Lecture Notes in Electrical Engineering

Volume 776

The book series *Lecture Notes in Electrical Engineering* (LNEE) publishes the latest developments in Electrical Engineering - quickly, informally and in high quality. While original research reported in proceedings and monographs has traditionally formed the core of LNEE, we also encourage authors to submit books devoted to supporting student education and professional training in the various fields and applications areas of electrical engineering. The series cover classical and emerging topics concerning:

- Communication Engineering, Information Theory and Networks
- Electronics Engineering and Microelectronics
- Signal, Image and Speech Processing
- Wireless and Mobile Communication
- Circuits and Systems
- Energy Systems, Power Electronics and Electrical Machines
- Electro-optical Engineering
- Instrumentation Engineering
- Avionics Engineering
- Control Systems
- Internet-of-Things and Cybersecurity
- Biomedical Devices, MEMS and NEMS

For general information about this book series, comments or suggestions, please contact leontina.dicecco@springer.com.

To submit a proposal or request further information, please contact the Publishing Editor in your country:

China

Jasmine Dou, Editor (jasmine.dou@springer.com)

India, Japan, Rest of Asia

Swati Meherishi, Editorial Director (Swati.Meherishi@springer.com)

Southeast Asia, Australia, New Zealand

Ramesh Nath Premnath, Editor (ramesh.premnath@springernature.com)

USA, Canada:

Michael Luby, Senior Editor (michael.luby@springer.com)

All other Countries:

Leontina Di Cecco, Senior Editor (leontina.dicecco@springer.com)

**** This series is indexed by EI Compendex and Scopus databases. ****

More information about this series at http://www.springer.com/series/7818

Sourav Dhar · Subhas Chandra Mukhopadhyay ·
Samarendra Nath Sur · Chuan-Ming Liu
Editors

Advances in Communication, Devices and Networking

Proceedings of ICCDN 2020

Springer

Editors
Sourav Dhar
Department of Electronics
and Communication Engineering
Sikkim Manipal Institute of Technology
Rangpo, Sikkim, India

Samarendra Nath Sur
Electronics and Communications
Engineering Department
Sikkim Manipal Institute of Technology
Rangpo, India

Subhas Chandra Mukhopadhyay
School of Engineering (E6B1.11)
Macquarie University
Sydney, NSW, Australia

Chuan-Ming Liu
Department of Computer Science
and Information Engineering
National Taipei University of Technology
Taipei, Taiwan

ISSN 1876-1100 ISSN 1876-1119 (electronic)
Lecture Notes in Electrical Engineering
ISBN 978-981-16-2910-5 ISBN 978-981-16-2911-2 (eBook)
https://doi.org/10.1007/978-981-16-2911-2

This Springer imprint is published by the registered company Springer Nature Singapore Pte Ltd.
The registered company address is: 152 Beach Road, #21-01/04 Gateway East, Singapore 189721, Singapore

Preface

ICCDN 2020, the 4th International Conference on Communication, Device and Networking, was held at the Department of Electronics and Communication Engineering (ECE) of Sikkim Manipal Institute of Technology (SMIT), Sikkim Manipal University (SMU), Sikkim, December 19–20, 2020. The proceedings of ICCDN 2020 is published by Springer Nature in the book series LNEE (Scopus Indexed).

The aim of the conference was to provide a platform for researchers, engineers, academicians, and industry professionals to present their recent research works and to explore future trends in various areas of engineering. The conference also brought together both novice and experienced scientists and developers, to explore newer scopes; collect new ideas; establish new cooperation between research groups; and exchange ideas, information, techniques, and applications in the field of Electronics, Communication, Devices, and Networking. The ICCDN-2019 Committees rigorously invited submissions of manuscripts from researchers, scientists, engineers, students, and practitioners across the world related to the relevant themes and tracks of the conference. The call for papers of the conference was divided into six tracks as mentioned, Track 1: Electronics Devices & Nano-technology, Track 2: Signal Processing, Track 3: Microwave and millimetre wave engineering, Track 4: AI and its Application, Track 5: Energy, Power and Control, Track 6: Communication and Networking

The conference was enriched with five keynote speeches each of 1-hour duration by eminent professors: Dr. Subhas Mukhopadhyay, Professor of Mechanical/Electronics Engineering, Macquarie University, Australia; Dr. S. R. Mahadeva Prasanna, Professor in the Department of Electrical Engineering, and Dean (Faculty Welfare, Research and Development) at IIT Dharwad; Dr. Marius M. Balas, Professor at, "Aurel Vlaicu" University of Arad, Romania; Dr. Sunit Kumar Sen, Ex-Professor of the Department of Applied Physics, Instrumentation Engineering, University of Calcutta; Dr. Dinh-Thuan Do, Assistant Professor of Department of Computer Science and Information Engineering, College of Information and Electrical Engineering Asia University, Taichung city, Taiwan; Dr. Angsuman Sarkar, Professor of Electronics and Communication Engineering in Kalyani Government Engineering College, West Bengal; Dr. Tanweer Ali, Assistant Professor-Senior Scale in the

Department of Electronics & Communication Engineering at Manipal Institute of Technology, Manipal Academy of Higher Education, Manipal.

A total of 113 papers were received; out of which, 50 papers (including 3 invited papers) were accepted in the conference. Participants came from different parts of the country as well as across different countries. All these efforts undertaken by the Organizing Committees led to a high-quality technical conference program, which featured high-impact presentations from keynote speakers and from paper presenters. The significant technical gathering in the conference really carried the message within the state of Sikkim, where people were being conveyed about the global progression toward smart homes, smart campuses, and smart cities. On behalf of the ICCDN organizing committee, we would like to thank Springer for the kind cooperation. We would like to thank the Chief Patrons, Patron General Chair, and Program chair for their continuous support and encouragements. Finally, we would also like to express our appreciation to all contributors for submitting highly accomplished work, to all reviewers for their time and valuable comments, and to all members of the conference committees for giving generously of their time over the past year.

Rangpo, India — Sourav Dhar
Sydney, Australia — Subhas Chandra Mukhopadhyay
Rangpo, India — Samarendra Nath Sur
Taipei, Taiwan — Chuan-Ming Liu

Contents

Editors and Contributors

About the Editors

Dr. Sourav Dhar is currently a Professor and Head of the ECE Department, SMIT. His current research interests include IoT, WSN, remote sensing, and microwave filter design. He is a member of IEEE, the IEEE-GRSS society and IEI, India. He has published more than 30 papers in SCI/Scopus indexed international journals and at conferences. He also serves as a reviewer for Wireless Personal Communication, IEEE Transactions on Vehicular Technology and several other journals and conferences.

Dr. Subhas Chandra Mukhopadhyay (M'97, SM'02, F'11) holds a B.E.E. (gold medallist), M.E.E., Ph.D. (India) and Doctor of Engineering (Japan). He has over 30 years of teaching, industrial and research experience.

Currently he is working as a Professor of Mechanical/Electronics Engineering, Macquarie University, Australia and is the Discipline Leader of the Mechatronics Engineering Degree Programme. His fields of interest include Smart Sensors and sensing technology, instrumentation techniques, wireless sensors and network (WSN), Internet of Things (IoT), wearable sensors, numerical field calculation, electromagnetics etc. He has supervised over 40 postgraduate students and over 100 Honours students. He has examined over 60 postgraduate theses.

He has published over 400 papers in different international journals and conference proceedings, written nine books and forty two book chapters and edited seventeen conference proceedings. He has also edited thirty books with Springer-Verlag and twenty four journal special issues. He has organized over 20 international conferences as either General Chairs/co-chairs or Technical Programme Chair. He has delivered 356 presentations including keynote, invited, tutorial and special lectures.

He is a Fellow of IEEE (USA), a Fellow of IET (UK), a Fellow of IETE (India), a Topical Editor of IEEE Sensors journal, and an associate editor of IEEE Transactions on Instrumentation and Measurements. He is a Distinguished Lecturer of the IEEE Sensors Council from 2017 to 2022. He is the Founding Chair of the IEEE Sensors Council New South Wales Chapter.

Dr. Samarendra Nath Sur is an Assistant Professor at the ECE Department, Sikkim Manipal Institute of Technology, India. His current research interests include broadband wireless communication, advanced digital signal processing, and remote sensing. He was the recipient of the University Medal & Dr. S.C. Mukherjee Memorial Gold Centered Silver Medal from Jadavpur University in 2007.He is a member of IEEE, IEEE-IoT, IEEE-SPS, IEI, India and IAENG. He has published more than 50 papers in SCI/Scopus indexed international journals and at conferences. He also serves as a reviewer for the International Journal of Electronics, IET Communication, Ad Hoc Networks, and IEEE Transactions on Signal Processing.

Dr. Chuan-Ming Liu is a Professor at the Department of CSIE, National Taipei University of Technology, Taiwan, where he was the Department Chair from 2013 to 2017. Currently, he is the Head of the Extension Education Center at the same school. He has published more than 80 papers in various prestigious journals and at international conferences. His current research interests include big data management and processing, uncertain data management, data science, spatial data processing, data streams, ad hoc and sensor networks, and location-based services.

Contributors

Anindita Adhikary Department of Management Studies, Sikkim Manipal Institute of Technology–Sikkim Manipal University, Gangtok, Sikkim, India

Amrita Agarwal SMIT, SMU, Gangtok, Sikkim, India

Neetu Agrawal Electronics and Communication Department, GLA University, Mathura, UP, India

Rekib Uddin Ahmed Department of Electronics and Communication Engineering National Institute of Technology Meghalaya, Shillong, Meghalaya, India

Srimanta Baishya National Institute of Technology, Silchar, Cachar, Assam, India

Garima Bakshi School of Engineering and Technology, Sushant University, Gurgaon, India

Ramashis Banerjee EE Department, NIT Silchar, Silchar, India

Saloni Bansal GLA University, Mathura, Uttar Pradesh, India

Rabindra K. Barik School of Computer Applications, KIIT Deemed To Be University, Bhubaneswar, India

Karabi Baruah National Institute of Technology, Silchar, Cachar, Assam, India

Ranjit Kumar Behera National Institute of Science and Technology, Berhampur, India

Rajdeep Bhattacharjee Electronics and Communication Engineering Department, Sikkim Manipal Institute of Technology, Gangtok, Sikkim, India

Sucharita Bhattacharyya Department of Applied Science and Humanities, Guru Nanak Institute of Technology, Kolkata, India

Amrita Biswas Electronics and Communication Engineering Department, Sikkim Manipal Institute of Technology, Gangtok, Sikkim, India

Anindya Bose Department of Physics, The University of Burdwan, Burdwan, India

Tanushree Bose Department of Electronics and Communication Engineering, Sikkim Manipal Institute of Technology—Sikkim Manipal University, Gangtok, Sikkim, India

Sudipta Chakrabarty Department of MCA, Techno India, Salt Lake, Kolkata, West Bengal, India

Swastika Chaktraborty Sikkim Manipal Institute of Technology, Sikim Manipal University, Majitar, Sikkim, India

Saikat Chatterjee EEE Department, Sikkim Manipal Institute of Technology, Sikkim Manipal University, Gangtok, India;
EEE Department, Sikkim Manipal Institute of Technology, Sikkim Manipal University, Majitar, Sikkim, India

Somenath Chatterjee Department of Electronics and Communication Engineering, Sikkim Manipal Institute of Technology (SMIT), Sikkim Manipal University (SMU), Majitar, Sikkim, India

Chandan Chetri SMIT, SMU, Gangtok, Sikkim, India

Keshab Ch Gogoi EEE Department, Sikkim Manipal Institute of Technology, Sikkim Manipal University, Majitar, Sikkim, India

Sukabya Dan Department of ECE, National Institute of Technology Mizoram, Chaltlang, Aizawl, Mizoram, India;
Department of Physics, The University of Burdwan, Burdwan, India

Pritam Das Electronics & Communication Engineering, Dibrugarh University, Dibrugarh, India

Puja Archana Das School of Computer Engineering, Kalinga Institute of Industrial Technology (KIIT) Deemed to be University, Bhubaneswar, Odisha, India

Sanat Kumar Das Sikkim Manipal Institute of Technology, Sikkim Manipal University, Majhitar, Rangpo, East Sikkim, Sikkim, India

Saumya Das Department of Electronics and Communication Engineering, Sikkim Manipal Institute of Technology, Sikkim Manipal University, Sikkim, India

Saurabh Debabrata Das Electronics and Communication Engineering Department, Sikkim Manipal Institute of Technology, Gangtok, Sikkim, India

Aritri Debnath CMRIT, Bangalore, Karnataka, India

Barnali Dey SMIT, SMU, Gangtok, Sikkim, India;
Electronics and Communication Engineering Department, Sikkim Manipal Institute of Technology, Gangtok, Sikkim, India

Sumit Dey Department of Physics, The University of Burdwan, Burdwan, India

Rudra Sankar Dhar Department of Electronics and Communication Engineering, National Institute of Technology, Aizwal, Mizoram, India

Sourav Dhar Department of ECE, Sikkim Manipal University, Sikkim, India

Dinh-Thuan Do Department of Computer Science and Information Engineering, College of Information and Electrical Engineering, Asia University, Taichung city, Taiwan

Ritam Dutta Department of Electronics and Communication Engineering, Surendra Institute of Engineering and Management, Maulana Abul Kalam Azad University of Technology, Kolkata, West Bengal, India;
Department of Electronics and Communication Engineering, Sikkim Manipal Institute of Technology, Sikkim Manipal University, Gangtok, Sikkim, India

Kritika Garg Department of Computer Science and Engineering, Sikkim Manipal Institute of Technology, Sikkim Manipal Univeristy, Gangtok, Sikkim, India

Kiran Gautam Department of Computer Science and Engineering, Sikkim Manipal Institute of Technology, Sikkim Manipal University, Gangtok, India

Pritam Kumar Gayen Kalyani Government Engineering College, Kalyani, Nadia, West Bengal, India

Manu George PES University, Bengaluru, India

Kamanashis Goswami Electronics Engineering Department, I.I.T. (I.S.M), Dhanbad, India

Mousumi Gupta Sikkim Manipal Institute of Technology, SMU, East Sikkim, India

Shivam Gupta Electronics and Communication Engineering Department, Sikkim Manipal Institute of Technology, Gangtok, Sikkim, India

Sandeep Gurung Sikkim Manipal Institute of Technology, Sikkim Manipal University, Sikkim, India

Nadine Hasan School of Electronics Engineering, Kalinga Institute of Industrial Technology, Deemed to Be University, Bhubaneswar, India

Mir Wajahat Hussain Department of Computer Science and Engineering, National Institute of Technology, Meghalaya, Shillong, Meghalaya, India

Muhammad Fazal Ijaz Department of Intelligent Mechatronics Engineering, Sejong University, Seoul, South Korea

Hashinur Islam Department of Electronics and Communication Engineering, Sikkim Manipal Institute of Technology, Sikkim Manipal University, Sikkim, India

Md. Ruhul Islam Department of CSE, SMIT, Rangpo, India

Ajeya Jha Department of Management Studies, Sikkim Manipal Institute of Technology—Sikkim Manipal University, Gangtok, Sikkim, India;
Department of Management Studies, Sikkim Manipal Institute of Technology, Majhitar, Sikkim, India

Saket Kumar Jha Department of Electronics and Communication Engineering, Sikkim Manipal Institute of Technology (SMIT), Sikkim Manipal University (SMU), Majitar, Sikkim, India

Sanjib Kabi Sikkim Manipal Institute of Technology, Sikkim Manipal University, Majhitar, Rangpo, East Sikkim, Sikkim, India

K. Kaviyarasi Electronics and Communication Engineering, Government College of Engineering Bargur, Bargur, Tamilnadu, India

Tawal Kumar Koirala Department of Computer Science and Engineering, Sikkim Manipal Institute of Technology, Sikkim Manipal University, Gangtok, India

Chaitali Koley Department of ECE, National Institute of Technology Mizoram, Aizawl, Mizoram, India

Abhishek Kumar UG Student, Sikkim Manipal Institute of Technology, SMU, East Sikkim, India

Anuj Kumar Assistant Professor, Apeejay School of Management, Dwarka, New Delhi, India

Harekrishna Kumar GLA University, Mathura, Uttar Pradesh, India

Jitendra Kumar Department of Management Studies, Sikkim Manipal Institute of Technology, Majhitar, Sikkim, India

Nitesh Kumar SMIT, SMU, Gangtok, Sikkim, India;
Department of Electronics and Communiation Engineering, Sikkim Manipal Institute of Technology, Sikkim Manipal Univeristy, Gangtok, Sikkim, India

Rubi Kumari Electrical and Electronics Engineering Department, Sikkim Manipal Institute of Technology, Sikkim Manipal University, Sikkim, India

Indranil Kushary Kalyani Government Engineering College, Kalyani, Nadia, West Bengal, India;
JIS College of Engineering, Kalyani, Nadia, West Bengal, India

Chi-Bao Le Faculty of Electronics Technology, Industrial University of Ho Chi Minh City, Ho Chi Minh City, Vietnam

Somnath Mahato Department of Physics, The University of Burdwan, Burdwan, India

Jai Mangal ECE, VIT Bhopal University, Bhopal, India

Vijay Kumar Mehta Department of Community Medicine, Sikkim Manipal Institute of Medical Sciences—Sikkim Manipal University, Gangtok, Sikkim, India

Ayaskanta Mishra School of Electronics Engineering, Kalinga Institute of Industrial Technology, Deemed to Be University, Bhubaneswar, India

Mohit Misra National Institute of Science and Technology, Berhampur, India

Ambik Mitra School of Electronics and Computer Engineering, Kalinga Institute of Industrial Technology (KIIT) Deemed To Be University, Bhubaneswar, Odisha, India

Shubham Mittal Sikkim Manipal Institute of Technology , SMU, Gangtok, Sikkim, India

Debasis Mohanty iNurture Education Solutions Private Limited, Bengaluru, Karnataka, India

Haraprasad Mondal Electronics & Communication Engineering, Dibrugarh University, Dibrugarh, India

Vivekananda Mukherjee Sikkim Manipal Institute of Technology, Sikkim Manipal University, Majhitar, Rangpo, East Sikkim, Sikkim, India

Subhas C. Mukhopadhyay School of Engineering, Macquarie University, Sydney, Australia

Subhas Chandra Mukhopadhyay School of Engineering, Macquarie University, Sydney, NSW, Australia

Nitai Paitya Department of Electronics and Communication Engineering, Sikkim Manipal Institute of Technology, Sikkim Manipal University, Gangtok, Sikkim, India

Moumi Pandit Electrical and Electronics Engineering Department, Sikkim Manipal Institute of Technology, Sikkim Manipal University, Sikkim, India

Mansi Parmar ECE, VIT Bhopal University, Bhopal, India

Amrut Patro National Institute of Science and Technology, Berhampur, India

Bishal Paudel Electronics and Communication Engineering Department, Sikkim Manipal Institute of Technology, Sikkim, India;
Sikkim Manipal Institute of Technology, Sikim Manipal University, Majitar, Sikkim, India

Kushal Pokhrel Department of Electronics and Communication Engineering Sikkim, Sikkim Manipal Institute of Technology, Sikkim Manipal University, Gangtok, Sikkim, India

Sumedha N. Prabhu School of Engineering, Macquarie University, Sydney, Australia

Nitisha Pradhan Department of Computer Science and Engineering, Sikkim Manipal Institute of Technology, Sikkim Manipal University, Gangtok, Sikkim, India

P. C. Pradhan Department of Electronics & Communication Engineering, SMIT, SMU, Majitar, India

Amulay Nidhi Puranik Amity Global Institute, Singapore, Singapore

Shankar Purbey Development Management Institute, Patna, Bihar, India

Prativa Rai Department of Computer Science and Engineering, Sikkim Manipal Institute of Technology, Sikkim Manipal University, Gangtok, Sikkim, India

S. Ramya Electronics and Communication, MIT, MAHE, Manipal, India

Abu ul Hassan S. Rana Department of Intelligent Mechatronics Engineering, Sejong University, Seoul, South Korea

Anand Ranjan Sikkim Manipal Institute of Technology, SMU, Gangtok, Sikkim, India

R. Ranjith Electronics and Communication Engineering, Government College of Engineering Bargur, Bargur, Tamilnadu, India

Vaibhav Rathore Sikkim Manipal Institute of Technology, SMU, Gangtok, Sikkim, India

Kaushik Ravichandran PES University, Bengaluru, India

Arun Kumar Ray School of Electronics Engineering, Kalinga Institute of Industrial Technology, Deemed to Be University, Bhubaneswar, India

Bijoyeta Roy Sikkim Manipal Institute of Technology, SMU, East Sikkim, India

Diptendu Sinha Roy Department of Computer Science and Engineering, National Institute of Technology Meghalaya, Shillong, India

Rupsa Roy Department of ECE, Sikkim Manipal University, Sikkim, India

Prabir Saha Department of Electronics and Communication Engineering National Institute of Technology Meghalaya, Shillong, Meghalaya, India

Saibal Kumar Saha Department of Management Studies, Sikkim Manipal Institute of Technology—Sikkim Manipal University, Gangtok, Sikkim, India

Satyajit Saha Kalyani Government Engineering College, Kalyani, Nadia, West Bengal, India

Atanu Santra Department of Physics, The University of Burdwan, Burdwan, India

Bishnu Prasad Sapkota Sikkim Manipal Institute of Technology, Sikkim Manipal University, Majhitar, Rangpo, East Sikkim, Sikkim, India

Samta Sapra School of Engineering, Macquarie University, Sydney, NSW, Australia

Nandita Sarkar Electronics and Communication Engineering Department, Sikkim Manipal Institute of Technology, Sikkim, India;
Sikkim Manipal Institute of Technology, Sikim Manipal University, Majitar, Sikkim, India

Swarup Sarkar Department of ECE, Sikkim Manipal University, Sikkim, India

Hiren Kumar Deva Sarma Department of IT, SMIT, Rangpo, India

Soumya Sen Department of ECE, Seacom Engineering College, Howrah, West Bengal, India

Azhar Shaik PES University, Bengaluru, India

Abhimanyu Sharma Department of Computer Science and Engineering, Sikkim Manipal Institute of Technology, Sikkim Manipal University, Gangtok, India

Anjali Sharma Department of Instrumentation and Control Engineering, Dr. B.R. Ambedkar NIT Jalandhar, Jalandhar, India

Anurag Sharma School of Computer Engineering, Kalinga Institute of Industrial Technology (KIIT) Deemed to be University, Bhubaneswar, Odisha, India

Anushka Sharma School of Computer Engineering, Kalinga Institute of Industrial Technology (KIIT), Deemed To Be University, Bhubaneswar, Odisha, India

Bikash Sharma Department of Electronics and Communiation Engineering, Sikkim Manipal Institute of Technology, Sikkim Manipal Univeristy, Gangtok, Sikkim, India

Kalpana Sharma Sikkim Manipal Institute of Technology, Sikkim Manipal University, Sikkim, India

Prashant Sharma Department of Electronics and Communication Engineering, Sikkim Manipal Institute of Technology, Sikkim Manipal University, Majitar, Rangpo, East Sikkim, India

Rajeev Sharma EEE Department, Sikkim Manipal Institute of Technology, Sikkim Manipal University, Majitar, Sikkim, India

K. S. Sherpa Electrical and Electronics Engineering Department, Sikkim Manipal Institute of Technology, Sikkim Manipal University, Sikkim, India

Aasheesh Shukla Department of Electronics and Communication, GLA University, Mathura, India

Ankit Singh Symbiosis Institute of Health Sciences, Symbiosis International (Deemed University), Pune, Maharashtra, India

Dilbag Singh Department of Instrumentation and Control Engineering, Dr. B.R. Ambedkar NIT Jalandhar, Jalandhar, India

Jayant Singh Department of Mechanical Engineering, Sikkim Manipal Institute of Technology, Sikkim Manipal Univeristy, Gangtok, Sikkim, India

Mandeep Singh Faculty, Department of ECE, National Institutes of Technology, Srinagar, Uttarakhand, India

Nira Singh Department of Electronics and Communication, GLA University, Mathura, India

Diptendu Sinha Roy Department of Computer Science and Engineering, National Institute of Technology, Meghalaya, Shillong, Meghalaya, India

Akshara Subramaniasivam PES University, Bengaluru, India

Samarendra Nath Sur Department of Electronics and Communication Engineering, Sikkim Manipal Institute of Technology, Sikkim Manipal University, Majitar, Rangpo, East Sikkim, India

Sweta UG Student, Sikkim Manipal Institute of Technology, SMU, East Sikkim, India

Jitendra Singh Tamang Department of Electronics and Communication Engineering, Sikkim Manipal Institute of Technology (SMIT), Sikkim Manipal University (SMU), Majitar, Sikkim, India;
Department of Electronics and Communication Engineering, National Institute of Technology, Aizwal, Mizoram, India

Sheba Diamond Thabah Department of Electronics and Communication Engineering National Institute of Technology Meghalaya, Shillong, Meghalaya, India

Adeep Thakuria Electronics & Communication Engineering, Dibrugarh University, Dibrugarh, India

Anup Kumar Thander Department of Applied Science and Humanities, Guru Nanak Institute of Technology, Kolkata, India

Abinash Thapa Department of Electronics & Communication Engineering, SMIT, SMU, Majitar, India

V. K. Tomar GLA University, Mathura, Uttar Pradesh, India

Subhranshu Sekhar Tripathy Department of Computer Science and Engineering, National Institute of Technology Meghalaya, Shillong, India

Biraj Upadhyaya Sikkim Manipal Institute of Technology, Sikkim Manipal University, Sikkim, India

Kaustubh Wani Electronics and Communication, MIT, MAHE, Manipal, India

Design and Implementation of Multi-operand $2^n - 1$, 2^n, and $2^n + 1$ Modulo Set Adder

Prabir Saha, Rekib Uddin Ahmed, and Sheba Diamond Thabah

Abstract In this paper, multi-operand modulo adders have been proposed and the adders have been designed based on a new 5:3 compressor. Gate-level design of such compressor has been carried out through the simplification of the K-map. The functionalities of the proposed modulo adders have been verified in the Xilinx 14.7 design environment implemented through VHDL coding. The performance parameters such as power consumption, delay, and power-delay product (PDP) have been obtained from the simulation using the device Virtex-6.

Keywords 5:3 compressor · modulo adders · Virtex-6

1 Introduction

In computer applications, the modulo operation finds the remainder or signed remainder after a division of one number by another. The computation through modulo arithmetic operations have a versatile applications in cryptography [1], residue number system [2], fault-tolerant computer system [3], digital signal processors [4], etc. The computation with the help of moduli set, viz., $2^n - 1$, 2^n, and $2^n + 1$ are the most popular approach due to the presence of the efficient combinational converter to and from binary systems [5]. Such special moduli set also have some advantages like low-cost efficient hardware for conversion, reduction as well as reverse conversion [6]. Enormous algorithms and designs for the additions have been found in literature for individual modulo adder which are based on either $2^n - 1$ [7, 8] and/or $2^n + 1$ [5, 9–11]. In $2^n - 1$ adder design, the researchers have been utilizing adder block either in parallel prefix adder or carry save adder, which are able to

P. Saha (✉) · R. U. Ahmed · S. D. Thabah
Department of Electronics and Communication Engineering National Institute of Technology Meghalaya, Shillong, Meghalaya, India
e-mail: sahaprabir1@gmail.com

R. U. Ahmed
e-mail: rekib@nitm.ac.in

S. Dhar et al. (eds.), *Advances in Communication, Devices and Networking*,
Lecture Notes in Electrical Engineering 776,
https://doi.org/10.1007/978-981-16-2911-2_1

handle the carry stage along with an extra stage for end-around carry [12]. On the other hand, some researchers also observed that the end-around carry can be computed on the prefix level itself [7]. Despite the utilization of the prefix adder, modulo $2^n - 1$ addition holds regular and structured format for addition with the penalty of the routing delay [7]. Modulo $2^n + 1$ adder design is much complex from its counterpart, i.e., $2^n - 1$ design and also has a higher hardware complexity [7]. Toward the efficient implementation of the modulo $2^n + 1$ design, Leibowitz [13] proposed the diminished-one number system, which has been utilized so far for such modulo computation. However, the diminished-one number system adder is also based on the end-around carry where the parallel adders are also required. In addition to this, most of the proposed modulo adders are based on two operands, viz., A and B.

In this paper, the multi-operand modulo adders namely $2^n - 1$, 2^n, and $2^n + 1$ have been proposed which can operate upon five operands (A, B, C, D, and E), each operand is of 4-bits, i.e., $n = 4$. The gate-level designs of the proposed adders have been simulated in Xilinx 14.7 using the device Virtex-6.

2 Proposed Methodology

The components used in the gate-level design of the modulo adders are 5:3 compressor, half adder, full adder, and 4-bit carry look ahead adders. Figure 1 shows the flowchart representation of the design of modulo 2^n adder which has been elaborated with an example illustrated in Fig. 2. In the example, five operands, viz., 12, 13, 14, 13, and 15 are considered (each of four bits) whose summation $(S_1) = 67$ which is represented as "01000011" in binary system. To compute 67 *mod* 16, the last four digit of the binary number is considered, i.e., "0011" ($Res = S_1[3:0]$) which represents "3" in decimal equivalent.

Figure 3 shows the flowchart representation of the modulo $2^n - 1$ adder illustrated with the help of the example shown in Fig. 4. In modulo $2^n - 1$ adder, the summation S is split into two vectors: P and Q, where, $P = S_1[3:0]$ and $Q = S_1[,:4]$. Referring to the example in Fig. 4, P ="0011" and Q ="0100" and the $Res = P + Q$, thus giving the result "0111" representing "7" in decimal equivalent. Flowchart shown in Fig. 3 works for almost all the 4-bit operands with a few exceptions, e.g., in cases where S_1 =30. Since, 30 *mod* 15 = 0, by following the example (shown in Fig. 4) the result comes to be =15 as shown in Fig. 5 which is not valid for modulo $2^n - 1$ adder. In this condition, Res is set to zero ("0000"). Fig. 6 shows the flowchart representation of the modulo $2^n + 1$ adder which is elaborated with the example shown in Fig. 7. In this adder, the S_2 is obtained by $P - Q$. Referring to the example in Fig. 6, the S_2 is a negative number, i.e., -1. In such case when S_2 is negative, Res is evaluated as $2^n + 1 + S_2$.

Figure 8 shows the addition schematic used in the proposed designs where the components such as 5:3 compressor, half adders, full adders, and 4-bit carry look ahead adders (CLA) have been used. Initially, the 5:3 compressor has been used to count the number of "ones" present in the input vectors ($A[i]$ $B[i]$ $C[i]$ $D[i]$ $E[i]$

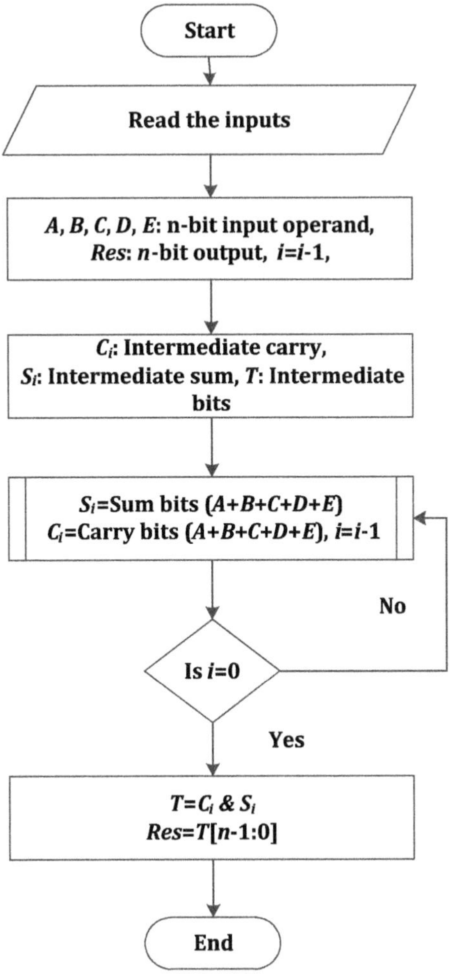

Fig. 1 Flowchart used for the design of the proposed 2^n modulo adder

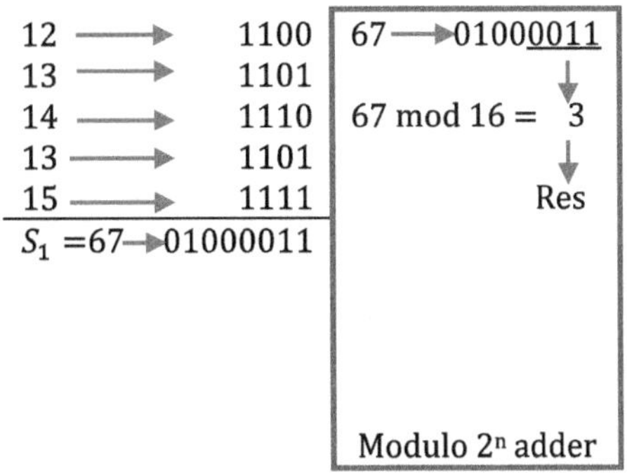

Fig. 2 Pictorial elaboration showing the steps in addition procedure of 2^n modulo adders

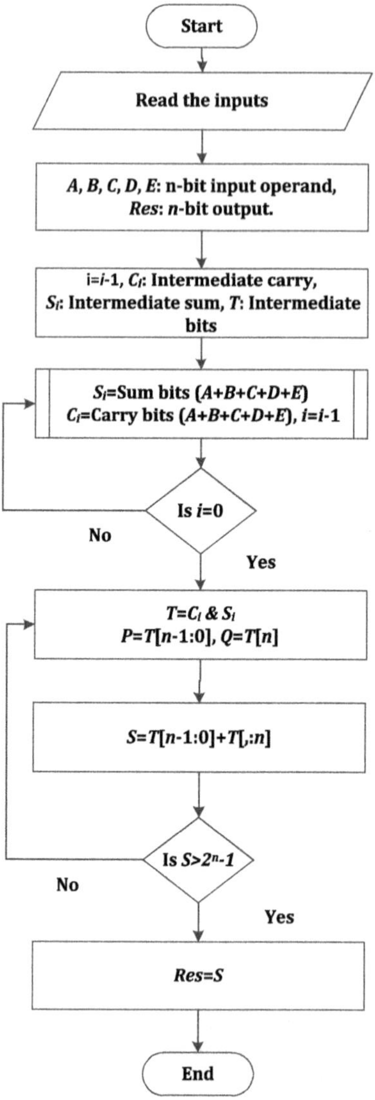

Fig. 3 Flowchart used for the design of the proposed $2^n - 1$ modulo adder

where $i = 0$ to 3). The Boolean functions for the output of 5:3 compressor are

$$X_i(2) = BCDE + ACDE + ABDE + ABCD \tag{1}$$

$$\begin{aligned} X_i(1) &= AC\overline{D} + A\overline{D}E + AB\overline{D} + C\overline{D}E + A\overline{B}\,\overline{C}D \\ &+ \overline{B}CD\overline{E} + \overline{A}BC\overline{E} + B\overline{C}D\overline{E} + \overline{A}B\overline{C}E \\ &+ \overline{A}\,\overline{B}DE \end{aligned} \tag{2}$$

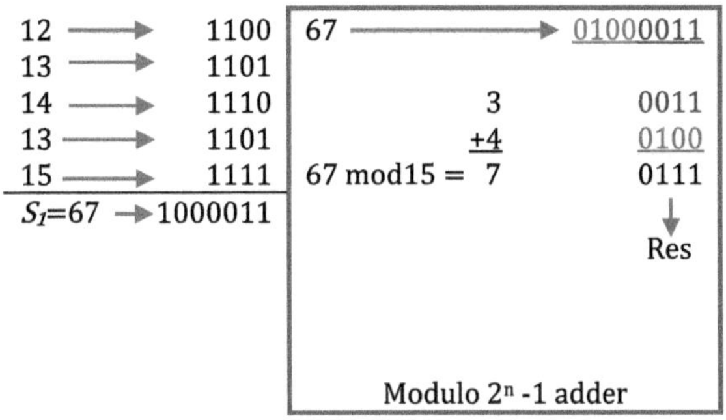

Fig. 4 Pictorial elaboration showing the steps in addition procedure of $2^n - 1$ modulo adders

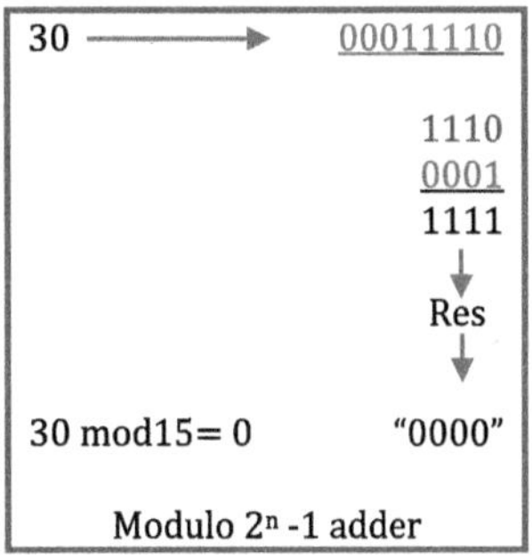

Fig. 5 Steps showing the steps involved in the operation 30 *mod* 15

$$
\begin{aligned}
X_i(0) = {} & BC\overline{D}E + AC\overline{D}E + AB\overline{D}E + ABC\overline{D} + ABCE \\
& + \overline{A}\,\overline{B}\,\overline{C}\,\overline{D}E + \overline{A}\,\overline{B}C\overline{D}\,\overline{E} + \overline{A}\,\overline{B}\,\overline{C}\,\overline{D}\,\overline{E} \\
& + \overline{A}B\overline{C}DE + \overline{A}B\overline{C}DE + \overline{A}BCD\overline{E} + A\overline{B}CD\overline{E} \\
& + A\overline{B}\,\overline{C}\,\overline{D}\,\overline{E} + A\overline{B}\,\overline{C}\,\overline{D}E + A\overline{B}\,\overline{C}DE \\
& + A\overline{B}CD\overline{E} + AB\overline{C}D\overline{E}
\end{aligned}
\tag{3}
$$

3 Results and Discussion

Gate-level design of the modulo adders is implemented through VHDL coding, and the functionalities are verified through simulation performed in Xilinx ISE 14.7 design environment using the device Virtex-6. The delay and power consumption are obtained from the synthesis and power report obtained from the simulation. The total number of look-up tables (LUTs) present in the Virtex-6 is 474240 [14], and its static power dissipation obtained from simulation is 4.447 W. Table 1 summarizes the performance parameters of the modulo adders. The power consumption is calculated using the relation:

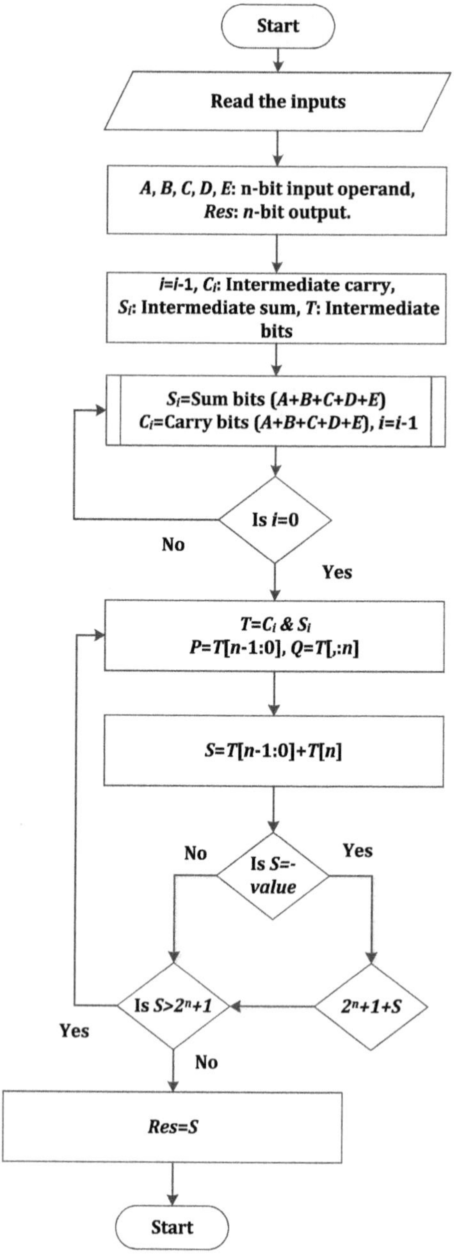

Fig. 6 Flowchart used for the design of the proposed $2^n + 1$ modulo adder

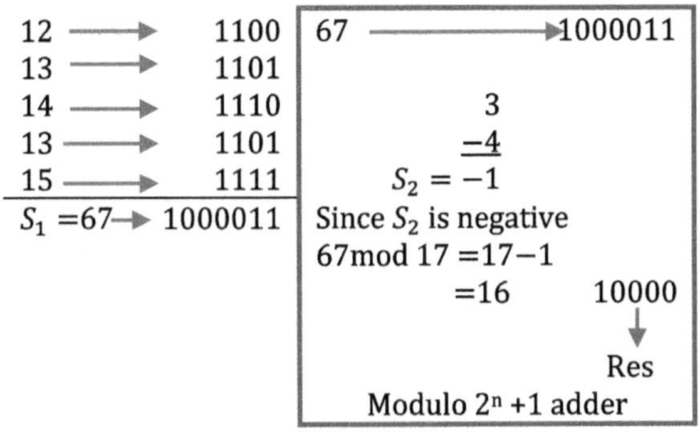

Fig. 7 Pictorial elaboration showing the steps in addition procedure of $2^n + 1$ modulo adders

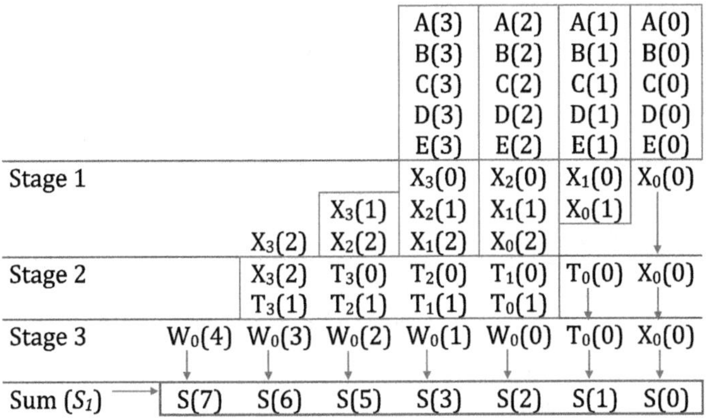

Fig. 8 Addition scheme used to compute the sum (S_1) of the five operands $ABCDE$

Table 1 Summary of power consumption, delay, and PDP of the modulo adders obtained from simulation in the device Virtex-6

Modulo adders	Number of LUTs used	Power (μW)	Delay (ns)	PDP (fJ)
2^n	13	121.90	2.901	353.63
$2^n - 1$	30	281.31	5.272	1483.06
$2^n + 1$	30	281.31	5.135	1444.53

$$\text{Power} = \frac{\text{Static Power}}{\text{Total number of LUTs}} \times \text{Number of LUTs used.} \quad (4)$$

4 Conclusion

This paper presents the design of the multi-operand modulo adders namely 2^n, $2^n - 1$, and $2^n + 1$. The simulation of the modulo adders have been carried out considering five operands (A, B, C, D, and E) each of 4-bits. The number of "ones" present in each column of the operands is counted using the 5:3 compressors. The modulo adders are implemented through VHDL coding that makes the design suitable for systematic application in high-performance modulo arithmetic units. The expected outputs are verified in the Xilinx 14.7 tool along with power consumption, delay, and PDP are obtained using the device Virtex-6.

Acknowledgements The work was supported by the project grant with a file no. NITMGH/TEQIP-III/MP/2019-20/252.

References

1. Lai X, Massey JL (1990) A proposal for a new block encryption standard. In: Advances in cryptology- EUROCRYPT'90, vol 473. Lecture notes in computer science. Springer, Berlin, Germany, pp 389–404
2. Soderstrand MA, Jenkins WK, Jullien GA, Taylor FJ (1986) Residue number system arithmetic: modern applications in digital signal processing. IEEE Press, New York
3. Rao TRN, Fujiwara E (1989) Error control coding for computer systems. Prentice-Hall, New Jersey, USA
4. Gholami E, Farshidi R, Hosseinzadeh M, Navi K (2009) High speed residue number system comparison for the moduli set ($2^n - 1$,2^n, $2^n + 1$). J Commun Comput 6(3):40–46
5. Vergos HT, Bakalis D, Efstathiou C (2008) Efficient modulo $2^n + 1$ multi-operand adders. In: 15th IEEE international conference on electronics, circuits and systems. IEEE Press, St. Julien's, pp 694–697
6. Vergos HT, Dimitrakopoulos G (2012) On modulo $2^n + 1$ adder design. IEEE Trans Comput 61(2):173–186
7. Kalampoukas L et al (2000) High-speed parallel-prefix modulo $2^n - 1$ adders. IEEE Trans Comput 49(7):673–680
8. Efstathiou C, Vergos HT, Nikolos D (2003) Modulo $2^n \pm 1$ adder design using select prefix blocks. IEEE Trans Comput 52(11):1399–1406
9. Skavantzos A (1989) Design of multioperand carry-save adders for arithmetic modulo ($2^n + 1$). Electron Lett 25(17):1152–1153
10. Jaberipur G, Parhami B (2009) Unified approach to the design of modulo-($2^n \pm 1$) adders based on signed-LSB representation of residues. In: 19th IEEE symposium on computer arithmetic. IEEE Press, Portland, pp 57–64
11. Lin S-H, Sheu M-H (2008) VLSI design of diminished-one modulo $2^n + 1$ adder using circular carry selection. IEEE Trans Circuits Syst II: Express Briefs 55(9):897–901
12. Dimitrakopoulos G, Nikolos DG, Vergos HT, Nikolos D, Efstathiou C (2005) New architectures for modulo $2^N - 1$ adders. In: 12th IEEE international conference on electronics, circuits and systems. IEEE Press, Gammarth, pp 1–4
13. Leibowitz LM (1976) A simplified binary arithmetic for the Fermat number transform. IEEE Trans Acoustics Speech Signal Process 24(5):356–359
14. Stavinov E (2011) 100 power tips for FPGA Designers, CreateSpace, Paramount, CA

Analysing the Behaviour of 14 nm, 10 nm, 7 nm FinFET and Predicting the Superiority Among the Lot

Soumya Sen and Mandeep Singh

Abstract The advancement of technology leads to the discovering of fresh avenues. Things took to bit challenging after the regular planar MOSFET dished out various shortcomings when they were subjected to downscaling. Few issues such as velocity saturation, drain-induced barrier lowering (DIBL), surface scattering, impact ionization etc. came into the picture. These are known as the short channel effects [1]. With the introduction of FinFET in the semiconductor market, it had brought with it a lot of advantages such as a notable improvement in the switching speed to the power consumption on one side to better controlling of leakage current also with significant switching speed. Also, with more added advantages an appreciable I_{ON}/I_{OFF} ratio is maintained. FinFET had given significant control of the channel by the devices for the usage of more than one gate. As a result, the short channel effects can be controlled without shooting high the carrier concentration. For all these, the researchers have dared to scale down the devices more to a significant extent. As the size gets lower there is a much healthier performance with very high efficiency. This very paper not only digs deep into the FinFET technology but also discusses the evolution of 14 nm, 10 nm, 7 nm technology for FinFETs and their features which takes the edge over their counterpart.

Keywords FinFET · Short channel effects · 14 nm · 10 nm · 7 nm · NanoHub

1 Introduction

MOSFET or the metal-oxide semiconductor field-effect transistor is a four-terminal device having the source, gate, drain and the body as the fourth terminal. The role of the gate is like a controlling tap that controls the flow of electrons. The distance between the source and the drain is the channel. The driving force of the I_d is the V_g.

S. Sen (✉)
Department of ECE, Seacom Engineering College, Howrah, West Bengal, India

M. Singh
Faculty, Department of ECE, National Institutes of Technology, Srinagar, Uttarakhand, India

S. Dhar et al. (eds.), *Advances in Communication, Devices and Networking*,
Lecture Notes in Electrical Engineering 776,
https://doi.org/10.1007/978-981-16-2911-2_2

The lowering of the channel length leads to downscaling of the device size and the diminishing of the gate voltage. This in turn makes the control of the gate low over the channel, which in turn invites the short channel effects [1, 2]. The device in turn becomes smaller to denser and the complexity of the circuit increases. The various short channel effects lead to leakage current to high power consumption issues which in turn damages the device. So in order for better gate control, the MugFETs have come into the picture. The FinFET falls into this category; the multiple-gate of the FinFET [3, 4] has been controlled by a single electrode and can be considered as one single gate. A major catch of the dual gate of the FinFET is that as the doping concentration is made to remain constant, the I_{ON} is two times that of the planar MOSFET which in turn is much beneficial according to the electrical integrity of the devices. The devices will have much improved carrier mobility and a lowered leakage current issues. If we explore deeper into the short channel effects of the devices, which include the *drain induced barrier lowering (DIBL), velocity saturation, hot electron effects, impact ionization*, the DIBL occurs due to the increase of the potential of the drain where the increase of the drain potential leads to the lowering of the barrier. The depletion region of the drain moves into the bulk. As a result, what happens is a huge rush of current flows through the subthreshold region. The major master of the device was the gate whose responsibility has been taken down by the drain. Now, as the barrier is lowered there is a current flow which is known as *leakage current* which would be very high and cannot be controlled by the gate [1]. Therefore, the device could not be turned off. It is seen that the entire depletion region from the drain would be entirely *punched through* to the source. So, there would be no depletion region around and the drain current would not be controlled by the gate, thus concluding that the gate loses its control over the device. Like this, many more like *velocity saturation* when the electric field will have no effect on the increased velocity. Figure 1a, b shows the simulated graphs in *NgSPICE* [5] of I_d versus V_{gs} and I_d versus V_{ds}, where in the aspect ratio W/L, the L is kept minimum. The aspect ratio is kept the same as 1.5. For the smallest value of L, the current is minimum

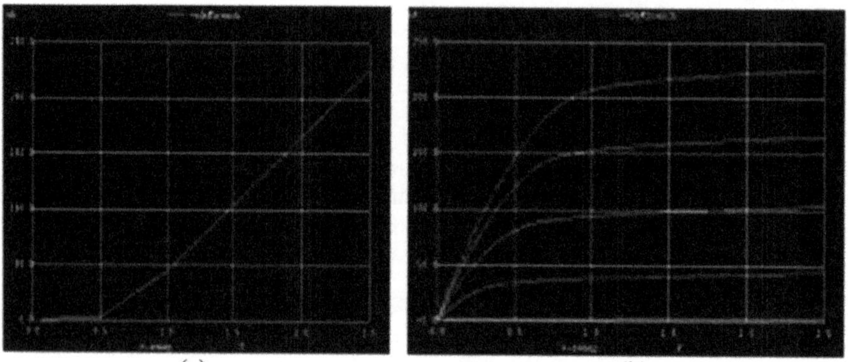

(a) (b)

Fig. 1 **a** The Id versus Vgs of 0.25 um MOSFET; **b** Simulation results for Id versus Vds due to velocity saturation condition in short channel device where the velocity gets saturated early before it reaches the pinch off point.

The next effect we can highlight is the *hot electron effect* where the electrons get ionized and upon entering into the oxide they increase the temperature of the oxide, hence the VT also shoots up degrading the overall performance of the device. Including all, we are having the *impact ionization effect* where one charge carrier will have sufficient energy to knock on another creating an e-h pair. With the short channel effects, the other problems which arise are the stray and parasitic capacitances and resistances. The major reason is the downscaling of the transistor. As much as the transistor sizes are shrinking down, the more the issues arise.

1.1 FinFET

The FinFET [3, 4] was designed to answer the hindrances of leakage current by wrapping the gate electrodes around the channel. Now, in the traditional transistor, we have it on the top of the channel. Here, the silicon fins act as the channel. The uncovered regions of the fins(gate) are the source and the drain. The source and the drain are surrounded by the silicon fin which has an undoped nature. This is further surrounded by an extension implant and poly oxide. There is also a high-K dielectric and metal gate. Energizing the gate electrode will lead to excellent control over the channels and the channel surrounds the gate. After the gate electrode, the region of the fin located beneath the electrode is inverted. It forms a path that will conduct between the source and the drain. FinFET is a fully depleted device and the conduction happens in the outermost edges of the fin. The double gate of the FinFET reduces the I_{OFF} (see Fig. 2).

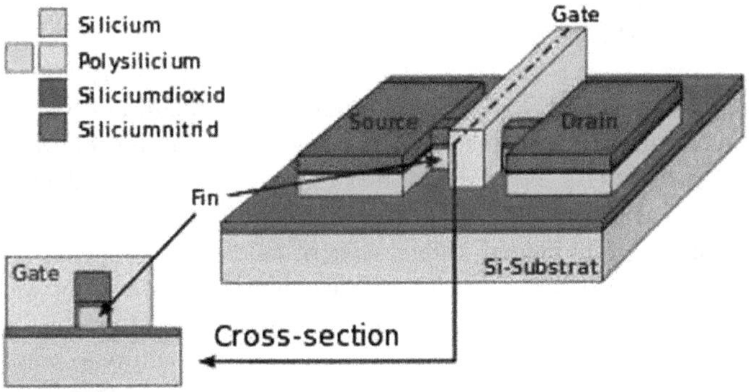

Fig. 2 FinFET device schematic [4]

1.2 FinFET Characteristics

- In order to improve the current drive, the source and the drain are raised, leading to the reduction of parasitic resistance and capacitance.
- The short channel effects are being suppressed by the development of silicon fin.
- FinFET is a type of MugFET [6]; it can achieve larger channel width by the usage of multiple fins. The number of fins is directly proportional to the increase in current.
- High performance can be achieved for symmetric gates, and asymmetric gates can also be built which focus on V_T.

The major foe of the semiconductor devices, short channel effects could be easily dealt with. FinFET is actually from the MuGFET [6] family which can be any fin-oriented multiple-gate transistors. The very law of Gordon Moore is very much satisfied when we see the fabrication technology revolution of *14 nm to 10 nm to 7 nm.* The theory of nanometres is the very story of processors which are used and the memories.

2 The FinFET Technology Nodes (14 nm, 10 nm, 7 nm)

The 14 nm was released in the year 2015, keeping in mind the battery lifetime of the cell phones. The switching speed was also high from the counterparts of SOI MOSFETs. The density of the Intel 14 nm FinFET is about 44 MTr/mm^2. Miniature the size, the cost explosion is also high. This will also be explored further as the paper explores more into the 10 nm and then to 7 nm FinFET technologies. The major usage is in the Intel's 5th to 9th generation processors. The 14 nm technology outsmarts the 12 nm as well as the 16 nm FinFET technology [7, 8].

10 nm came into the picture nearly about in the year 2017 with a density of about 99.00 MTr/mm^2 for Intel, and as for TSMC, it is about 60 MTr/mm^2. The 10 nm FinFET technology is nearly about three times denser than the 14 nm process. The major parameter under the scanner is their clock speed, and for this very reason, it is not been considered for the processor for desktops. The power reduction is nearly 60% and its 20% is faster than the TSMC's 12 nm/16 nm.

7 nm is the most highlighted technologies of modern times having a density of about 1.6 times that of 10 nm of TSMC, even a dominating power reduction over 45% over the 10 nm technology of FinFET. The density of the 7 nm process as depicted by Samsung is about 95.00 MTr/mm^2 (approx.) having an ability to present a much lightweight and slimmer version of smart phones; adding to that, there is a significant high switching speed in the 7 nm FinFET technology for the increased drive current and leakage current reduction. The thinner gate oxides lead to the requirement of lesser input voltage supply and finally lesser voltage swings. The FinFET technology outsmarts the normal planar MOSFET by their gate around approach around the

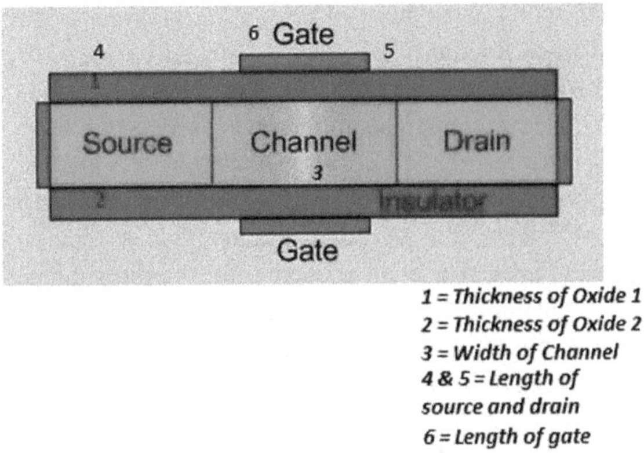

Fig. 3 Cross-section of FinFET simulated in NanoHub

channel, which in turn gives a better gate control, and as we shrink the size much more gate control is needed.

Despite all of these, we have a major challenge under the scanner, that is, the quantum tunnelling. This comes into the picture when the size shrinks and the transistors are in a fully turned on mode and cannot be turned down (see Fig. 3).

3 Results and Discussion

For the simulation purpose, we have used the *NgSPICE* as well as the *Nanohub.org* [8] in order to simulate the MOSFET as well as the FinFET in accordance to their dimension. In order to simulate the FinFETs, we have used the dimensions in accordance to Table 1. Here, the different technology nodes which are 14 nm, 10 nm and 7 nm, respectively, are shown. The default dielectric constant used in the insulator

Table 1 Parameters and technology nodes

Parameter	Technology nodes and testing conditions		
L_g	14 nm	10 nm	7 nm
L_s	8 nm	6 nm	4 nm
L_d	8 nm	6 nm	4 nm
O_s	2 nm	2 nm	2 nm
O_d	2 nm	2 nm	2 nm
W_{ch}	5 nm	4.5 nm	3.5 nm
T_{ox1}	1 nm	0.8 nm	0.65 nm
T_{ox2}	1 nm	0.8 nm	0.65 nm

L_g: Length of gate, L_s: Length of source, L_d: Length of drain, O_s: Gate overlap to source, O_d: Gate overlap to drain, W_{ch}: Width of channel, T_{ox1}: Thickness of oxide1, T_{ox2}: Thickness of oxide2 is 3.9, which is because of SiO_2, *silicon dioxide.* The given bias points are 3, to get the values for the DIBL, *drain-induced barrier lowering*, which will be discussed later in the paper. The thickness of the oxide or Tox is decided on the value of the dielectric constant. The gate type is used as the metal. The *MuGFET* [6] technology is used for testing conditions. The parameters are taken as such in consideration with the current technology nodes. For the simulation purpose, Geometry—X and Geometry—Y were considered, as the analysis is done on the 2-D device. The simulator used is *Padre.* The critical current for the threshold voltage is 1e−4. This is done for the logarithmic scale for the calibration of the drain current.

It is observed from Fig. 4a that the 14 nm FinFET [9] has good short channel controlling abilities. When compared with Fig. 4b and c, it can be observed clearly that the V_T is decreasing from 14 nm > 10 nm > 7 nm. The leakage current also is getting decreased. Lesser leakage current will also lead to low power consumption and high device efficiency (see Figs. 5 and 6).

As it is known that the current starts driving forward when it crosses the V_T, in the ideal case $I_D = 0$. But if we see practically, there is some current in the zone where there is a negligible gate voltage. This part is the subthreshold region, and the steeper

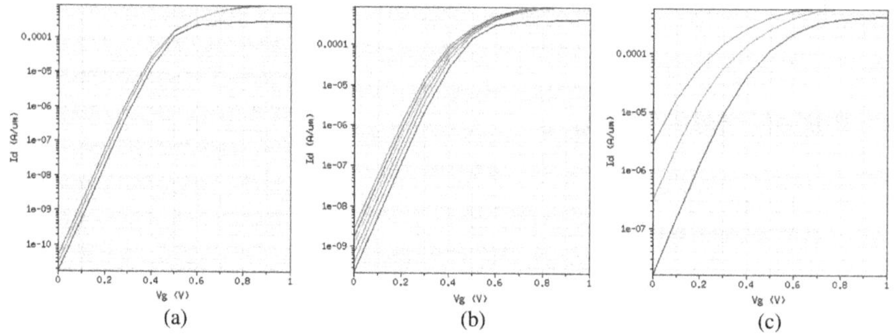

Fig. 4 **a** The log I_d versus V_{gs} of 14 nm FinFET; **b** Simulation results for log I_d versus V_{gs} of 10 nm FinFET. **c** Transfer characteristics for 7 nm FinFET

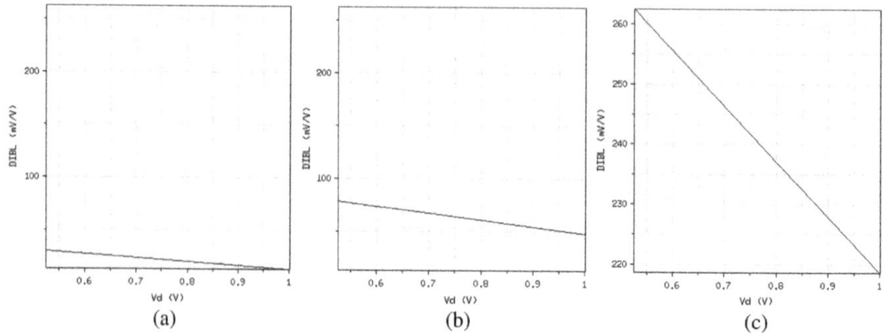

Fig. 5 **a** Drain-induced barrier lowering of 14 nm FinFET; **b** Simulation results for DIBL of 10 nm FinFET. **c** DIBL for 7 nm FinFET

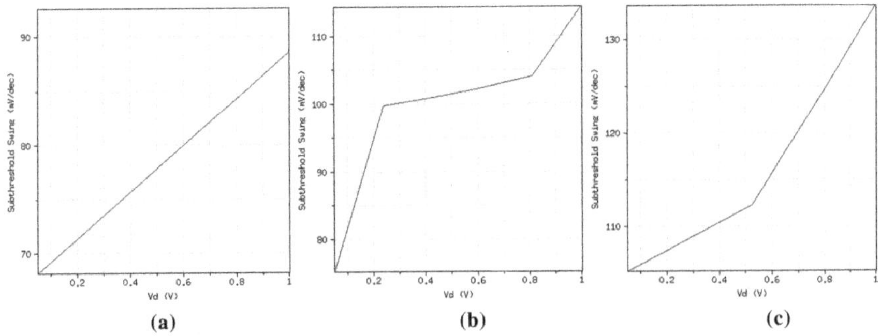

Fig. 6 **a** Simulation in nanohub.org for subthreshold swing for 14 nm FinFET; **b** Simulation results for subthreshold swing of 10 nm FinFET. **c** Subthreshold swing simulation for 7 nm FinFET

the *subthreshold swing* the lesser is the I_{OFF} and the more is the I_{ON}/I_{OFF}. The drastic decrease of the drain-induced barrier lowering can be seen clearly as the devices get shrunk down, and finally being used for cell phones or desktop processors for their longer battery lifetime and significant low V_T, to provide the same encouraging circuit designs with low power circuits.

4 Conclusion

This paper uses the 2-D models for examining the characteristics of the 14 nm, 10 nm and 7 nm technology nodes. It can be well enough concluded that though the 7 nm of Intel is under the development mode the TSMC has launched its 7 nm, and the current technology node of 7 nm is quite superior in terms of density. As examined from the overall characteristics though even after a bit of leakage current, there are abundant superior characteristics for the 7 nm FinFET technology to outsmart the others. The blessing is that the advance of research is such that the 5 nm and 3 nm technologies are not very far away from commercial use.

References

1. Taur Y, Ning TH (1998) Fundamentals of modern VLSI devices, 2nd edn. Cambridge Univ. Press, New York
2. Narendar V, Narware P, Bheemudu V et al (2020) Investigation of short channel effects (SCEs) and analog/RF figure of merits (FOMs) of dual-material bottom-spacer ground-plane (DMBSGP) FinFET. Silicon 12:2283–2291. https://doi.org/10.1007/s12633-019-00322-2
3. Pal RS, Sharma S, Dasgupta S (2017) Recent trend of FinFET devices and its challenges: a review. In: 2017 conference on emerging devices and smart systems (ICEDSS), Tiruchengode, pp 150–154. https://doi.org/10.1109/icedss.2017.8073675

4. Computer Hope (2017) Accessed on May 31, 2020. https://www.computerhope.com/jargon/f/finfet.htm
5. Maszara WP, Lin M (2013) FinFETs—technology and circuit design challenges. In: 2013 Proceedings of the ESSCIRC (ESSCIRC), Bucharest, pp 3–8. https://doi.org/10.1109/esscirc.2013.6649058
6. Li J, Gauthier R, Li Y, Mishra R (2014) ESD device performance analysis in a 14 nm FinFET SOI CMOS technology: fin-based versus planar-based. In: Electrical overstress/electrostatic discharge symposium proceedings 2014, Tucson, AZ, pp 1–6
7. Xie Q, Xu J (2016) Recent research development of FinFETs. Sci China Phys Mech Astron 59: https://doi.org/10.1007/s11433-016-0394-5
8. Klimeck G (2007) NanoHUB.org tutorial: education simulation tools. In: 2007 2nd IEEE international conference on nano/micro engineered and molecular systems, Bangkok, pp nil41. https://doi.org/10.1109/nems.2007.351992
9. Jurczak M, Collaert N, Veloso A, Hoffmann T, Biesemans S (2009) Review of FINFET technology. In: 2009 IEEE international SOI conference, Foster City, CA, pp 1–4. https://doi.org/10.1109/soi.2009.5318794

Delta-Doped Layer-Based Hetero-Structure DG-PNPN-TFET: Electrical Property and Temperature Dependence

Karabi Baruah and Srimanta Baishya

Abstract In this work, we have analyzed the electrical properties and the impact of temperature variation on the characteristics of delta-doped layer-based hetero-structure DG-PNPN-TFET. The basic electrical properties of the proposed device are compared with the conventional double-gate TFET, and enhanced results are obtained in terms of ON current, ON/OFF current ratio, and subthreshold swing (SS). It is familiar that temperature variation influences the device characteristics. The paper shows that the device characteristics have weak temperature dependence at high gate voltage, i.e., the ON-state current of the proposed device is almost temperature independent. However, the dependence is more when the gate voltage is low, resulting in an increase of leakage current at upraised temperatures. A wide range of temperatures 250–500 K has been considered to analyze the performance of the device using the 2D Sentaurus TCAD device simulation tool.

Keywords Band-to-band tunneling (BTBT) · Cut-off frequency (f_T) · Gain-bandwidth product (GBP) · PNPN-TFET · Subthreshold swing (SS) · Tunnel FET (TFET)

1 Introduction

Continuous scaling of CMOS technology results in severe problems in device characteristics, including degradation of leakage current and subthreshold swing, and enhancement of some short channel effects [1]. A tunnel field-effect transistor (TFET) is a promising candidate in this field, for providing low OFF current and steep subthreshold slope [2–4]. The carrier transport mechanism in TFET is different from conventional MOSFET, which is mainly due to band-to-band quantum tunneling (BTBT) of electrons [5, 6]. However, this tunneling behavior of TFET results in low ON current, which is a major flaw of tunnel FET. Several works have been carried

K. Baruah (✉) · S. Baishya
National Institute of Technology Silchar, Silchar, Cachar, Assam, India

S. Dhar et al. (eds.), *Advances in Communication, Devices and Networking*,
Lecture Notes in Electrical Engineering 776,
https://doi.org/10.1007/978-981-16-2911-2_3

out by the researchers using various techniques such as optimizing device geometry, using a dual material gate, high-k gate dielectric, low bandgap material, source pocket engineering [7–9], etc. to achieve higher ON current in TFET.

Thermal reliability is one of the critical issues faced by the semiconductor industry nowadays. As the transistor's number on a chip increases gradually, heat dissipation on the chip also increases remarkably. For this reason, it is mandatory to analyze the device performances at various ambient temperatures [10, 11]. In this work, we have proposed a delta-doped layer-based DG-PNPN-TFET structure, with SiGe as pocket material. When a highly doped delta layer is inserted in the source area, the tunneling volume increases, resulting in a higher drain current compared to the conventional TFET [12]. The double-gate TFET structure provides better electrostatic control than the single-gate TFET structure. When a pocket is introduced in the source side of a TFET (PNPN-TFET), it increases the lateral field, and hence, ON current. In this work, the temperature sensitivity analysis of the proposed device is studied in terms of DC and analog/RF parameters considering various ambient temperatures ranging from 250 to 500 K.

2 Parameters and Models Used for Device Simulation

The 2D cross-sectional view of the proposed device is shown in Fig. 1. The device body thickness is fixed at 10 nm and 2 nm and HfO_2 is used as high-k gate dielectric material (EOT = 0.37 nm). The other device parameters used in the simulation process are listed in Table 1. Low bandgap material, Germanium (bandgap 0.66 eV at room temperature), is used as a source material to improve the ON current of the device [13]. Constant profile doping is considered for both conventional and proposed devices. Source doping is considered higher than drain doping to suppress the ambipolar conduction which is an inherent property of tunnel FET. $Si_{1-X}Ge_X$ (mole fraction, $X = 0.3$) is introduced as pocket material in the source-channel junction [14]. An optimized width of the delta-doped layer (L_2) is present at an optimized distance (L_1) from the source channel junction in our proposed device. An optimum value of gate metal work function, 4.25 eV, has been considered for

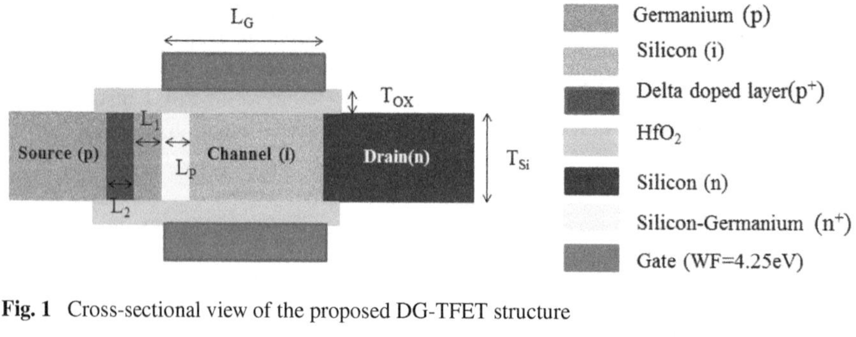

Fig. 1 Cross-sectional view of the proposed DG-TFET structure

Table 1 Parameters used in the simulation for the proposed device

Parameters	Dimensions
Gate length (L_g)	25 nm
Gate work function	4.25 eV
Source doping (N_S)	4×10^{19} cm^{-3}
Delta layer doping (N_D)	1×10^{20} cm^{-3}
Pocket doping (N_P)	2×10^{19} cm^{-3}
Drain doping (N_D)	5×10^{18} cm^{-3}
channel doping (N_{CH})	1×10^{16} cm^{-3}
Pocket length (L_P)	2 nm
Distance of delta layer from source channel interface (L_1)	2 nm
Length of delta-doped layer (L_2)	2 nm

improved characteristics. All simulation is performed using the SENTAURUS TCAD device simulator [15].

Various models are used for simulation such as the non-local band-to-band tunneling model to calculate the generation rate at each mesh point in the tunneling region, bandgap narrowing model, Fermi Dirac statistics, concentration and field-dependent mobility model, and SRH model for carrier recombination. The TCAD models were calibrated with the experimental results showed in [16], as shown in Fig. 2.

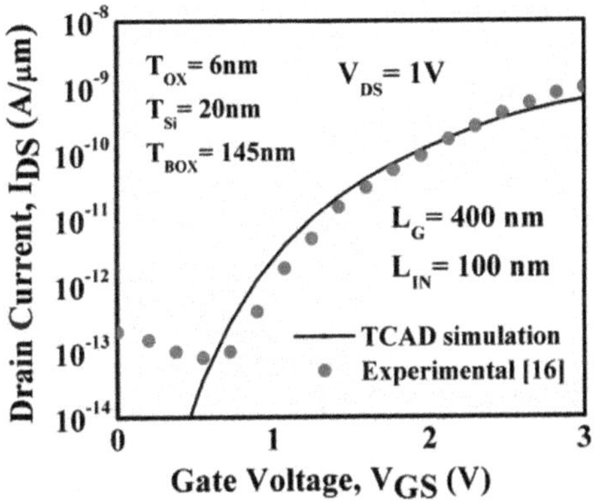

Fig. 2 The TCAD simulated models are calibrated with the experimental result [16]

3 Results and Discussion

A comparative view of transfer characteristics of the proposed device as well as conventional TFET is shown in Fig. 3a. An improved I_{ON} (0.12 mA) and steeper subthreshold slope (32 mV/decade) are achieved in the case of the proposed TFET due to the use of the Germanium and delta-doped layer in the source side. However, OFF current degrades than the conventional one because of the use of high k gate dielectric material in the proposed structure as well as Germanium at the source. There is a tradeoff between ON current and OFF current when Germanium is used as source material. The proposed structure provides an I_{ON}/I_{OFF} ratio in the order of 10^9. To achieve better performance, we have optimized the device in terms of drain voltage, taking three different drain voltages as shown in Fig. 3b. From the figure it can be seen that there is a slight increase in ON current as well as OFF current, with an increase in V_{DS}; however, the undesirable ambipolar current, which is an inherent property of TFET, also increases tremendously with drain voltage in all the three cases. Hence the optimized value 0.7 V of V_{DS} is considered for the rest of the work.

The energy band diagram of the proposed device at ON-state condition as well as at OFF-state condition is shown in Fig. 4a. During OFF-state condition ($V_{GS} = 0$, $V_{DS} = 0.7$ V) the channel region conduction band lies at a higher position than the source region valance band. No electron can tunnel through the source channel junction barrier as the barrier width is large. When we apply some voltage in the gate terminal, the conduction band in the channel moves downward. Now carriers can easily tunnel from the source valance band to the channel conduction band through the narrow barrier. Due to the presence of the pocket which is n-type, near the source which is p-type, the conduction band at the channel region inclines closer toward the valance band edge of the source region. The ON-state energy band diagram of the proposed device at three different temperatures of 250, 350 and 500 K is shown in Fig. 4b. From the figure, it can be seen that there is a little variation of the energy band with different temperatures.

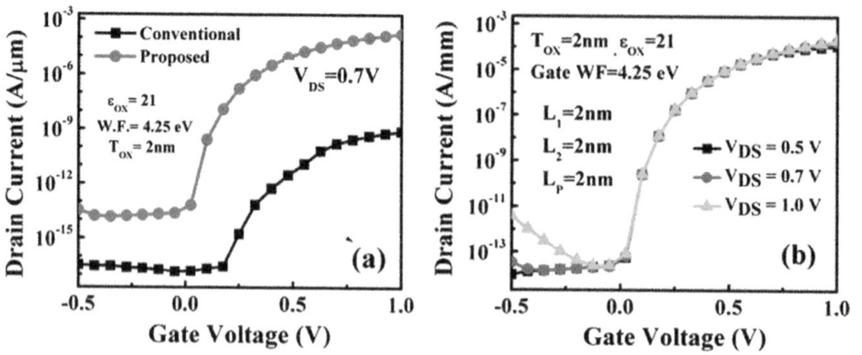

Fig. 3 **a** Comparative view of transfer characteristics of conventional DG-TFET and the proposed TFET, **b** Transfer characteristics of the proposed device at different V_{DS}

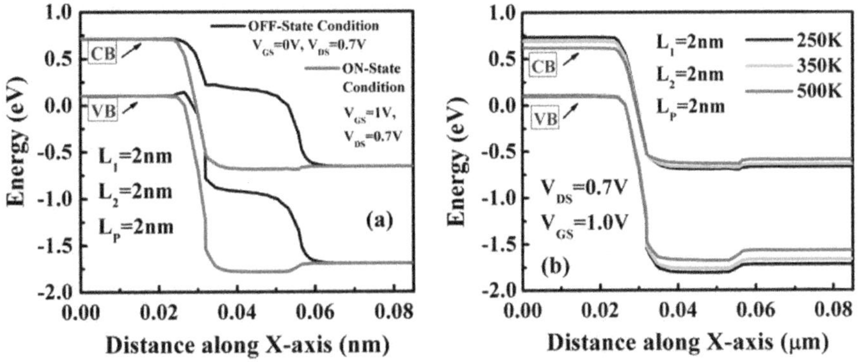

Fig. 4 **a** Energy band of the proposed device at OFF-state and ON-state condition, **b** ON-state energy band of the proposed TFET at three different temperatures

Temperature is a physical parameter that indirectly influences device performance by affecting the energy bandgap [8, 17]. The temperature has an inverse relationship with the energy bandgap which can be understood by the equation, $E_g(T) = E_g(0) - \frac{\alpha T^2}{T+\beta}$, where $E_g(T)$ is the bandgap at temperature T, $E_g(0)$ is the bandgap at zero Kelvin, and α, β are the fitting parameters [18]. Figure 5a shows the I_D-V_{GS} characteristics of the proposed device for different temperatures ranging from 250 to 500 K at $V_{DS} = 0.7$ V. The figure reveals that the transfer characteristics depend on temperature and is a strong function of gate bias [11]. At lower gate voltage the drain current is almost constant at all temperatures except 250 K. The leakage current is measured in this region at $V_{GS} = 0$ V, which increases from the order of 10^{-15} to 10^{-8} with an increase in temperature from 250 to 500 K. This is a result of carrier generation in the reverse-biased junction. After that, the drain current increases exponentially with gate voltage up to some V_{GS}. The average SS progressively degrades from 250 to 500 K in this region. At higher gate voltage there

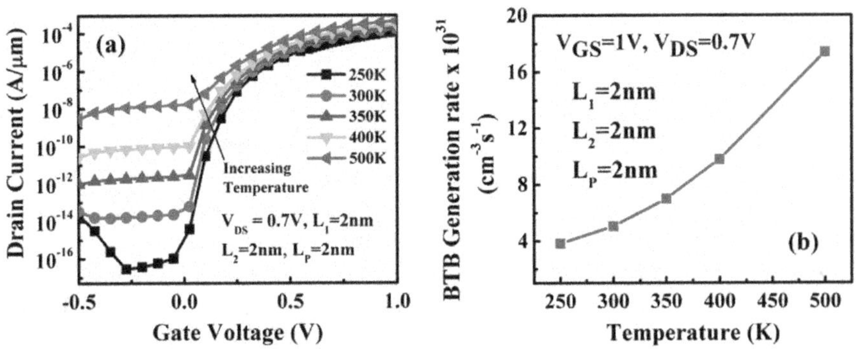

Fig. 5 **a** Transfer characteristics, **b** Band-to-band tunneling rate of the proposed structure at different temperatures ranging from 250 to 500 K, respectively

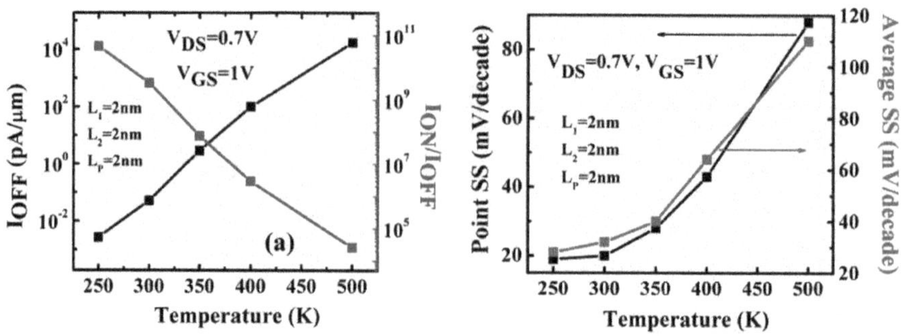

Fig. 6 **a** Variation of OFF current and ON/OFF ratio with the temperature at constant V_{DS} and V_{GS}, **b** Variation of SS with temperature

is a slight change of I_{ON} with temperature because the device depends more on the band-to-band tunneling rather than the temperature. Figure 5b presents the variation in the BTBT rate at various temperatures. It is clear from the figure that the BTBT rate is reliant on temperature. With an increase in temperature the bandgap of semiconductor decreases, thus narrower bandgap increases the tunneling of the carriers, hence increases the BTBT rate.

The variation of I_{OFF} and $I_{\mathrm{ON}}/I_{\mathrm{OFF}}$ with temperature is analyzed in Fig. 6a. It can be observed that the OFF current increases with temperature due to thermally generated charge carriers in the depletion region. However, as the increment in the ON current is very less compared to the OFF current, the ON/OFF current ratio decreases as the temperature increases from 250 to 500 K. Subthreshold swing is an important parameter that is defined as the change in gate voltage required causing a one-decade change in drain current in the subthreshold region [19]. Figure 6b shows the variation of SS with temperature. A slight variation of SS with temperature can degrade the performances of the device. Point SS and average SS are shown in the same figure for the different temperatures at 0.7 V of V_{DS}. It is found that the average SS of the device is larger than the point SS for all temperatures.

Transconductance, $g_{m,}$ plays an important role in achieving a high value of f_T. A higher value of g_m gives a higher value of cut-off frequency. Figure 7a shows the transconductance (g_m) variation with gate voltage at different temperatures. The increasing value of the transconductance is observed at lower gate voltage due to the increase in drain current which is a result of the application of low bandgap material with a delta-doped layer at the source side. As V_{GS} increases further, the value g_m starts decreasing due to mobility degradation [20]. Further, it can be seen that variation of g_m with gate voltage is small at lower temperatures (up to 400 K) compared to that in high temperatures. In Fig. 7b we have analyzed the reliance of biased-dependent parasitic capacitance C_{gd} on ambient temperatures. In the case of TFET, the total gate capacitance is majorly composed of gate-drain capacitance rather than gate-source capacitance [21]. Hence the variation in C_{gd} is almost the same as C_{gg} with respect to drain voltage. It can be observed from the figure that C_{gd} increases with the rise in temperature, i.e., the value of the capacitances degrades

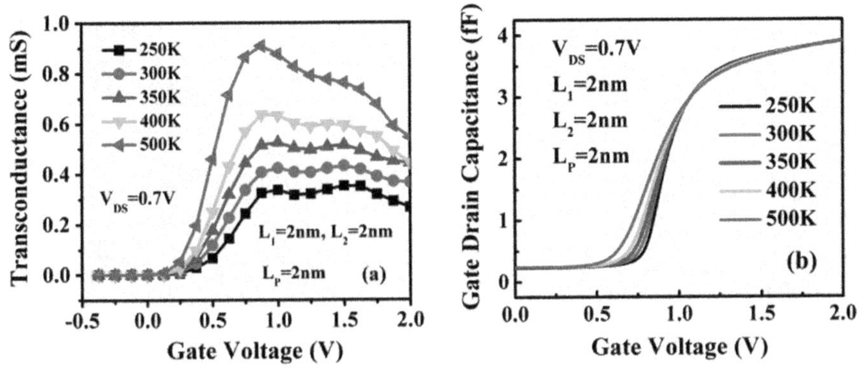

Fig. 7 **a** Variation of transconductance, **b** Variation of gate to drain capacitance with gate voltage at different temperatures, respectively

with an increase in temperature, leading to high leakage current, thus making the device less suitable in a high-temperature application.

Another most important parameter to check the RF performances of a device is the cut-off frequency [22]. It is the frequency at which the current gain becomes unity and is defined as [17], $f_T = \frac{g_m}{2\pi C_{gg}}$. Figure 8 shows the variation of f_T with gate voltage at different temperatures ranging from 250 to 500 K, when V_{DS} is 0.7 V. From the figure, it can be seen that f_T almost follows the pattern g_m. Up to around 1 V of V_{GG}, g_m increases exponentially due to the increase in ON current, hence f_T also shows the enhancement up to some V_{GG}. When V_{GG} increases beyond 1 V (approx..), a decrement in cut-off frequency is observed due to the degraded value of g_m which is because of mobility degradation and higher value of total gate capacitance [20]. Table 2 shows the comparison chart of some electrical properties of the proposed device with respect to some practical sources [23–25].

Fig. 8 Cut-off frequency variation with gate voltage for different temperatures at $V_{DS} = 0.7$ V

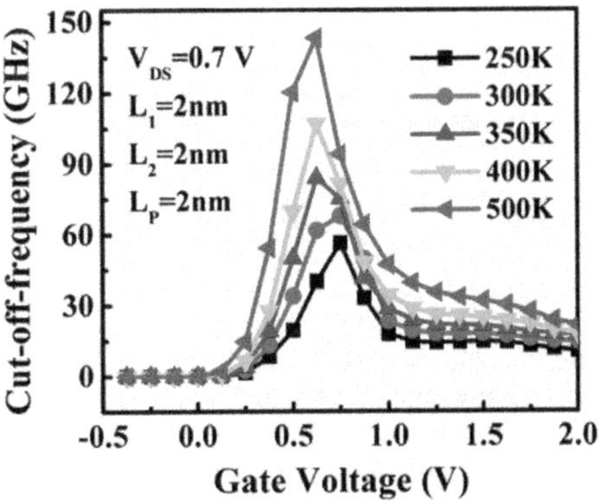

Table 2 Comparison of electrical properties of the proposed TFET with some practical authentic sources

	Ref. [23]	Ref. [24]	Ref. [25]	This work
	Ge-TFET	Si-Ge PAI pocket	SiGe/Si heterojunction	Delta doped with pocket
T_{ox} (nm)	3(SiO_2)	SiO_2	HfO_2/SiO_2	2(HfO_2)
L_G (nm)	5000	2000	1000	22
I_{ON} (A/μm)	0.42×10^{-6}	14.5×10^{-6}	20×10^{-6}	0.12×10^{-3}
I_{OFF} (pA/μm)	0.12	0.1	10	0.6
I_{ON}/I_{OFF}	3E6	1.45E8	2E6	1E9
SS (mV/Decade)	~70	80	>120	32
V_D (V)	0.5	1	1.2	0.7

4 Conclusion

In this work, we have investigated the performance of a delta-doped layer-based hetero-structure DG-PNPN-TFET with temperature variation. We have achieved improved results in terms of I_{ON}, I_{ON}/I_{OFF}, and SS for the proposed TFET. It is observed that the device characteristics are less affected by the temperature at the high gate voltage in contrast to the low gate voltage. The device shows a huge variation of OFF current (10^{-15} to 10^{-8} A/μm) and SS (28-110 mV/decade) for the temperature range of 250–500 K. On the other hand, the variation in I_{ON} (1.23 $\times$ 10^{-4} to 4.57 $\times$ 10^{-4} A/μm) is negligible for different temperatures. The study reveals that there is more distinguished degradation in subthreshold characteristics in comparison to the superthreshold characteristics. The proposed DG-PNPN-TFET shows a positive temperature coefficient for the entire gate bias range. In summary, the characteristics of the proposed device degrade at high temperatures, indicating that the device is more suitable for low and medium-temperature applications.

References

1. Ionescu AM, Riel H (2011) Tunnel field-effect transistors as energyefficient electronic switches. Nature 479(7373):329–337
2. Seabaugh AC, Zhang Q (2010) Low-voltage tunnel transistors for beyond CMOS logic. Proc IEEE 98(12):2095–2110
3. Ram MS, Abdi DB (2014) Single grain boundary tunnel field effect transistors on recrystallized polycrystalline silicon: proposal and investigation. IEEE Electron Device Lett 35(10):989–991
4. Vishnoi R, Kumar MJ (2014) Compact analytical model of dual material gate tunneling field effect transistor using interband tunneling and channel transport. IEEE Trans Electron Device 61(6):1936–1942
5. Naik R, Gupta R (Apr 17) An introduction double gate tunnel field effect transistor. Int J Res Dev Technol 7(4)

6. Thorat KS, Turkane SM (Feb 2014) A comparative analysis of TFET characteristics for low power digital circuits. Int J Curr Eng Technol 4(1)
7. Krishnamohan T, Kim D, Nguyen CD, Jungemann C, Nishi Y, Saraswat KC (2006) High-mobility low band-to-band-tunneling strained-germanium double-gate heterostructure FETs: simulations. IEEE Trans Electron Devices 53(5):1000–1009
8. Boucart AM (2007) Double-gate tunnel FET with high-k gate dielectric. IEEE Trans Electron Devices 54(7):1725–1733
9. Jhaveri R, Nagavarapu NV, Woo JCS (2011) Effect of pocket doping and annealing schemes on the source-pocket tunnel fieldeffect transistor. IEEE Trans Electron Devices 58(1):80–86
10. Madan J, Chaujar R (2017) Numerical simulation of n+source pocket PIN-GAA-Tunnel FET: impact of interface trap charges and temperature. IEEE Trans Electron Devices 64(4):1482–1488
11. Mookerjea S, Mohata D, Mayer T, Narayanan V, Datta S (2010) Temperature-dependent I–V characteristics of a vertical $In_{0.53}Ga_{0.47}As$ tunnel FET. IEEE Electron Device Lett 31(6):564–566
12. Panda S, Dash S, Behera SK, Mishra GP (July 2016) Delta-doped tunnel FET (D-TFET) to improve current ratio (ION/IOFF) and ON-current performance. J Comput. Electron
13. Krishnamohan T, Kim D, Raghunathan S, Saraswat K (2008) Double-gate strained-ge heterostructure tunneling FET (TFET) with record high drive currents and <60 mV/dec subthreshold slope. In: 2008 IEEE international electron devices meeting, San Francisco, CA, pp 1–3
14. Nagavarapu V, Jhaveri R, Woo JCS (2008) The tunnel source (PNPN) n-MOSFET: A novel high performance transistor. IEEE Trans Electron Devices 55(4):1013–1019
15. Sentaurus Device User Guide, Version G-2012.06, June 2012
16. Biswas A, Dan SS, Royer CL, Grabinski W, Ionescu AM (2012) TCAD simulation of SOI TFETs and calibration of non-local band to band tunneling model. Microelectron Eng 98:334–337
17. Shrivastava V, Kumar A, Sahu C, Singh J (Mar 2016) Temperature sensitivity analysis of dopingless charge-plasma transistor. Solid-State Electron 117:94–99. ISSN 0038-1101
18. Sze SM, Ng KK (2011) Physics of semiconductor devices, 3rd edn. Wiley India Edition
19. Tsividis Y, McAndrew C (2011) Modeling and operation of MOS transistor. Oxford University Press
20. Nigam K, Kondekar P, Sharma D (Apr 2016) DC Characteristics and Analog/RF performance of novel polarity control GaAs-Ge based tunnel field effect transistor. Superlattices Microstruct 92:224–231. ISSN 0749-6036
21. Yang Y, Tong X, Yang L-T, Guo P-F, Fan L, Yeo Y-C (2010) Tunneling field-effect transistor: capacitance components and modeling. IEEE Electron Device Lett 31(7):752–754
22. Pandey S, Sahu C, Singh J (Jan 2016) A highly linear RF mixer using gate-all-around junctionless transistor. Int J Electron Lett (Taylor & Francis), 1–8
23. Kim SH, Kam H, Hu C, Liu TJK (2009) Germanium source Tunnel field effect transistors with record high I_{ON}/I_{OFF}. In: Symposium on VLSI technology, IEEE, pp 178–179
24. Cheng W, Liang R, Xu G, Yu G, Zhang S, Yin H, Zhao C, Ren TL, Xu J (Apr 2020) Fabrication and characterization of a novel Si line tunneling TFET with high drive current. IEEE J Electron Devices Soc (8):336–340
25. Walke AM, Vandooren A, Rooyackers R, Leonelli D, Hikavyy A, loo R, Verhulst AS, Kao KH, Huyghebaert C, Groeseneken G, Rao VR (Mar 2014) Fabrication and analysis of a $Si/Si_{0.55}Ge_{0.45}$ heterojunction line tunnel FET. IEEE Tans Electron Devices 61(3):707–715

Performance Study of Ambipolar Conduction Suppression for Dual-Gate Tunnel Field-Effect Transistors

Ritam Dutta and Nitai Paitya

Abstract The continuous device structure degradation in nanoscaled device modeling restricts conventional tunnel field-effect transistor (TFET) to meet the best low-power applications. In this paper, dual-gate tunnel field-effect transistor (DG-TFET) with ultra-thin pocket doping has been thoroughly investigated and the performance analysis is studied. The incorporation of 2 nm of source and drain pocket doping results in substantial changes in electric field distribution and surface potential variation along the channel (two-dimensional). Nonlocal band-to-band tunneling (BTBT) model is used for simulation work. At supply voltage (V_{DD}) of 0.5 V, the source pocket-doped tunnel FET (SP-TFET) produces better leakage current ($I_{\rm OFF}$) control, i.e., ambipolar conduction suppression, validated by analytical modeling. SP-TFET provides 30.08 mV/decade of subthreshold swing (SS) recorded best, which further improves $I_{\rm ON}/I_{\rm OFF}$ switching ratio. Two-dimensional numerical device simulator is used for the simulation work.

Keywords DG-TFET · Band-to-band tunneling (BTBT) · Subthreshold swing (SS) · Source pocket doping · Drain pocket doping · Switching ratio

1 Introduction

Continuous miniaturization in device parameters limits the metal oxide field-effect transistor (MOSFET) operation on various factors. Since the device parameters are scaled down rapidly from millimeter to nanometer range, therefore short channel effects (SCE) play a vital role due to which the MOSFET operations are not providing

R. Dutta (✉)
Department of Electronics and Communication Engineering, Surendra Institute of Engineering and Management, Maulana Abul Kalam Azad University of Technology, Kolkata, West Bengal, India

R. Dutta · N. Paitya
Department of Electronics and Communication Engineering, Sikkim Manipal Institute of Technology, Sikkim Manipal University, Gangtok, Sikkim, India
e-mail: nitai.p@smit.smu.edu.in

S. Dhar et al. (eds.), *Advances in Communication, Devices and Networking*,
Lecture Notes in Electrical Engineering 776,
https://doi.org/10.1007/978-981-16-2911-2_4

the best switching solutions. The subthreshold swing (SS) provided by the conventional MOSFET is now getting limited to 60 mV/decade [1]. This problem is well identified by several researchers and the new device physics has been introduced named quantum tunneling. Using band-to-band quantum tunneling, the tunnel field-effect transistors (TFET) achieves better control over subthreshold swing (less than 60 mV/decade) [2]. Although having a better subthreshold swing, tunnel field-effect transistors face the real challenges for its low ON-state current (I_{ON}) and ambipolar conduction. In this article, we have thoroughly studied the recent literature works [3, 4], which addressed the low drive current and ambipolar conduction of tunneling FETs.

F. Hosenfield et al. investigated the major drawback of ambipolar conduction and introduced a thin pocket layer beside the source-channel interface regions of TFET [5]. P. Kondekar et al. emphasize work function engineering [6] as their approach for suppressing ambipolarity in their research work. S. Saurabh et al. discussed the limitations of ambipolarity issues, which can raise the unwanted leakage current. This leakage current is also known as OFF-state current (I_{OFF}). The researchers enlightened the disadvantages of growing leakage current for scaled TFET structures [7]. Besides these, the gate–drain overlapping and source–gate overlapping have been introduced by M.J Kumar et al. in their research work [8] for controlling ambipolar current in TFET devices.

2 Device Structures with Thin Pocket Layer Doping

2.1 Device Model Specifications

The conventional structure of dual-gate tunnel FET (DG-TFET) has been studied and modeled using a numerical device simulator [9–12]. The middle portion between source and drain regions is termed as the intrinsic region in this TFET structure. Since the middle region is less doped, therefore the quantum tunneling can be produced between the source–channel interface for the *n*-channel TFET and drain–channel interface for the *p*-channel TFETs.

Figure 1 shows the conventional structure of dual-gate TFET where the intrinsic channel region is less doped and all three regions' materials are assumed as silicon (Si) for ease of simulation.

A thin pocket doping layer (source–pocket) of 2 nm has been introduced at the source–channel junction area (Fig. 2). Source–pocket layer length denoted as (L_{SP}) is introduced for better control in ambipolar current suppression.

Similarly, in Fig. 3, we have placed a thin pocket doping layer (drain–pocket) of 2 nm introduced at the drain–channel junction area. The length of the drain–pocket layer denoted as (L_{DP}) is introduced for better control in ambipolar current suppression. However, in all three device structures, *n*-channel is modeled by a device simulator.

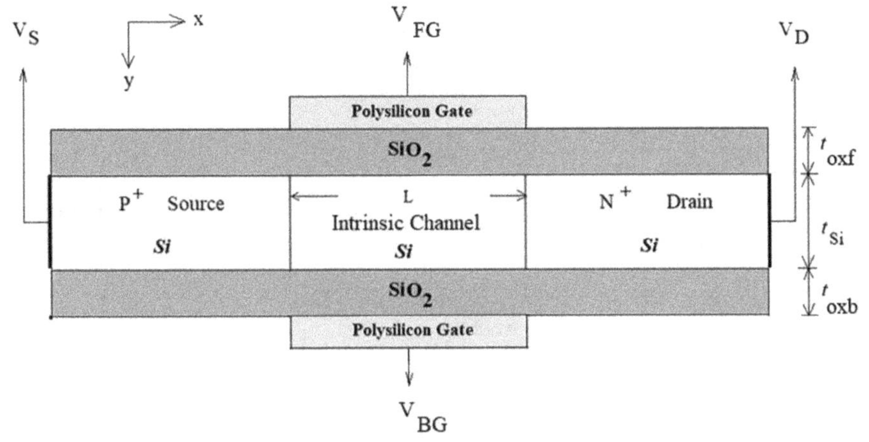

Fig. 1 Schematic model of dual-gate tunnel FET (DG-TFET)

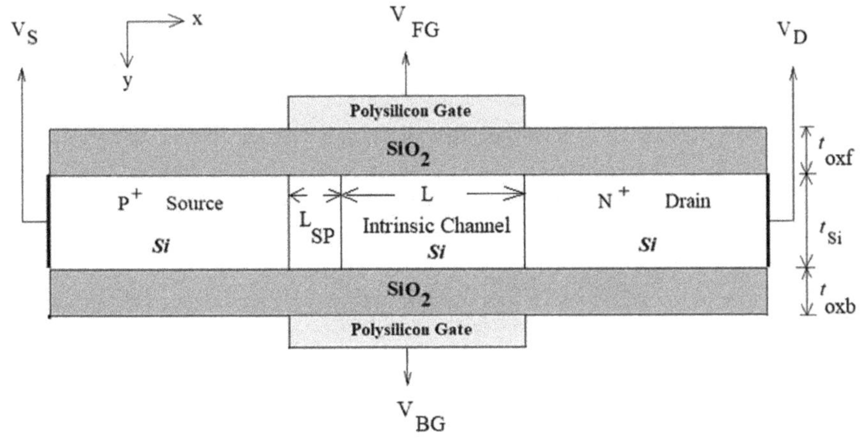

Fig. 2 Schematic model of dual-gate source pocket tunnel FET (DG-SP-TFET)

2.2 *Device and Electrical Parameter Analysis*

The DG-SP-TFET and DG-DP-TFET are designed by varying the device and electrical parameters to obtain better I_{ON}. The typical parameters for the above-mentioned tunnel FET structures are summarized in Table 1. The device parameters used in this work are purely based on the recent literature survey of both simulation and experimental results.

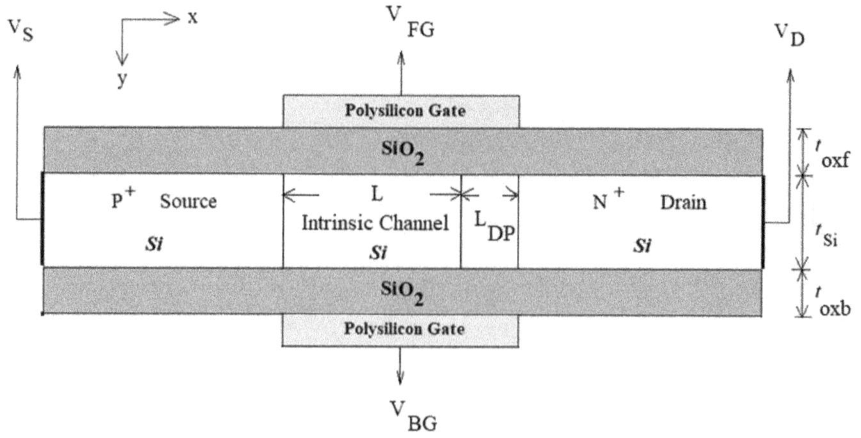

Fig. 3 Schematic model of dual-gate drain pocket tunnel FET (DG-SP-TFET)

Table 1 Typical parameters of various dual-gate tunnel FET structures

Parameters	Dual-gate tunnel FET with *Si* channel
Intrinsic channel length (L)	50 nm
Source-doped region length (L_{SP})	2 nm
Drain-doped region length (L_{DP})	2 nm
Front gate oxide thickness (t_{oxf})	2 nm
Back gate oxide thickness (t_{oxb})	2 nm
Body thickness (t_{si})	10 nm
Doping at source area (N_s)	1×10^{20} cm^{-3}
Doping at drain area (N_d)	5×10^{18} cm^{-3}
Channel doping (N_i)	1×10^{17} cm^{-3}
Work function of metal used	4.8 eV
Gate voltage (V_{GS})	1 V
Supply voltage (V_{DD})	0.5 V

3 Proposed Methodology for Analytical Modeling

A detailed numerical analysis is made to validate our simulation result. Therefore, at first, the regions have been categorically identified; then individual boundary conditions are considered.

Two-dimensional nonlinear Poisson equation is used to determine the surface potential [13].

$$\frac{\partial^2\psi(x,y)}{\partial x^2}+\frac{\partial^2\psi(x,y)}{\partial y^2}=\frac{qN_i}{\varepsilon_{Si}} \quad for \quad 0\leqslant x\leqslant L, \quad 0\leqslant y\leqslant t_{Si} \tag{1}$$

where $\psi(x, y)$ = electrostatic potential, q = charge, ε_{Si} = permittivity of material used and N_i = intrinsic doping along the channel (L) and t_{si} = thickness of channel.

For the electric field modeling, Yong's formula is used for electric field calculations of E_x and E_y (lateral and vertical) [14]:

$$E_x = -\frac{\partial\psi(x,y)}{\partial x} \text{ and } E_y = -\frac{\partial\psi(x,y)}{\partial y}. \tag{2}$$

So, the total electric field can be denoted as:

$$E = \sqrt{{E_x}^2 + {E_y}^2}. \tag{3}$$

Now, the tunneling current (I_{tun}) can be determined [15] using Kane's model [16],

$$I_{\text{tun}} = q\iint G_{BTBT}\ \text{d}x\ \text{d}y \tag{4}$$

$$G_{BTBT} = A_K\frac{|E|^D}{\sqrt{E_g}}\exp\left(-\frac{B_K E_g^{3/2}}{|E|}\right). \tag{5}$$

where G_{BTBT} is the generation rate of BTBT tunneling, A_K and B_K are process parameters. I_{tun} can now be computed from Eq. (4), and E_g is the energy gap.

4 Simulation Results with Contour Plots

The contour plot of the electric field distribution (Fig. 4) using nonlocal BTBT model is investigated [17–20].

The contour plot of surface potential is obtained, Fig. 5, by keeping the supply voltage $V_{DD} = 0.5$ V and $V_{GS} = 0.5$ V.

In Fig. 6, the transfer characteristics is plotted where the 2 nm thick source–pocket dual-gate TFET provides better drain current (I_{ON}) and better suppressed ambipolar current (I_{OFF}). The DG-SP-TFET provides subthreshold swing (SS) as 30.08 mV/decade which is much lower than the conventional TFET (43.74 mV/decade) without any pocket doping. This further results in better switching.

From the graphical analysis of Fig. 7, we can see that $I_{\text{OFF}} = 6.12 \times 10^{-16}$ A/μm and $I_{\text{ON}} = 2.12 \times 10^{-5}$ A/μm for DG-SP-TFET. From this, we also can find the subthreshold swing (SS) mentioned in Table 2.

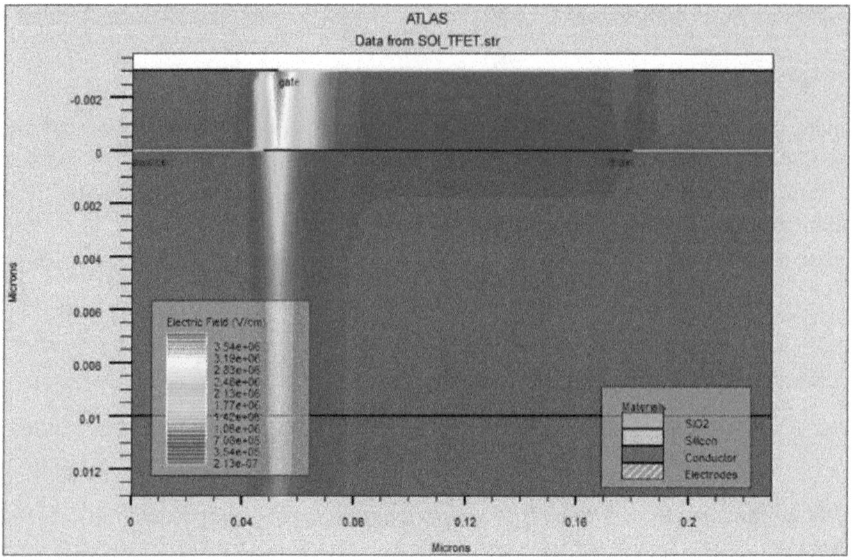

Fig. 4 Contour plot of electric field distribution of dual-gate TFET

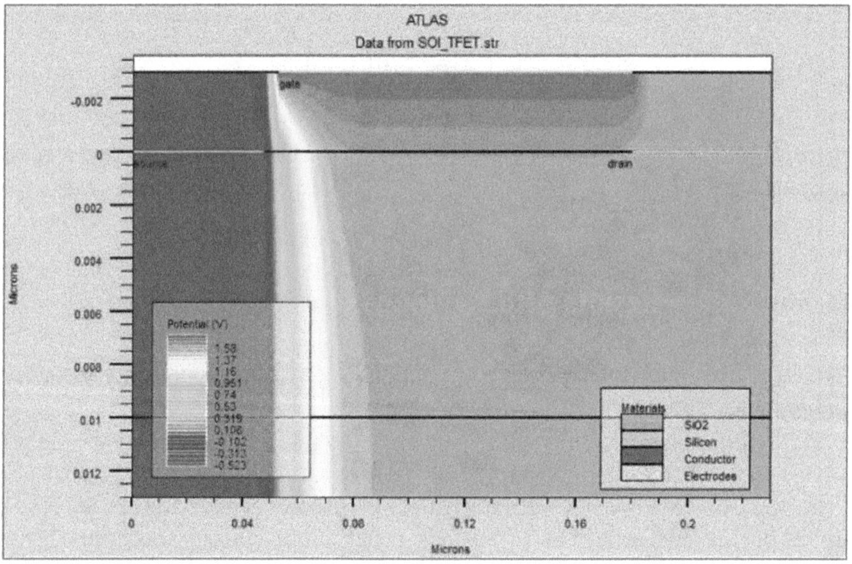

Fig. 5 Surface potential contour of dual-gate TFET

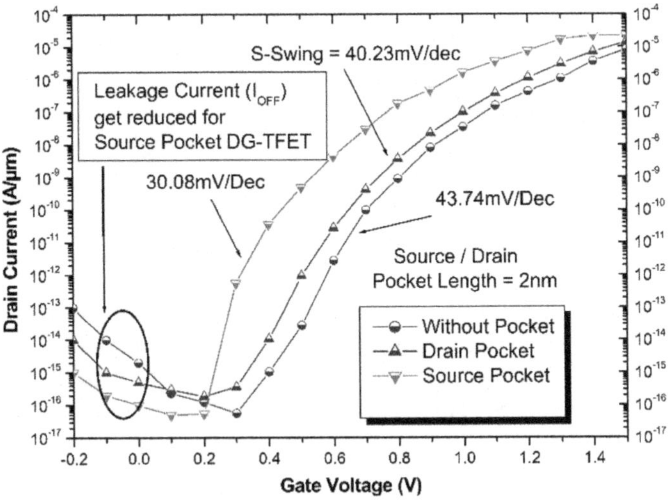

Fig. 6 Transfer characteristics of different dual-gate TFETs at the supply voltage $V_{DD} = 0.5$ V

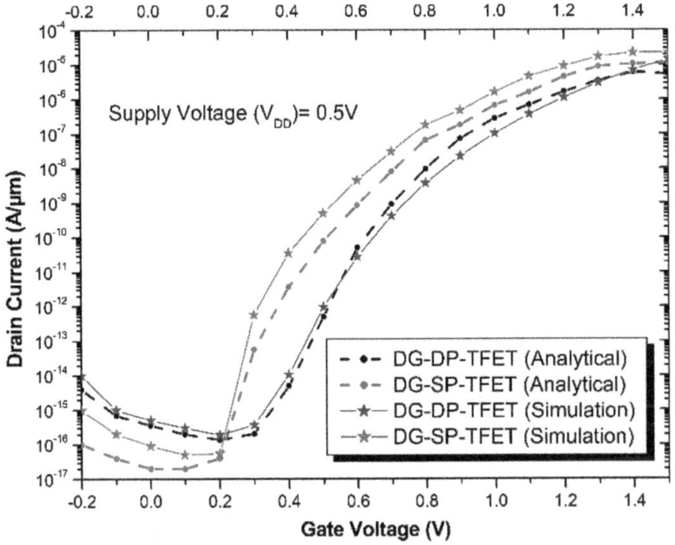

Fig. 7 Simulation and analytical data comparison at $V_{DD} = 0.5$ V

5 Conclusion

From this survey work, it can be concluded that though tunnel FET is the correct alternative of MOSFET for better subthreshold swing (SS), but ambipolar conduction can be well controlled by thin pocket layers at tunneling junctions. Here we have used

Table 2 Subthreshold swing of all DG-TFET (*n*-channel) models

Device models	Subthreshold swing (SS)
Conventional DG-TFET without pocket doping layers	43.74 mV/decade
Dual-gate drain pocket (DG-DP-TFET)	40.23 mV/decade
Dual-gate source pocket (DG-SP-TFET)	30.08 mV/decade

the *n*-channel TFET for simulation purposes, therefore the BTBT tunneling performs at the source–channel interface, due to which the I_{ON} has been raised to a considerable amount. In addition to this, the ambipolarity can also be well controlled by the dual-gate source pocket (DG-SP-TFET), keeping I_{OFF} at a lower value compared to all possible device structures.

References

1. Seabaugh C, Zhang Q (2010) Low-voltage tunnel transistors for beyond CMOS logic. Proc IEEE 98(12):2095–2110
2. Thomas N, Philip Wong HS (2017) The end of Moore's Law: a new beginning for information technology. IEEE J Comput Sci Eng 19(2):41–50
3. Anghel C, Gupta A, Amara A, Vladimirescu A (2011) 30-nm tunnel FET with improved performance and reduced ambipolar current. IEEE Trans Electron Devices 58(6):1649–1654
4. Narang R, Saxena M, Gupta RS, Gupta M (2012) Assessment of ambipolar behavior of a tunnel FET and influence of structural modifications. J Semicond Technol Sci 12(4)
5. Abdi DB, Kumar MJ (2014) Controlling ambipolar current in tunneling FETs using overlapping gate-on-drain. IEEE J Electron Devices Soc 2(6):187–190
6. Saurabh S, Kumar MJ (2016) Fundamentals of tunnel field-effect transistors. CRC Press, 292 pages. ISBN 9781498767132
7. Nigam K, Kondekar P, Sharma D (2016) Approach for ambipolar behavior suppression in tunnel FET by work function Engineering. Micro Nano Lett 11(8):460–464
8. Graef M, Hosenfeld F, Horst F, Farokhnejad A, Hain F, Iñíguez B, Kloes A (2018) Advanced analytical modeling of double-gate tunnel-FETs-a performance evaluation. Solid-State Electron 141:31–39
9. Das GD, Mishra GP, Dash S (2018) Impact of source-pocket engineering on device performance of dielectric modulated tunnel FET. Superlattices Microstruct 124:131–138
10. Dutta R, Paitya N (2019) Electrical characteristics assessment on heterojunction tunnel FET (HTFET) by optimizing various high-κ materials: HfO_2/ZrO_2. Int J Innov Technol Explor Eng 8(10):393–396
11. Garg S, Saurabh S (2018) Suppression of ambipolar current in tunnel FETs using drain-pocket: proposal and analysis. Superlattices Microstruct 113:261–270
12. Dutta R, Paitya N (2020) Effect of pocket intrinsic doping on double and triple gate tunnel field effect transistors. In: Proceedings of the 2nd international conference on communication, devices and computing. Lecture notes in electrical engineering, vol 602. Springer
13. Shen C, Ong SL, Heng CH, Samudra G, Yeo YC (2008) A variational approach to the two-dimensional nonlinear Poisson's equation for the modeling of tunneling transistors. IEEE Electron Device Lett 29(11):1252–1255

14. Young KK (1989) Short-channel effect in fully depleted SOI MOSFETs. IEEE Trans Electron Devices 36(2):399–402
15. Verhulst AS, Leonelli D, Rooyackers R, Groeseneken G (2011) Drain voltage dependent analytical model of tunnel field-effect transistors. J Appl Phys 110(2):024510-1–024510-10
16. Kane EO (1961) Theory of tunneling. J Appl Phys 32(1):83–91
17. TCAD Atlas Manual, Silvaco, Inc., CA 95054, USA (2015)
18. Dutta R, Konar SC, Paitya N (2020) Influence of gate and channel engineering on multi-gate tunnel FETs: a review. In: Maharatna K, Kanjilal M, Konar S, Nandi S, Das K (eds) Computational advancement in communication circuits and systems. Lecture notes in electrical engineering, vol 575. Springer
19. Dutta R, Paitya N (2019) TCAD performance analysis of P-I-N tunneling FETS under surrounded gate structure. SSRN-Elsevier
20. Dutta R, Paitya N, Majumdar A (2020) Ambipolar reduction methodology for SOI tunnel FETs in low power applications: a performance report. Int J Recent Technol Eng 8(5)

Physical Design and Implementation of Multibit Multilayer 3D Reversible Ripple Carry Adder Using "QCA-ES" Nanotechnique

Rupsa Roy, Swarup Sarkar, and Sourav Dhar

Abstract Quantum dot cellular automata with electro-spin ("QCA-ES") nanotechnology is an effective alternative to today's popularly used "CMOS" technique. Owing to the advancement of the proposed "QCA-ES" which is a more effective transistor-less nanotechnology, the optimization of area occupation, delay, energy dissipation and device complexity becomes easier. In this work, the "QCA-ES" nanotechnology is proposed for the design of an advanced 4-bit reversible adder, which is also known as ripple-carry-adder ("RCA"). Here, the proposed advanced adder is designed with multilayer 3D platform using multilayer inverter gate. A comparative study among the different other methods of full-adder design is shown in the paper, which shows that the delay, cell complexity and area occupation of multibit advanced full-adder structure can be minimized more by forming a multilayer 3D reversible "RCA" design, where multilayer inverter gates are used. Further, a comparative study is also presented between the proposed advanced multibit multilayer "RCA" design and other advanced multibit adder designs using the selected "QCA-ES" technology by varying the quantum cell size (18–16 nm). A brief discussion of high-temperature tolerance with proper output strength of the proposed advanced multibit multilayer "QCA-ES"-based design is also presented in this paper.

Keywords CSA · CLA · Multilayer · QCA-ES · RCA · Reversible

R. Roy (✉) · S. Sarkar · S. Dhar
Department of ECE, Sikkim Manipal University, Sikkim, India

S. Sarkar
e-mail: swarup.s@smit.smu.edu.inb

S. Dhar
e-mail: sourav.d@smit.smu.adu.inc

S. Dhar et al. (eds.), *Advances in Communication, Devices and Networking*,
Lecture Notes in Electrical Engineering 776,
https://doi.org/10.1007/978-981-16-2911-2_5

1 Introduction

Arithmetic and logic unit or ALU is the key component of any processor in today's digital world and full-adder circuit is an important part of this key component, which can be used to represent the addition-result of digital numbers. "CMOS" technology is widely used to design this important component till now because this technology supports Moore's law [1–3] to optimize device area and delay with time. But, the increment of device number in a single chip with time creates trouble to properly maintain the device size and speed with the proper fault-free operation. When the component size is scaled down, it increases the design complexity as well as power consumption and dissipation. So, a low-power, high-speed, less-complex, nano-sized and effective replacement of the above technology is required to design the components of ALU in this modern era. This paper chooses an advanced "QCA-ES" nanotechnology to design a multibit advanced full-adder circuit ("RCA") using a novel reversible logic in multilayer 3D platform.

In this work, a novel full-adder structure is designed at first, where a three-input majority gate with only four quantum cells are used and then this design is compared with the other previously published designs of 2019–2020 to represent the advancement of our proposed structure to optimize the occupied area, cell complexity and delay. After proving the betterment of the proposed "QCA-ES"-based full-adder design, it is compared with the previously published full-adder designs, and this paper focuses on the optimization of the occupied area, power, delay, complexity and cost of the proposed multibit adder design by using reversible gate in multilayer 3D platform. Modified "3:3 TSG Gate" is applied over here to avoid the energy dissipation per single bit and to compensate for the high-temperature effect due to layer number increment by using adiabatic logic [4].

The rest of the paper is presented as follows. After introducing the main focus, the proposed component, technology, logic of this presented paper, and the background of the proposed technology with proposed logic are discussed in the Sect. 2. After that, the Sect. 3 of this paper presents the literature review of the proposed circuitry and the Sect. 4 shows the presented work of this paper. Section 5 shows the simulated outcomes with required comparisons and proper discussion. After all the above stages, a brief conclusion about this proposed paperwork is presented with proper future-scope discussion in the sixth section.

2 Theoretical View of the Proposed Technique and Logical Expression

2.1 *"QCA-ES" Cell Configuration, Gate Structure and Clocking Scheme*

The proposed "QCA-ES" nanotechnology is mainly utilized to configure the physical-level structural design of any digital component in this nanotechnical era, and these components become an effective replacement of "CMOS"-based components in this recent nanoelectronic era, which is already discussed in the introduction section of this paper. This proposed advanced technology "QCA-ES" is a four-quantum dot-based technology, where the quantum cells are placed one after another to form a quantum wire and the information flow from input to output can be possible by the interaction of neighboring quantum cells. The discussed four-quantum dots are placed in each quantum cell, where two diagonally placed dots are occupied by the electron. When the electrons become excited they start to move through the tunnel in a spintronic direction and this electron-spin concept helps to move information from one cell to another cell. The polarization of quantum cells is another important factor of this technology, which depends on the above concept. When electrons start to move in the clockwise direction it shows positive polarization, meaning binary "1", and when the movement is in the anti-clockwise direction, the cell polarization becomes negative, meaning the cells act as binary "0".

Figure 1a, b represent the binary 0 and binary 1 cell design (Fig. 1 is "90°" cell design and Fig. 2 is "45°" cell design), where the two electrons are placed in

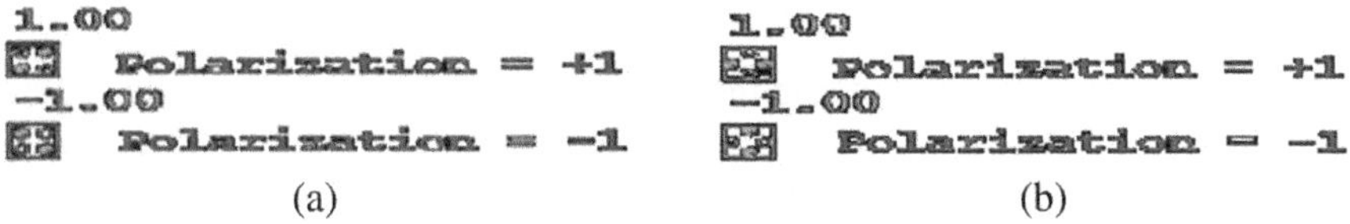

Fig. 1 **a** The "90°" binary 0 cell design and binary 1 cell design in QCA. **b** The "45°" binary 0 cell design and binary 1 cell design in QCA

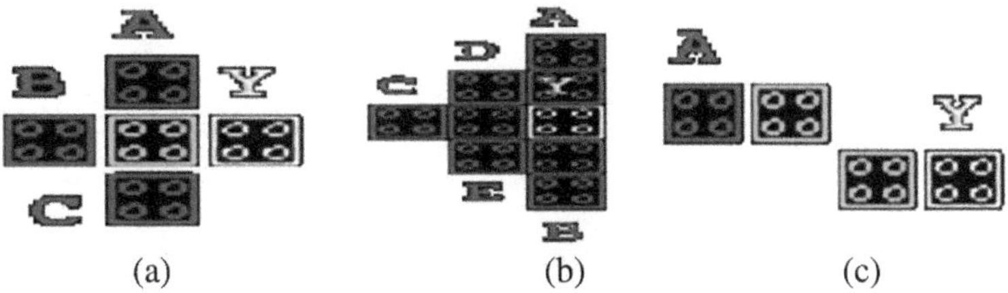

Fig. 2 **a** Three-input majority gate by using quantum dot cells. **b** Five-input majority gate by using quantum dot cells. **c** A basic inverter gate structure of QCA technology

two separate diagonal positions due to the electrostatic repulsion between the same charged carriers (electrons).

In this portion, some important gates in QCA electro-spin technique are properly discussed which are rapidly used to form any binary expression in the digital field. "90°" quantum cells are used in these presented gate designs due to less energy consumption [5] and higher output amplitude level compared to "45°" rotated quantum cells. Three-input majority gate, five-input majority gate and inverter gate are some of the most effective and rapidly used logic gates in this selected advanced nanotechnique. The input polarity of three-input majority gate is the same as the output polarity. Let a, b and c are the three inputs of this above gate structure, then the output looks like Eq. (1). This gate also can be used to design "AND" gate and "OR" gate by changing the polarity (-1 and $+1$) of one of the used three inputs in three-input majority gate, which are also given in Eqs. (2) and (3), respectively. The three-input majority gate is presented in Fig. 2a. In the introduction part, it is clearly revealed that this proposed technology is more advanced than the "CMOS" technology. If in this part, the "CMOS" designs of these above gates are compared with "QCA-ES" design, then it can be said that the number of used cells is less than the number of used transistors to form all those above gate structures, such as more than 25 transistors are required to design a three-input majority gate (MG) and six transistors are required to design each "AND" gate and "OR" gate.

$$M\,(A, B, C) = AB + BC + CA \tag{1}$$

$$M\,(A, B, 1) = AB + B.1 + A.1 = A + B \tag{2}$$

$$M\,(A, B, 0) = AB + B.0 + A.0 = A.B \tag{3}$$

The five-input majority gate with the proper truth table is also represented in this part in Fig. 2b and Table 1. As it is known that for any binary representation the "NOT" gate plays an important component role, it can be easily designed in QCA technology as an inverter gate. Fig. 2c presents an inverter gate in single-layer form, but when it is converted into multilayer form the number of used cells is 50% reduced with high output strength. This acts as another advantage of the proposed technology to form low-sized and low-power digital logic gates.

Table 1 Truth table of the five-input majority voter gate

Sum of (A, B, C, D, E)	Majority of (A, B, C, D, E)
0	0
1	0
2	0
3	1
4	1
5	1

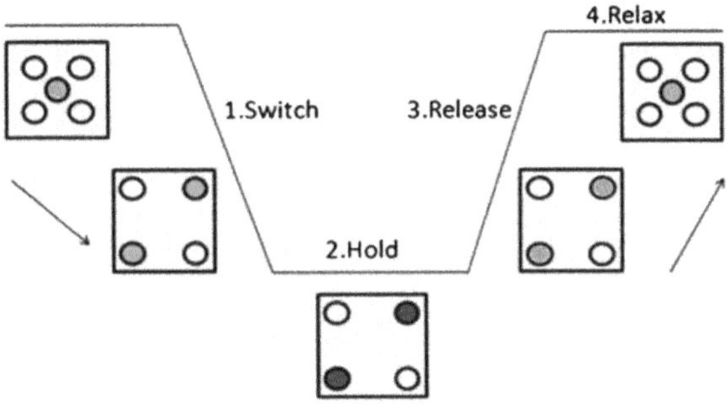

Fig. 3 "QCA-ES" clocking scheme

The "QCA-ES" technology-based design can be performed in two ways by clocking phenomenon, such as zone clocking and continuous clocking. In zone clocking the change of the clock zone in a quantum path is required to change the potential barrier between the quantum dots to get the propagation delay of output [4, 6]. Thus, the change in the clock zones needs to be controlled, because the rapid change in clock zones means a continuous change of barrier potential, which can increase the power consumption and power dissipation also. An advanced clocking scheme, named "Bennett Clocking Scheme" is already introduced [7], where the information can be saved before erase, which reduced the amount of dissipated power per bit without compromising the delay reduction of the proposed design.

If we discuss the basics of the clock zones, then it is important to discuss the fort clock phases of each clock zone. The four clock phases are switch, hold, release and relax, which are clearly shown in Fig. 3. In the switch phase, the cells are switched to work. Then the hold phase helps to flow the information by electron localization in the quantum dots based on the polarization of neighbor cells. In the release phase, the relocation of electrons takes place and complete relocation is possible in the relax phase. The phase difference between the two clock phases is 90°.

2.2 *Basics of Reversible Gate and Used Multilayer Structure*

Previously discussed "Bennet Clocking Scheme" is mainly used in reversible logic, which follows adiabaticity to maintain the extra heat generation with layer increment. This reversible logic gate is used in this work to design an arithmetic unit and logic unit combination (an important part of the core component, named "Arithmetic and Logic Unit" or "ALU", of any digital processor). In a conventional gate only input-based outputs are achieved, but in the reversible gate (presented in Fig. 4) inputs can also be represented by outputs and vice versa [8]. It can reduce the area, delay and complexity of the design. Briefly, it can be said that the advancement of selected "QCA-ES"-based design can be explored by applying reversible logic [9–12].

In our proposed structure an advanced wire crossing is used to achieve power-efficient, high-dense and high-speed circuit easily and this is multilayer wire crossing,

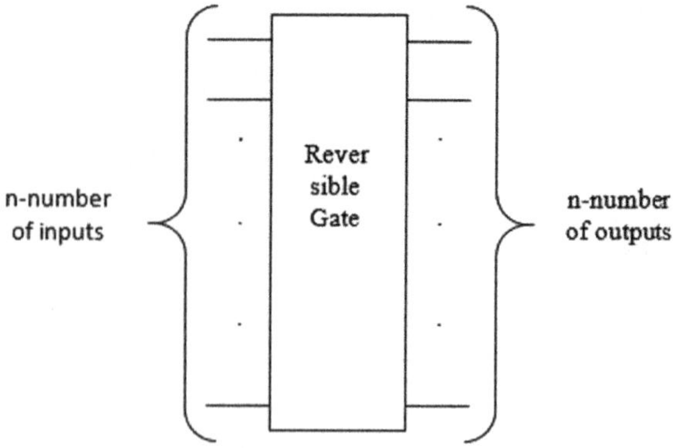

Fig. 4 Block diagram of reversible gate

which can form a 3D design by specifying different cells in different layers. A connecter layer between two different layers is also required in this type of design.

The multilayer structural improvement of quantum cell automata technique with electron-spin phenomenon is utilized in this paper, where a three-dimensional structure (presented in this paper in Fig. 5) can be formed from a two-dimensional structure for a unit area and delay reduction, and multilayer 3D inverter gates are used to form the proposed components with 16 nm quantum cell. Another advantage of this multilayer structure is an increase in the output strength. When a single-layer structure needs a huge number of quantum cells, at the same time, the design complexity is also maximized. But, it can be reduced by using this multiyear crossover technique [13].

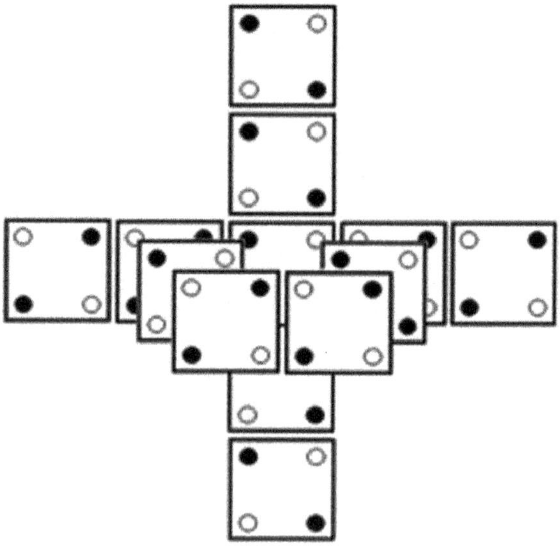

Fig. 5 " 3D" structure using multiple layers in QCA design

3 Literature Review

A QCA-based novel full-adder structure is designed in this paper before the "RCA" formation. This proposed adder design is used to form a 4-bit "RCA" in this work, which is discussed in the next section. In this section, some recently published QCA-based full-adder designs are shown one after another, which helps to make a comparison between the published works and our proposed work. In 2019, Fahimeh Danedaran et al. designed a fault-tolerant full-adder with 0.02 μm^2 area, 0.5 clock phases, 32 quantum cell complexity and 0.07305 eV dissipated energy for 1 E_k using five-input majority gate [14]. In this year also, a reversible full-adder subtractor design is formed by Moein Sarvaghad Moghaddam [15] with 58 quantum cells, 2 clock zones and 0.04 μm^2 to add another subtraction operation with reversibility. Next year, Nuriddin Safoev et al. designed a QCA-based 8-bit full-adder subtractor circuitry using a novel structure of coplanar full-adder with 49 quantum cells, 0.04 μm^2 area and 4 clock phases delay [16]. An incrementer/decrementer design is also formed this year by using the full-adder structure with 56 cells and the output strength calculation is also given in that paper [17].

Marshal Raj et al. presented a novel adder design where the clock pulse, area and cell number are increased, but this design presents an extra operation (subtraction) with adder operation [18]. A novel "Arithmetic and logic Unit" using QCA technology is designed in 2020 by Seyed-Sajad Ahmadpour, and as we know the full-adder is an important component in "Arithmetic and logic unit". This paper [5] presented a novel fault-tolerant design of full-adder circuitry by QCA. But in that, if a fault-tolerant full-adder design is used, the number of cells increased and the area is also increased, which creates a limitation of that presented circuitry. The above limitations of that previously published paper are already reduced in the same year by Behrooz Parhami et al. [19], where new majority-voter gate-based QCA-full-adder with 55 cells, 0.037 μm^2 area and 0.75 clock-cycle delay is used to form different types of 8-bit carry generator. Another novel full-adder structure is presented in the same year by M.M Abutaleb, where 38 quantum cells are used with 0.03 μm^2 area and 0.75 clock-cycle delay to form and compare two different types of "RCA" circuitry [20]. The used cell complexity of full-adder design is reduced (4 cells are decreased) in a published paper of Nuriddin Safoev et al., with less delay and the same area is compared with the previous one, by making the proposed component's formation in multilayer QCA platform [21]. The cell-reduction technique of QCA-based full-adder design is continued by Subhanjan Subhasis Das et al., where the delay is the same as the previously discussed paper but the area is reduced down to 0.02 μm^2. This paper also calculates the energy dissipation, which is $2.39e^{-003}$ eV for 1 E_K [22].

The 13 number of cells to design a QCA-based full-adder circuit is designed with the same clock phase delay in a published work of Jeyalakshmi Maharaj et al. [23], which is utilized to design a multibit "RCA" circuit. The discussed designs are compared with our proposed circuit which is presented in the next part of this paper through a parametric comparison table. There are various types of advanced multibit full-adder designs already discussed in the paper [24], such as ripple-carry-adder or "RCA", carry-save-adder or "CSA" and carry-look-ahead-adder "CLA". The various types of "QCA-ES"-based designs of these above-advanced adder circuits are presented in previously published papers. In this research work, the parameters of some previously published multibit advance adder designs are summarized, which helps to make a clear comparison of different types of proposed advanced adder designs of different papers with our presented reversible multibit design. In [25], V. Pudi et al.'s advance multilayer 4-bit adder design took 698 quantum cells with 0.618 μm^2 area occupation, 4 clock-cycle delay and 9.888 quantum cost, which were reduced by a multilayer 4-bit "CSA" design of paper [13] written by D. De and J. C. Das, where cell complexity, area occupation, delay and quantum cost were 525, 0.55 μm^2, 2.5 clock-cycle and 3.44, respectively. Then Sarvarbek Erniyazov et al. presented a paper [26] where the advancement of quantum cell-based 4-bit "CSA" over the fastest advanced 4-bit adder "CLA" using quantum cell was discussed after reducing the quantum cell size. The advancement of the presented "CSA" was discussed in the above paper, where the occupied area, complexity, delay and cost reduction can be possible compared to the proposed "CLA" design in a single-layer platform. But, an advanced adder design compared to the above published "CSA" design is presented in this work, where the previously discussed parameters of proposed circuitry are optimized more by using reversible logic in a multilayer 3D "QCA-ES" platform with comparatively small cell size and proper output strength.

4 Proposed Design

As we know, a novel reversible "QCA-ES"-based "RCA" design is ultimately presented in this paper in multilayer platform, but 1-bit full-adder structure is also represented in this part of this paper, where a clear parametric comparison with other previously published (discussed in the previous part of this paper) QCA-based full-adder circuit is represented in this portion through Table 2. This proposed full-adder structure is shown in Fig. 6.

Table 2 proves the advancement of our "QCA-ES"-based full-adder compared with other discussed designs. This structure is mainly utilized to form the reversible multilayer 4-bit "RCA", which is discussed below (see Table 3).

A novel seventh-layer reversible 4-bit "RCA" design using quantum cells are presented in Fig. 7, where the used full-adder designs with extra "XOR" operations are shown in layers 1, 3, 5 and 7, and layers 2, 4 and 6 present the communication line between every two layers. In Fig. 7, a new four-cell three-input majority gate [5, 23] with multilayer inverter gates is used by specifying one layer above another layer to reduce the unit area and delay, and the reversible logic reduces power dissipation of the proposed multilayer 3D design. "3:3 TSG Gate" is used over here as a reversible gate where three inputs are used (let A, B and C) and the three outputs are (1) direct output of input A, (2) "XOR" of A, B and C and (3) "XOR" of A and B. The truth table of the used reversible gate is also presented in the following sections.

Table 2 Parametric comparison among different adder designs

Referred papers	Cell complexity	Occupied area (μm^2)	Latency (ps)
[5]	32	0.02	0.5
[27]	58	0.04	0.5
[16]	49	0.04	1
[17]	56	0.047	1
[18]	75	0.09	0.75
[2]	85	0.082	2
[19]	55	0.037	0.75
[20]	38	0.03	0.75
[21]	34	0.03	0.5
[22]	32	0.02	0.5
[28]	26	0.02	0.5
[29]	16	0.006	0.5
[15]	13	0.009	2
Proposed (18 nm)	12	0.008	0.5
Proposed (16 nm)	12	0.006	0.4

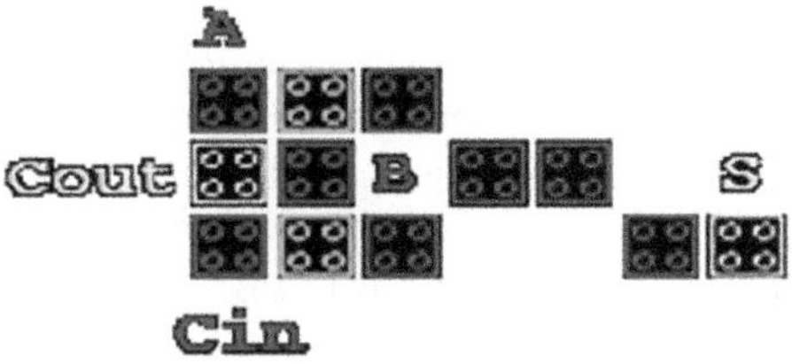

Fig. 6 Physical-level design of the proposed 1-bit full-adder design

Table 3 Truth table of "3:3 TSG Gate"

Input			Output		
A	B	C	A'	P = A XOR B XOR C	Y = A XOR B
0	0	0	0	0	0
0	0	1	0	1	0
0	1	0	0	1	1
0	1	1	0	0	1
1	0	0	1	1	1
1	0	1	1	0	1
1	1	0	1	0	0
1	1	1	1	1	0

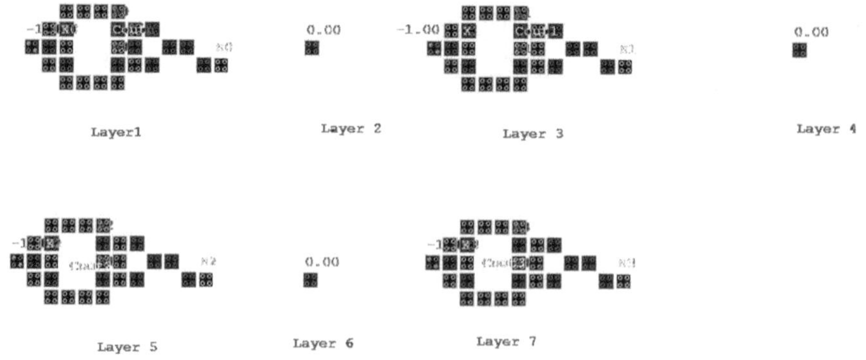

Fig. 7 Physical-level design of reversible multilayer 4-bit "RCA"

5 Simulated Outputs and Discussion

In this portion, the direct output of reversible full-adder structure is given, where the previously discussed proposed 1-bit full-adder circuit is used. After the design-simulation, the used "Full-Adder" design to form the proposed reversible multibit multilayer "RCA" is simulated in advanced "QCA-ES" platform and the simulated result is presented below in Fig. 8 separately. These "Full-Adder" designs are used in this work to form the proposed structure of 16 nm quantum cell-based multilayer 3D 4-bit reversible "RCA" design and the simulated output of this proposed circuitry is presented in this paper in Fig. 9.

In the above outcomes, A0 to A3 and B0 to B3 are the inputs of 4-bit "RCA" and the carry-out of the previous "Full-Adder" acts as the carry-input of next "Full-Adder". S0 to S3 are the sum-output of the proposed design and output X0 to X3 are

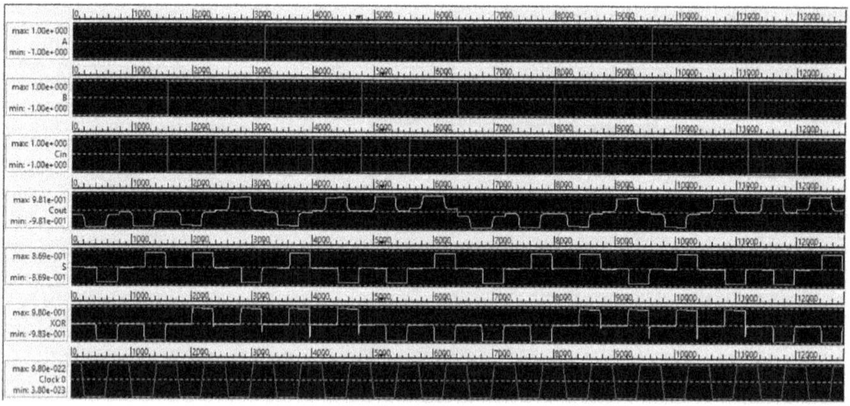

Fig. 8 QCA output of reversible "full-adder" design with "XOR" operation

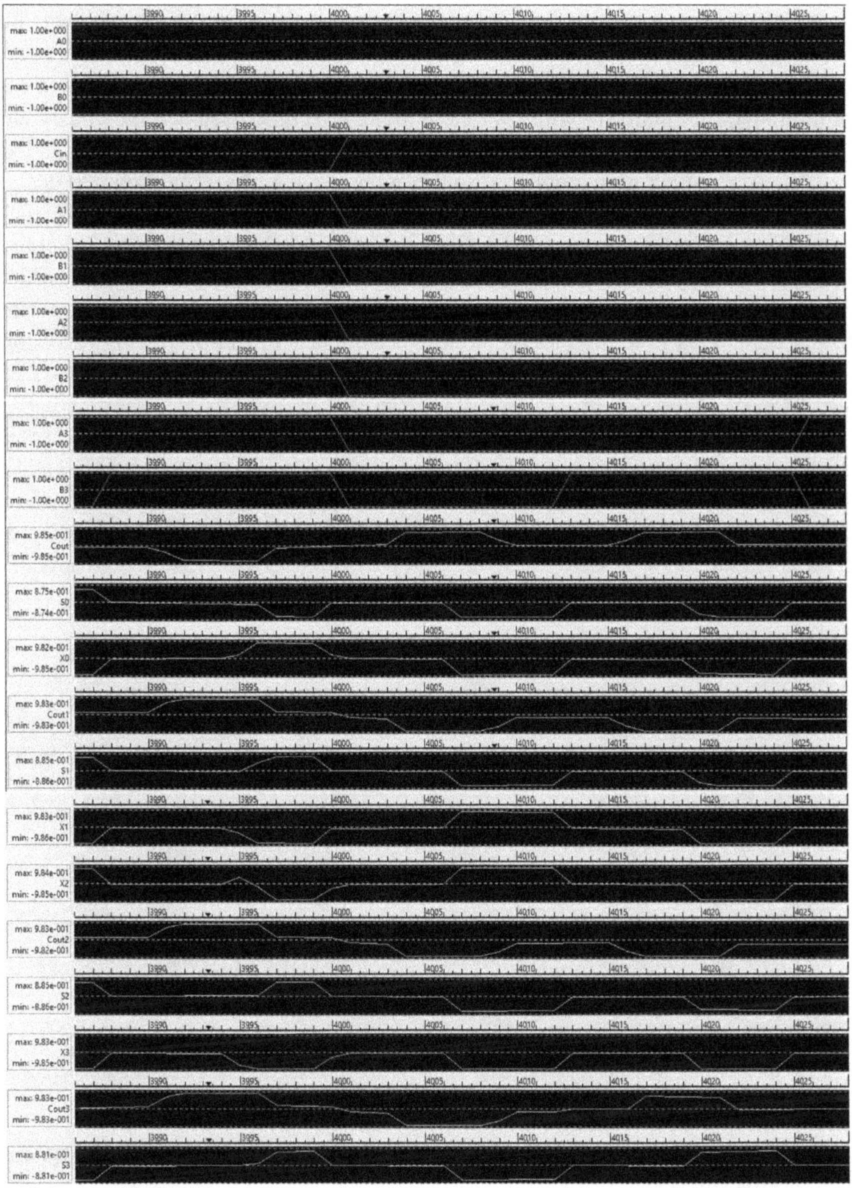

Fig. 9 Output of four-bit reversible multilayer "RCA" design with "XOR" operation

the “XOR” outputs which are another logical expression of the proposed reversible 4-bit “RCA”.

This proposed more advanced nanotechnology than “CMOS” is used in this work in multilayer 3D platform with reversible logic, which is discussed previously and this proposed multilayer “RCA” design is better than the “QCA-ES”-based “CLA” and “CSA” configuration, published in the previous year [26]. The proposed design of our paper can reduce the area occupation, area utilization factor (AUF) [27] (ratio of area occupation of the proposed design and area occupied by used quantum cells), delay, complexity and cost with proper output strength (9.3 at 3 K temperature, which is graphically presented in Fig. 10) and low power dissipation. Another comparison table, Table 4, is presented to show the advancement of multilayer 3D reversible “RCA” compared to previously published “QCA-ES”-based “CSA” and “CLA” structure.

Basically, if the advancement of adder-circuitry based on the given parameters in this paper is discussed, then it can be said that “CSA” is the fastest advanced adder among other advanced adder structures, which are formed in a single-layer platform. But, Table 4 clearly shows that in a multilayer 3D platform with the help of a selected reversible gate the proposed “RCA” circuitry can optimize the required area, delay,

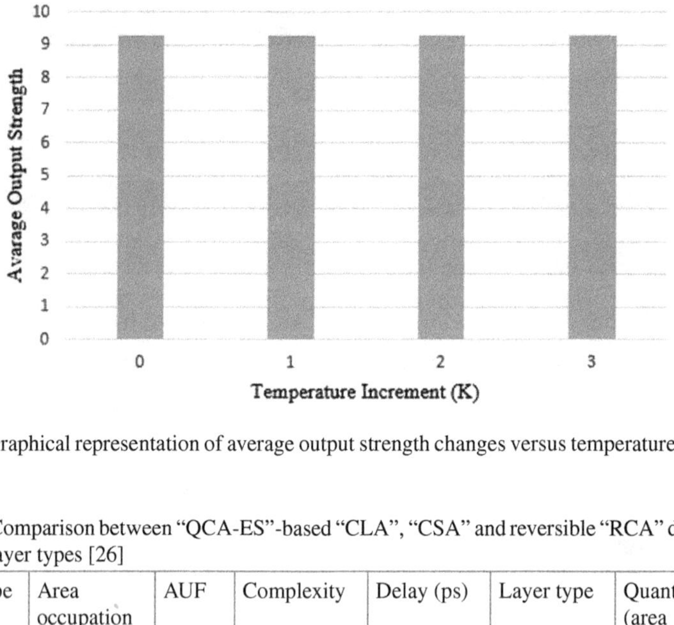

Fig. 10 Graphical representation of average output strength changes versus temperature increment

Table 4 Comparison between “QCA-ES”-based “CLA”, “CSA” and reversible “RCA” design with different layer types [26]

Adder type	Area occupation (μm^2)	AUF	Complexity	Delay (ps)	Layer type	Quantum cost (area × latency^2)
“CLA”	0.92	11.5	777	2.75	Multilayer	6.95
“CSA”	0.66	9.42	696	2.25	Coplanar	3.34
Reversible “RCA”	0.03	1.5	111	1.75	Multilayer	0.091

AUF, complexity and cost of the design more than previously published single-layer "CSA" circuitry.

The output strength is reduced at the time of cell diameter reduction due to the electron-scattering problem, but the multilayer design can reduce this problem by reducing the chance of quantum wire crossing. Thus, multilayer design can maintain the output strength of "QCA-ES"-based digital designs at the time of cell diameter reduction. Owing to the above discussion, our proposed design can also give proper output with better output strength (up to 9.94) at the time of layer separation gap reduction up to 5 nm.

6 Conclusion

A low-power, cost-effective, nano-sized, less complex, high-speed and temperature-tolerant advanced multibit-adder design is the primary objective of this work. The proposed design leads to a better arithmetic component of ALU compared to the current CMOS transistor-based components, since the CMOS technology suffers from scalability problems and high interlapping complexities at the time of integrated circuit formation. Extending our previous work [25], further optimization of "CSA" and "CLA" has been achieved in this work by using QCA-ES technology in addition to multilayer reversible "RCA" formation. In our previous work [25], the advancement of "CSA" compared to "CLA" has been proved, and in this work, it is shown that the use of multilayer reversible "RCA" is capable of optimizing the area occupation up to 95.4%, AUF up to 84%, complexity 84%, delay 22.22% and quantum-cost 97.3% with proper output strength. As a future scope of this work, the proposed design may be used as a combination of the arithmetic unit and the logic unit in ALU with proper fabrication and proper validity checking (checked by "FPGA Spartan 3E" board).

References

1. Tougaw PD et al (1994) Logical devices implemented using quantum cellular automata. J Appl Phys 75:1818–1825
2. Labrado C, Thapliyal H (2016) Design of adder and subtractor circuits in majority logic-based field-coupled QCA nano computing. Electron Lett 52(6):464–466
3. Sasamal TN, Sing AK et al (2016) An optimal design of full adder based on 5-input majority gate in coplanar quantum-dot cellular automata. Opt Int J Light Electron Optics 127:8576–8591
4. Pidaparthi SS, Lent CS (2018) Exponentially adiabatic switching in quantum-dot cellular automata. J Low Power Electron Appl 8:1–15
5. Ahmadpour S-S, Mosleh M, Heikalabad SR (2020) The design and implementation of a robust single–layer QCA ALU using a novel fault–tolerant three–input majority gate. J Supercomput. https://doi.org/10.1007/s11227-020-03249-3

6. Abedi D, Jaberipur G, Sangsefidi M (2015) Coplanar full adder in quantum-dot cellular autmata via clock-zone-based crossover. IEEE Trans Nanotechnol. https://doi.org/10.1109/tnano.2015.2409117
7. D'Souza N, Atulasimha J, Bandyopadhyay S (2012) An energy-efficient bennett clocking scheme for 4-state multiferroic logic. IEEE Trans Nano Technol. https://doi.org/10.1109/TNANO.2011.2173587
8. Waje MG, Dakhole P (2013) Design implementation of 4-bit arithmetic logic unit using quantum dot cellular automata. IEEE, IACC, pp 1022–1029
9. Yelekar PR, Chiwande SS (2011) Introduction to reversible logic gates & application. In: NCICT, IJCA, pp 5–9
10. D'Souza N, Atulasimha J et al (2012) Energy-efficient bennett clocking scheme for four-state multiferroic logic. IEEE Trans Nanotechnol 11:418–425
11. Singh MK, Nakkeeran R (2017) Design of novel reversible logic gate with enhanced traits. In: ICICI. IEEE, pp 202–205
12. Adelnia Y, Rezai A (2018) A novel adder circuit design in quantum-dot cellular automata technology. Int J Theor Phys. https://doi.org/10.1007/s10773-018-3922-0
13. De D, Das JC (2017) Design of novel carry save adder using quantum dot-cellular automata. J Comput Sci 22:54
14. Danehdaran F, Angizi S, Khosroshahy MB, Navi1 K, Bagherzadeh N (2019) A combined three and five inputs majority gate-based high performance coplanar full adder in quantum-dot cellular automata. Int J Inf Tecnol. https://doi.org/10.1007/s41870-019-00365-z
15. Ali MS-M, Orouji A (2019) New symmetric and planar designs of reversible full-adders/subtractors in quantum-dot cellular automata. Eur Phys J D 73:125
16. Safoev N, Jeon J-C (2020) A novel controllable inverter an adder/subtractor in quantum-dot cellular automata using cell interaction based XOR gate. Microelectron Eng 222:
17. Safoev N, Jeon J-C (2020) Design of high-performance QCA incrementer/decrementer circuit based on adder/subtractor methodology. Microprocess Microsyst 72:
18. Raj M, Gopalakrishnan L, Ko S-B (2020) Design and analysis of novel QCA full adder-subtractor. Int J Electron Lett. https://doi.org/10.1080/21681724.2020.1726479
19. Parhami B, Abedi D, Jaberipur G (2020) Majority-Logic, its applications, and atomic-scale embodiments. Comput Electr Eng 86:
20. Abutaleb MM (2020) Utilizing charge reconfiguration on quantum-dot cells in building blocks to design nanoelectronics adder circuits. Comput Electr Eng 86:
21. Safoev N, Jeon J-C (2020) Design and evaluation of cell interaction based vedic multiplier using quantum-dot cellular automata. Electronics 9:1036
22. Das SS, Singh R (2020) Design of efficient adders and subtractors based on quantum dot cellular automata (QCA). IJRASET 8:2562–2571
23. Maharaj J, Muthurathinam S (2019) Effective RCA design using quantum dot cellular automata. Microprocess Microsyst. https://doi.org/10.1016/j.micpro.2019.102964
24. Sasamal TN, Singh AK, Ghanekar U (2018) Efficient design of coplanar ripple carry adder in QCA. IET Circuits Devices Syst 12(5):594–605
25. Pudi V, Sridharan K (2012) Low complexity design of ripple carry and Brent–kung adders in QCA. IEEE Trans Nanotechnol 11:105
26. Erniyazov S, Jeon J-C (2019) Carry save adder and carry look ahead adder using inverter chain based coplanar QCA full adder for low energy dissipation. Microelectron Eng 211:37–43
27. Patidar M, Gupta N (2020) An efficient design of edge–triggered synchronous memory element using quantum dot cellular automata with optimized energy dissipation. J Comput Electron 19:529–542
28. Gudivada AA, Sudha GF (2020) Design of Baugh–Wooley multiplier in quantum–dot cellular automata using a novel 1–bit full adder with power dissipation analysis. SN Appl Sci 2:813
29. Heikalabad SR, Salimzadeh F, Barughi YZ (2020) A unique three-layer full adder in quantum-dot cellular automata. Comput Electr Eng 88:106735

Computational Study of the Electrical Properties of LD-LaSrMnO$_3$ for Usage as Ferromagnetic Layer in MTJ Memory Device

Abinash Thapa, P. C. Pradhan, and Bikash Sharma

Abstract Colossal magnetoresistance (CMR) was observed in a few ferromagnetic materials (FM) such as $La_{0.71}Sr_{0.29}MnO_3$ (LSMO). LSMO bearing greater magnetoresistance effect has the potential application in high-density information storage, spin valve and sensor technology. The atomistic computational analysis was performed for LSMO using quantum ATK. From LSMO bandstructure plots bandgap was observed at 0 eV bandgap at E_f (Fermi level). Overlap bandstructure predicted its ferromagnetic property. Effective mass [m*(m$_e$)] was calculated as 0.025 and −0.539 at 671 band index (−0.134 eV energy) and 672 band index (−0.124 eV energy). DOS plots of LSMO show a higher number of states at the Fermi level (by observing the number of spikes), depicting its application as a high-density device. The highest peak at the valance band was observed at −18 eV energy. *p* and *d* orbitals contributed to a larger number of states in DOS plots. Computation of transmission spectrum helped in predicting TMR by providing conductance value of the material used. Mulliken population observation leads to the prediction of atomic conjunction in the material. Thus, the electrical and electronic properties depicted superior properties of LSMO as FM. Implementing LSMO-based ferromagnetic layers in MTJ memory device is of high interest.

Keywords CMR · LSMO · MTJ · TMR

1 Introduction

Electrons, neutrons and protons have quantum mechanical property called "spin" along with electron charge [1]. Landau and Lifsbitz depicted spin to be quantum theory, not classical interpretation [2]. Stern-Gelach experiment provided a watershed event in the history of spin [1]. Earlier researchers Kronig, Uhlenbeck and Goudsmit thought space quantization concerned only quantum number *m* (quantum number), *l* (orbitals) and *n* (magnetic), but not the spin concept [1]. "Spin" plays

A. Thapa · P. C. Pradhan · B. Sharma (✉)
Department of Electronics & Communication Engineering, SMIT, SMU, Majitar, India

S. Dhar et al. (eds.), *Advances in Communication, Devices and Networking*,
Lecture Notes in Electrical Engineering 776,
https://doi.org/10.1007/978-981-16-2911-2_6

an important role in describing the multiplicity of atomic spectra [1]. The physical origin of magnetism is explained by "spin" depicted by the Bloch model, Heinsberg model and Stoner model [1]. Spin was also used for digital information encoded processing as binary bits 0 and 1 [3, 4].

Today, the magnetoelectronics field of spintronic deals with the magnetic or magnetoresistive (MR) effect for sensing and storing information such as read/write head in HDDs for sensing purpose [1], non-volatile MRAM [5], programmable spintronic logic gate [6], position control device in robotics [7], high current monitoring device for power system [1], etc. In 1980, the discovery of GMR led to the revolution in spintronic, during the investigation of spin-polarized electric current in FM/PM multilayer [8, 9]. In the quantum mechanism phase coherence of "spin" has longer retention than "charge" [1]. Thus it is important to build a solid-state scalable quantum logic processor and memory [1].

The prominent spin-based memory is magnetic tunnel junction (MTJ). It is a basic storage element in MRAM [10]. MTJ is FM/tunnel barrier/FM stack with a relative change in resistance that occurs depending on the comparative magnetic orientation of FM layers [10]. Modern MTJ is a spin valve with fixed and free FM layers, where the fixed FM layer has a fixed magnetic direction compared to the free FM layer according to the input field/current and store information [10]. Spin-dependent tunneling involving electron transport between majority and minority spin states in nanostructure [10]. In 1975, Jullier discovered TMR of 14% in Fe/Ge/Co junction at low temperature [11]. Similarly, in 1995, TMR of 10% at RT is observed in amorphous Al-oxide (Al-O) barrier MTJ [12]. Since then, the TMR values are increasing year by year, and the TMR ratio is given by 2P1P2/(1-P1P2) as spin polarization [11]. The TMR parameter depends on (a) spin-valve structure for stabilization of AP configuration [13], (b) optimization of FM/electrode materials [14], (c) magnetic field annealing [15], (d) oxidization method [16] and (e) etching technique [17].

A new concept called colossal magnetoresistance (CMR) has come up recently in MTJ memory devices [18]. The CMR was found in a few FM materials and has huge scope in high-density information storage, spin valve and sensor technology applications [19, 20]. $La_{0.71}Sr_{0.29}MnO_3$ (LSMO) demonstrates stable magnetic property at high temperature and practical application for data storage [21, 22]. CMR observed as electron transport takes place in e_g electron of Mn^{3+} ions by applying large magnetic field and could suppress the thermal-magnetic disorder [19]. The main drawback of CMR is that it requires a very large field (several Tesla) for operation [18]. Thus, we opted for low-field magnetoresistance (LFMR) found in LSMO [1, 23–25]. LSMR is associated with a large number of grain boundaries having noncollinear spin structure and high saturation magnetic moment [1, 26]. Better LFMR was observed by adding artificial grain boundaries in FM manganite/spacer/FM manganite trilayers such as LSMO/$SrTiO_3$/LSMO thin film junction with the highest LFMR of 45–50% at 4.2 K [27, 28].

2 Methodology

The X-ray diffraction (XRD) pattern is used for the computation of LSMO structure using the Rietveld refinement method [18]. The LSMO unit cell as Rhombohedra with R3c space group is considered and shown in Table 1.

A 2 × 2 × 2 layer in x, y and z of LD-LSMO bulk structure is considered for computation using quantum ATK. A semi-empirical calculator was used instead of LCAO-based DFT calculator [29]. The semi-empirical calculator is appended for a faster and accurate SCF calculation of bulk elements [29]. In a semi-empirical calculator extended Huckel is used, with Hoffmann parameter, density mesh cut-off as 70 Hartree, occupation as Fermi-Dirac, broadening at 1200 K and *k*-points sampling as 2 × 2 × 50. Further bandstructure, the density of states (DOS), effective mass, transmission spectrum and Mulliken population were calculated (Tables 2 and 3).

Table 1 Sample space, lattice parameters, angles and cell volume [18]

Space group = 167 R3c (Rhombohedral)					
a =	5.5058	α =	90°		
b =	5.5058	β =	90°	b/a =	1
c =	13.3597	γ =	120°	c/a =	2.42648

Table 2 Atomic position (x, y and z) [18]

	Element	x	y	z
36f	La	0	0	1/4
36f	Sr	0	0	1/4
36f	Mn	0	0	0
36f	O	0.131	0.3333	0.0833

Table 3 Occupation number [18]

Element	Occupation no.
La	0.71
Sr	0.29
Mn	1
O	1

3 Results and Discussion

The low-dimensional (LD) LSMO atomistic computation was performed using the quantum ATK tool. Subsequently, crucial electrical and electronic properties were calculated such as effective mass, DOS, bandstructure, Mulliken population and transmission spectrum (Fig. 1).

The LSMO XRD structure was considered by Bhattacharyya et al. NEGF and DFT-based spin transport simulations were performed using the quantum ATK tool [29].

Similar to Fe_3GeTe_2 and Ni_3GeFe_2 bandstructure from Chettri et al. [30], the LSMO bandstructure plots shown in Fig. 2 depict the overlap of bandstructure. The

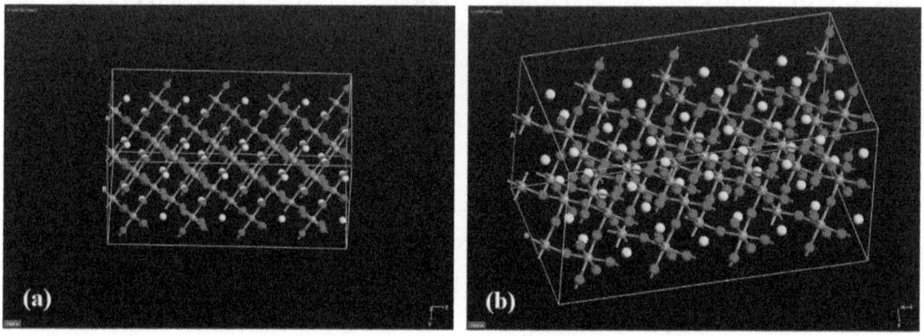

Fig. 1 **a** Top and **b** side view LD-LSMO structure (blue indicates La and Sr atom, purple indicates Mn atom and red indicates O_3 atom)

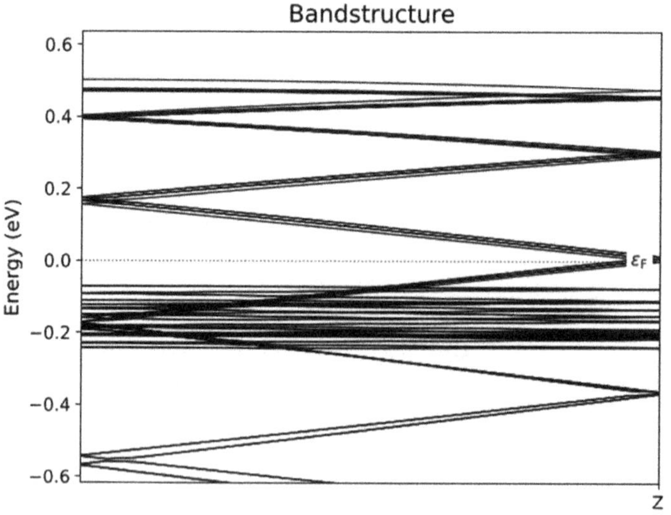

Fig. 2 Bandstructure of LD-LSMO material

Table 4 Band index, energy and effective mass of LSMO

Band index	Energy (eV)	Effective mass ($m^*(m_e)$)
671	−0.134	0.025
672	−0.124	−0.539

0 eV bandgap represents LSMO indicating ferromagnetic (FM) property equivalent FM materials, such as Fe, Co, CrO_2 etc.

The effective mass calculation was performed for LD-LSMO, as shown in Table 4. An effective mass of 0.025 was observed for 2 × 2 × 2 LSMO FM material with a band index of 671 and −0.134 eV energy.

Figure 3 depicts the LD-LSMO density of states (DOS) plots at *s*, *p*, *d* and *f* orbitals. DOS plots show the number of states in a memory device by observing the number of peaks obtained. For LSMO considerable peak is observed at E_f (Fermi level) depicting the high number of electrons transport. The highest peak at the valance band was observed at −18 eV energy. On further observing, the maximum peak was contributed from *s* and *p* orbitals (Fig. 4).

Mulliken population (MP) analysis determines the electronic charge distribution in a molecule [31], and also its bounding or non-bonding nature of molecular orbitals for particular atoms [31]. It reports the population of atoms at each shell and resolves it on individual orbitals [31]. Fig. 5 shows the Mulliken population analysis results

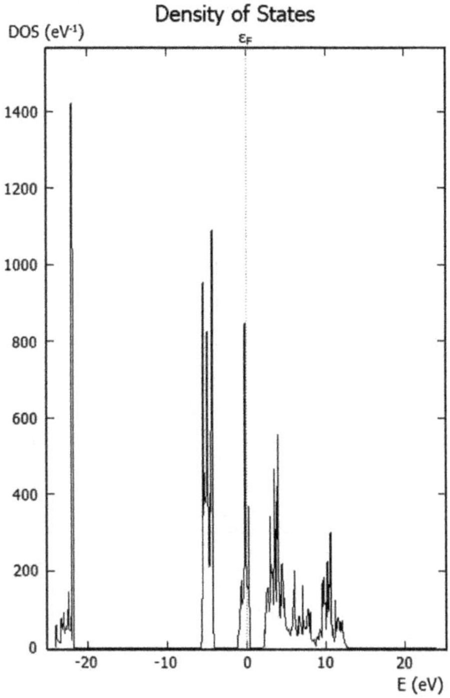

Fig. 3 DOS plots of LSMO (*s*, *p*, *d* and *f* orbitals)

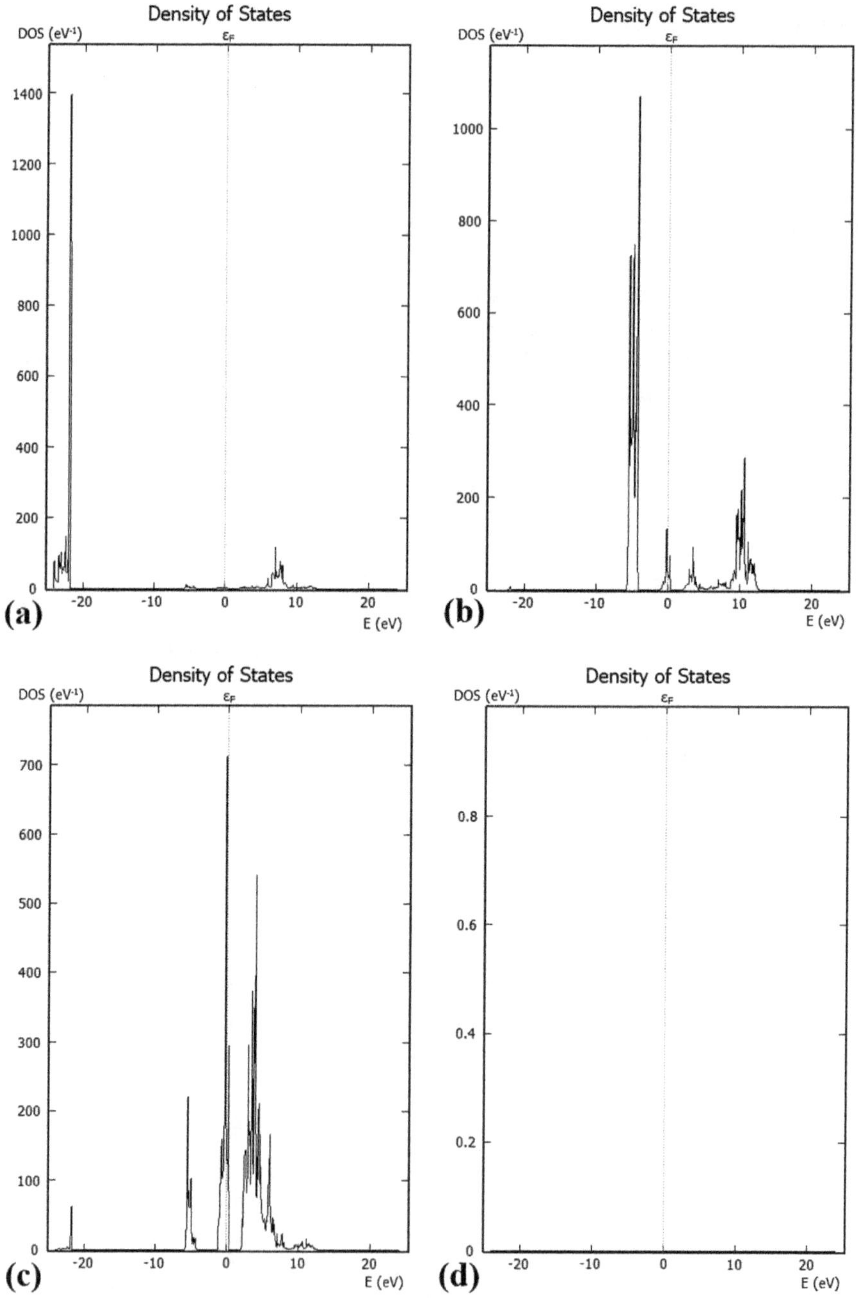

Fig. 4 DOS plots of LSMO at **a** *s* orbital, **b** *p* orbital, **c** *d* orbital and **d** *f* orbital

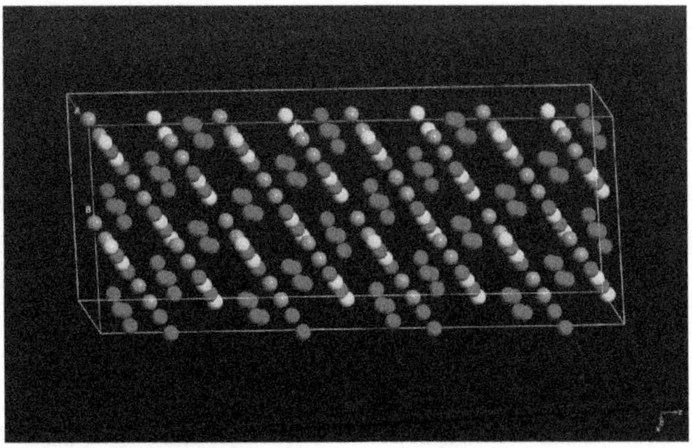

Fig. 5 Mulliken population analysis (blue indicates La and Sr atom, purple indicates Mn atom and red indicates O_3 atom)

for each atom in LSMO. MP verifies that each atom in the system has anticipated the spin polarization [31].

The MP of orbitals M_i [32] is given as

$$M_i = \sum_j D_{ij} S_{ji} \tag{1}$$

The MP of atoms M_μ [32] is given as

$$M_\mu = \sum_{i \in \mu} \sum_j D_{ij} S_{ji} \tag{2}$$

The MP of bonds $M_{\mu v}$ [32] is given as

$$M_{\mu v} = (2 - \delta_{\mu v}) \sum_{i \in \mu} \sum_{j \in v} D_{ij} S_{ji} \tag{3}$$

where D is the density matrix and S is the overlap matrix summed over all orbitals in the system.

The electrical conductance property of LSMO material can be calculated with the help of the transmission spectrum [31]. The transmission spectrum analysis in Fig. 6 depicted high conductance of 17.5 T(E) at E_f for both total spin and up/down spin at 0 V bias. In transmission coefficient plots, better conductance property of LSMO material was observed at 0.00 k_B and 0.00 k_A.

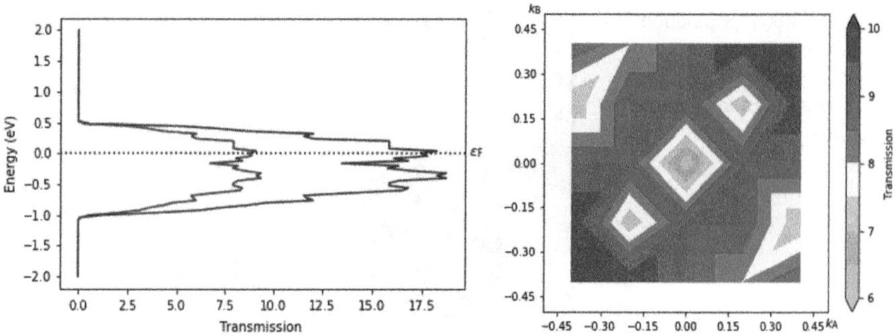

Fig. 6 Transmission spectrum and transmission coefficient of LSMO (red indicates total spin and blue indicates up/down spin)

4 Conclusion

CMR is a state-of-the-art concept in an MR-based memory device. LSMO, a new CMR material with better ferromagnetic property, was studied. It has wider applications in the case of high-density information storage, spin valve and sensor technology. Quantum ATK tools with NEGF and DFT were used for computation analysis of LSMO material. A 2 × 2 × 2 Rhombohedra unit cell LSMO structure was designed with the R3c group. Further, bandstructure, DOS, Mulliken population and transmission spectrum were analyzed. The LSMO bandgap from bandstructure plots was observed at 0 eV, depicting its ferromagnetic nature. In the DOS plot, the highest peak was observed at −18 eV energy and a better number of peaks observed anticipated a higher number of states in the memory device. The maximum peak was contributed from *s* and *p* orbitals in the DOS plot. The Mulliken population of LSMO structure was calculated depicting each atom bond, anti-bond and position. The transmission spectrum depicted high conductance of 17.5 T(E) at E_f for both total spin and up/down spin at 0 V bias. Therefore, the implementation of LSMO as ferromagnetic material in MTJ memory devices is of high interest in the near future.

Acknowledgments This work was supported by All India Council for Technical Education (AICTE) Govt. of India under Research Promotion Scheme for North-East Region (RPS-NER) vide ref.: File No. 8-139/RIFD/RPS-NER/Policy-1/2018-19.

References

1. Bandyopadhyay S, Cahay M (2015) Introduction to spintronics. CRC Press (2015 Sept 18)
2. Bird RB (1964) The Feynman lectures on physics, Richard P. Feynman, Robert B. Leighton, and Matthew Sands, Addison-Wesley, Reading, Mass, Volume I, II (1964); Volume III (1965). AIChE J. 10(6):794
3. Awschalom DD, Flatté ME, Samarth N (2002) Spintronics. Sci Am 286(6):66–73

4. Wolf SA, Treger DM (2003) Scanning the issue-special issue on spintronics. Proc IEEE 91(5):647–51
5. Freitas PP, Silva F, Oliveira NJ, Melo LV, Costa L, Almeida N (2000) Spin valve sensors. Sens Actuators A Phys 81(1–3):2–8
6. Wang J, Meng H, Wang JP (2005) Programmable spintronics logic device based on a magnetic tunnel junction element. J Appl Phys 97(10):10D509
7. Freitas PP, Costa JL, Almeida N, Melo LV, Silva F, Bernardo J, Santos C (1999) Giant magnetoresistive sensors for rotational speed control. J Appl Phys 85(8):5459–61
8. Baibich MN, Broto JM, Fert A, Van Dau FN, Petroff F, Etienne P, Creuzet G, Friederich A, Chazelas J (1988) Giant magnetoresistance of (001) Fe/(001) Cr magnetic superlattices. Phys Rev Lett 61(21):2472
9. Binasch G, Grünberg P, Saurenbach F, Zinn W (1989) Enhanced magnetoresistance in layered magnetic structures with antiferromagnetic interlayer exchange. Phys Rev B 39(7):4828
10. Ikeda S, Hayakawa J, Lee YM, Matsukura F, Ohno Y, Hanyu T, Ohno H (2007) Magnetic tunnel junctions for spintronic memories and beyond. IEEE Trans Electron Devices 54(5):991–1002
11. Julliere M (1975) Tunneling between ferromagnetic films. Phys Lett A 54(3):225–6
12. Miyazaki T, Tezuka N (1995) Giant magnetic tunneling effect in $Fe/Al_2O_3/Fe$ junction. J Magn Magn Mater 139(3):L231–4
13. Lu Y, Altman RA, Marley A, Rishton SA, Trouilloud PL, Xiao G, Gallagher WJ, Parkin SS (1997) Shape-anisotropy-controlled magnetoresistive response in magnetic tunnel junctions. Appl Phys Lett 70(19):2610–2
14. Kano H, Bessho K, Higo Y, Ohba K, Hashimoto M, Hosomi M (2002) MRAM with improved magnetic tunnel junction material. In: Proceedings of INTERMAG conference, 2002, p BB-04
15. Sato M, Kikuchi H, Kobayashi K (1999) Effects of interface oxidization in ferromagnetic tunnel junctions. IEEE Trans Magn 35(5):2946–8
16. Sun JJ, Soares V, Freitas PP (1999) Low resistance spin-dependent tunnel junctions deposited with a vacuum break and radio frequency plasma oxidized. Appl Phys Lett 74(3):448–450
17. Tsunoda M, Nishikawa K, Ogata S, Takahashi M (2002) 60% magnetoresistance at room temperature in Co–Fe/Al–O/Co–Fe tunnel junctions oxidized with Kr–O 2 plasma. Appl Phys Lett 80(17):3135–7
18. Sadhu A, Bhattacharyya S (2014) Enhanced low-field magnetoresistance in $La_{0.71}Sr_{0.29}MnO_3$ nanoparticles synthesized by the nonaqueous sol–gel route. Chem Mater 26(4):1702–1710
19. Prinz GA (1998) Magnetoelectronics. Science 282(5394):1660–1663
20. Moodera JS, Mathon G (1999) Spin polarized tunneling in ferromagnetic junctions. J Magn Magn Mater 200(1–3):248–273
21. Jin S, McCormack M, Tiefel TH, Ramesh R (1994) Colossal magnetoresistance in La-Ca-Mn-O ferromagnetic thin films. J Appl Phys 76(10):6929–6933
22. Urushibara A, Moritomo Y, Arima T, Asamitsu A, Kido G, Tokura Y (1995) Insulator-metal transition and giant magnetoresistance in La(1–x)Sr(x)MnO3. Phys Rev B 51(20):14103
23. Chen A, Bi Z, Tsai CF, Lee J, Su Q, Zhang X, Jia Q, MacManus-Driscoll JL, Wang H. Tunable low-field magnetoresistance in $(La_{0.7}Sr_{0.3}MnO_3)_{0.5}:(ZnO)_{0.5}$ self-assembled vertically aligned nanocomposite thin films. Adv Funct Mater 21(13):2423–2429
24. Staruch M, Gao H, Gao PX, Jain M (2012) Low-Field Magnetoresistance in La0. 67Sr0. 33MnO3: ZnO Composite Film. Adv Funct Mater 22(17):3591–3595
25. Mathur ND, Burnell G, Isaac SP, Jackson TJ, Teo BS, MacManus-Driscoll JL, Cohen LF, Evetts JE, Blamire MG (1997) Large low-field magnetoresistance in $La_{0.7}Ca_{0.3}MnO_3$ induced by artificial grain boundaries. Nature 387(6630):266–268
26. Balcells L, Fontcuberta J, Martinez B, Obradors X (1998) High-field magnetoresistance at interfaces in manganese perovskites. Phys Rev B 58(22):R14697
27. Lu Y, Li XW, Gong GQ, Xiao G, Gupta A, Lecoeur P, Sun JZ, Wang YY, Dravid VP (1996) Large magnetotunneling effect at low magnetic fields in micrometer-scale epitaxial $La_{0.67}Sr_{0.33}MnO_3$ tunnel junctions. Phys Rev B 54(12):R8357
28. Sinha UK, Das B, Padhan P (2020) Interfacial reconstruction in $La_{0.7}Sr_{0.3}MnO_3$ thin films: giant low-field magnetoresistance. Nanoscale Advances

29. https://docs.quantumatk.com/tutorials/tutorials.html
30. Chettri B, Sharma B, Thapa A, Chettri P, Sharma B (2020) Performance analysis of Ni_3GeFe_2/Fe_3GeTe_2 composites as Ferromagnetic layer in MTJ memory devices. In: 2020 IEEE VLSI device circuit and system (VLSI DCS) 18 Jul 2020. IEEE, pp 494–499
31. https://docs.quantumatk.com/tutorials/fe_mgo_fe/fe_mgo_fe.html?highlight=mulliken
32. https://docs.quantumatk.com/manual/Types/MullikenPopulation/MullikenPopulation.html

Realization of Ultra-Compact All-Optical Logic AND Gate Based on Photonic Crystal Waveguide

Kamanashis Goswami, Haraprasad Mondal, Pritam Das, and Adeep Thakuria

Abstract In this article, we have proposed a design of all-optical logic AND gate in the two-dimensional photonic crystal of triangular lattice structure with silicon rods in air background medium. Introducing line defect in the crystal, a Y-shaped waveguide is created and no nonlinear material is used to construct the device. The radiuses of few rods at the junction area are optimized to realize the operation of logic AND gate. The size of the proposed device is very small with a dimension of 77 μm^2 and it provides a contrast ratio of 6.9 dB. It offers a response time of 0.193 picoseconds and a data rate of 2.8 Tbit/second. Plane wave expansion method and finite difference time domain method have been applied to calculate the band structure and to analyze the performance of the simulation, respectively.

Keywords Photonic crystal · Photonic integrated circuit · Finite difference time domain · Plane wave expansion · AND gate

1 Introduction

Among various optical platform, like Semiconductor Optical Amplifiers (SOAs) [1], Photonic Crystals (PhC) [2–4], ring resonators [5], and Periodically Poled Lithium Niobate (PPLN) waveguides [6] for designing all-optical devices, photonic crystal is the most promising and popular platform of designing ultra-compact, ultrafast all-optical switches [7–9] and logic devices [10–12] because of its unique characteristic known as Photonic Band Gap (PBG). The photonic crystal structure is an arrangement of periodically modulated refractive index of dielectric materials. This periodic arrangement of dielectric material establishes the photonic band gap, which provides information about the reflectance and transmittance of the Electromagnetic

K. Goswami
Electronics Engineering Department, I.I.T. (I.S.M), Dhanbad, India

H. Mondal (✉) · P. Das · A. Thakuria
Electronics & Communication Engineering, Dibrugarh University, Dibrugarh, India

S. Dhar et al. (eds.), *Advances in Communication, Devices and Networking*,
Lecture Notes in Electrical Engineering 776,
https://doi.org/10.1007/978-981-16-2911-2_7

(EM) wave through it. Depending on the principle of operation, the PhC can be categorized into two groups, namely, linear and nonlinear.

The main advantages of linear optics, compared to nonlinear optics, are low operating power, small size, fast response time, and high extinction ratio. In the recent past, so many designs of all-optical switches/logic devices [13] have been reported in linear domain using PhC. As compared to other logic gates, AND gate plays an important role for photonic integrated circuits. In recent years, few research works have been reported on all-optical AND gate. For example, Rani et al. [14] proposed Y-shaped all-optical logic AND gate on two-dimensional PhC (2D-PhC) structure of triangular lattice, where one junction rod was optimized to realize the operation. But in their work response time, data rate and contrast ratio were reported as significantly low as 1.024 picoseconds, 0.976 Tbit/s, and 4.3 dB, respectively. Kiazand [15] proposed a T-shaped AND gate in the 2D-Phc structure of a square array where two extra rods were incorporated in input waveguides at the junction and optimized the radius to realize its operation. The extinction ratio was reported as low as 6 dB and the device can't perform well in cascading mode because 50% output power appears when both inputs to be considered are logic 1. Inspired by these previous works, authors have proposed a new design of all-optical AND gate where the aforesaid drawbacks have been overcome to a high extent.

In this paper, we have presented a simple design of all-optical AND gate based on two-dimensional rods in air photonic crystal with a triangular lattice structure. The beauty of this device is its ultra-compact size of 77 μm^2. The simulation results and performances of the device such as transmittance for various input signal wavelengths, contrast ratio, response time, and data rate have been calculated by using the Finite Difference Time Domain (FDTD) method [16].

The remaining portion of the paper is arranged into five sections. Section 2 elaborates the structural design of the device and band structure of the PhC. In Sect. 3 of this paper, working principle and performance analysis have been described. Section 4 includes simulation results for various input conditions of the device. Finally, the conclusion is drawn in Sect. 5.

2 Structural Design and Band Analysis

A new structure of all-optical logic AND gate is proposed with Y-shaped channel waveguide in the two-dimensional photonic crystal platform which is shown in Fig. 1a. The hexagonal lattice array of 11 × 15 is used in the form of rods in air to design the device. Circular solid silicon rods with dielectric constant ε of 11.42, i.e. refractive index of 3.38 are used in the background medium air, considering the refractive index 1. Line defect is introduced by removing rods in Γ-M direction of the crystal so as to form Y-shaped channel waveguide. The proposed device has two input ports (input-A, input-B) and one output port (output-C). The size of the device is very small which is in the order of 77 μm^2. Lattice constant 'a' of the structure and diameter of silicon rods are chosen as 750 nm and 480 nm, respectively. Point

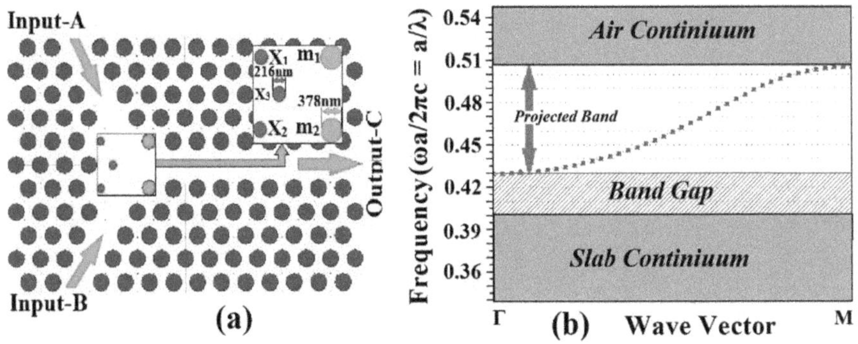

Fig. 1 The diagram of **a** two-dimensional photonic crystal AND gate. **b** Projected band of the photonic crystal line defect waveguide at TM Mode

defects are introduced in the junction of the three waveguides to make a cavity which controls the flow of light wave towards the output port, hence the operation of the device as logic AND gate can be realized. The radius of five rods x_1, x_2, x_3, m_1, and m_2 are optimized, by successive approximation method, to obtain the maximum output power at the output port C for desired input logic combinations. The radius of the defect silicon rods xi (where i = 1, 2, and 3) and m_j (where j = 1 and 2) are 108 nm and 189 nm, respectively.

The Plane Wave Expansion (PWE) algorithm [17] has been applied to calculate the Photonic Band Gap (PBG) of the PhC in Transverse Magnetic (TM) mode. For TM mode, band gap is found within the normalized frequency 0.4–0.51 a/λ, i.e. wavelengths ranging from 1470 to 1875 nm. The PWE method is also applied to calculate the projected band diagram in the defect mode of the PhC structure. Figure 1b shows that only one guided band has been found within the normalizing frequency range from 0.43 to 0.505 a/λ, which lies within PBG. However, the projected band depicts that the 1550 nm standard wavelength for optical communication lies within the projected band.

3 Working Principle and Performance Analysis

Optimized silicon rods play a vital role to perform the operation of a AND gate. When injected continuous Electromagnetic (EM) wave at port A reaches the junction of three waveguides, then rods x_1 and x_3 deliver ¾ of incoming signal towards port B and ¼ of incoming signal towards output port C. Similarly, when applied continuous light wave at port B reaches the junction, then rods x_2 and x_3 forward ¾ of the incoming signal towards the port A and ¼ of incoming signal towards output port C. Finally, when applied signals at both the input ports reach to the junction and interact with each other, then it creates a new mode which has maximum power transfer towards output port.

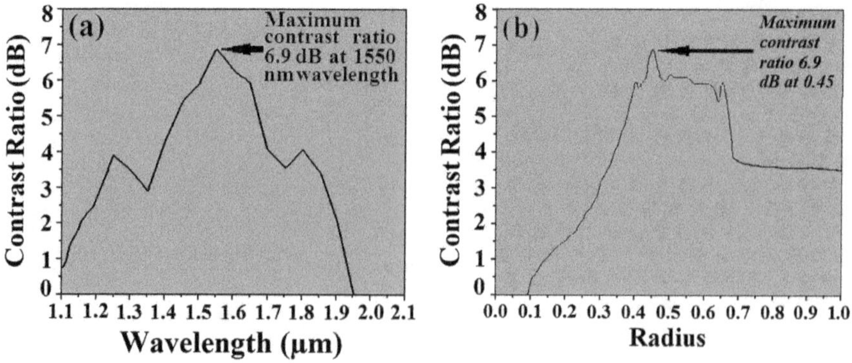

Fig. 2 Contrast ratio **a** for various operating wavelengths. **b** with various radiuses of defect rod x_3

The performance of the proposed device has been analyzed by measuring some performance matrixes like contrast ratio, transmittance, response time, and data rate which are discussed in the following subsections.

3.1 Contrast Ratio (CR)

The contrast ratio is defined as a ratio of the output power at 'logic-1'and the output power at 'logic-0'; mathematically it can be expressed as follows:

$$CR = 10\log_{10}\frac{P_1}{P_0} \tag{1}$$

where P_1 is the output power level at logic-1 and P_0 is the output power level at logic-0. İn Fig. 2a, the contrast ratio of the device for various wavelengths is shown which establishes that the maximum contrast ratio of 6.9 dB has been obtained at 1550 nm wavelength. This large CR contributes to the low Bit Error Rate (BER) because BER is inversely proportional to CR. The radius of the defect rods (X_1, X_2, X_3) is fixed to 108 nm, i.e. 0.45r (where 'r' is the radius of silicon rods) to obtain maximum power at the output port. Figure 2b shows the variation of contrast ratio with the variation of radius of X_3. From the figure, it is prominent that the maximum contrast ratio (6.9 dB) is obtained when the radius of X_3 is 0.45r.

3.2 Transmittance (T)

Transmittance (T) is defined as the ratio of the signal power received at the output port (C) to the signal power applied at the input terminals (A, B) for various wavelengths.

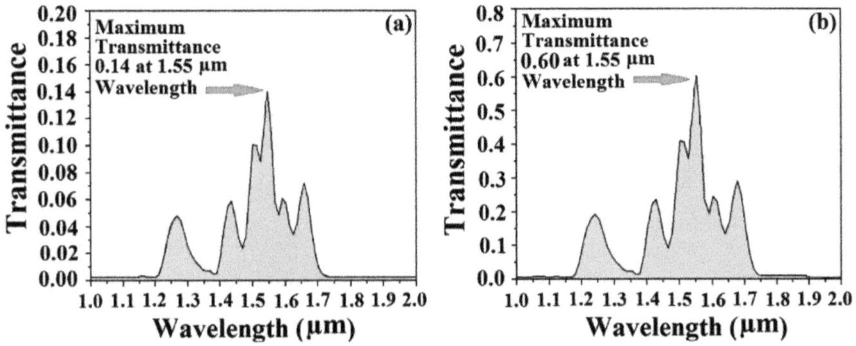

Fig. 3 Transmittance at the output **a** when the signal is applied at single input. **b** When the signals are applied at both the inputs

Figure 3a shows the transmittance characteristics of the proposed AND gate when only one input waveguide is excited with TM signal which represents the logic '01' or logic '10'. The maximum transmittance of 0.14 is obtained for logic '01' as well as for logic '10' at 1550 nm wavelength. Figure 3b represents the transmittance characteristics of the proposed device for '11' input logic state.

3.3 *Response Time and Data Rate*

The data rate and response time of the proposed AND gate have been calculated using time evolving graph which is shown in Fig. 4. Response time is known as the time required to reach from 0 to 90% of the average output power of the final steady state. It is shown that response time cT = 31 μm (where 'c' is the speed of light in vacuum), i.e. 0.193 ps. The response time is divided into two parts, namely transition time (t_{tr}) and steady-state time (t_{ss}) which have also been calculated as 0.103 ps and 0.09 ps, respectively. Due to operation in the linear optical domain, the required

Fig. 4 Time-evolving graph for the output power

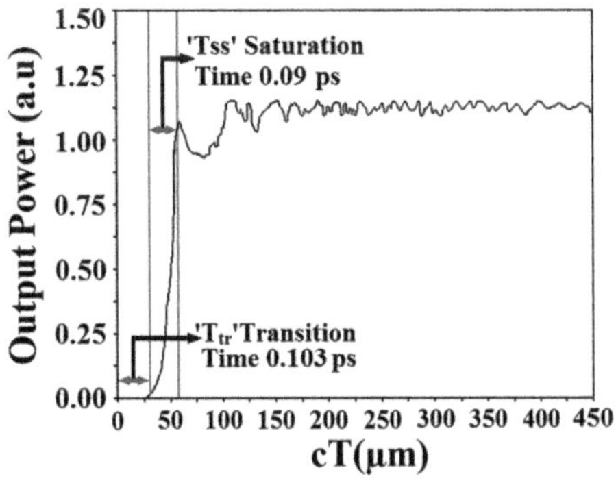

falling time from maximum to 10% of average output power is approximately the same as the steady-state time, i.e. 0.09 picoseconds, and hence the width of a narrow pulse is the sum of rise time and fall time which is equal to $2t_{ss}$, i.e. 0.18 picoseconds which is considered as ON time for a signal. For a signal of duty cycle 50%, the time period can be considered as $4t_{ss}$, i.e. 0.36 picoseconds. Therefore, the data rate has been calculated as 2.7 Tbit/s.

4 Simulation Result Analysis

The imulation result for the proposed AND Gate is shown in Fig. 5. When no signal is applied at the input ports, which represents '00' logic combination, no signal is obtained at output port C. When an input signal of power Pi is injected either at port A or Port B that represents '10'or '01'logic combination, a signal of 0.24Pi only is obtained at output port C. Such a low power signal is considered as logic '0'. The TM field propagations are shown in Fig. 5a and b.

Upon application of input signal of same power at both the input ports simultaneously, an output signal of power 1.18 Pi is achieved at port C, which is considered as logic '1'. The electric field propagation for '11' input logic combination is shown in Fig. 5c. The above result satisfied the conditions of a two-input AND gate for all possible input logic combinations. Output power for different input logic combinations is tabulated in Table 1.

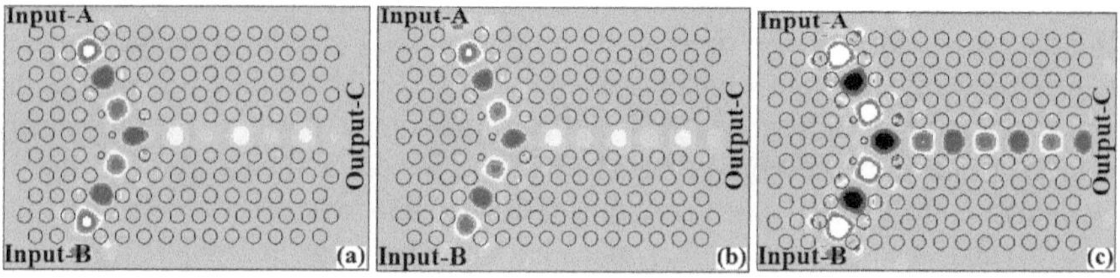

Fig. 5 **a** TE mode wave is applied to the input port B depicting logic '01'. **b** TE mode wave is applied to the input port- A depicting logic '10'. **c** TE mode waves are applied to both inputs depicting logic '11'

Table 1 Truth table and output power level of AND gate

Input-A	Input-B	Output-C	Power level at the output-C
0	0	0	0
0	1	0	$0.24P_i$
1	0	0	$0.24P_i$
1	1	1	$1.18P_i$

5 Conclusion

In this paper, a design of two-input all-optical logic AND gate in 2D-PhC platform of triangular lattice has been proposed. The design is based on Y-shaped channel waveguide, where radiuses of five junction rods are optimized. PWE method is used to analyze the normal and projected band diagram, and FDTD technique is used to analyze the performance of the simulation results. The footprint of the device is as small as 77 μm^2 and the device is made of linear optical material which requires very low operating power. The proposed AND gate provides a high data rate of 2.8 Tbit/s and response time of 0.193 picoseconds with contrast ratio as high as 6.9 dB. The design of the device, the performance of operation, and simulation results establish that the proposed logic AND gate can be applied for the chip-level integration.

References

1. Zhang X, Wang Y, Sund J, Liu D, Huang D (2004) All-optical AND gate at 10 Gbit/s based on cascaded single-port-couple SOAs. Opt Express 12:361–361
2. Mondal H, Goswami K, Prakash C, Sen M (2018) An all-optical ultra-compact 4-channel wavelength de-multiplexer. In: Proceedings of ICMAP, Dhanbad, India, pp 1–2
3. Gogoi D, Das K, Mondal H, Talukdar P, Hussain K (2016) Design of ultra-compact 3-channel wavelength demultiplexer based on photonic crystal. In: Proceedings of ICADOT, Pune, India, pp 590–593
4. Prakash C, Mondal H, Goswami SM (2018) Investigation of optimum position of interface between strip waveguide and PhC slot waveguide for maximum power coupling. In: Proceeding of ICMAP, Dhanbad, India, pp 1–2
5. Serajmohammadi S, Alipour-Banaei H, Mehdizadeh F (2015) All optical decoder switch based on photonic crystal ring resonators. Opt Quantum Electron 47:1109–1115
6. Wang J et al (2008) PPLN-based flexible optical logic and gate. IEEE Photonics Technol Lett 20:211–213
7. Prakash C, Sen M, Mondal H, Goswami K (2018) Design and optimization of a TE-pass polarization filter based on a slotted photonic crystal waveguide. J Opt Soc Am B 35:1791–1798
8. Mondal H, Sen M, Goswami K (2019) Design and analysis of all-optical 1 to 2-line decoder based on linear photonic crystal. IET Optoelectron 13:191–195
9. Mondal H, Sen M, Goswami K (2019) Design and analysis of a 0.9 Tb/s six-channel WDM filter based on photonic crystal waveguides. J Opt Soc Am B 36:3181–3188
10. Mondal H, Chanda S, Sen M, Datta T (2015) All optical AND gate based on silicon photonic crystal. In: Proceedings of ICMAP, Dhanbad, India, pp 1–2
11. Mondal H, Sen M, Prakash C, Goswami K, Sarma C (2018) Impedance matching theory to design an all optical AND gate. IET Optoelectron 12:244–248
12. Mondal H, Chanda S, Gogoi P (2016) Realization of all-optical logic AND gate using dual ring resonator. In: Proceedings of ICACDOT, Pune, India, pp 553–556
13. Sen M, Das MK (2015) High-speed all-optical logic inverter based on stimulated Raman scattering in silicon nanocrystal. Appl Opt 54:9136–9142
14. Rani P, Kalra Y, Sinha RK (2015) Design of all optical logic gates in photonic crystal waveguides. Optik 126:950–955
15. Kiazand F (2016) Design and simulation of linear logic gates in the two-dimensional square-lattice photonic crystals. Optik 127:4669–4674

16. Huang W, Chu S, Chaudhuri S (1991) A semi-vectorial finite difference time-domain method (optical guided structure simulation). IEEE Photonic Technol Lett. 3:803–806
17. Johnson SG, Joannopoulos JD (2001) Block-iterative frequency domain methods for Maxwell's equations in a plane wave basis. Opt Express 8:173–190

Investigation of Optical Properties of Ag-Doped Zinc Oxide Thin Film Layer for Optoelectronic Device Applications

Bishnu Prasad Sapkota, Sanat Kumar Das, Vivekananda Mukherjee, and Sanjib Kabi

Abstract We propose theoretical modeling to interpret the experimental transmittance as a function of wavelength (λ) of amorphous or crystalline semiconductors and dielectric thin films for optoelectronic device applications. The presented model reproduces the values of effective single oscillator energy (E_0) of the undoped and Ag-ZnO thin films that are in good agreement with other theoretical models. We found that effective single oscillator energy (E_0) of Ag-ZnO has increased from 6.313 eV to 6.46 eV as doping concentration Ag in ZnO films is increased from 0 to 5%. Furthermore, we executed the newly derived model to reproduce other optical parameters.

Keywords Optoelectronic device · Refractive index · Effective single oscillator energy · Dispersion energy · Nonlinear optical susceptibility · Doping

1 Introduction

Zinc Oxide (ZnO) thin films and nanostructures are potential candidates for modeling and synthesis for solar cell application due to their large exciton binding energy (≈60 meV) [1–3]. For the next generation of optoelectronic and microelectronic devices, using ZnO appears at the top of the frontier list due to its high thermal, mechanical, and chemical stability [1].

In this work, we primarily emphasize on the investigation of the dependency of optical properties and dispersion of the undoped and Ag-doped ZnO thin films on various Ag concentrations for different applications such as lasers, sensors, absorption, and field emission devices [2, 3].

B. P. Sapkota · S. K. Das (✉) · V. Mukherjee · S. Kabi
Sikkim Manipal Institute of Technology, Sikkim Manipal University, Majhitar, Rangpo, East Sikkim 737136, Sikkim, India
e-mail: sanat.d@smit.smu.edu.in

S. Dhar et al. (eds.), *Advances in Communication, Devices and Networking*, Lecture Notes in Electrical Engineering 776,
https://doi.org/10.1007/978-981-16-2911-2_8

2 Opto-Electronic Parameters and Dispersion Models Studies

Wemple–DiDomenico (WDD) single effective oscillator model is employed to determine the effective single oscillator energy (E_0) and dispersion energy (E_d) [rf]. WDD model is based on three types of energies, namely E_0, E_d, and the incident photon energy (h ϑ) [4, 5]. WDD model introduced in Eq. (1) gives a physical interpretation of the measured quantities.

$$\left(n^2 - 1\right) = \frac{E_d E_o}{E_o^2 - h\vartheta^2} \tag{1}$$

$$\left(n^2 - 1\right)^{-1} = \frac{E_o}{E_d} - \frac{1}{E_d E_o} h\vartheta^2 \tag{2}$$

$$\varepsilon_o = n_0^2 = 1 + \frac{E_d}{E_o} \tag{3}$$

The parameters E_d and E_0 have a key role in pinpointing the characteristics of optical materials and allow the estimate of required factors for the design of the spectral dispersion and optical communication devices. The obtained values of E^0 and E_d can be applied to ascertain the zero-frequency dielectric constant ε_0 and zero-frequency refractive index n_0.

We can additionally calculate the moments M_{-1} and M_{-3} of the optical spectra of thin films by using the relations:

$$E_o^2 = \frac{M_{-1}}{M_{-3}} \tag{4}$$

$$E_d^2 = \frac{M_{-1}^3}{M_{-3}} \tag{5}$$

After rearranging all the equation, it becomes

$$\left(n^2 - 1\right) = \frac{S_o \lambda_o^2}{\left(1 - \lambda_o^2\right)/\lambda^2} \tag{6}$$

The analyzed $\chi^{(1)}$ (linear optical susceptibility) values are given in Table 1. The third-order nonlinear optical susceptibility $\chi^{(3)}$ and the nonlinear refractive index are articulated as [6, 7]

$$x^{(3)} = 6.82 \times 10^{-15} \frac{E_d^4}{E_o^4} \tag{7}$$

Table 1 Estimation of some essential optical parameters of the undoped and Ag-doped ZnO thin films for various Ag concentrations

Sl. No.	Parameter	Undoped ZnO	Ag-ZnO 1%	Ag-ZnO 2%	Ag-ZnO 3%	Ag-ZnO 4%	Ag-ZnO 5%
1	Effective single oscillator, E_0 (eV)	6.227	6.313	6.408	6.557	6.681	6.646
2	Dispersion energy, E_d (eV)	10.452	12.433	14.743	17.143	20.189	22.234
3	Optical Moments, M_{-1}	1.378	1.398	1.489	1.512	1.587	1.614
4	Third-order nonlinear optical susceptibility, $\chi^{(3)}$ 10^{-14} (esu)	2.513	4.785	7.549	12.324	35.687	84.289

$$n^2 = \varepsilon' = \varepsilon_\infty - \frac{1}{4\pi^2\varepsilon_o}\left(\frac{e^2}{c^2}\right)\left(\frac{N_e}{m^*}\right)\lambda^2 \tag{8}$$

$$\varepsilon'' = \frac{1}{4\pi^3\varepsilon_o}\left(\frac{e^2}{c^3}\right)\left(\frac{N_e}{m^*}\right)\left(\frac{1}{\tau}\right)\lambda^3 \tag{9}$$

where e is the electronic charge, Nc is the charge carrier density, c is the light speed, ε'' is the imaginary part of the dielectric constant, and m* is the effective mass of the carrier.

3 Results

See (Figs. 1 and 2).

4 Conclusion

We found that E_0 is increased from 6.227 to 6.646 eV as Ag concentration incorporated in ZnO thin films is increased from 0% to 5%. This is expected due to the formation of stronger bonds among the constituents of the Ag-doped thin films. Additionally, the dispersion energy E_d also increased from 10.452 to 22.234 eV. The calculated values are presented in Table 1. It is worth noting that the obtained values of the zero-frequency refractive index are in good agreement with reported theoretical and experimental values of the normal refractive index. The zero-frequency dielectric constant ε_0 and zero-frequency refractive index n_0 of undoped ZnO thin film

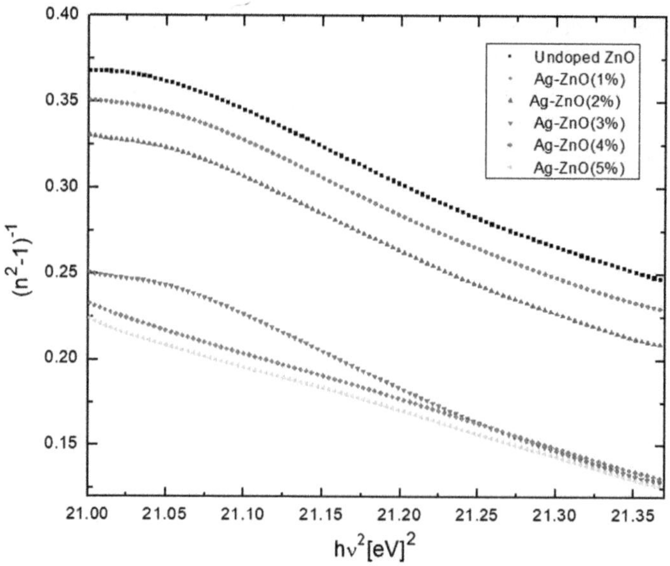

Fig. 1 $(n^2 - 1)^{-1}$ versus $h\vartheta^2$ of undoped and Ag-doped ZnO thin films for various Ag concentrations

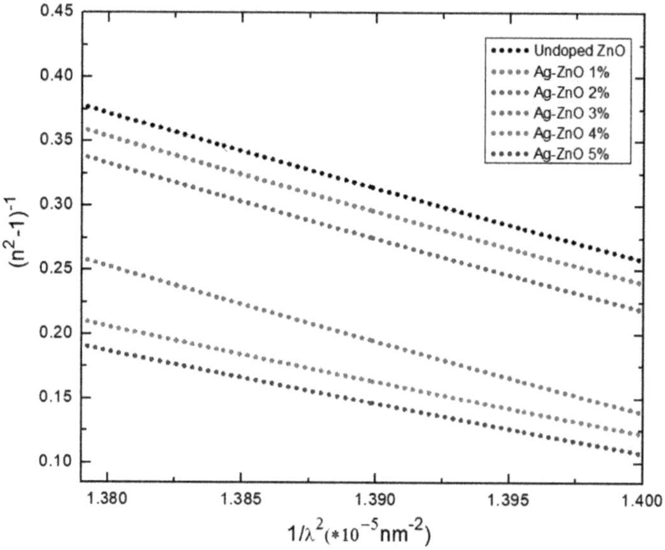

Fig. 2 $(n^2 - 1)^{-1}$ versus λ^{-2} of undoped and Ag-doped ZnO thin films for various Ag concentrations

are found to be 2.123 and 1.613, respectively. As the Ag concentration is increased from 0 to 5%, both the parameters are increased significantly to 3.176 and 2.138, respectively.

References

1. Özgür Ü et al (2005) A comprehensive review of ZnO materials and devices. J Appl Phys 98:11
2. Wrobel P et al (2019) Plasmon-enhanced absorption in heterojunction n-ZnO nanorods/p-Si solar cells. International Society for Optics and Photonics, Metamaterials XII
3. Hassanien AS, Akl AA (2015) Influence of composition on optical and dispersion parameters of thermally evaporated non-crystalline Cd50S50−xSex thin films. J Alloys Compd 648:280–290
4. Kumarasinghe P et al (2017) Effect of post deposition heat treatment on microstructure parameters, optical constants, and composition of thermally evaporated CdTe thin films. Mater Sci Semicond Process 58:51–60
5. Güneri E, Kariper A (2012) Optical properties of amorphous CuS thin films deposited chemically at different pH values. J Alloys Compd 516:20–26
6. Badran HA et al (2016) Determination of optical constants and nonlinear optical coefficients of violet 1-doped polyvinyl alcohol thin film. Pramana 86:1
7. Okutan M et al (2005) Investigation of refractive index dispersion and electrical properties in carbon nano-balls' doped nematic liquid crystals. Phys B Conden Matter 362:180–186

Bibliometric Analysis of Home Health and Internet of Health Things (IoHT)

Ankit Singh, Jitendra Kumar, Ajeya Jha, and Shankar Purbey

Abstract A bibliometric analysis on Home Health together with Internet of Health Things (IoHT) is not done earlier. The objective of this study is to carry out a bibliometric analysis of Home Health with the Internet of Health Things (IoHT). The analyzed parameters include top ten institutions by affiliation, top ten authors, top ten countries, top ten funding sponsors, top ten sources and subject-wise classification. In addition to that, the network analysis is done for keywords, sources citation analysis, author's co-citation analysis and country-wise co-authorship. The data is taken from the Scopus database on 23 September 2020. The search string results have shown a total of 1000 documents, after limiting for language as English and document type as articles. The results have revealed that the key themes in the future could be a congruence of assisted living of older patients with the health monitoring devices. Congruence of smart homes with smart devices is based on Body Area Networks (BANs) and Personal Area Networks (PANs). Moreover, the research in this area is expected to increase in the coming years and there is a requirement of enhanced collaboration between developed and developing countries.

Keywords Home health · Internet of health things · Bibliometric · Network analysis

---

A. Singh
Symbiosis Institute of Health Sciences, Symbiosis International (Deemed University), Pune, Maharashtra, India

J. Kumar · A. Jha (✉)
Department of Management Studies, Sikkim Manipal Institute of Technology, Majhitar, Sikkim, India
e-mail: ajeya.jha@smit.smu.edu.in

S. Purbey
Development Management Institute, Patna, Bihar, India

S. Dhar et al. (eds.), *Advances in Communication, Devices and Networking*,
Lecture Notes in Electrical Engineering 776,
https://doi.org/10.1007/978-981-16-2911-2_9

1 Introduction

Rise in the elderly population, changing disease pattern, the increased representation of women in the labour force, rising demand for care are some of the reasons pushing the growth of home healthcare services in Europe [6]. However, this trend is also seen in other countries and the demand for home healthcare is expected to rise globally. Home healthcare fills a void in the healthcare system by providing a mechanism to render care at all the three levels, i.e. primary, secondary and tertiary of pre-hospitalization and post-hospitalization. The current pandemic of COVID-19 has also brought the much-needed attention towards home healthcare as this disease spreads through the "droplet transmission" [9] and due to the increased risk of transmission, there is a shift in the patient (consumer) behaviour for healthcare services consumption. The adoption of "Home Health" and "Telemedicine" is recommended and has shown an uprising trend [16]. Home care or home healthcare can be defined as "an array of health and social support services provided to clients in their residence. Such co-ordinated services may prevent, delay or be a substitute for temporary or long-term institutional care". Home Healthcare which was earlier confined predominantly to the USA [14] and European countries [7] has now started to gain relevance in developing countries such as Iran [8], India [18] and Jordon [2]. This trend can also be attributed to the phenomenal technological development, which has happened in the last decade and the increased adoption of wearable devices; serving the purpose of constant monitoring of the patients has increased exponentially [16]. This trend is again fuelled by the COVID-19 situation, in a study it was highlighted that around 38 per cent of the chief executive officers have admitted that they had not included any digital component in their overall strategic plan; but now even the United States government has announced that to push technological adoption they will not put penalty against non HIPPA compliant technology [10]. Internet of Things (IoT) can be understood as a framework where objects can communicate with each other and process the vital parameters of the individual in a partially or fully automatized manner [5]. Similarly, Internet of Health things can be explained as "a framework in which the objects exchange and monitor a patient's health status". The application of the Internet of Health Things has increased in the last decade [15]. The bibliometric analysis is considered as a very useful tool for generating single-source reference for multiple scholars interested in any field of work, the strength of this tool is that it introduces a measure of objectivity in the evaluation of scientific literature and promotes the research rigour [20]. There is plenty of individual researches happening in the various domains of knowledge, however, for every new researcher in any field, it takes significant efforts and energy to find out the relevant sources, authors and articles in the interested field. The bibliometric analysis makes this job easier and provides synthesized information which assists the growth of the particular field of the study. Bibliometric methods are also considered as superior to narrative reviews because of the possibility of researcher bias [20]. This paper presents a comprehensive assessment of the chosen domain of "Home Health" and "Internet of Health Things", starting with a pool of over 1,000 published studies from the Scopus database on 23 September 2020. The pool was further filtered to document type as articles and language as English. The initial sections of this article identify the top ten authors, sources, funding sponsors, publication by countries and the publication trend. The second section covers the comprehensive network analysis based on author keywords, sources and countries.

2 Methods

Documents were analyzed according to their type, authors, sources, citations, co-citations, country-wise bibliographic coupling.

2.1 Initial Search Results

Total 1000 documents were shown for the entered search string in the Scopus database, See Table 1.

2.2 Refinement of Search Results

The results were further refined by limiting the source type to articles and the language to English.

The data were analyzed with the help of MS Excel 2016 and the VOSviewer version 1.6.

Table 1 Search String

Primary keywords	"Home Health" OR "Home Care"
Secondary keywords	AND "Internet of Health Things" OR "IoHT"
	OR "Internet of Medical Things" OR "IoMT"
	OR "Internet of Things" OR "IoT"
	OR "Smart Health" OR "Wearable Devices"

2.3 Institution by Affiliations

The top 10 institutions, which have published articles in this domain are represented in Fig. 1. It can be seen that the King Saud University together with Chinese academy

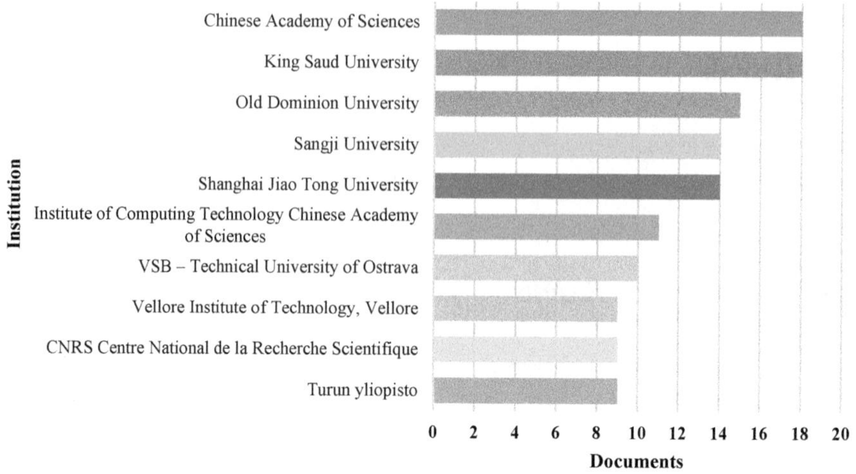

Fig. 1 Top 10 institutions by affiliations

of sciences has published the maximum number of documents (18), followed by Old Dominion University (15) and Sanji University (14). See Fig. 1.

2.4 Authors

The analysis of the authors has revealed that Chung K has published the maximum number of the articles (12), followed by Vanus J (9) and others. See Fig. 2

2.5 Country

The analysis of the articles based on the publication country has revealed that the maximum number of the article is published in the United States of America [170] followed by China [145] and others. See Fig. 3.

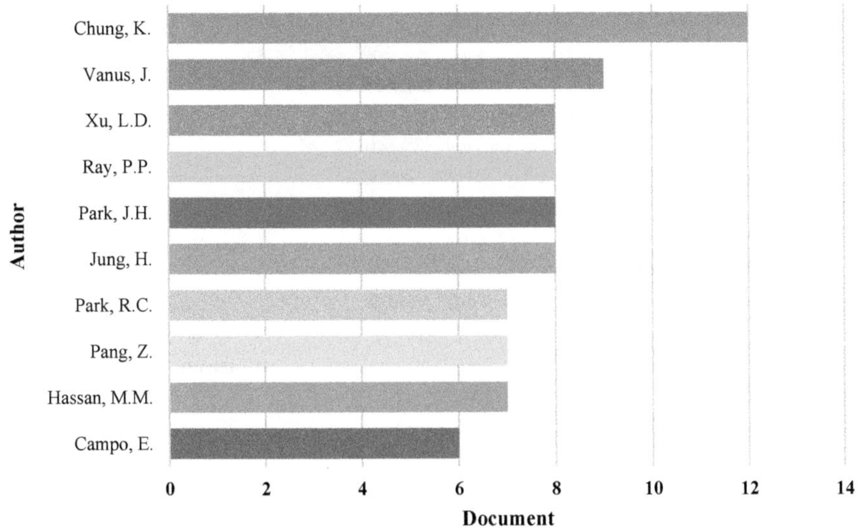

Fig. 2 Top 10 authors

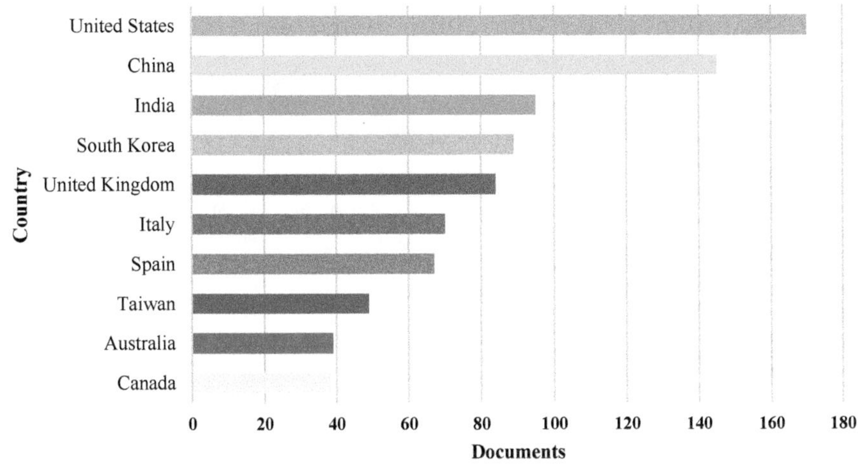

Fig. 3 Top 10 countries by publications

2.6 *Funding Sponsor*

The top 10 funding institutions are mentioned in Fig. 4. National research science foundation of China has sponsored a maximum number of published articles (59) followed by National research foundation of china (32) and National Science Foundation (24). See Fig. 4.

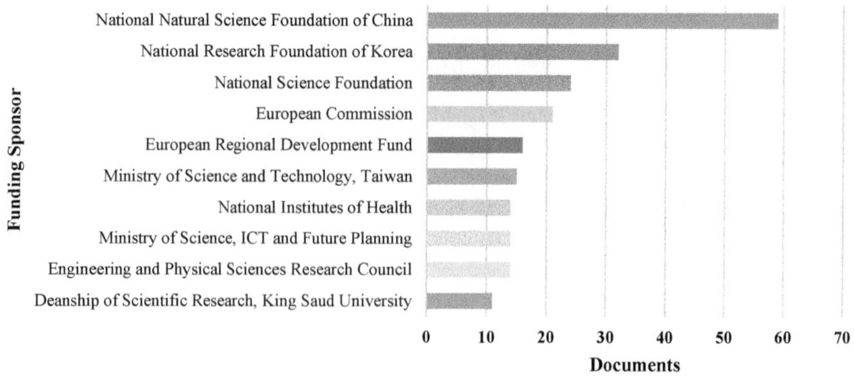

Fig. 4 Top 10 funding sponsor institutions

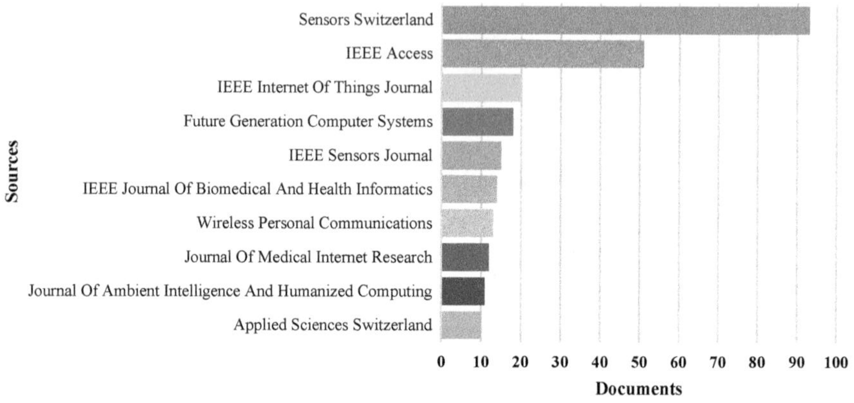

Fig. 5 Top 10 sources

2.7 Sources

The analysis of the sources has revealed that the maximum number of articles were published in the Journal Sensors Switzerland (93), followed by IEEE Access (51) and IEEE Internet of things Journal (20). See Fig. 5.

2.8 Subject

The analysis of the articles by the subject-wise has revealed that the maximum number of articles were published in the domain of computer science (527) followed by engineering (468) and Medicine (195). See Fig. 6.

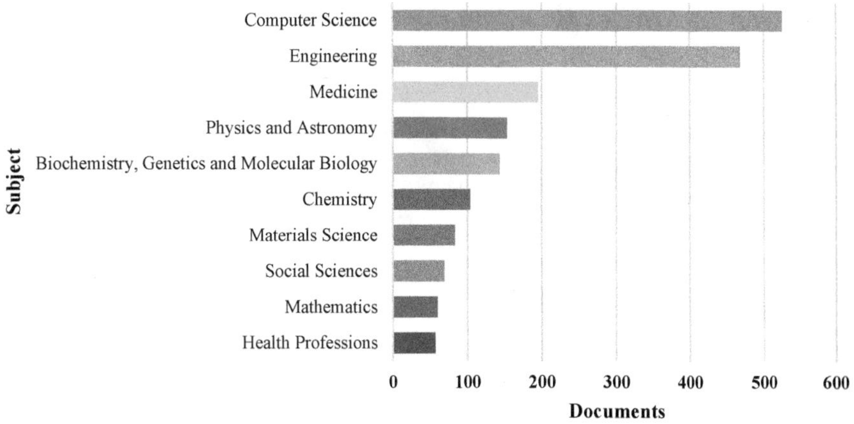

Fig. 6 Subject-wise distribution

2.9 *Trend Analysis*

The trend analysis of the research articles publications reveals that the focus towards the discipline of home health and Internet of things started in early 2000. In the years between 2012 and 2019, the number of articles published in this domains has increased exponentially, which highlights that it is one of the current favoured research areas and in the coming years the numbers of articles published in this interjection of the internet of things and healthcare can be expected to rise further. See Fig. 7.

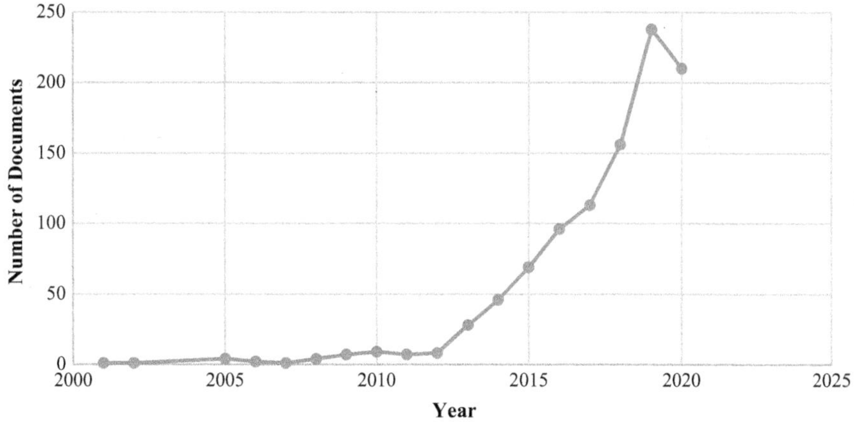

Fig. 7 Publication trend analysis, year wise

2.10 Top 10 Articles Based on the Citation

The analysis of the top 10 articles based on the citation reveals that the article "A Survey on ambient assisted living tools for older adults" has got the highest citations, i.e. 609, the article "A review of smart homes—present states and future challenges" with 603 citations and others. See Table 2.

2.11 Keyword Analysis

The VoSViewer is used for analyzing the co-occurrence of the author keywords, with the full counting method. The minimum occurrence for the keyword was chosen as 5, and out of 3190 keywords, 108 keywords met the threshold. In this density visualization map, the colour ranges from blue to green to yellow. The more closure the colour is towards yellow the more weight that particular item has. In this density visualization map, the Internet of Things, machine learning, ambient-assisted network, context awareness, telemedicine, e-health and home care came out as prominent author keywords. Another highlight is that the wireless sensor networks, wireless sensor network, wearable sensor, and wearable keywords together with health monitoring and heart rate and diabetes are also significant, which can be interpreted that there is an increased application of wearable devices for heart and diabetes-related ailments. See Fig. 8.

2.12 Citation Analysis: Sources

The citations analysis for the sources was done, the minimum number of documents of sources was selected as 3 and the minimum number of citations of sources was also selected as 10, out of 440 sources, 65 sources met the threshold see Fig. 8. This analysis reveals that there is a strong network between Sensors Switzerland and IEEE access, Sensors Switzerland and IEEE journal of biomedical and informatics. See Fig. 9.

2.13 Co-citation Analysis: Authors

The co-citation analysis of cited authors based on full counting method was carried out, the minimum number of citations of an author was selected as 20 and out of 86463 authors, and 657 authors met the threshold the results are presented in Fig. 10.

Table 2 Top 10 articles based on the citation

Publication year	Document title	<2016	2016	2017	2018	2019	2020	Subtotal	>2020	Total
2013	A survey on ambient-assisted living tools for older adults	120	97	122	111	94	64	488	1	609
2008	A review of smart homes-Present state and future challenges	367	59	51	55	47	24	236	0	603
2012	Smart wearable systems: Current status and future challenges	102	63	83	91	82	43	362	1	465
2016	2016 European Guidelines on cardiovascular disease prevention in clinical practice the Sixth Joint Task Force of the European Society of Cardiology and Other Societies on Cardiovascular Disease Prevention in Clinical Practice (constituted by representatives of 10 societies and by invited experts) Developed with the special contribution of the European Association for Cardiovascular Prevention and Rehabilitation (EACPR)	0	21	118	154	107	61	461	0	461
2013	A survey on fall detection: Principles and approaches	107	68	88	66	87	39	348	0	455
2014	Unobtrusive sensing and wearable devices for health informatics	42	64	86	94	74	37	355	1	398
2014	IoT-Based intelligent perception and access of manufacturing resource toward cloud manufacturing	18	69	62	91	85	54	361	0	379

(continued)

Table 2 (continued)

Publication year	Document title	<2016	2016	2017	2018	2019	2020	Subtotal	>2020	Total
2013	A survey on ambient intelligence in healthcare	81	80	46	57	49	36	268	1	350
2014	Data mining for internet of things: A survey	22	48	59	67	91	47	312	0	334
2009	ESPEN Guidelines on Parenteral Nutrition: Non-surgical oncology	191	35	28	32	26	21	142	0	333

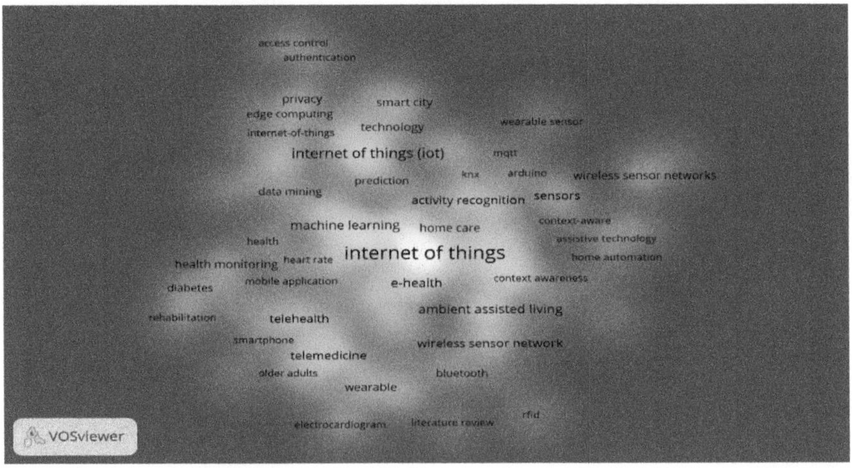

Fig. 8 Author keywords density visualization

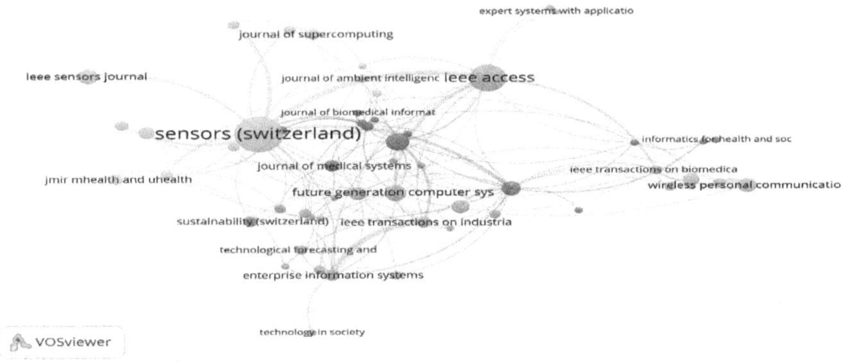

Fig. 9 Network visualization of sources based on citation

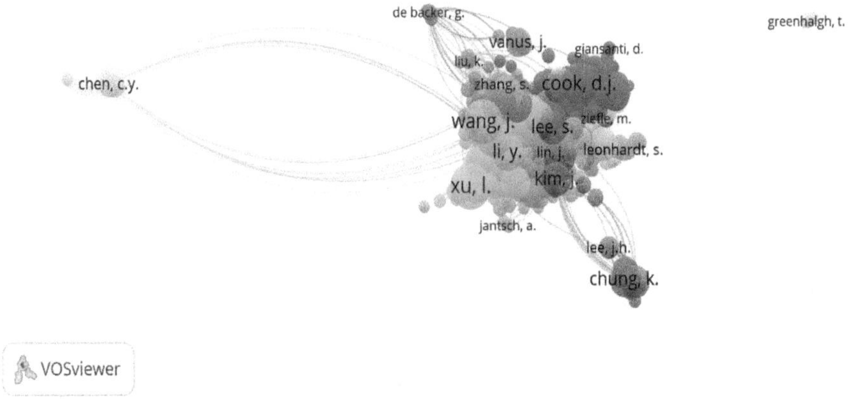

Fig. 10 Network visualization of authors based on co-citation

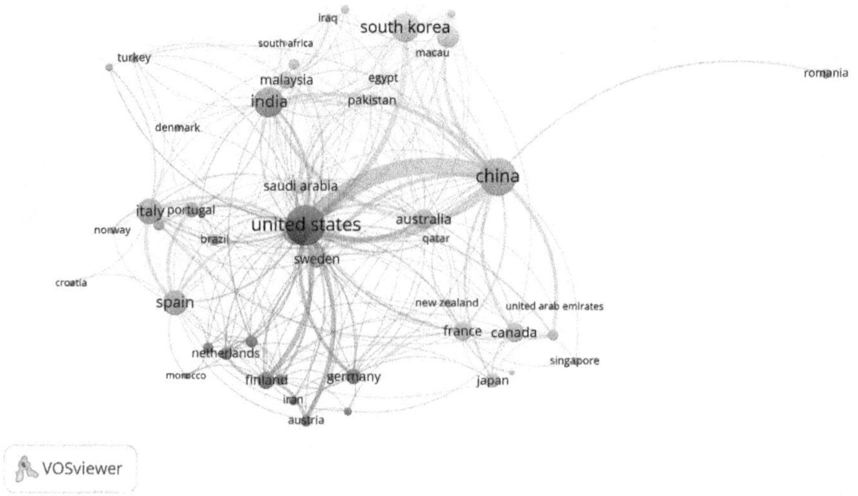

Fig. 11 Network analysis of co-authorship based on countries

2.14 Co-authorship, Country-Wise

To analyze the authors based on co-authorship and country-wise, the minimum number of documents by a country was selected as 3 and the minimum number of citations of a country was selected as 5. Out of 81 countries, only 50 countries met the threshold. The analysis reveals that the majority of the work in the field of home care and internet of health things is taking place in United States, China, South Korea and India followed by Italy, Spain, Australia and Canada. There is a very strong network between the United States of America Authors and the Authors of China. Moreover, Authors of United states have also significant linkage with the authors of the countries such as South Korea, India, Brazil, Sweden, Spain, Germany and Italy. See Fig. 11.

3 Discussion

Application of Internet of things is going to increase in the health care industry in the coming days. There are some evident trends which hint that the future of the Internet of Health things has a concurrence with smart homes [3], and assisted living. Moreover, the application of ambient intelligence has the potential to support the people affected by various physical or mental disabilities or chronic diseases [1]. Similarly, the possibility of integrating internet of health things with the clinical guidelines for medical specialities and sub-specialities will also be an area of research which will be explored in the future [12]. In the coming days, the impetus will be on low-cost wearable, non-invasive alternative for 24 × 7 monitoring of activity, health and mental status with the help of Body area Network "BAN" and Personal area network "PAN" [4]. The application Body area networks can be found in the monitoring of physiological parameters such as blood glucose, blood pressure, CO_2 gas sensor, ECG sensor and pulse oximetry [1]. The integration of wearable devices with health informatics would also dominate the research focus in the coming days, the health informatics provides the unique benefit of early prediction and treatment of major diseases [19]. The utility of Internet of Health Things is more for the elderly population and this will also be a key area for research in the coming days as some of the studies have attempted to detect the fall incidents for older patents and this is one of the important cause of injury to the old age patients [11]. Similarly, a PROSAFE project was carried out, which was developed to monitor the motor behaviour of older patients [3, 13].

The Majority of the research is happening in the developed countries such as The United States, China, Germany, Italy and Japan where the life expectancy of the individuals is higher and the adoption and availability of technology are also better. However, there are some developing countries such as India and Brazil where also the research in this domain has started to pick up and this is happening because of the demographic shift happening in these countries which are drawing the attention of the researchers towards old age population of the respective countries [17].

References

1. Acampora G, Cook DJ, Rashidi P, Vasilakos AV (2013) A survey on ambient intelligence in health care NIH public access. Proc IEEE Inst Electr Electron Eng 101(12):2470–2494. https://doi.org/10.1109/JPROC
2. Ajlouni MT, Dawani H, Diab SM (2015) Home Health Care (HHC) managers perceptions about challenges and obstacles that hinder HHC services in Jordan. Global J Health Sci 7(4). https://doi.org/10.5539/gjhs.v7n4p121
3. Chan M, Estève D, Escriba C, Campo E (2008) A review of smart homes-present state and future challenges. Comput Methods Programs Biomed 91(1):55–81. https://doi.org/10.1016/j.cmpb.2008.02.001
4. Chan M, Estève D, Fourniols JY, Escriba C, Campo E (2012) Smart wearable systems: current status and future challenges. Artif Intell Med 56(3):137–156. https://doi.org/10.1016/j.artmed.2012.09.003
5. Da Costa CA, Pasluosta CF, Eskofier B, da Silva DB, da Rosa Righi R (2018) Internet of health things: toward intelligent vital signs monitoring in hospital wards. Artif Intell Med 89(March 2017):61–69. https://doi.org/10.1016/j.artmed.2018.05.005
6. Genet N, Boerma WG, Kringos DS, Bouman A, Francke AL, Fagerström C, Melchiorre M, Greco C, Devillé W (2011) Home care in Europe: a systematic literature review. BMC Health Serv Res 11. https://doi.org/10.1186/1472-6963-11-207
7. Genet N, Boerma W, Kroneman M, Hutchinson A, Saltman RB (2012) Home care across Europe-current structure and future challenges. World Health Organ 156
8. Janati B, Khalesi T, Gholizadeh M (2010) Iranian professional's perception about advantages of developing home health care system in Iran. Hakim Res J 13(2):71–79

9. Jayaweera M, Perera H, Gunawardana B, Manatunge J (2020) Transmission of COVID-19 virus by droplets and aerosols: a critical review on the unresolved dichotomy. Environ Res, June 19–21. https://doi.org/10.1007/s00134-020-05991-x.Bizzarro
10. Koven S (2020) Engla, Journal-2010-New engla nd journal. New Engl J Med 82(1):1–2. https://doi.org/10.1056/NEJMp2009027
11. Mubashir M, Shao L, Seed L (2013) A survey on fall detection: principles and approaches. Neurocomputing 100:144–152. https://doi.org/10.1016/j.neucom.2011.09.037
12. Piepoli MF, Hoes AW, Agewall S, Albus C, Brotons C, Catapano AL, Cooney MT, Corrà U, Cosyns B, Deaton C, Graham I, Hall MS, Hobbs FDR, Løchen ML, Löllgen H, Marques-Vidal P, Perk J, Prescott E, Redon J, Richter Dimitrios J, Sattar Naveed, Smulders Yvo, Tiberi Monica, Bart van der Worp H, van Dis Ineke, Monique Verschuren WM, Binno Simone, Gale C (2016) 2016 European Guidelines on cardiovascular disease prevention in clinical practice. Eur Heart J 37(29):2315–2381. https://doi.org/10.1093/eurheartj/ehw106
13. Rashidi P, Mihailidis A (2013) A survey on ambient-assisted living tools for older adults. IEEE J Biomed Health Inform 17(3):579–590. https://doi.org/10.1109/JBHI.2012.2234129
14. Kisa S, Kisa A, Younis MZ (2009) The adoption of American home health care systems and its developing market in Turkey: What do physicians think? A pilot study in Turkey. Int J Health Promot Educ 47(2):44–50. https://doi.org/10.1080/14635240.2009.10708158
15. Santos MAG, Munoz R, Olivares R, Filho PPR, Del Ser J, de Albuquerque VHC (2020) Online heart monitoring systems on the internet of health things environments: a survey, a reference model and an outlook. Inf Fusion, 53(May 2019):222–239. https://doi.org/10.1016/j.inffus.2019.06.004
16. Seshadri DR, Davies EV, Harlow ER, Hsu JJ, Knighton SC, Walker TA, Voos JE, Drummond CK (2020) Wearable sensors for COVID-19: a call to action to harness our digital infrastructure for remote patient monitoring and virtual assessments. Front Digit Health 2(June):1–11. https://doi.org/10.3389/fdgth.2020.00008
17. Singh A (2017) Home Health Care: the missing link in health delivery system for Indian elderly population—a narrative review. Int J Res Found Hosp Healthc Adm 5(December):89–94. https://doi.org/10.5005/jp-journals-10035-1082
18. Singh A, Jha A, Purbey S (2020) Perceived risk and hazards associated with home health care among home health nurses of India. Home Health Care Manag Practice 32(3):134–140. https://doi.org/10.1177/1084822319895332
19. Zheng YL, Ding XR, Poon CCY, Lo BPL, Zhang H, Zhou XL, Yang GZ, Zhao N, Zhang YT (2014) Unobtrusive sensing and wearable devices for health informatics. IEEE Trans Biomed Eng 61(5):1538–1554. https://doi.org/10.1109/TBME.2014.2309951
20. Zupic I, Čater T (2015) Bibliometric methods in management and organization. Organ Res Methods 18(3):429–472. https://doi.org/10.1177/1094428114562629

Target Detection from Brain MRI and Its Classification

Bijoyeta Roy, Mousumi Gupta, Abhishek Kumar, and Sweta

Abstract Early detection of brain tumor is a very exigent task for radiologists. Delineating the boundaries of brain tumors from magnetic resonance imaging is an important task for brain cancer research. Throughout the most recent decade, various strategies have just been proposed for automated brain tumor detection. In this paper, we have introduced a novel strategy to characterize a given MR cerebrum image as normal or having unusual characteristics. The proposed strategy incorporates automated segmentation using K-means clustering, extraction of textural features using discrete wavelet transformation trailed by applying principle component analysis (PCA) to reduce the dimension of elements of highlights. The diminished highlights were submitted to a kernel support vector machine (KSVM) which was tested using three different kernels. It could be applied to the field of MR brain image characterization and can assist the specialists to analyze the degree of abnormality and severity of the brain tumor. Experimental findings indicate that the proposed approach offers an efficient and promising tool for highlighting and classifying brain tumors from MR images with an average accuracy of 79%.

Keywords Magnetic resonance image · Discrete wavelet transform · Principal component analysis · Kernel SVM · Brain tumor

B. Roy (✉) · M. Gupta
Sikkim Manipal Institute of Technology, SMU, East Sikkim, India
e-mail: bijoyeta.r@smit.smu.edu.in

M. Gupta
e-mail: mousumi.g@smit.smu.edu.in

A. Kumar · Sweta
UG Student, Sikkim Manipal Institute of Technology, SMU, East Sikkim, India

S. Dhar et al. (eds.), *Advances in Communication, Devices and Networking*, Lecture Notes in Electrical Engineering 776,
https://doi.org/10.1007/978-981-16-2911-2_10

1 Introduction

The brain is significantly an explicit and unstable organ of the human body [1]. Brain tumor grows very quickly; its average size doubles in just 25 days [2]. It is an intracranial mass comprised of abnormal improvement of tissue in the brain or around the cerebrum. Tumor is an irregular uncontrolled development and division of the cell which results in increased pressure inside the skull. Tumors can wreck neural connections or damage them by causing irritation, compacting various bits of the brain, provoking brain edema, brain necrosis, and various other brain-related abnormalities [3]. The abnormal brain tumor tissues differ in characteristics from the normal part of the brain such as gray matter, white matter, and the cerebrospinal fluid. A brain tumor can be either called benign which is considered to be noncancerous or malignant when there is no uniformity in structure and it contains active cancerous cells [2]. Benign tumors appear like normal cells in the brain and they expand gradually. Tumors are tested for their position, orientation, shape, density, magnitude, and margin description. Higher density is a typical indicator of malignancy.

There are various restorative diagnostic imaging techniques including MRI, CT, and PET used by clinicians to plan their treatment by assessing the location and severity of brain tumors. Magnetic Resonance Imaging (MRI) is considered as one of the most developed therapeutic imaging strategy which brings about high-resolution images enabling the specialists in distinguishing the shape, size, and morphology of the tumor [4]. Regular methodologies for viewing and diagnosing the ailments rely upon perceiving the proximity of explicit incorporates by a human spectator. Segmentation of brain MRI is a very essential step having significant applications in neurology, radiology, etc. Traditionally, segmentation is done manually in an operator-dependent clinical setting. Due to the vast volume of MRI evidence to be processed, the manual segmentation process by surgical specialists turns into a very tedious and time-consuming process and is therefore arbitrary and vulnerable to human error. It is difficult to produce sound results in MRI brain segmentation due to inhomogeneity, noise, low contrast, and several other artifacts of acquisition. Wavelet change is a viable means for extracting and highlighting meaningful features from MR cerebrum pictures since it permits an investigation of MR brain pictures at different degrees of resolution because of its multi-resolution analytical property [3]. In this paper, brain tumor is segmented using the unsupervised K-means clustering approach. The discrete wavelet transform (DWT) approach is used for extracting meaningful features which is accompanied by Principal Component Analysis (PCA) for reducing the dimension of the extracted features. Experts have proposed a great deal of approaches using support vector machine (SVM) for the classification of biomedical images. SVMs have noteworthy inclinations of high precision, numerical tractability, and direct geometric comprehension [4]. In this piece of work, we have used kernel support vector machine (KSVM) which can deal with datasets having linear as well as nonlinear decision boundary. Out of various available kernels, Linear, Radial Basis Function (RBF), and Polynomial kernels are used for transforming the feature space to a higher dimension.

2 Previous Related Work

Brain MRI segmentation is an important step that has numerous applications in neurology such as theoretical and quantitative analysis, operational planning, and functional imaging. According to the reports from National Brain Tumor Foundation (NBTF), over the past three decades the number of deaths due to brain tumor has risen by 300% [5, 6]. Therefore, various researchers have devoted their work on the brain tumor segmentation of data obtained from MRI using different segmentation techniques. Islam et al. [7] in 2013 proposed a technique using Fractal and Fractional Brownian Motion and got fair results. In [8], the authors have tried three segmentation techniques, namely modified gradient magnitude region growing technique (MGMRGT), level set segmentation (LSS), and marker controlled watershed segmentation (MCWS) on 9 different types and sizes of MRI brain datasets containing malignant cells. The results have shown that MGMRGT performed better than LSS and MCWS with respect to the measurement of area computed and compared with the ground truth image. Thaha et al. [9] in 2019 worked on an optimization-based segmentation technique using Enhanced Convolutional Neural Network. The loss function is optimized by using the BAT algorithm for segmenting the MRI brain portion in an efficient and automated way. With respect to performance parameters like precision, accuracy, and recall, the proposed algorithm has shown significant good results as compared to some of the other existing techniques. Kalavathi et al. [10] outstretched the standard Fuzzy C Means algorithm for clustering by incorporating the Gray-Level Co-occurrence Matrix (GCLM) for extracting features in order to construct a robust technique. It is further evaluated and validated with various FCM-based methods with varying levels of noise and attained better classification results. In 2015, using Artificial Bee Colony algorithm and Fuzzy C Means Clustering method, Neeraja et al. presented a hybrid technique for segmentation of MRI brain images [11].

3 Target Detection and Classification

Therapeutic image segmentation for distinguishing proof of brain tumors from the Magnetic Resonance (MR) pictures or other medicinal imaging modalities is a noteworthy strategy for picking the appropriate treatment at the right time. The various techniques that are utilized to achieve the desired results are the following.

3.1 Otsu Thresholding for Preprocessing

An extent of locale homogeneity is change (i.e. zones with high homogeneity will have low contrast). Otsu's strategy picks the point of confinement by restricting

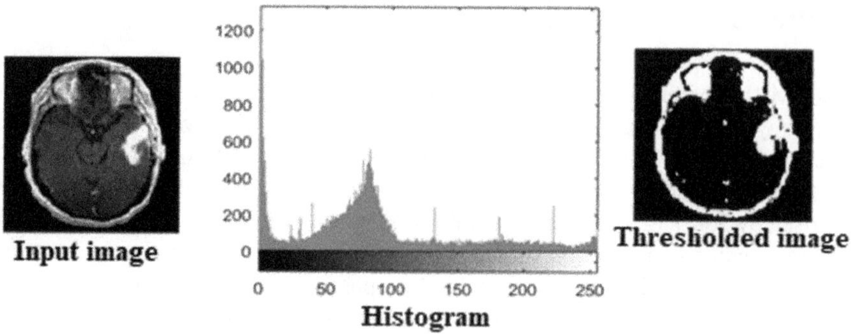

Fig. 1 Thresholded image

within class change of the two get-togethers of pixels detached by the thresholding value. Thresholding method scales images through the creation of a binary partition based on the intensities of each pixel of the image. Using the threshold value, the region between the brain segment and the skull portion can be distinguished. It will help to focus on the brain tissue part where the tumor is to be located further by segmentation (Fig. 1).

3.2 Segmentation with K-Mean Clustering

The main aim of this segmentation technique is to divide an image into mutually exclusive regions such that the intra-region differences are minimized and the inter-region differences are maximized. The calculation will order the things into k gatherings of likeness. In K-means algorithm, to ascertain that likeness, the Euclidean separation is utilized as estimation. It is an unsupervised technique involving the following steps:

(a) A value N is chosen randomly. N represents the total number of clusters.
(b) For each cluster, assign cluster centers randomly.
(c) Centroid is calculated for each cluster.
(d) Distance between each pixel and each cluster center is computed. Here, the Euclidean distance is used as the distance metric.
(e) Assign each pixel to a cluster where the distance between the corresponding pixel and the cluster center is minimum.
f) Recompute the cluster centroid and repeat the same process till the cluster center does not change.

3.3 Feature Extraction Using Discrete Wavelet Transform

After the MRI scan, the brain image is partitioned into exhausted regions and it is very crucial to extract detailed quantitative information in order to extract and analyze the various coherent features of an image. Various image features include shape, color, texture, contrast, etc. [12]. The main aim of applying DWT is to extract wavelet coefficient from MRI scanned brain images. This helps in localizing signal function frequency information [12]. Although Fourier transform (FT) is a commonly used tool for signal analysis, it is associated with a disadvantage that it excludes the time details from the signal. Unlike Discrete Cosine Transform (DCT) and Discrete Fourier Transform (DFT) which use sinusoidal waves as basic functions, DWT uses small waves known as wavelets each having variation in frequencies [13]. It approximates wavelet coefficients and can also analyze an image signal at various levels of resolution. In 2D-DWT, wavelet transform is applied in each dimension separately, and it converts the image to approximation coefficient (LL) and detailed coefficient (LH, HL, and HH). Features can be extracted from the approximation coefficient by applying decomposition in three levels. There are a variety of wavelets like Daubechies, Biorthogonal wavelets, Symplets 1, etc. [13]. In our study, we have worked on Daubechies wavelets. In Fig. 2, the various levels of decomposition of approximation coefficient is shown. Analysis of various statistical features like mean, standard deviation, entropy, skewness, kurtosis, smoothness, homogeneity, correlation, etc. also plays a vital role to understand the in-depth texture of an image. These parameters were assessed using gray-level co-occurrence matrix (GLCM) in this work. Functions in GLCM depict an image's texture by measuring how frequently pairs of pixels with the same values and within a given spatial relationship occur in an image, generating a matrix from where the statistical measurements can be captured. Mean (M) refers to the overall image intensity average value and determines the overall illumination of the image. Thus, images with better illumination have higher mean value. Contrast (C) finds a measure of the difference in intensities between a pixel and its neighboring area over the entire image. Higher the contrast, higher is the standard deviation (SD) of the image. Homogeneity (H) feature gives a measure of how closely the elements of an image are distributed. Other parameters like Variance (V), Kurtosis (K), Skewness (S), Entropy (E), and Correlation (CORR) are also measured. Therefore, the study of all these statistical features will contribute to help

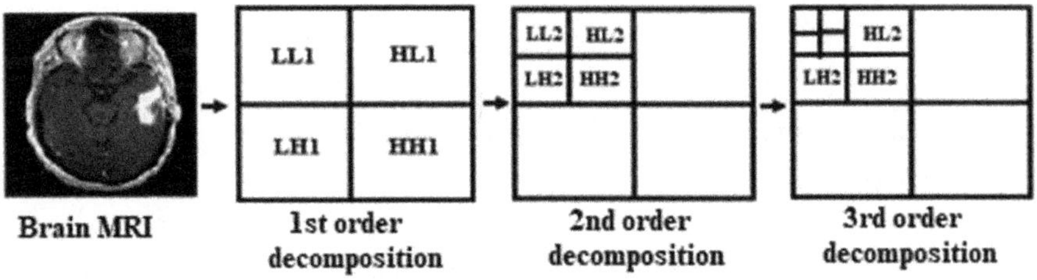

Fig. 2 Phases of DWT

Table 1 GLCM features

Features	Image 1	Image 2	Image 3	Image 4	Image 5
M	0.00324	0.00206	0.00321	0.00423	0.00341
SD	0.0897	0.0897	0.0897	0.0897	0.0807
E	3.577	3.518	3.37	3.551	2.994
V	0.00801	0.00804	0.00806	0.00803	0.00805
S	0.9324	0.8845	0.9314	0.9403	0.9209
K	6.273	6.767	7.35	6.061	7.582
SK	0.6331	0.4412	0.635	0.5104	0.5317
CORR	0.10067	0.09911	0.09327	0.1072	0.1294
H	0.9206	0.9365	0.9328	0.9287	0.9344

in differentiating normal and anomalous brain tissues for human visual perception. The result of statistical features obtained when applied on the testing images is shown is Table 1.

3.4 Principle Component Analysis for Feature Reduction

Medical images consist of enormous amount of information that are hard to preserve and process taking a great deal of computing energy. PCA is one of the most accepted multivariate and linear techniques for feature reduction [14]. It mainly contributes in approximating the original data with feature vectors of lower dimension and thus minimizes the feature space. The main idea is to project the original features of the brain image into Eigenvectors and then rearranging features according to their variance by considering the largest Eigenvectors from the correlation matrix [15]. In order to extract the PCA features of any training image '$\mathbf{I_1}$' having dimension **R x C** (R: no. of rows; C: no. of columns) into a vector $\boldsymbol{\lambda_1}$, we need to concatenate each of the rows into a single vector. The main aim of PCA is to transform the vector $\boldsymbol{\lambda_1}$ to another vector $\boldsymbol{\Omega_1}$ such that the dimension (d) of the vector $\boldsymbol{\Omega_1}$ is reduced (**d < < MxN**) without losing the important details of the image. The same is repeated for each training image: '**Ii**' and the corresponding vector $\boldsymbol{\Omega}$ of new dimension is computed and stored for further processing

3.5 Kernal Support Vector Machine (KSVM) for Classification

One of the significant applications of medical image classification is the identification of the type of tumor in abnormal MR brain images. Support Vector Machine which

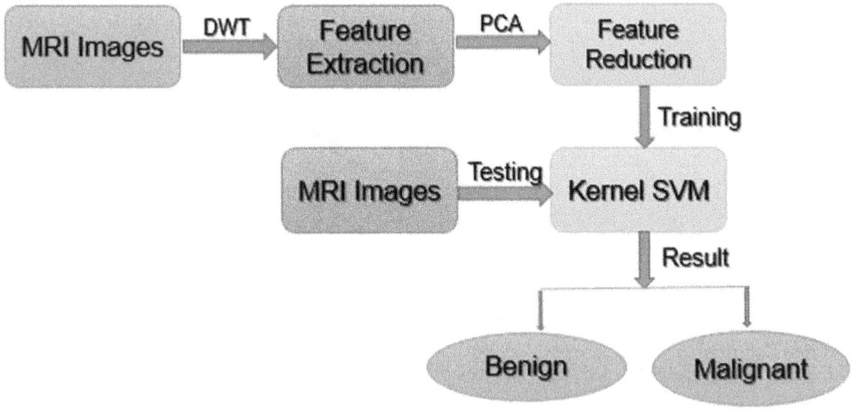

Fig. 3 Overall steps involved in the proposed work of segmentation and classification

is a popular pattern recognition tool for classification cannot consider neighboring pixel labeling intensities. The basic idea of finding a hyperplane for separating d-dimensional data into two groups often does not fit well for linearly nonseparable data. The introduction of a set of mathematical functions called kernel in SVM can contribute toward casting the data into a greater dimensional space making the data separable. Kernel SVM has a wide range of kernels such as linear, radial basis function, and polynomial which makes it a good choice for classification (Fig. 3).

4 Results and Discussions

In this proposed work, machine learning concepts were implemented using MATLAB 2014a in a system with 4 GB RAM and i5 processor and in Windows 10 operating system. In this study, two datasets are considered, one is the trained dataset and the other one is the test dataset. The datasets were taken from www.kaggle.com. For the test, a total of 70 images were considered out of which 42 images were in the training dataset and 28 images were in the testing dataset. Since the image has 3 colors, we will create 3 clusters using K-means clustering and measure the distance using Euclidean Distance Metric. After segmentation, two-dimensional discrete wavelet transform is applied on the MRI images which are resized to 200×200 to extract coefficients of wavelets which will help in giving more detailed information of various directions to provide a better characterization of brain tumor. First-level decomposition resulted in a feature matrix of size 103×103, second level decomposition produced 55×55 feature matrix, and finally the after third decomposition the final feature vector reduced to $31 \times 31 = 961$ features. DWT also filters the image by passing it through high pass as well as low pass filter. Since all the features extracted are not relevant for tumor detection, PCA is applied to reduce the dimensionality of data to make it more

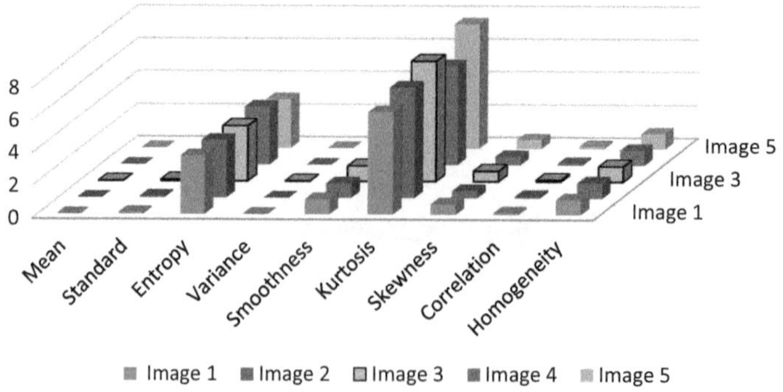

Fig. 4 Graph representing statistical features of brain MRI images

decipherable without having any significant information loss. After reducing the dimension of the MRI images by applying PCA, gray-level co-occurrence matrix (GLCM) is computed whose dimension resulted in a 8 × 8 matrix. Various relevant statistical features like skewness, kurtosis, correlation, homogeneity, variance entropy, etc. are also being calculated for each of the images using GLCM which will be helpful in classification. The statistical GLCM features obtained are shown in Table 1 and their corresponding graphical representation is depicted in Fig. 4.

This algorithm is applied to a number of MR images and the corresponding accuracies are evaluated. In case of making the hyperplane decision boundary between the classes, the linear, radial basis function (RBF), and polynomial kernels are clearly distinct. The kernel functions are used to transform the original dataset which can be linear or nonlinear in a high-dimensional space with a view to rendering it to a linear dataset. The RBF kernel is a very popular kernel function used in machine learning techniques like support vector machine. Considering N and N' as two feature vectors in any input space, and 'σ ' as a free parameter, the RBF kernel using squared Euclidian distance can be understood as (Fig. 5)

$$K(N.N') = exp\left(-\frac{||N - N'||^2}{2\sigma^2}\right) \tag{1}$$

A linear kernel can be used on any two observations by computing a standard dot product where the values obtained by multiplying each pair of input values are summed up together.

$$K(N, N_i) = \sum_{i=1}^{n}(N * N_i) \tag{2}$$

A polynomial kernel is represented by the degree (d) of the polynomial which needs to be input manually in the learning algorithm and a value of d = 1 can be thought of as a linear transformation.

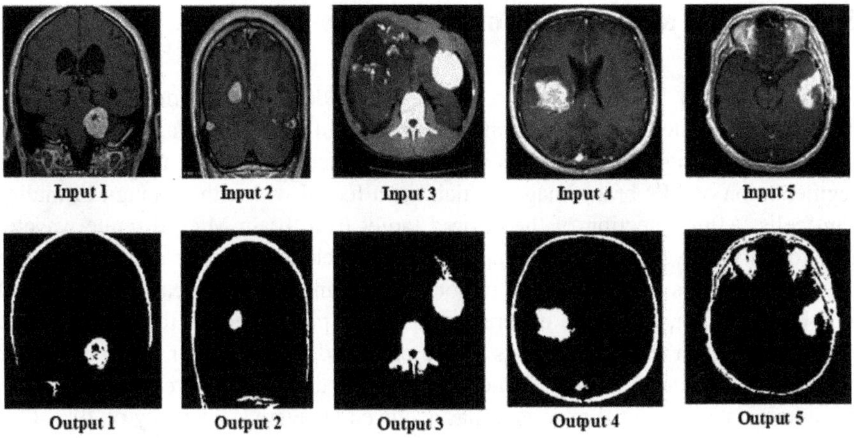

Fig. 5 Segmented output of five different input brain MR images

Table 2 Classification result

Classification result	
No. of brain tumor images	54
Benign	32
Malignant	22

$$K(N, N_i) = 1 + \sum_{i=1}^{n} (N * N_i) \,^\wedge d \tag{3}$$

In KSVM, the kernel parameters are to be adjusted before the data training process starts. In our study, we have considered RBF, Linear and kernel and found linear kernel to have the best performance. Table 2 and Table 3 depict the result of classification and kernel accuracies that are being computed.

Table 3 Kernel accuracy

Kernel accuracy	
RBF	79
Linear	80
Polynomial	77

5 Summary and Conclusion

Brain tumor may be considered to be one of the deadliest and most indomitable diseases. Automated and quick segmentation of brain tumors for diagnosis and prognosis is a challenging task. In this paper, it is aimed to perform automated tumor segmentation of MR brain images which are affected with either benign or malignant cells. After detection of the desired target from brain MRI, discrete wavelet transforms followed by kernel support vector machine are utilized to extract the important features and classify the tumor stage by analyzing the feature vectors and area of the tumor. From the experimental results performed on the different MR images, it is clear that the analysis of the proposed technique for the brain tumor detection is fast and accurate when compared with the manual detection performed by radiologists or clinical experts. It has achieved an average accuracy of 79%. Our experimental results show that the proposed approach can aid in the accurate and timely detection of brain tumor along with the identification of its exact location. The experimental results show different accuracies demonstrating the effectiveness of the proposed technique in different kernels for identifying the tumor as benign and malignant. In future, to improve the accuracy of the classification of the present work, further investigation can be done on the selective scheme of the classifier by combining more than one classifier and feature selection techniques to get more precise results. Also, focus can be given to finding out the textural features to analyze the stage of the tumor.

References

1. Ramteke RJ, Monali KY (2012) Automatic medical image classification and abnormality detection using k-nearest neighbor. Int J Adv Comput Res 2(4):190
2. Amin J, Sharif M, Yasmin M, Fernandes SL (2017) A distinctive approach in brain tumor detection and classification using MRI. Pattern Recognit Lett
3. Kharrat A, Gasmi K, Messaoud MB, Benamrane N, Abid M (2010) A hybrid approach for automatic classification of brain MRI using genetic algorithm and support vector machine. Leonardo J Sci 17(1):71–82
4. Praveen GB, Agrawal A (2015) Hybrid approach for brain tumor detection and classification in magnetic resonance images. In: 2015 Communication, control and intelligent systems (CCIS). IEEE, pp 162–166. (Nov 2015)
5. Logeswari T, Karnan M (2010) An improved implementation of brain tumor detection using segmentation based on hierarchical self-organizing map. Int J Comput Theor Eng 2(4):591
6. El-Dahshan ESA, Mohsen HM, Revett K, Salem ABM (2014) Computer-aided diagnosis of human brain tumor through MRI: a survey and a new algorithm. Expert Syst Appl 41(11):5526–5545
7. Islam A, Reza SM, Iftekharuddin KM (2013) Multifractal texture estimation for detection and segmentation of brain tumors. IEEE Trans Biomed Eng 60(11):3204–3215
8. Dubey RB, Hanmandlu M, Vasikarla S (2011) Evaluation of three methods for MRI brain tumor segmentation. In: 2011 eighth international conference on information technology: new generations. IEEE, pp 494–499. (Apr 2011)

9. Thaha MM, Kumar KPM, Murugan BS, Dhanasekeran S, Vijayakarthick P, Selvi AS (2019) Brain tumor segmentation using convolutional neural networks in MRI images. J Med Syst 43(9):294
10. Kalavathi P, Ilakkiyamuthu R (2017) Feature extraction based hybrid method for segmentation of brain tumor in MRI brain images. Int J Comput Sci Trends Technol (IJCST) 5(1):95–100
11. Menon N, Ramakrishnan R (2015) Brain tumor segmentation in MRI images using unsupervised artificial bee colony algorithm and FCM clustering. In: 2015 international conference on communications and signal processing (ICCSP). IEEE, pp 0006–0009. (Apr 2015)
12. Mathew AR, Anto PB (2017) Tumor detection and classification of MRI brain image using wavelet transform and SVM. In: 2017 international conference on signal processing and communication (ICSPC). IEEE, pp 75–78. (July 2017)
13. Shree NV, Kumar TNR (2018) Identification and classification of brain tumor MRI images with feature extraction using DWT and probabilistic neural network. Brain Inform 5(1):23–30
14. Gaikwad SB, Joshi MS (2015) Brain tumor classification using principal component analysis and probabilistic neural network. Int J Comput Appl 120(3)
15. Abdalla SA, Mustafa ZA, Abrahim BA (2019) Brain tumor classification using principal component analysis and artificial neural network. J Clin Eng 44(2):70–75

Gender Based HRV Changes Occurring in ANS During Graded Head-Up Tilt and Head-Reverse Tilt

Anjali Sharma and Dilbag Singh

Abstract The tilt table test was performed on 20 healthy individuals, who voluntarily participated in this study, out of which the data of 19 volunteers has been used to determine the gender related differences occurring in HRV because 1 volunteer was making movements and was continuously talking. The selected subjects were in the age ranging from 20 to 30 years and had a mean age of 23.894 ± 2.183 years. The subjects considered for this study were B.Tech and M.Tech students of NIT Jalandhar. The individuals considered in this study are all Indians belonging from different states. The tilt table test was performed at different angles of tilt which were 0°, 20°, 40°, 60°, 40°R, 20°R, and 0°R. The data was recorded in the biomedical instrumentation lab, NIT Jalandhar, which was later analyzed and assessments were drawn out between males and females. The study aimed to investigate the effect of gender on HRV using tilt table testing particularly head-up tilt and head-reverse tilt. The frequency-domain indices were calculated and comparison graphs were made to illustrate the changes occurring in HRV of men and women at different tilt angles. Frequency-domain analysis conveys how the energy of a signal is distributed over a range of frequencies. The tilt table was not brought back to 0° after every recording session at every angle, which means the angles were continuously increased after acquiring data for 7 min per specified degree tilt. The results convey that the sympathetic activity increases during head-up tilt and the parasympathetic activity takes over during head-reverse tilt. The outcome of this study demonstrates that healthy females in the age group of 20–30 years have sympathetic dominance at different angles of tilt, whereas healthy males of the same age group illustrate parasympathetic dominance at different tilt angles.

Keywords Autonomic Nervous System · Head-Reverse Tilt · Head-up Tilt · Parasympathetic Response · Sympathetic Response

A. Sharma (✉) · D. Singh
Department of Instrumentation and Control Engineering, Dr. B.R. Ambedkar NIT Jalandhar, Jalandhar, India

S. Dhar et al. (eds.), *Advances in Communication, Devices and Networking*,
Lecture Notes in Electrical Engineering 776,
https://doi.org/10.1007/978-981-16-2911-2_11

1 Introduction

Tilt Table Test is a basic, simple, and a non-invasive procedure to determine unexplained fainting spells or syncope [1]. It has clinical, medical, and research relevance. There is no need for external stimuli because tilt table itself induces the natural stimulus in the individual's body which is controllable. There are no severe complications in this testing procedure. Heart rate does not completely convey about the cardiovascular mortality, therefore, HRV is taken. HRV is advantageous because it is a non-invasive, quantitative, and a helpful research tool which is used to examine autonomic effects on heart. Furthermore, HRV is used to designate the physiologic reactions in normal individuals and patients with cardiovascular disorders [2]. The variation in time intervals of continuous cardiac cycles is referred to as heart rate variability. HRV illustrates that it is a totally non-invasive means of study utilized for research purposes in cardiology [3]. It is used to examine the influence of autonomic functions on heart. HRV describes physiological responses in healthy, normal, and in cardiovascular disordered patients as well. HRV describes instantaneous heart rate and RR intervals. HRV is recently used as an indicator of epileptic seizures [4]. Every time, a low HRV does not arise as a consequence of noise, but it can be severe as it expresses a greater chance of arrhythmia and mortality rate in patients with cardiovascular disorders, heart related diseases, and unhealthy heart [4]. RR intervals or inter-beat intervals can be distinguished among the intrinsic sources of HRV, which are ULF, VLF, LF, HF, and LF/HF, as each of the indices have different ranges. Short-term HRV is defined as the HRV of an individual observed for small time, whereas long-term HRV illustrates large time HRV observations of an individual. The past research have discovered that parasympathetic dominance was seen in women as contrasted with men during short and long-term HRV recordings [5]. The past investigations conclude that females have high vagal tone as a result of which they are being protected against arrhythmias, sudden cardiac death, and possess lower chance of these conditions as contrasted with males [6]. Some prior studies have also concluded that women have high sympathetic dominance than men or men have high vagal tone [7]. Even during fetal check-up, the heart rates are the gender predicting factors of the baby. It was determined in past investigations and experimentation that baby girls demonstrate higher heart rates close to 140 bpm or above, but the baby boys show heart rates less than 140 bpm [7]. But in some cases it was found that baby girls have shown lower heart rates than baby boys. Resting HRV of females is found to be more than males. Sympathetic responses cause increased blood pressure, increased heart rate, pupil dilation, tachycardia, and increases the risk for cardiac mortality. The previous researches also indicate that exercise, physical workout, yoga, and other body-mind relieving activities should be performed by an individual in order to keep the heart and other bodily systems healthy [8]. There are mainly two influencing factors that determine HRV of healthy population, which are age and exercise [8]. Young population depict a higher HRV due to active autonomic modulations. Youngsters depict higher HRV than older people because some investigations show that HRV decreases with increasing age, but it is also stated that a higher HRV is always not a healthy heart indicator [9]. As age increases, the HRV decreases, and due to certain diseased conditions such as diabetes, stress, anxiety, hypertension, heart diseases, etc [10, 11]. Data is recorded using MP-36 Biopac Student Lab system software. The obtained data has ECG on its 1st channel, which is processed to extract required features and then HRV is plotted.

2 Methodology

(a) **Inclusion–Exclusion criteria**: The subjects participated voluntarily and were electrocardiographically healthy. The subjects who voluntarily participated in present study were physically examined and their resting ECG was found to be normal. There were no cardiovascular complaints. Completely healthy subjects who do not smoke and drink were considered for this study. None of the subjects had neurological or psychiatric disease. Diabetic people were not considered for this study. The volunteers were verbally explained about the recording procedure and a written consent was being signed by them as well.

(b) **Subject recruitment**: The subjects were requested not to drink caffeinated beverages at least two hours prior the procedure. Since everyone knows that coffee is a stimulant which elevates blood pressure, heart rate, may lead to other variations in other bodily systems as a result of which the baroreceptor activity is increased. The individuals aged between 20 and 30 years and there mean was 23.894 ± 2.183 years. The subjects considered for this study are B.Tech and M.Tech students of NIT Jalandhar. The subject was made to relax for 10 min. prior the recording process. During this time, Ag/AgCl electrodes were attached at the specific body sites of the volunteer, i.e., near the right wrist, medial surface of left and right ankle of foot. BIOPAC ECG lead set consists of three leads of different colors, which are black, red, and white. The black colored lead specifies ground/ earth, red lead is for positive polarity, and white lead is for negative polarity. Only positive and negative polarity leads will not capture the signals, therefore, for this reason ground lead is connected for referencing and to record good signal. The ECG lead set was pinched on the volunteer's cloth in order to reduce cable strain or tension. The subjects should not be in contact with any metallic object during the recording session. Motorized automatic tilt table was being used to record the data at different degrees of tilt. The subject was asked to keep his/ her foot on the cushioned foot board to prevent slipping. The individuals were fixed on the tilt table by holding them with the 4″ positioning straps tied near chest, abdomen, and lower leg area. The positioning straps were not tied tightly as it would lead to changes in blood pressure and heart rate or it may suffocate the volunteer as well.

(c) **Signal recording**: The signals were recorded in the biomedical instrumentation laboratory at NIT Jalandhar during 9 am to 5 pm. Recording of data took place in an isolated room where there was normal ambient light and pressure, temperature were kept constant. The temperature ranged from 16to 25 °C. 3-

lead electrocardiography was performed on all the subjects. The ECG signal was recorded at different tilt angles for 7 min on each angle. The angles selected for this study are 0°, 20°, 40°, 60°,40°R, 20°R, and 0°R. BIOPAC student lab system software was utilized to record the ECG data. The middle 5 min data was selected for analyzing ECG/EKG recording and 1 min from starting and ending was truncated in order to analyze a good set of data so that variations can be seen in the data. The variations in our body arise minimum after 2 min of postural change or angle change. ECG signal is sensitive to motion artifacts, EMG, power line transmission, and external electrical noise. Talking too may alter the recordings, so all the volunteers were asked not to talk, laugh, and move. They were asked to stay still during the recording session on the tilt table with eyes closed but not sleeping. It was made sure that there occurs no technical failure, no artifacts or insufficient data recording or monitoring.

(d) **Proposed methodology**: ECG, age, and gender data was collected from 20 individuals without any cardiovascular complaints. Subjects were divided by gender into two sides for HRV analysis. Short-term HRV was recorded using Biopac student lab system software. The data was recorded at different tilt angles for 7 min each. The first and last minute data were clipped, only middle 5 min data was considered for the analysis process. For offline HRV evaluation using MATLAB, the digitized real-time ECG recordings were taken into removable hard disk. HRV is a marker of beat-to-beat intervals drawn from ECG recordings by distinguishing R-waves and constructing RR interval information. This is done for the examination of HRV in frequency domain.
Frequency-domain analysis conveys the energy of the signals distributed over a range of frequencies. It is obtained by doing the Fourier transformation of its time domain function. Frequency measurements can be done in two ways, which are absolute and relative powers. Measurements of frequency components like VLF, HF, and LF are done in absolute values of power (ms^2), but the measurements of LF and HF can also be done in normalized units (n.u.). The **absolute power** is calculated as ms2/ Hz, which means millisecond squared divided by cycles per second. While the **relative power** is estimated as percentage of total HRV power or n.u. (normal units). Normal units divides absolute power for a specific or particular frequency band by addition of absolute power of LF and HF bands. LF and HF measurements in n.u. represent organized and stable behavior of two branches of autonomic nervous system. HF signifies a measure of parasympathetic activity, whereas LF is a measure of sympathetic activity with vagal modulation. LF/HF signifies vagal tone and is a measure of instant sympathovagal balance. LF and HF ratio is a marker of SNS and PNS activity ratio under well-ordered conditions.

(e) **Spectral Estimation**: It forms the origin for distinguishing and tracing the signals of concern in the existence of noise and for extracting information from the received data. Fourier spectrum analysis is the foundation for almost every spectral estimation equipment including FFT, compressive spectrum analyser, etc. Fourier spectrum analysis makes assumptions regarding data outside the observation interval these impractical assumptions diminish the

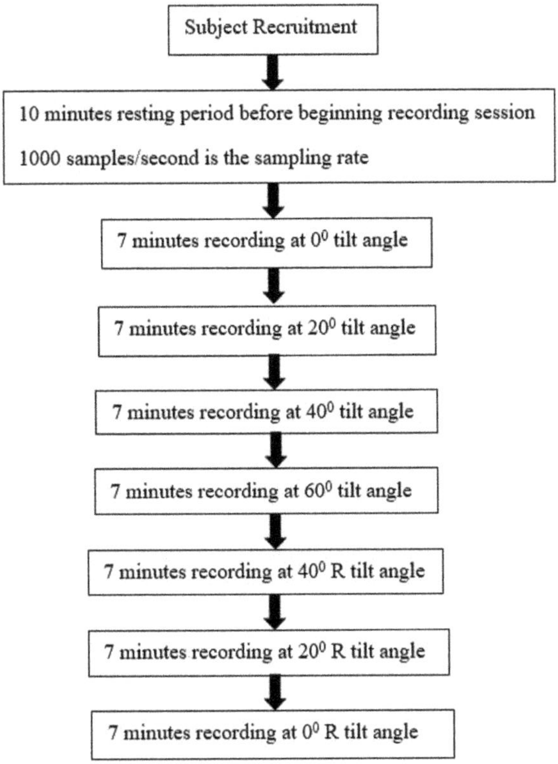

Fig. 1 Sequence of ECG data recording

quality of the evaluations. The frequency reflects the period of time over which rhythm occurs. It forms the origin for distinguishing and tracing the signals of concern in the existence of noise and for extracting information from the received data. Fourier spectrum analysis is the foundation for almost every spectral-estimation equipment including FFT, compressive spectrum analyser, etc. Welch's method is an approach to spectral density estimation. It is used to estimate power of a signal at different frequencies. This method is based on the concept of converting a signal from time to frequency domain using the periodogram spectrum estimates. Welch's method reduces noise in the estimated power spectra rather than the frequency resolution and its way better than the standard periodogram spectrum estimating method. Fourier spectrum analysis makes assumptions regarding data outside the observation interval, these impractical assumptions diminish the quality of the evaluations. The analysis was performed in 5 min. sliding window with a one-second increment. It is the minimum length recommended to completely cover and capture spectral cardiac dynamics [12].

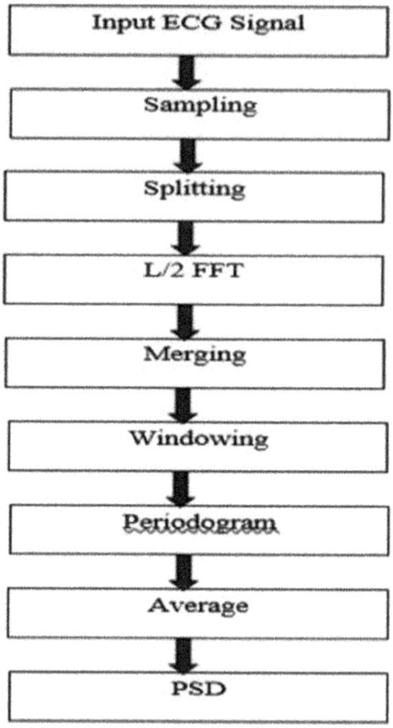

Fig. 2 Sequence of PSD extraction using Welch's Periodogram

Table 1 FD indices of HRV. LF: Low Frequency, HF: High Frequency, LF/HF: Ratio of LF and HF

Table for FD range values of HRV			
S. No	FD indices of HRV	Frequency Range (Hz)	Defined by
1	LF	0.04–0.15 Hz	Parasympathetic and Sympathetic Influence
2	HF	0.15–0.4 Hz	Parasympathetic Influence
3	LF/ HF		Sympatho-vagal Balance

$$\text{Normalized LF} = \frac{LFp}{Np} \quad (1)$$

$$\text{Normalized HF} = \frac{HFp}{Np} \quad (2)$$

where LFp is the power of LF band, HFp is the power of HF band and Np = 0.04–0.15 Hz.

The results are estimated using Welch's periodogram. For the utilization of Welch's periodogram, there should be prior knowledge of discrete fourier transform

(DFT), spectral estimation, basic and modified periodogram [7]. It is a non-parametric method which consists of the periodogram having a benefit of possible execution by means of Fast Fourier Transform. If data length samples are selected optimally, the periodogram technique based on Welch method is proficient in providing good resolution. The periodogram is an elementary method of estimating power spectral density of a time-series and is given by the below equation

$$\mathrm{DFT(f)} = \sum_{n=0}^{n} X(n)e^{-2\pi fn} \tag{3}$$

$$\mathrm{P(f)} = \frac{1}{N}\left|\sum_{n=0}^{N-1} X(n)e^{\frac{2\pi fk}{l}}\right|^2 \ldots \tag{4}$$

k = 0,1,……….L-1.

The modified periodogram is given by

$$\mathrm{PM(f)} = \frac{1}{MU}\left|\sum_{n=0}^{M-1} X(n)e^{-2\pi fn}\right|^2 \ldots \tag{5}$$

i = 0,1,……….L-1; where $\mathrm{U} = \frac{1}{M}\sum_{n=0}^{M-1} w^2(n)$

PSD by Welch Periodogram is given by the following equation:

$$\mathrm{PW(f)} = \frac{1}{N + \sum_{n=0}^{N-1}(P_{Mi}(f))} \ldots \tag{6}$$

$P_{Mi}(f)$ is the ith modified periodogram from the data series.

3 Results and Discussion

In low frequency domain, on graded HUT from 0° to 20°, female data depicted a gradual increase in the LF values then the trend sharply increased till 60°. During HRT from 60° to 40°R, the trend gradually decreased then it sharply decreased till 0°R. This shows that the sympathetic activity in females shot when the tilt table was raised from 0° to 60°, and the values decreased when the tilt angle was lowered. In case of males, during HUT from 0° to 20°, there was a slight decrement observed in the pattern of LF. As the tilt angle was elevated from 20° to 40°, sharp increment in the pattern was observed. As the angle of tilt was further increased, a sharp increment in the trend was witnessed. As the tilt table was lowered from 60° to 0°R, the data associated with the graphical pattern illustrated a slow decrementing trend.

HF shows a complete inverted trend as compared to LF. The rising trends of LF are now replaced by falling trends and vice versa as can be seen in the graph. The graphical pattern illustrates that HF is the converse of LF. In case of females, during HUT from 0° to 20°, there was slow decrease in HF index which further decremented sharply till 60°. During HRT from 60° to 40°, there was slow increase in HF value which increased further as angle of tilt decreased to 20R and then to 0R. So, it can be concluded that HF decreased as the tilt angle was increased and HF was found to be increasing when the angle of tilt was reduced. This means that parasympathetic activity decreases as tilt angle or the natural stimulation in the body increases. On the other hand, parasympathetic activity increases as the tilt angle or the natural stimulation in the body decreases. This shows that at rest or at the stages near to rest, the parasympathetic activity was found to be dominating. It can be clearly seen from the graphical pattern that the HF indices of males are higher than the HF indices of females, which means parasympathetic activity in males is higher than in females as only healthy individuals were considered for this study. During HUT, the HF of males slightly increases as the tilt angle increased from 0° to 20° and then it fell sharply when tilt table was raised from 20° to 40° then to 60°. During HRT, that is, when the tilt table was lowered from 60° back to 40°, HF increased sharply and then there was slow increase in HF when tilt table was lowered to 20°. HF in males was found to be increased when the tilt table was brought back to 0°. It's witnessed that the value of HF at 0°R nearly returned to pre-tilt value of 0°. When the male and female results were compared, it was found that HF indices of males was higher at every tilt angle than the HF indices of females at the corresponding tilt angles. This signifies that HF or the parasympathetic activity in males is higher which means that their sympathetic activity is lower than females.

LF/HF signifies vasovagal balance. It depicts the balance between the two branches of ANS, which are Sympathetic and Parasympathetic Nervous System. During HUT from 0° to 20°, there was a slow increment in the LF/HF of females, then there was sharp increment observed in it when the tilt table was raised to 40° and then further to 60°. LF/HF was observed to be maximum at 60° tilt. When the tilt angle was reduced back to 40°, then there was slow decrement observed in LF/HF of females. When the angle was further lowered to 20°, then there was a sharp decrease observed, which was further observed until the tilt table was brought back to 0°. It is witnessed that the LF/HF value returned to nearly pre-tilt value at 0°R when compared with 0°. In case of males, during HUT from 0° to 20°, there was slow increment in the LF/HF index. A sharp increment in LF/HF index was observed when the tilt table was raised to 40° and then to 60°. LF/HF was found maximum at 60°. When the tilt tale was lowered back to 40°, then LF/HF decremented sharply, which was further reduced when the tilt angle was lowered back to 20° and then to 0°. Rising trend was observed from 0° to 60°, and from there, a falling trend was observed till 0°R.

When the LF/HF of females and males were distinguished, then it was found that there was slow increase in the indices during HUT, but there was sharp decrement in the LF/HF indices during HRT. This concludes the fact that males have high parasympathetic activity or simply it can be said that the sympathetic activity is less dominant in males. It can also be seen from the table and the graphical trend that parasympathetic activity is more dominant during HRT, that is, when the tilt table was lowered back to 0°. The results also show that females have sympathetic activity dominant in themselves and a less dominant parasympathetic side than in males.

4 Tables and Corresponding Graphs

It can be witnessed from the table and the graphical pattern that there was a slow and sharp increment in LF/HF values during HUT, but there was a slow decrement in the values as compared to men when the tilt table was lowered back to normal. There's another witnessing factor that, in case of women, the index values returned to nearly pre-tilt values at 0°R, but in case of men, the 0°R value is quite higher than the value at 0° (initial point) (Figs. 3, 4, 5 and Tables 2, 3, 4).

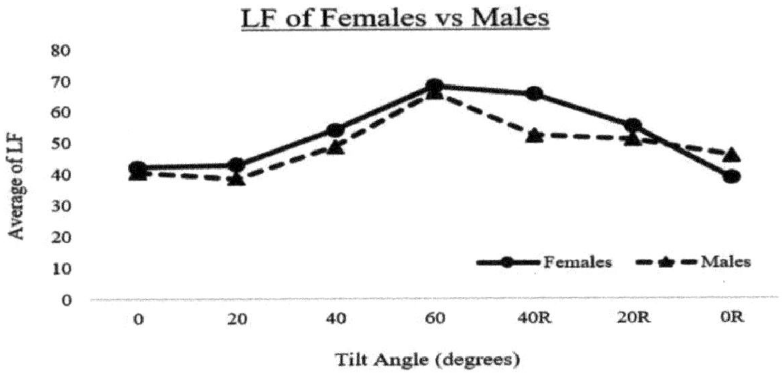

Fig. 3 LF of female vs males

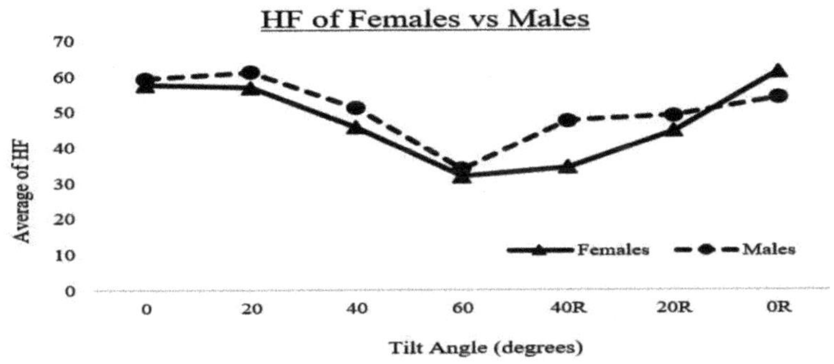

Fig. 4 HF of female vs males

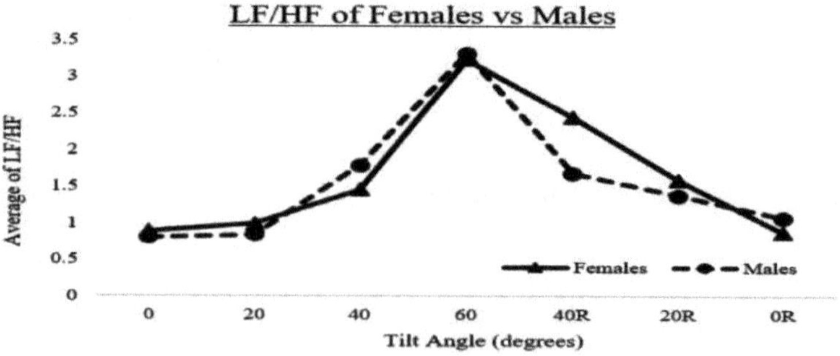

Fig. 5 LF/HF of female vs males

Table 2 Depicts LF of females and males at different tilt angles

Females	0	20	40	60	40R	20R	0R
1.	44.089	50.96	55.635	47.905	56.515	53.111	40.691
2.	51.814	25.115	33.026	56.849	39.017	36.018	11.162
3.	45.749	26.219	36.113	84.105	77.678	78.101	22.182
4.	7.6737	23.144	58.615	36.455	53.762	49.011	31.22
5.	62.402	64.098	68.918	80.541	76.338	63.348	70.672
6.	64.189	35.611	75.321	86.902	78.369	58.05	32.271
7.	13.869	12.148	20.181	44.079	61.197	14.491	25.358
8.	59.26	72.592	65.036	69.285	85.527	55.074	55.542
9.	42.291	60.76	62.215	82.645	60.272	70.746	44.965
10.	32.594	46.715	53.898	75.848	75.391	52.638	33.291
11.	41.861	56.489	67.276	84.476	56.972	75.797	57.459
AVG	42.3447	43.07736	54.20309	68.09909	65.54891	55.12591	38.61936
SD	18.37975	19.74198	17.23302	18.48907	14.03103	18.27359	17.42595
Males	0	20	40	60	40R	20R	0R
1.	57.021	67.715	73.953	86.05	59.774	50.355	56.044
2.	41.474	25.572	31.847	26.239	8.6597	40.481	34.906
3.	28.256	13.831	30.637	60.403	47.134	27.422	25.623
4.	24.259	25.848	19.997	43.32	34.229	49.259	37.695
5.	57.226	45.672	44.606	84.044	78.448	70.3	58.831
6.	21.237	22.488	54.791	61.227	44.606	26.049	19.878
7.	35.261	43.223	47.273	79.979	65.95	69.278	62.181
8.	60.791	65.106	88.049	87.102	79.779	74.912	71.821
AYG	40.69063	38.68188	48.89413	66.0455	52.32246	51.007	45.87238
SD	15.93416	20.09629	22.91541	22.41717	23.89834	19.16444	18.83601

Table 3 Depicts HF of females and males at different tilt angles

Females	0	20	40	60	40R	20R	0R
1.	55.737	48.994	44.336	51.979	43.412	46.799	59.194
2.	48.12	74.824	66.964	43.104	60.97	63.817	88.729
3.	54.171	73.743	63.503	15.868	22.299	21.879	77.793
4.	92.31	76.836	41.356	63.536	46.22	50.792	68.69
5.	37.535	35.764	31.069	19.418	23.6	36.592	29.305
6.	35.804	64.367	24.667	13.021	21.612	41.903	67.691
7.	86.087	87.758	79.719	55.846	38.648	85.277	74.339
8.	40.667	27.391	34.951	30.703	14.47	44.918	44.451
9.	57.448	39.026	36.819	16.526	39.243	28.956	54.896
10.	67.209	53.209	46.068	24.12	24.591	47.315	66.627
11.	57.94	43.493	32.693	15.492	42.92	24.149	42.378
AYG	57.548	56.855	45.64955	31.783	34.36227	44.76336	61.28118
SD	18.37996	19.75656	17.22581	18.5461	14.00356	18.23428	17.4048
Males	0	20	40	60	40R	20R	0R
1.	42.909	32.281	26.007	13.922	40.179	49.609	43.918
2.	58.486	73.854	67.89	73.681	90.869	59.471	64.712
3.	71.717	86.153	69.275	39.581	52.856	72.557	74.34
4.	75.731	74.081	79.98	56.664	65.75	50.719	62.179
5.	42.741	54.298	55.345	15.915	21.473	29.679	41.161
6.	78.698	77.484	45.165	38.763	55.371	73.783	80.007
7.	64.699	56.766	52.725	19.967	34.029	30.677	37.8
8.	39.147	34.667	11.951	12.83	20.196	25.081	28.12
AYG	59.266	61.198	51.04225	33.91538	47.59038	48.947	54.0296
SD	15.94299	20.07654	22.87997	22.41588	23.79221	19.13571	18.78608

5 Conclusion

This study aimed at documenting the frequency-domain parameters of HRV. The study was performed to investigate the effect of gender on HRV using tilt table testing particularly head-up tilt and head-reverse tilt. HRV values have been documented in normal population and the sample size was 20 (n = 20). Age, gender, race, timing, before eating, after eating, physical workout, performing yoga and meditation, etc., are some of the influencing factors that rule HRV. There can be age mismatching as compared to racial differences. Healthy students with no cardiovascular disorders voluntarily participated in the current study. There are certain limitations of this paper, which are the sample size for gender differences considered was small, the subjects selected for the study were young and healthy. All the participants belong to different states of India. The outcome of this study demonstrates that healthy

Table 4 Depicts LF/HF of females and males at different degrees of tilt

Females	0	20	40	60	40R	20R	0R
1.	0.79102	1.0401	1.2548	0.92163	1.3018	1.1349	0.68742
2.	1.0768	0.33565	0.49319	1.3189	0.63993	0.56439	0.1258
3.	0.84453	0.35555	0.56869	5.3002	3.4835	3.5696	1.28514
4.	0.083129	0.30121	1.4173	0.57376	1.1632	0.96492	0.45451
5.	1.6625	1.7923	2.2182	4.1478	3.2346	1.7312	2.4116
6.	1.7928	0.55324	3.0536	6.6741	3.6261	1.3853	0.47673
7.	0.16111	0.13843	0.25315	0.78929	1.5835	0.16993	0.34111
8.	1.4572	2.6503	1.8608	2.2566	5.9107	1.2261	1.2495
9.	0.73616	1.5569	1.6898	5.0008	1.5359	2.4432	0.8191
10.	0.48497	0.87795	1.17	3.1446	3.0657	1.1125	0.49966
11.	0.72248	1.2988	2.0578	5.4528	1.3274	3.1387	1.3559
AVG	0.892064	0.990948	1.457939	3.234589	2.442939	1.585522	0.882406
SD	0.563767	0.778019	0.836592	2.191223	1.565858	1.055159	0.652781
Males	0	20	40	60	40R	20R	0R
1.	1.3289	2.0977	2.8436	6.1807	1.5835	1.015	1.2761
2.	0.70912	0.34625	0.4691	0.35612	0.095299	0.6807	0.5394
3.	0.39399	0.16054	0.44224	1.5261	0.89175	0.37793	0.34468
4.	0.32034	0.34898	0.25002	0.76451	0.5206	0.97122	0.60624
5.	1.3389	0.84115	0.80596	5.2808	3.6533	2.3687	1.4293
6.	0.26985	0.29022	1.2131	1.5795	0.80558	0.35305	0.24845
7.	0.54501	0.76142	0.89659	4.0055	1.9381	2.2584	1.645
8.	1.5529	1.878	7.3677	6.7888	3.9503	2.9868	2.5541
AVG	0.807376	0.840525	1.786039	3.310254	1.679804	1.376475	1.080409
SD	0.519041	0.747787	2.397999	2.566059	1.432557	1.012701	0.792963

females in the age group of 20–30 years have sympathetic dominance at different angles of tilt, whereas healthy males of the same age group illustrate parasympathetic dominance at different tilt angles. The difference in the frequency-domain parameters suggests that gender has a significant effect on RR interval, heart rate, normalized LF, normalized HF, and low frequency–high frequency ratio. The total power remains nearly the same in both the cases, but the heart rate changes. That might be the reason that women have thinner blood, their body's blood temperature is higher than men, and have higher heart rates than men. Sympathetic reactions is the basis for increased blood pressure, increased heart rate, pupil dilation, tachycardia, increases the risk for cardiac mortality. The previous researches have depicted that age has a greater impact on HRV rather than gender [13]. Exercise, physical workout, yoga, and other relaxing activities tend to relax the bodily systems and hence result in good HRV [8]. A higher HRV is not always a good indicator of heart health, but could be a matter of worry [9, 14]. Lack of physical fitness may also become a factor for low HRV in young age [8]. The outcomes of this study might be useful in healthcare and research to determine ranges and differences in HRV of different genders under different experimental and clinical environments.

6 Future Scope

It is a vast area of research. New tools and methods can be utilized to distinguish among the genders. New diseases and other related effects can be predicted from the differences within the same gender as well.

Acknowledgements The study was performed in Biomedical Instrumentation Lab of Instrumentation and Control Engineering Department at Dr. B.R. Ambedkar NIT Jalandhar, Punjab. The recordings were particularly recorded in the month of December, 2019. The authors assume scientific responsibility for this study.

References

1. Barón-Esquivias G, Martínez-Rubio A (2003) Tilt table test: state of the art. Indian Pacing Electrophysiol J 3(4):239–252
2. Heart rate variability: standards of measurement, physiological interpretation and clinical use. Task force of the European society of cardiology and the North American society of pacing and electrophysiology (1996). Circulation 93(5):1043–1065
3. Pumprla J, Howorka K, Groves D, Chester M, Nolan J (2002) Functional assessment of heart rate variability: Physiological basis and practical applications. Int J Cardiol 84(1):1–14. https://doi.org/10.1016/S0167-5273(02)00057-8
4. Gender-Related Differences in Heart Rate Variability of Epileptic Patients Soroor Behbahani, PhD1, Nader Jafarnia Dabanloo, PhD2, Ali Motie Nasrabadi, PhD3, and Antonio Dourado, PhD, DSc4
5. Bigger JT Jr, Fleiss JL, Steinman RC, Rolnitzky LM, Kleiger RE, Rottman JN (1992) Frequency domain measures of heart period variability and mortality after myocardial infarction. Circulation 85:164–171
6. Bigger JT Jr, Fleiss JL, Steinman RC, Rolnitzky LM, Schneider WJ, Stein PK (1995) RR variability in healthy, middle-aged persons compared with patients with chronic coronary heart disease or recent acute myocardial infarction. Circulation 91:1936–1943
7. Evans JM, Ziegler MG, Patwardhan AR, Ott JB, Kim CS, Leonelli FM, Knapp CF (2001) Gender differences in autonomic cardiovascular regulation: spectral, hormonal, and hemodynamic indexes. J Appl Physiol 91:2611–2618
8. Rossy LA, Thayer JF (1998) Fitness and gender-related differences in heart period variability. Psychosom Med 60:773–781
9. Umetani K, Singer DH, McCraty R, Atkinson M (1998) Twenty-four-hour-time domain heart rate variability and heart rate: Relations to age and gender over nine decades. Am Coll Cardiol 31:593–601
10. Hoyer D, Friedrich H, Zwiener U, Pompe B, Baranowski R, Werdan K, Muller-Werdan U, Schmidt H (2006) Prognostic impact of autonomic information flow in multiple organ dysfunction syndrome patients. Int J Cardiol 14:359–69.8.

11. Sato A, Sato Y, Schmidt RF (1984) Changes in blood pressure and heart rate induced by movements of normal and inflamed knee joints. Neurosci Lett 52:55–60
12. Malik M, Bigger JT, Camm AJ, Kleiger RE, Malliani A, Moss AJ et al (1996) Heart rate variability: standards of measurement, physiological interpretation, and clinical use. Circulation 93:1043–1065
13. Moodithaya SH, Avadhany ST (2012) Gender differences in age-related changes in cardiac autonomic nervous function. J Aging Res 679345. https://doi.org/10.1155/2012/679345
14. Montano N, Ruscone TG, Porta A, Lombardi F, Pagani M, Malliani A (1994) Power spectrum analysis of heart rate variability to assess the changes in sympathovagal balance during graded orthostatic tilt. Circulation 90:1826–1831
15. Kuo TB, Lin T, Yang CC, Li CL, Chen CF, Chou P (1999) The effect of aging on gender differences in neural control of heart rate. Am J Physiol 277:H2233–H2239
16. Choi JB, Hong S, Nelesen R, Bardwell WA, Natrajan L, Schubert C, Dimsdale JE (2006) Age and ethnicity differences in short-term heart-rate variability. Psychosom Med 68:421–426
17. Mandic S, Fonda H, Dewey F, Le V, Stein R, Wheeler M, Froelicher VF (2010) Effect of gender on computerized electrocardiogram measurements in college athletes. Phys Sportsmed 2:1–9
18. Uusitalo ALT, Vanninen E, Levälahti E, Battie MC, Videman T, Kaprio J (2007) The role of genetic and environmental influences on heart rate variability in middle-aged men. Am J Physiol 293:H1013–H1022
19. Task Force of the European Society of Cardiology & the North American Society of Pacing and Electrophysiology (1996) Heart rate variability: standard of measurement, physiological interpretation, and clinical use. Circulation 93:1043–1065

Impact of Electrophoresis on Normal and Malignant Cell: A Mathematical Approach

Saikat Chatterjee and Ramashis Banerjee

Abstract Electroporation is a noninvasive technique for artificial drug delivery systems. This technique uses electric pulses to create transient pores in the cell membrane, which promote the delivery of biologically active molecules into cells. According to the variation of field intensity membrane potential, pore radius, pore density, and effective area of pore also change. Here we focus on the effect of electrophoresis on normal cell and malignant cell (blood cell) based on cell parameters. A comparative study is shown based on the mathematical model of a single cell.

Keywords Electrophoresis · Artificial drug delivery · Active molecules · Mathematical modeling · Transmembrane potential (TMP)

1 Introduction

Electroporation is a very useful technique to transfer drug molecules through a specific path (pores) to the exact destination of the cell. Pulse electric field effects become the focus of electroporation because of vast biomedical applications. The characteristics of the cell membrane change when it is exposed to electric pulses in electroporation. In the presence of an external electric field of 40 kV/m for 1 ms duration electroporation is performed through three stages. Activate or charging of cell membrane (0–0.51 μs), generation of pores (0.51–1.43 μs), and analysis of pore radius (1.43 μs–1 ms). The external field creates 97.8% small radius pores (1 nm) and 2.2% are large. The penetration level varies according to the intensity of the external electric field. The electric field is applied to the cell from different positions. A number of studies observed on the effect of pulsed electric fields. The change of shape of the cell due to an external field is also observed [1]. The unusual shape of

S. Chatterjee (✉)
EEE Department, Sikkim Manipal Institute of Technology, Sikkim Manipal University, Gangtok, India

R. Banerjee
EE Department, NIT Silchar, Silchar, India
e-mail: ramashisgnitee@rediffmail.com

S. Dhar et al. (eds.), *Advances in Communication, Devices and Networking*,
Lecture Notes in Electrical Engineering 776,
https://doi.org/10.1007/978-981-16-2911-2_12

the cell is explained by using Gielis' super formula [2]. Different test cases on human cells have been examined. For each type of cell, the effect of the relevant shape, the dielectric nature, and the external electric pulse effect on the electroporation method have been studied [2]. It shows that at huge fields, the spherical cell geometry can be considerably changed, and even ellipsoidal forms would be unacceptable to explain the variation. The transmembrane potential of the cell changes with the external applied field. This TMP can also be measured with a finite element method [3]. A comparative mathematical model is shown here between normal and malignant single blood cell basis of pore radius, the effective area of pores, membrane potential, and pore density.

2 Mathematical Model of Cell

E is the electric field, P is a point on the membrane, a_c, a_n, a_{mem}, and a_{ne} are the cell radius, nucleus radius, cell membrane thickness, and nuclear envelope thickness, respectively. C_{m}, C_{mem}, C_c, C_{ne}, and C_{np} are the media conductivity, cell membrane conductivity, cytoplasm conductivity, nuclear envelope conductivity, and nucleoplasm conductivity, respectively. M_{m}, M_{mem}, M_c, M_{ne}, and M_{np} are the media permittivity, cell membrane permittivity, cytoplasm permittivity, nuclear envelope permittivity, and nucleoplasm permittivity. The external-induced field changes its position at 15° interval. Single, monophasic, biphasic pulses are applied to the single cell (Figs. 1 and 2).

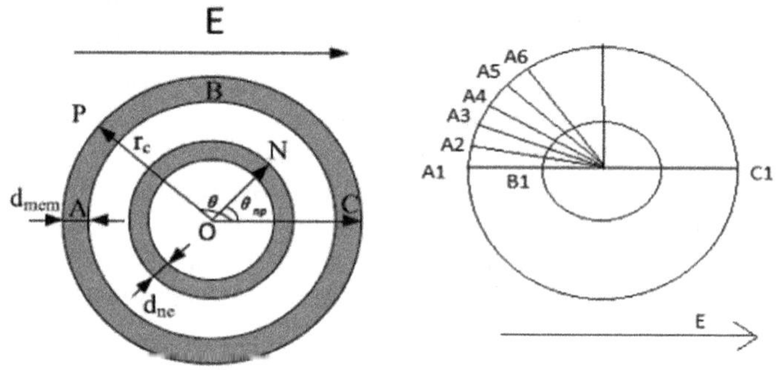

Fig. 1 Dielectric model of the cell [1, 2]

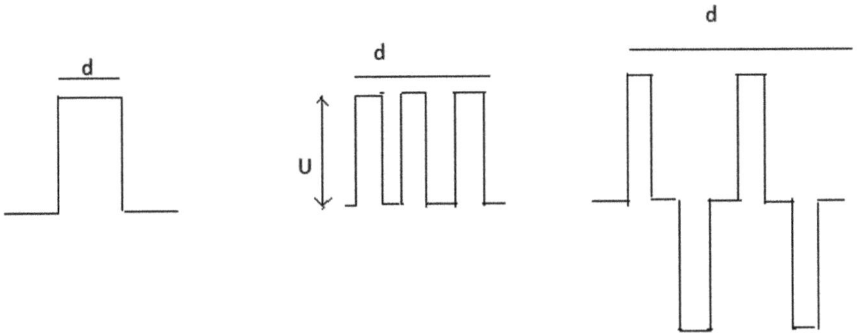

Fig. 2 Single pulse, monophasic and biphasic shock used as external field

2.1 Model of Electroporation

We consider a cell exposed to an electric field of strength E_{in} and an underground field with a conductive medium. The electric potential P inside and outside the cell can be computed using the electromagnetic field theory.

The induced transmembrane voltage (ΔP) was calculated [4] as the difference between electric potentials on both sides of the membrane

$$\Delta \mathrm{P} = \mathrm{P}_i(t) - \mathrm{P}_0(t) \tag{1}$$

$$\Delta \mathrm{P} = 1.5\mathrm{E}a_c \cos\theta\left(1 - \mathrm{e}^{-\frac{1}{\tau}}\right) + V_{\mathrm{rest}} \tag{2}$$

where θ represents angle of external electrode [1, 2]; τ = charging time constant; V_{rest} = resting potential of cell. Transient hydrophilic pores in the lipid bilayer of the plasma membrane are formed upon exposure to an external electric field.

2.2 Mathematical Model of Pore Radius and Effective Area

The pores that are initially created with radius a* (minimum radius of hydrophilic pores) change their size to minimize the energy of the entire lipid bilayer. For a cell with a total number of pores (K), the rate of change of their radius (a_j) is determined by a set of ordinary differential equations

$$\frac{daj}{dt} = (\mathrm{a}, \mathrm{Vm}, \mathrm{A_{eff}}) \tag{3}$$

A_{eff} is the combined area of all pores existing on the cell, where U is the advection velocity.

$$A_{eff}(W_p) = \frac{2\sigma' - \sigma_0}{\left(1 - \frac{W_p}{W}\right)^2} \quad (4)$$

$$W_p = N\left(\pi a*^2\right) \quad (5)$$

$$W = \pi a^2 \quad (6)$$

where σ_0 is the tension of the membrane without pores, σ' is the energy per area of the hydrocarbon–water interface, W_p and W are the effective surface area [2, 3] of the pore and the cell, respectively.

3 Different Parameters of Normal and Cancer Cell [5, 6]

Different parameters are shown in Table 1. In the presence of an external electric field, the values of different parameters are observed for normal and malignant cells.

Table 1 Different parameters of normal and cancer cell

Parameters	Normal cell	Cancer cell
Resting potential of cell membrane (mV)	−0.08	−0.026
Cell radius (μm)	3.3	5.2
Membrane capacitance (μF/cm^2)	1e-2	1.6
Cell conductivity (S/m)	5.6*1e-7	9.1*1e-6
Cell membrane permittivity	12*8.85*e-12	9.8*8.85*e-12
Nucleoplasm conductivity (S/m)	2.04	1.07
Nuclear envelope conductivity (S/m)	1.11*e-2	4.4*e-3
Nuclear envelope permittivity	106	60.3
Radius of nucleus (μm)	2.8	4.4
Thickness of cell membrane (μm)	0.007	0.007
Thickness of nuclear envelope (μm)	0.04	0.04

4 Results and Discussion

The variations of different parameters due to an external field are observed and abnormalities of the cells are identified. It is our goal to identify the abnormalities of cell according to the change of parameters. These abnormalities may arise due to tumor cell or malignant cell. Here, a comparative study between normal and malignant cells (cancer cells) is shown [6] based on conductivity, transmembrane potential and pore density, pore radius, and effective pore area, respectively (Fig. 3).

4.1 Conductivity

In the presence of an external electric field (the electric field range: 0.25–5.00 kV, position of electrode: 30°), the value of cell conductivity is observed [7]. In the case of the normal cell (^) it always maintains the stable condition (homogeneous transfer of ions does not affect conductivity) and remains unchanged (pH 7.4) but when there is a store of ions (Na^+, Cl^-, Ca^{++}) or the flow of ions (through iontophoresis, electrophoresis) varies, conductivity will vary (at 10 μs). It is also observed that conductivity of cancer cell decreases (2–6 μs) initially (due to a decrease of pH value of 4.6) but later due to more transfer of ions the conductivity increased (at 7 μs) (store of ions on cell membrane and nucleomembrane) and when the opposite polarity ions cross or store (due to neutralization of opposite polarity charges) the conductivity decreased.

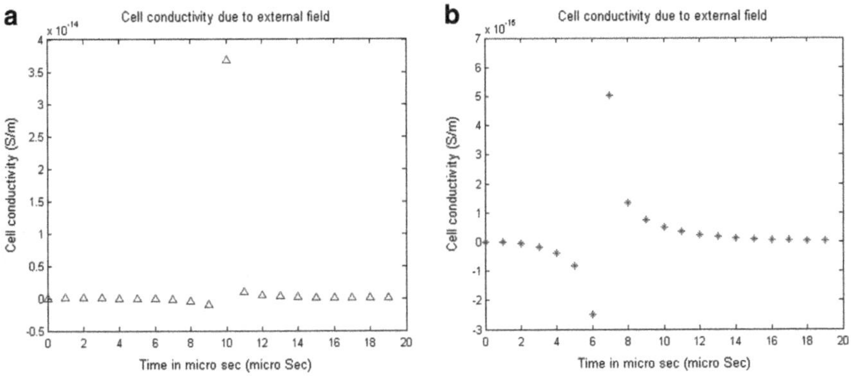

Fig. 3 Cell conductivity of **a** normal cell and **b** cancer cell [5] due to an external field (30°)

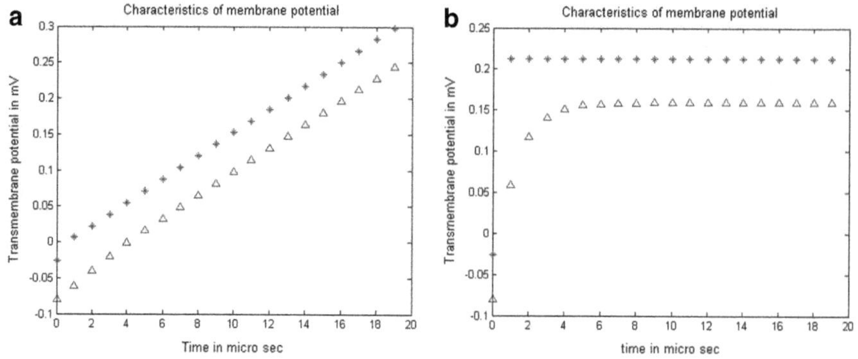

Fig. 4 Membrane potential for **a** variable external field [5] and **b** fixed external field [1, 2] due to an external field (30°)

4.2 Transmembrane Potential

In the presence of an external electric field (0.25–5 kV), the rate of transfer of ions changed. It is observed that due to this flow of ions through pores the transmembrane potential also changes (Fig. 4a, b). In the presence of the variable external electric field, the rate of increase of membrane potential of cancer cell (*) is higher compared to the normal cell (^). For fixed field (3.5 kV) (Fig. 4b) the transmembrane potential of both cells increased initially but later both remain constant. The membrane potential of a normal cell (0.15 mV at 5 μs) and cancer cell (0.22 mV at 1 μs) becomes constant after a certain time.

4.3 Pore Density

In the presence of an external electric field (0.25–5 kV at a certain position of electrode: 30°) [1], it is observed that the rate of generation of pores (pore density) is higher in the case of cancer cell (*) compared to normal cell (^) within a certain interval of time (0–19 μs). It is also observed that initially (0–10 μs) pore generation rate is low compared to the next interval of time (11–20 μs) (Fig. 5).

4.4 Pore Radius

In the presence of an external electric field (0.25–5 kV at a certain position of electrode: 30°) [1, 2], the pore radius changed with respect to time. Initially, it is observed that the pore radius is equal in both cases (at 4 μs) but later it decreases [due to repulsive force; Eq. (1, 2)]. It is observed that the rate of decrease of pore radius is higher

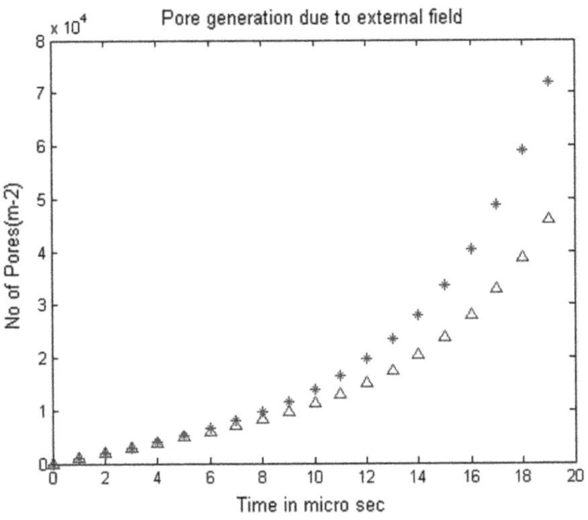

Fig. 5 Characteristics of pore density (m^{-2}) [7] due to external field (electrode position: 30°)

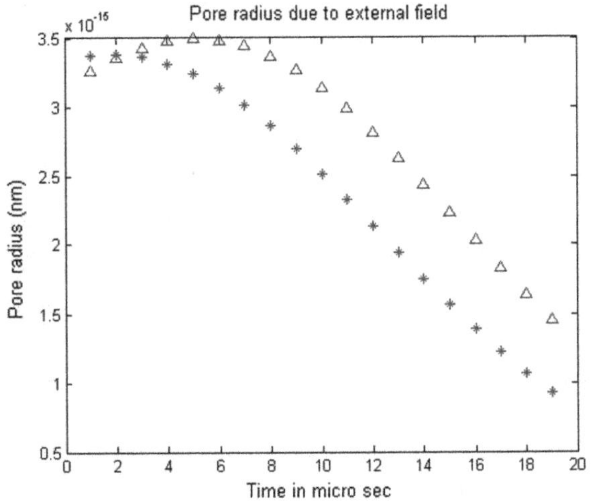

Fig. 6 Characteristics of pore radius [1] due to an external field (electrode position: 30°)

in the case of cancer cell (*) compared to normal cell (^) with a certain interval of time (Fig. 6).

4.5 Effective Area of Pore

In the presence of an external electric field (0.25–5 kV at a certain position of electrode: 30°) [1, 2], it is observed that the effective pore area of a normal cell is more compared to cancer cell initially (due to very less number of pores existing in the cancer cell), but with respect to time, it decreases. The rate of decreasing of

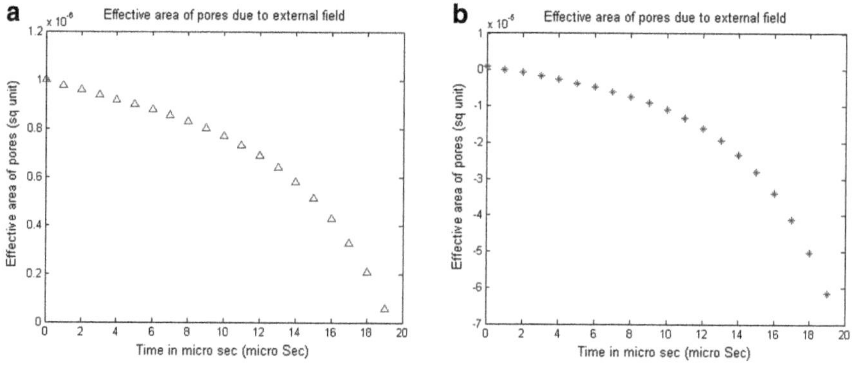

Fig. 7 Effective area of pores [8] for **a** normal cell and **b** cancer cell subjected to an external field (electrode position: 30°)

effective pore area is higher in the case of cancer cell (*) compared to normal cell because in the case of cancer cell the pore density is very low (effective pore area is negligible) (Fig. 7).

5 Conclusion

In the presence of an external field, the internal characteristics of single cells (normal as well as malignant cells) are observed.

(a) In the presence of an electric field, the concentration of H^+ ions is more in the malignant cell compared to the normal cell. It takes less time to reach a peak.
(b) Transmembrane potential cancer cell increases rapidly compared to a normal cell. In the case of the fixed external field, the transmembrane potential of the cancer cell reaches a saturation level faster than the normal cell.
(c) At a certain position of the external field (30°), the generation of pore (pore density) of a cancer cell is higher compared to the normal cell.
(d) Initially, the radius of the cell is the same for both cases, but later the rate of decrease of radius of the cancer cell is faster than the normal cell.
(e) The rate of change of the effective area of the pore is faster for the cancer cell due to reduction of cell radius.

References

1. Joshi RP, Hu Q, Schoenbach KH, Beebe SJ (2002) Simulations of electroporation dynamics and shape deformations in biological cells subjected to high voltage pulses. IEEE Trans Plasma Sci 30(4):1536–1546
2. Mescia L, Chiapperino MA, Bia P, Gielis J, Caratelli D (2018) Modeling of electroporation induced by pulsed electric fields in irregularly shaped cells. IEEE Trans Biomed Eng 65(2):414–423
3. Meny I, Burais N, Buret F, Nicolas L (2007) Finite-element modeling of cell exposed to harmonic and transient electric fields. IEEE Trans Magnet 43(4):1773–1776
4. Wanda K, Petar DF (2007) Modeling electroporation in a single cell. Biophys J 92(2):404–417
5. Yao C, Zhao Y, Dong, S (2017) Analysis of dynamic processes in single- cell electroporation and their effects on parameter selection based on the finite-element model. IEEE Trans Plasma Sci 45(5):889–900
6. Sree G, Udayakumar V, Sundararajan K (2009) electric field-mediated inactivation of tumor cells. Biomedical conference presentation, pp 30–34
7. DeBruin KA, Krassowska W (1999) Modeling electroporation in a single cell. II: Effects of ionic concentrations. Biophys J 77:1225–1233
8. Macqueen LA, Thibault M, Buschmann MD, Wertheimer MR (2012) Electro-manipulation of biological cells in microdevices. IEEE Trans Dielectr Electr Insul 19(4):1261–1268

DevChar: An Extensive Dataset for Optical Character Recognition of Devanagari Characters

Akshara Subramaniasivam, Azhar Shaik, Kaushik Ravichandran, and Manu George

Abstract The advent of cameras has only accelerated the need to digitize content as it helps prevent data corruption by natural processes and enables faster transfer of the data across communities. Handwritten documents and ancient manuscripts form a large part of this data as they call for a need to be translated from the local languages they were written in. The first step into solving this problem is the recognition of handwritten text. Existing handwritten datasets for the Devanagari script can be used for the recognition of individual characters, but they fail to perform well when the text contains matras and conjuncts created by joining character modifiers. This also introduces a dependency between the model and the data source due to required pre-processing for extracting characters recognized by the model from the word itself. These datasets also lack variation in their penmanship which is essential to encompass diversity in the writing style. We present an extensive dataset that addresses these issues. Our dataset has around 4 million characters of varying handwriting styles, complex characters and matras. Training a simple CNN on our data, to detect characters with matras, gave accuracies exceeding 98%. We also show that using this dataset allows a separation of the input data format from the model design, thus allowing researchers to focus on the latter. This dataset is made publicly available at DevChar2020.

Keywords Pattern Recognition (PR) · Optical Character Recognition (OCR) · Handwritten dataset · Devanagari script · Hindi dataset

1 Introduction

Since its inception in 1914, the first optical character recognition (OCR) model [1] was a mechanical statistical machine, and its intricacy has only exponentiated over time and so has its compactness. It is now a prime area of study in conjunction with complex deep learning models, which can easily be deployed on any home computer.

A. Subramaniasivam · A. Shaik · K. Ravichandran (✉) · M. George
PES University, Bengaluru, India

S. Dhar et al. (eds.), *Advances in Communication, Devices and Networking*, Lecture Notes in Electrical Engineering 776,
https://doi.org/10.1007/978-981-16-2911-2_13

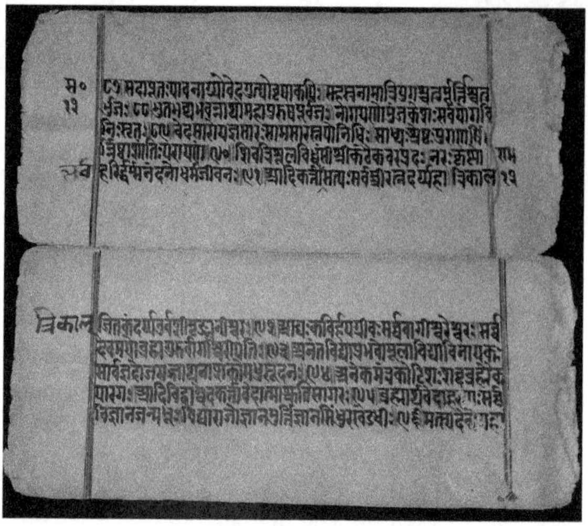

Fig. 1 Ancient manuscripts written in Devanagari script [5]

To simplify this process for the common man, advancements have been made with the release of open-source applications such as the Google Lens. The primary focus of these models is to convert handwritten text into digital documents, the importance of which is maintained by Nagy [2].

The study of OCR for printed text is not much of our concern as there are many novel methods proposed for the same, as mentioned by Impedovo et al. [3]. OCR for printed text is a fairly straightforward problem as the text is well defined in its parameters, such as the font used, which is specified and the size, which is kept uniform. The challenge lies in building a model for OCR for handwritten text. Writing any text by hand introduces a large variation in the text, as there is no clear definition of font, size and other such features. There can be a large variation within the font itself, and a constant size cannot be guaranteed due to anthropogenic factors.

One of the key elements required for the advancement of existing OCR technology is a good dataset. There are many large and extensive datasets for the English language such as NIST [4], but very few exist for Indian languages. It is known that most Indian and a huge number of Southeast Asian languages use the Devanagari script. In addition to this, the script is one of the oldest known scripts, which is being used since the first century CE. Some of the world's oldest manuscripts and scriptures, as shown in Fig. 1, are written in Sanskrit, which also uses the Devanagari script. This calls for a requirement for a large and extensive dataset for this script. The Devanagari script is used by over 120 Indo-Aryan languages and is a logical composition of its constituent symbols in a horizontal fashion from left to right, connected by a horizontal line, also known as the sirorekha. It has 14 vowels and 33 simple consonants. Besides the consonants and vowels, other constituent symbols in Devanagari are the set of vowel modifiers called Matras (placed to the left, right, above, or at the bottom of a character or conjunct) and pure consonants (also called half letters) which when combined with other consonants yield conjuncts. The presence of these vowel modifiers and pure consonants (half letters) often makes character recognition for the Devanagari script quite complex.

Some of the most widely used handwritten datasets in the Devanagari script include a printed and written dataset which was created by Yadav et al. [6]. ISM office fonts were used for the creation of printed characters, and the written dataset was collected from people of various age groups and professions. Holambe et al. [7] prepared a database of Hindi characters, which also follows the Devanagari script, totaling 77 different characters in 5 fonts, giving 385 characters which were used in training. An accuracy of 98.5% was obtained for the character-level recognition. The fallback of this dataset is that it is varied in terms of fonts, but not in terms of handwriting. Acharya et al. [8, 9] introduce a public image dataset for Devanagari script, containing 92 thousand segmented images of handwritten characters, of 46 different classes of which 72 thousand images were of consonants and 20 thousand were numerals. The characters were handwritten and then manually cropped. The processing stages included converting the image to grayscale followed by color inversion to produce white characters on a black background. One contribution of this paper is that the images were pre-processed well, but the dataset still lacks variation and is small in size (Fig. 2).

In this paper, we present an extensive dataset for the characters of the Devanagari script. Our large dataset comprises almost 4 million images with the following novelties:

- Our dataset contains 378 different character classes.
- Not only pure but also character-vowel and character-character combinations were considered.
- Our dataset has over five different handwriting styles.
- The dataset is pre-processed and ready to train.
- This dataset allows using simpler models for character recognition, which was not so earlier, as they had to accommodate combined characters within the model.
- The dataset is available in two sizes—DevChar-small and DevChar-large, with the same amount of variance in parameters.

The rest of this paper is structured as follows. The next two sections talk about the shortcomings of existing datasets in the Devanagari script and our contributions to overcome them, respectively. We then talk about the dataset creation in detail followed by the evaluation and results of the dataset.

2 Shortcomings of Existing Datasets

The base form of the consonants can be coalesced with vowels to form additional characters. There are 33 consonants and 14 vowels giving us a total of 462 different characters to detect and identify. So far, most available datasets address only recognition of individual characters, and not conjuncts created by joining two characters

Fig. 2 Problems with OCR for Devanagiri text

छ	६	Difference being horizontal line at top
ड	ङ	Difference being presence of single dot on right side
द	ढ	Difference being presence of small circle and small down stroke line

(a) Structural differences in characters

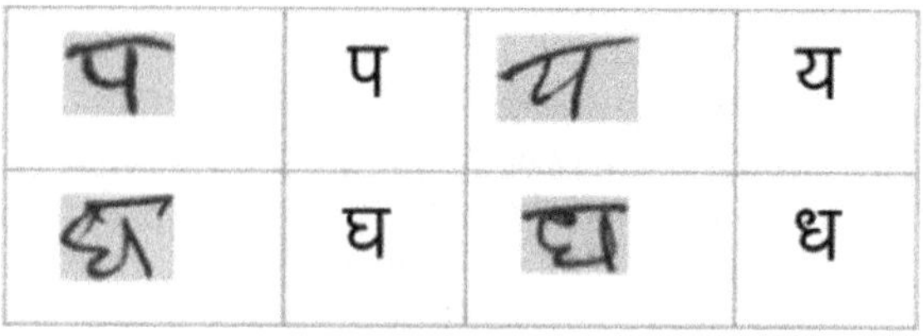

(b) Difference in writing styles

or a character and a vowel modifier, which is a common occurrence. Creating such an extensive dataset for handwritten text involves a lot of manual work. Neglecting these conjunct characters might seem like a simple solution, but comes with a giant overhead to deal with during the image processing stage, since separating the vowel from the consonant is not simple due to the cursive nature of the written text. This nature of the script makes the segmentation process of the script into its characters for recognition a lot more complex.

Since we are dealing with images of handwritten text collected from various different sources, each image needs to be pre-processed efficiently such that factors such as the ink used while writing, the paper it was written on and the thickness of the pen nib do not become considerably significant. All these factors make it hard for us to separate the foreground and the background in the image. Pre-processing is an often overlooked stage in the creation of a dataset, but one which carries significant weight in deciding the merit of a good dataset.

Segmenting sentences from a paragraph of text is fairly easy as there exists a punctuation mark to signify the end of a sentence, a vertical line. Segmenting the sentence into words is also a simple task as the characters combined to form a single word are all joined together by a continuous horizontal line connecting the tops of every character. Segmenting the words into individual characters is a complex undertaking. Existing methods identify the characters joined by the horizontal line and then remove the line to try and segment the characters, which gives rise to another issue where the few structural differences that distinguish the characters are lost.

3 Our Contributions

Ancient scripts have an irregular nature which is also an inherent characteristic of characters in the Devanagari script. Regardless of the language, segmenting a word using the horizontal line connecting the literals is not accurate since it is seldom present in ancient scripts and carvings. Due to the age of the Devanagari carvings and manuscripts, the characters, words and lines are ill-formed and require in-depth analysis and segmentation at the character level. By creating a huge variety of combinations of characters and vowels, rather than just the characters alone, we address the need to further segment characters into consonants and vowels for identification, which would otherwise be required.

The performance of a machine learning model depends heavily on the quality of the training dataset. The performance is the best when the dataset is diverse enough to truly represent the target concept. We have created a varied dataset representing the many circumstances of alignment, spacing, slope and shape of characters. There are many acceptable ways of writing the same character which may depend on the region, the font, the era of the script and most importantly, the author. A combination of all these varied factors gives rise to a heavily varied target distribution. The dataset was created by collecting the manually written and annotated characters from various people to encompass diversity in the writing style. In addition to this, we ensure that a large variation is captured by artificially augmenting the images by using a tool that allows us to perform elastic distortions.

One of the key stages during the development of OCR techniques is the pre-processing of images before training the model. Every image in our dataset is subjected to various phases of pre-processing in order to optimize and normalize each image. This saves time and resources as researchers can now focus on the architecture of the model rather than the uniformity of the input. The stages and details of the pre-processing steps are mentioned further in Sect. 4.2 of the paper.

4 Dataset Creation

The Devanagari script is composed of 47 primary characters, including 33 consonants and 14 vowels. The vowels are mostly used along with characters, and sometimes standalone, but never combined with each other. Compound characters can also be used, depending on the rules of the language. The dataset was created with the main focus being combinations of consonants and vowels. Since there are over 120 languages using this script, making a dataset with all combinations of primary characters is not easily feasible and hence has been avoided. The digits in Devanagari script have also been avoided as they are hardly seen in texts. This dataset includes 378 classes of 10000 images each, with the following classes being covered:

- Individual vowels;
- Individual characters;

- Characters combined with vowels;
- Characters combined with other characters.

This section lists the steps involved in creating the dataset. To give a thorough insight into the dataset creation, we have divided it into four subsections. The first subsection briefly goes over the collection of the images, followed by a subsection on the pre-processing involved to make the data simpler to train models. The following subsection covers the steps used to augment the dataset to increase its size. We finally conclude this section by presenting the metrics of our dataset compactly.

4.1 Data Collection

This dataset was collaborated by five people, to include a large variance in the handwriting. Each individual contributed to writing about 63 classes. This gave an equal distribution for each handwriting, thus increasing the variance in the dataset. We started with one class in one data sheet, which is a plain A4 size sheet of paper. Each data sheet was divided into 8 columns with 11 rows each, thus giving 88 cells per class, where each cell corresponds to one image, as shown in Fig. 3. This whole sheet was scanned at 300 dpi and then cut into 88 images, programmatically. We then organized this dataset into one folder per class, which makes it easier for generating more data, and training the model. The next few sections cover the pre-processing done and the steps involved in augmenting the images to generate a larger quantity of images.

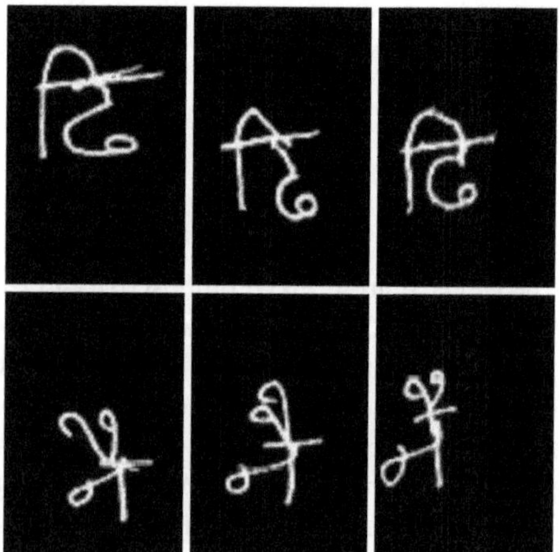

Fig. 3 Dataset images

4.2 Pre-Processing

Working with raw images adversely affects the training process as the foreground character blends into the background burdening the network to learn to differentiate between foreground and background first. To mitigate this undesirable overload, we first obtain the grayscale of the image and then perform image inversion. This results in an image where the white foreground character is placed upon a black background marking a clear distinction between the foreground and background aiding in the character recognition process. In addition to this, each character image is resized to 100×100 resolution as the model takes this resolution as input. We restrict the resolution of the input images to ensure that the training of the model happens in a reasonable time.

4.3 Data Generation

Once we obtain the dataset and perform the necessary pre-processing, we use Augmentor [10] to generate more data. Augmentor is a software package used for image augmentation, which provides support for many commonly used operations in image augmentation for machine learning problems. After loading the image, we add each operation we would like to perform on the image into a pipeline, where the operations are executed sequentially. Once all the operations are loaded in the pipeline, we sample the images to generate the specified number of output images. The output images may or may not be augmented, based on the probability.

We applied a random distortion to all our images, which means that any elastic distortion is done, with a specific probability. Two other parameters we provided were grid size and magnitude. The grid size controlled how fine and granulated the distortion will be, whereas the magnitude controlled the size of the distortion. An elastic distortion means that the aspect ratio of the image is maintained, but the horizontal or vertical lines in the images are distorted from their original position. Figure 4 shows augmented images for different distortion parameters for each image. Augmented images along with the original are shown in Fig. 4. The simple algorithm for augmenting is given in Algorithm 1.

Algorithm 1: Augmentor

Result: Set of Augmented Images

```
for classes i=0 to n do
    initialize pipeline;
    set random distortions;
    sample images for class i;
end
```

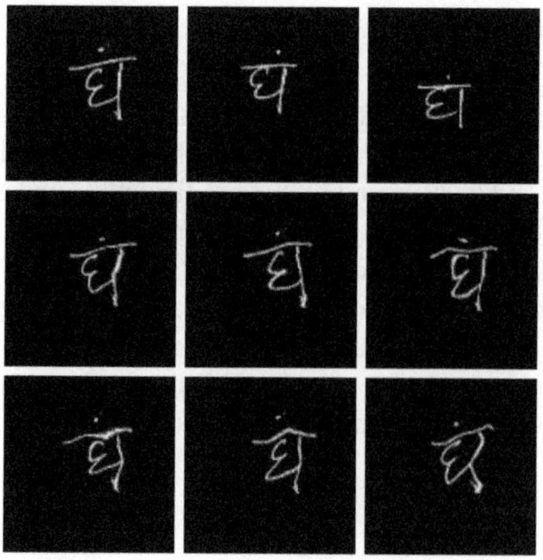

Fig. 4 Augmented Images Row 1: Original Images; Row 2: Grid size 6 × 6 and magnitude 3; Row 3: Grid size 8 × 8 and magnitude 4

Table 1 Dataset metrics

Characters	Images per class	Number of classes	DevChar-small	DevChar-large
Consonants	88	33	14850	330000
Vowels	88	14	6300	140000
Consonants + Vowels	88	329	148050	3290000
Consonants + Consonants	88	2	900	20000
Total	–	378	170100	3780000

4.4 Dataset Metrics

The different classes in the dataset have been covered in the first subsection of this section. The list of vowels, characters and combinations in the dataset are listed in Table 1. Each class has 450 images in the small dataset, and 10000 images in the large dataset.

5 Evaluation and Results

The goal of our evaluation is twofold. We first evaluate the performance of our dataset when trained on a simple CNN. We also compare this with the performance of the same model on the dataset obtained from the UCI Machine Learning Repository [8, 9].

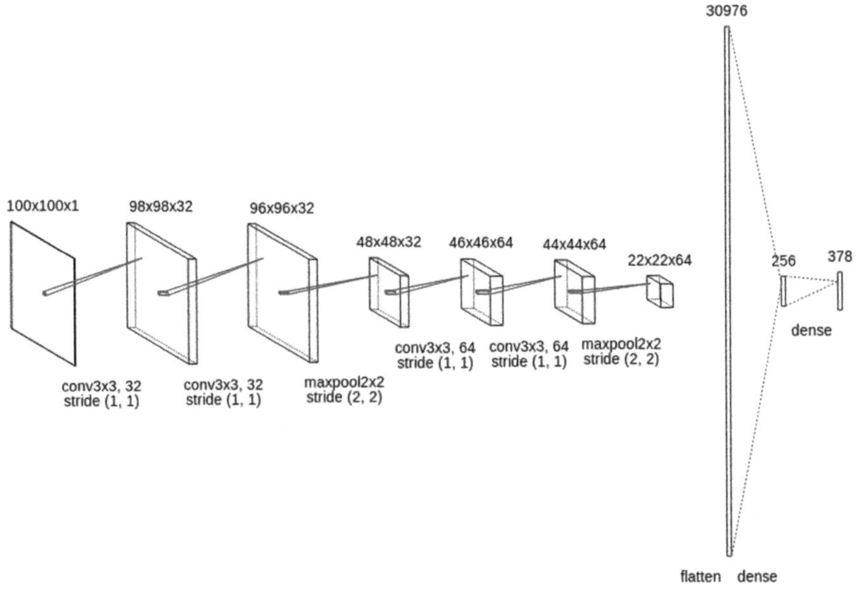

Fig. 5 CNN model used

The first subsection discusses the model we used in brief along with a presentation of the results obtained. In the second subsection, we discuss how this dataset makes building a model easier, in comparison with previously used models.

$$p = \frac{\text{True Positive}}{\text{True Positive + False Positive}} \tag{1}$$

$$r = \frac{\text{True Positive}}{\text{True Positive + False Negative}} \tag{2}$$

$$F_1 = 2 * \frac{p * r}{p + r} \tag{3}$$

The metrics [11] that were used are given below. Precision is the probability of false predictions actually being false, as given in Eq. (1). Recall is the probability of false being predicted false, and is calculated according to the formula in Eq. (2). F_1 is the harmonic mean of Precision and Recall, which is given by Eq. (3). We also report the accuracy of the model.

Table 2 Evaluation results

Dataset	Precision	Recall	F1	Accuracy
UCI	0.992	0.992	0.992	99.23
DevChar-small	0.929	0.949	0.939	92.19
DevChar-large	0.994	0.992	0.993	99.25

5.1 *Model*

The architecture of our Convolutional Neural Network is shown in Fig. 5 and is similar to VGG16 [12]. Instead of having 16 blocks of 2 Convolutional and 1 max-pooling layer, we have two such blocks followed by 2 dense layers. The final layer corresponds to the number of output classes which in our case is 378. We also use a dropout with p = 0.25 and p = 0.5 for the last layer. All activations are ReLU except for the last layer which uses the softmax activation function to give a probability distribution over the classes. We ran the model for 40 epochs. The batch size was kept constant at 64 for training and testing. The training was done at 1200 steps per epoch, and the testing was done at 215 steps per epoch. We ran our model on our datasets for 6 epochs, and used a batch size of 256, with 10000 steps per epoch for the first three epochs, and 2000 steps per epoch for the last three epochs. The evaluation was done on an Apple MacBook Pro 2017 Intel(R) Core(TM) i5-7360U CPU @ 2.30GHz, 8 GB RAM using Python v3.7.3. Our model, obtained an accuracy of 99.23%, on the UCI Machine Learning Repository Dataset and an accuracy of 99.25% on our large dataset, as shown in Table 2.

5.2 *Inference of Results*

Recognition of characters with matras and compound characters formed by the combination of two characters is a challenging task. Recognition of these characters often requires complex models to isolate the base character from its complex form and then use models like a classifier or a finite state automaton to detect the character. For example, [13] uses Zernike moment feature descriptor and proposes SVM and k-NN-based classification systems for the extraction of features from handwritten compound characters. Using our dataset, the same can be detected with a much simpler CNN model as mentioned in the previous subsection as the dataset itself contains characters along with their matras and compound forms. This helps us avoid the process of segmentation to isolate the base form of the character. Furthermore, simplifying the model also makes the process of optical character recognition much faster and also provides us with accuracies exceeding 98%. Hence, this dataset helps make character detection faster, simpler and more accurate.

6 Conclusion and Future Work

In this paper, we present a new dataset which is made publicly available for any research scholar to build models for Optical Character Recognition of characters of the Devanagari script. The experimental analysis illustrates that the dataset is robust and can be applied to solve the difficulty in character recognition of all Devanagari script-based languages. The characters were handwritten, manually cropped and scanned. Augmentation of the characters provided variations that would have otherwise required more time-consuming manual labor in writing and annotating. The addition of conjunct characters, i.e. consonants with all relevant vowels makes the dataset more detailed, extensive and allows us to achieve thorough and accurate results.

The dataset we created is extensive in that it covers most conjunct characters, but a lot of the images were augmented. Each class can be enlarged by writing more samples per class by hand, rather than augmenting. We have focused mostly on vowel-consonant combinations. There are many more allowed combinations of consonants with other consonants, depending on the language and its dialect. The dataset could also be enlarged in this direction as well. The network used to test robustness only provided a first look into what could be used for the process of recognition. A much more detailed study can be done to build our own network. We only show how character recognition is done, and avoid word-level and sentence-level recognition, which is where our dataset would perform the best. These two additional steps are essential to digitize a document, which is the end goal of our study. There are also efforts being made by the authors to donate the dataset to be a part of the UCI Machine Learning Repository [9].

References

1. Tauschek G (1929) Reading machine, United States Patent 2026329, May 27, 1929
2. Nagy G, 29 optical character recognition–Theory and practice. Classification pattern recognition and reduction of dimensionality, 621–649
3. Impedovo S, Ottaviano L, Occhinegro S, Optical character recognition: a survey. Int J Pattern Recogn Artif Intell 05(1–2):1–24
4. Grother PJ (1995) NIST special database 19 handprinted forms and characters database. National Institute of Standards and Technology
5. Research Group on Manuscript Evidence, Hindu Manuscript on Paper http://manuscriptevidence.org/wpme/sanskrit-and-prakrit-manuscripts/
6. Yadav D, Sánchez-Cuadrado S, Morato J (2013) Optical character recognition for hindi language using a neural-network approach. J Inf Process Syst 9(1)
7. Holambe AN, Thool RC, Jagade SM (2010) Printed and handwritten character & number recognition of Devanagari script using gradient features. Int J Comput Appl (0975- 8887) 2(9)
8. Acharya S, Pant AK, Gyawali PK (2015) "Deep Learning based large scale handwritten Devanagari character recognition. 9th International conference on software. Knowledge, Information Management and Applications (SKIMA)
9. Dua D, Graff C, UCI machine learning repository. http://archive.ics.uci.edu/ml

10. Bloice MD, Roth PM, Holzinger A (2019) Biomedical image augmentation using Augmentor. Bioinformatics 35(21):4522–4524
11. Chinchor N (1992) MUC-4 evaluation metrics. In: Proceedings of the fourth message understanding conference (MUC-4)
12. Simonyan K, Zisserman A, Very deep convolutional networks for large-scale image recognition. International conference on learning representations (ICLR)
13. Kale KV (2014) Zernike moment feature extraction for handwritten Devanagari (Marathi) compound character recognition

A Survey on Various Stemming Techniques for Hindi and Nepali Language

Biraj Upadhyaya, Kalpana Sharma, and Sandeep Gurung

Abstract Stemming is considered to be an important step in the field of natural language processing. Any word may be represented in various forms as per its use in a sentence or a paragraph. The various forms of the word are derived from one root word. This root word is called stem and the process of extracting the root word is known as stemming. As an example, words like singing and singer have the root word sing. The stemming technique particularly finds its use in the field of information retrieval where it contributes to efficient retrieval. The underlying algorithm used by various search engines like Google uses stemming for mapping a user's query which is put in various word forms to a base word or stem. This paper highlights several stemming techniques which have been developed for languages like Hindi and Nepali.

Keywords Stem · Root · Stemming · Information retrieval

1 Introduction

Stemming is the method by which the inflected or derived words are reduced to a stem, base or root form. The software which produces the stem or the root word is referred to as a stemmer [1]. A text in a natural language may contain several variants of the same word. These morphologically similar words have their meanings rooted in the stem or the root word [2, 3]. Today, several stemmers have been developed for many languages across the world, mainly for English and other European languages

B. Upadhyaya (✉) · K. Sharma · S. Gurung
Sikkim Manipal Institute of Technology, Sikkim Manipal University, Sikkim, India
e-mail: upadhayaya.biraj@smit.smu.edu.in

K. Sharma
e-mail: kalpana.s@smit.smu.edu.in

S. Gurung
e-mail: sandeep.gu@smit.smu.edu.in

S. Dhar et al. (eds.), *Advances in Communication, Devices and Networking*,
Lecture Notes in Electrical Engineering 776,
https://doi.org/10.1007/978-981-16-2911-2_14

as well [4]. However, not much work in the field of stemming has been carried out for South Asian languages like Hindi and Nepali [5].

Languages like Hindi and Nepali are written using a writing methodology known as Devanagari script which is considered to be a descendant of Brahmi script. It is written and read from left to right direction and there is no distinction between upper case and lower case letters [6]. The Hindi and the Nepali alphabets share many similar characteristics in the way they are written, as both are written using the Devanagari script. The Devanagari script is the fourth most widely adopted writing system in the world and is composed of 47 primary characters, including 14 vowels and 33 consonants [7]. In order to develop a stemmer for any language, it is very important to understand the word constructs pattern for that language. Nepali is an Indo-Aryan language written in Devanagari script. It follows a Subject + Object + Verb pattern in sentences which is different for languages like English. From the aspect of applicability, stemming techniques find their use in various information retrieval tasks used by various search engines like Google, Yahoo etc. The stemming techniques also find their use in finding domain vocabularies related to a particular domain of interest.

1.1 Types of Stemmer

Stemming algorithms can be broadly classified into the following three categories.

1.2 Rule-Based Stemming

It is a structural stemming approach that utilizes the structure of the words in a language as well as the morphological rules to identify the stem of each word. The rules are written based on the morphology of the language and its word derivation structure [8].

1.3 Statistical Stemming

Rule-based stemmers have the disadvantage of being reliant on a fixed set of rules for carrying out the stemming operation. Rule-based stemmers also require an adequate amount of language expertise [9]. On the other hand, statistical stemming approaches do not require language expertise and use statistical information from a large corpus of a given language to learn the morphology of words [10].

1.4 *Hybrid Stemming*

This approach combines the features of a rule-based stemmer and a statistical stemmer for stemming. Combining the features of both stemmer helps to increase the accuracy of the stemming algorithm [11].

2 Challenges in Stemming

Stemming algorithms are often challenged with two kinds of errors that occur frequently depending upon the nature of algorithms used by the stemmer. The types of resulting errors are given below.

2.1 *Over-Stemming*

Over-stemming happens when two words with different stems are derived from the same root. Over-stemming may also be known to be false-positive.

2.2 *Under-Stemming*

Under-stemming happens when two words that do not have separate stems are derived from the same root. It is possible to interpret under-stemming as false-negative.

3 History of Stemming

Initially, the only language where stemming was carried out was the English language. The first-ever algorithm [1] for stemming was proposed by Julie Beth Lovins in the year 1968 and published in the Journal of Mechanical Translation and Computational Linguistics. An inflected word is a result of a combination of the word with prefix or suffix or both. The stemmer was a rule-based stemmer basically aimed at extracting the word by suffix removal. The Lovins stemmer has 294 endings, 29 conditions and 35 transformation rules [1].

Inspired by the work of Lovins, an algorithm [2] for stemming was proposed by Martin Porter in the year 1980 and published in the journal named Program. This stemmer is the most widely used stemmer in the world and is the most cited paper on stemming [7]. It is a de facto standard for stemming. The algorithm used was a rule-based approach. However, the stemmer had a lesser number of rules as compared to Lovins stemmer when it was derived.

The third significant work on stemming was carried by Christopher D Paice in the year 1990. The paper [3] was published in the proceedings of the conference on Special Interest Group on Information Retrieval. The stemmer was a rule-based stemmer with an added advantage of the inclusion of a subroutine for index compression in the algorithm. It was faster but produced a relatively large number of over-stemming errors [3] compared to the previous algorithms on stemming.

4 Stemming Techniques for Hindi Language

Various algorithms have been proposed for stemming based on the Hindi language. Ramanathan and Rao [4] proposed a lightweight stemmer for Hindi which uses handcrafted set of suffix list and looks for longest match stripping. They have used the name "Light stemming" as the algorithm is used for tripping of a small set of either prefixes or suffixes or both, without trying to deal with infixes, or recognize patterns and find roots. Out of 35,977 words used as input to the algorithm, 4.6% of words were found to be under-stemmed while 13.8% were found to be over-stemmed [4].

Pandey and Siddiqui [5] proposed an unsupervised Hindi stemmer with the aim of improving the combining various prefix and suffix rules based on heuristics. This paper focuses on the development of an unsupervised stemmer for Hindi and the evaluation of the approach using manually segmented words. The algorithm was evaluated on 1000 words randomly extracted words from the Hindi WordNet-1 database [12]. The training data has been constructed by extracting 106,403 words extracted from EMILLE (Enabling Minority Language Engineering) 2 corpus [13]. The observed accuracy was found to be 89.9% after applying some heuristic measures. The F-score was 94.96% [5].

Ganguly et al. (6) proposed two separate rule-based stemmers for the Bengali and Hindi languages. In this paper, linguistic knowledge was used to manually craft the rules for removing the commonly occurring plural suffixes for Hindi and Bengali. A baseline was fixed by choosing words randomly from websites of news articles written in Hindi on the web. The improvement obtained with the incorporation of new rules for stemming by handling some exceptional words which were not a part of the baseline was 5.03% [6].

Mishra and Prakash [7] proposed a stemmer named "Maulik" for the Hindi language. This stemmer is purely based on Devanagari script and it uses the hybrid approach which combination of brute force and suffix removal approach. A lookup table with 15,000 words was maintained in the database. The average accuracy of the stemmer was obtained to be 91.59% [7].

Anand et al. [8] proposed a semi-supervised approach for stemming text written in the Hindi language. This paper uses an algorithm to find the stem of a word in Hindi. The proposed algorithm uses word2vec, which is a semi-supervised learning algorithm, for finding the 10 most similar words from a corpus. Also, a mathematical function is used to find the stem. The results are verified by selecting a set of 1000 Hindi words randomly taken from a corpus [8].

5 Stemming Techniques for the Nepali Language

Bal and Shrestha [9] proposed a morphological analyzer and a stemmer for the Nepali language. This earliest stemming technique did not handle words formed as a result of the combination of two free morphemes. This paper discusses the design and implementation issues as well as the linguistic aspects of a morphological analyzer and a stemmer for the Nepali language [9].

Sitaula [10] proposed a hybrid algorithm for stemming Nepali text. The hybrid Nepali stemming algorithm uses affix stripping in conjunction with a string similarity function and reports a recall rate of 72.1% on 1200 words. The handwritten rules comprised 150 suffixes and around 35 prefixes were considered. The accuracy of this hybrid algorithm is 70.10% [10].

Paul et al. [11] proposed an affix removal stemmer for the Nepali language. This work has a rule base of 120 suffixes and 25 prefixes and a root lexicon of over 1000 words and reports an overall accuracy of 90.48% [11].

Shrestha and Dhakal [14] proposed a new stemmer for the Nepali language and classifies suffixes into three categories and stem them according to different criteria. The proposed algorithm was implemented in Ruby and was tested in a data set of 5000 words, extracted from a corpus containing E-Kantipur news. The accuracy of the algorithm is obtained as 88.78% [14].

Koirala and Shakya [15] proposed a Nepali rule-based stemmer and analyzed its performance on different NLP applications. The corpus contained articles from various different areas, including news, sports, politics, literature etc. Corpus contained a total of 438 news articles with a total word count of 11,813 and a total unique word count of 11,346. Each news article, on average, contained 269 total words and 181 unique words. The F1 score was 0.79 [15].

References

1. Lovins JB (1968) Development of a stemming algorithm. J Mech Trans Comput Linguist, 22–31
2. Porter M (1980) An algorithm for suffix stripping Program. Program 14(3):130–137
3. Paice CD (1990) Another stemmer. Proc SIGIR Forum 24(3):56–61
4. Ramanathan A, Rao D (2003) A lightweight stemmer for hindi. Proceedings of workshop on computational linguistics for south asian languages, 10th conference of the European chapter of association of computational linguistcs., pp 42–48
5. Pandey AK, Siddiqui TJ (2008) An unsupervised hindi stemmer with heuristic improvements. In: Proceedings of the second workshop on analytics for noisy unstructured text data, vol 303, pp 99–105

6. Ganguly D, Leveling J, Jones GJF (2013) Rule-based stemmers for bengali and hindi. Post-Proceedings of the 4th and 5th workshops of the forum for information retrieval evaluation, New Delhi, India, December 4–6, 2013
7. Mishra U, Prakash C (2012) MAULIK: an effective stemmer for hindi language. Int J Comput Sci Eng, ISSN: 0975-3397, Vol 4 No 05 May 2012
8. Anand A, Chatterji S, Bhattacharya S (2019) Semi-supervised approach for hindi stemming. 8th International conference on natural language processing (NLP 2019), Vol 9, No 12, September 28–29, 2019, Copenhagen, Denmark
9. Bal BK, Shrestha P (2004) A morphological analyzer and a stemmer for Nepali. PAN Localization, working papers 2007, pp 324–31
10. Sitaula C (2013) A hybrid algorithm for stemming of Nepali text. Intelligent information management
11. Paul A, Dey A, Purkayastha BS (2014) An Affix Removal stemmer for natural language text in Nepali. Int J Comput Appl
12. https://catalog.ldc.upenn.edu/, LDC2008L02 Bhattacharyya, Pushpak, Prabhakar Pande, and Laxmi Lupu. Hindi WordNet LDC2008L02. Web Download. Philadelphia: Linguistic Data Consortium, 2008
13. http://www.emille.lancs.ac.uk/, The EMILLE (Enabling Minority Language Engineering) project, Lanchester University, United Kingdom (Website Last Accessed on 12.10.2020)
14. Ingroj Shrestha and Shreeya Singh Dhakal, “A new stemmer for Nepali language”, International Conference on Advances in Computing and Communication, 2016
15. Pravesh Koirala and Aman Shakya, “A Nepali Rule Based Stemmer and its performance on different NLP applications”, arXiv preprint arXiv:2002.09901, 2020 arxiv.org

Smart Face Recognition with Mask/No Mask Detection

Amrita Biswas, Bishal Paudel, and Nandita Sarkar

Abstract Ever since Covid-19 has affected our lives, the use of masks in public places has taken extreme importance. Be it restaurants, shopping malls, workplaces, airports or any other public place, wearing a mask is essential for the safety of oneself as well as the others. However, there are always some defaulters who will not wear a mask putting all at risk. In this paper, a mask/no mask detector has been implemented to identify such defaulters. Also, a face recognition system has been incorporated to identify those who are not wearing a mask. This setup could be used in offices or any other place of public gathering to ensure the safety of one and all.

Keywords Mask detection · Face recognition · Yolo v2 · HOG · Convolutional neural networks

1 Introduction

The world today is facing a huge crisis due to Covid-19 pandemic. It is highly infectious and as of now millions of people have been affected and thousands of people lost their lives. Covid-19 has drastically changed the way we work these days. Coronavirus being a novel virus, not much medical knowledge is available. Hence taking precautions is the utmost need of time, especially in closed spaces as the chances of getting infected is high. From working from home to social distancing in workplaces we have come a long way. One thing that has become an essential weapon in this fight against the coronavirus is the use of masks in all public places. As the virus spreads primarily through our mouth and nose, a face mask has become an important accessory. CCTV cameras are widely used these days in places of public gathering for security purposes. In this paper, we have proposed and implemented an intelligent mask detection system which can be easily incorporated in the surveillance systems to identify anyone who is not wearing a mask automatically and also recognize the

A. Biswas (✉) · B. Paudel · N. Sarkar
Electronics and Communication Engineering Department, Sikkim Manipal Institute of Technology, Sikkim, India
e-mail: amrita.a@smit.smu.edu.in

S. Dhar et al. (eds.), *Advances in Communication, Devices and Networking*, Lecture Notes in Electrical Engineering 776,
https://doi.org/10.1007/978-981-16-2911-2_15

defaulter and issue some warning to the individual for not wearing a mask. This type of system could be useful in organizations, shopping malls, hospitals or other places of public gathering to generate automated warnings to non-mask wearers without any human intervention. Since masks are a recent part of our lives, we did not come across formal publications on mask detection. However, some projects have been carried out on mask detection and are available online. We have included these projects in our review work. Roserbock [1] carried out mask/no mask detection in real time. He used SSD neural network for **face detection. Classification** for mask/no mask cases was carried out using another neural net (MobileNetV2). He used a synthetic mask dataset with the same type of masks superimposed artificially on the face images to test his algorithm. However, the actual images captured from the CCTV cameras have several issues, like images are smaller and often not clear, the faces may be at varying angles and a single still frame may not be sufficient for determining if a person is using a mask or not.

In [2] the authors detected the faces using pose estimation and used Kalman filter for tracking the faces. CIFAR was used for the classification of the detected faces.

In the proposed work, first Yolo v2 algorithm was used for face detection. Yolo v2 is an object detection system targeted for real-time processing [3]. HOG features have been used for feature extraction of the detected faces and support vector machines (SVM) have been used for the classification of mask and no mask cases. Once a no-mask situation is detected, to recognize the person who is not wearing a mask, the convolutional neural networks have been used.

2 Methodology

The system comprises three processes: face detection using the Yolo v2 algorithm, mask detection using the HOG feature descriptor and face recognition for non-masked faces using the convolutional neural networks. Each of the processes has been described below.

Continuous video input is fed to the proposed model. The Yolo v2 detector takes the video input and detects the head and inserts the bounding box framewise. The image is cropped as per the bounding box. This image is passed to the face mask recognizer to classify mask and non-mask faces. The HOG features are extracted and passed to the SVM classifier. If a mask is detected, it displays "Mask detected". If no mask is detected, the query image is fed to the face recognizer model. This model is trained with faces of 50 people. The model extracts the features from the query image using the CNN network and matches it with the training set and recognizes the person. The classification for face recognition has been done using SVM. The workflow has been shown in Fig. 1.

Yolo v2 as the name suggests (You Only Look Once) works swiftly in object detection applications. The Yolo v2 model runs a deep learning convolution neural network on the input image and produces the predicted output and generates the bounding box. The image is divided into a 13 $\times$ 13 matrix. Then it is convoluted

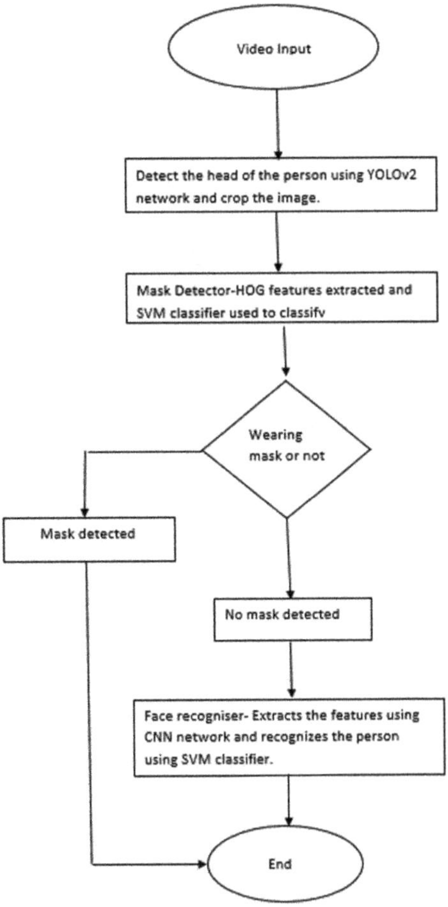

Fig. 1 Workflow of smart face recognition with mask/no mask detection

with anchor boxes to predict offsets and confidence score of the detected subject. The bounding box having a greater confidence score is retained and the rest is discarded. A 25-layer Yolo v2 network was used comprising primarily the convolution layer, batch normalization layer, Relu layer and Maxpool layer.

The network designed for our system comprises 25 layers. The complete details of the layers used have been shown in Fig. 2.

The performance of Yolo v2 layer is evaluated and Fig. 3 shows the error parameter. Based on the average precision of 0.96 and log average miss rate of 0.04, it can be concluded that Yolo v2 is well suited for the used face detection application.

For mask/no mask detection, the HOG features have been used. The full form of HOG is histogram of oriented gradients. It is a popular feature descriptor often used in object detection. The number of gradient orientations in localized parts of the image is counted. The image is split into small subimages and the histogram of the oriented gradients is computed for each subimage. The results are normalized

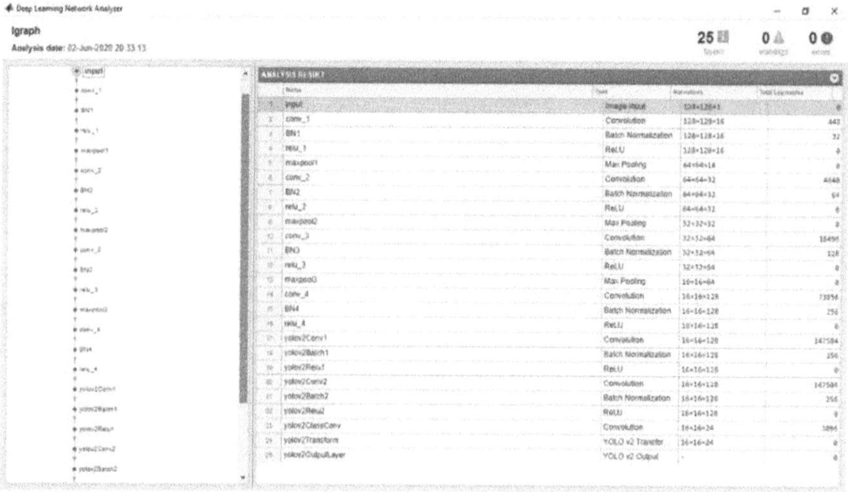

Fig. 2 Yolo v2 layers

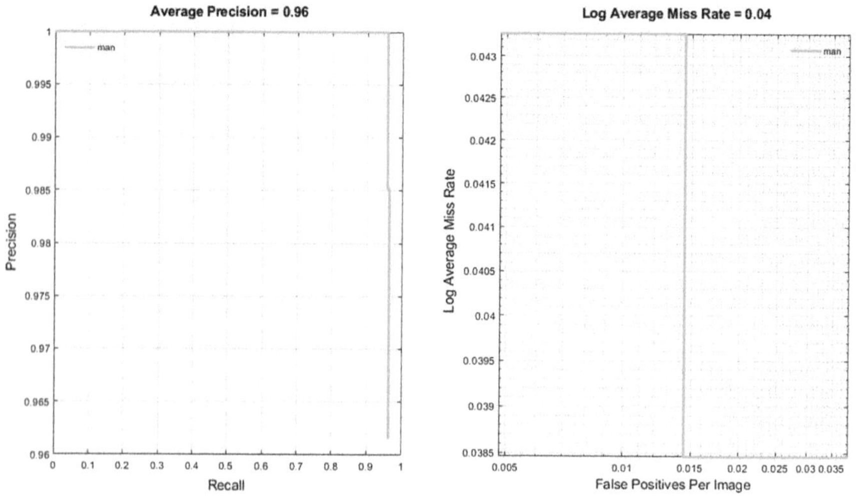

Fig. 3 Yolo v2 average precision error and log average miss rate

and the descriptor is returned. As it works on small cells it is robust to geometric and photometric transformations [4, 5]. The gradient for each pixel includes the magnitude and direction as shown below:

$$g = \sqrt{g_x^2 + g_2^2}$$
$$\theta = \arctan \frac{g_y}{g_x} \tag{1}$$

where g_x and g_y are the horizontal and vertical components of the change in the pixel intensity, respectively. We have used a window size of 150 × 150 for the face images because it matches the aspect ratio of human faces. Block size of 4 × 4 has been used for descriptor calculation. A total of nine bins have been used for each block. Normalization has been done over a 16 × 16 block to reduce the effect of lighting conditions. The detection threshold has been set to 60%. The feature vector size obtained is 1 × 46656. The percentage of correct face recognition was 99%.

The SVM model is trained using the HOG features. Support vector machines (SVMs) are used to classify data. SVMs are widely used in many classification applications. SVMs are based on the concept of hyperplanes that are used to separate multiple classes.

In case a no-mask situation is detected, the convolutional neural networks have been used for face recognition. CNN comes under the category of deep neural networks and is widely used in processing images. The CNN features are shift-invariant and hence also known as shift-invariant artificial neural networks (SIANN), as they shared weights architecture and translation invariant characteristics.

Convolutional networks have a connectivity pattern that is like the arrangement of the human visual cortex [6, 7].

The CNNs need lesser pre-processing compared to other image classification algorithms. There is no need of any prior feature extraction step, and directly feeding the query images to the network provides satisfactory results [8].

3 Results Obtained

The mask/no mask detector was used to test several videos with mask and no mask scenarios. It correctly recognized mask and no mask faces. The system works when the face detected by the bounding box has a pixel area of 6000 or more. The distance between the camera and the subject should be approximately 10 m or less for the face to be detected correctly. The snapshots of the results obtained for mask detection have been shown in Fig. 4.

The face recognition system was trained for 50 images and it correctly classified all face images which were without mask. Some snapshots of the obtained results have been shown in Fig. 5.

4 Conclusion and Future Scope

Since person recognizer uses personal data, the implementation must be with the consent of the people whose data is being used. Hence, it is ideal for corporate offices where the system can access personal data for societal benefit. The face recognizer can be trained for all the employees. Then the system can be deployed in the security cameras. It will continuously monitor the people in its range and record the people

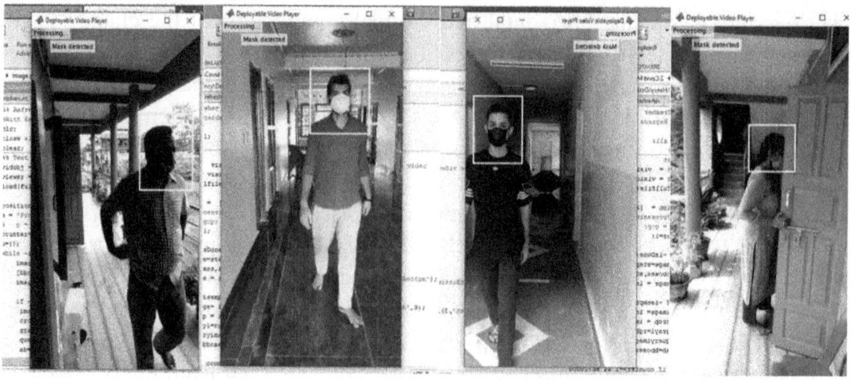

Fig. 4 Snapshots of mask detected

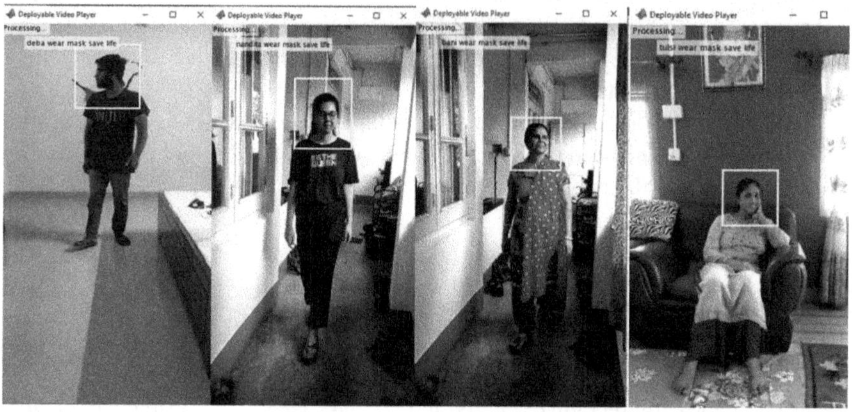

Fig. 5 Screenshots of the mask not detected scenario and face recognition done successfully for no mask cases

not wearing a mask. Actions can be taken against the defaulters with the help of the face recognition system. This project aims for better enforcement of wearing masks among the public.

The system works when the face covers a pixel area of 6000 in the bounding box which means the subject should be at least 10 m or closer to the camera. The classifier is large in size and hence it requires time to boot the system. But once it is loaded, real-time detection happens.

References

1. Rosebrock A (2020) Covid 19 face mask detector with open CV, Keras/tensor flow and deep learning, https://www.pyimagesearch.com/2020/05/04/covid-19-face-mask-detector-with-opencv-keras-tensorflow-and-deep-learning/
2. Braulio, Face Mask Detection in street camera video streams using AI: behind the curtain, https://tryolabs.com/blog/2020/07/09/face-mask-detection-in-street-camera-video-streams-using-ai-behind-the-curtain/, July 9,2020
3. https://medium.com/@jonathan_hui/real-time-object-detection-with-yolo-yolov2-28b1b9 3e2088
4. Dalal N, Triggs B (2005) Histograms of oriented gradients for human detection. 2005 IEEE computer society conference on computer vision and pattern recognition (CVPR'05), San Diego, CA, USA, pp 886–893 vol 1, https://doi.org/10.1109/cvpr.2005.177
5. Shu C, Ding X, Fang C (2011) Histogram of the oriented gradient for face recognition. Tsinghua Sci Technol 16(2):216–224. https://doi.org/10.1016/S1007-0214(11)70032-3
6. Fukushima K (1980) Neocognitron: a self-organizing neural network model for a mechanism of pattern recognition unaffected by shift in position. Biol Cybern 36:193–202. https://doi.org/10.1007/BF00344251
7. Matusugu M, Mori K, Mitari Y, Kaneda Y (2003) Subject independent facial expression recognition with robust face detection using a convolutional neural network. Neural Networks 16(5–6):555–559
8. Núñez-Marcos A, Azkune G, Arganda-Carreras I (2017) Vision based fall detection with convolutional neural networks, wireless communications and mobile computing, Article ID 9474806, 16 pages, https://doi.org/10.1155/2017/9474806

Matching Song Similarity Using F-Test Measure

Sudipta Chakrabarty, Md. Ruhul Islam, and Hiren Kumar Deva Sarma

Abstract The paper proposes an intelligent method to generate similarities among songs by finding the similarities of fundamental frequencies using the F-Test tool. The primary objective of this tool is to find whether two or more independent series of fundamental frequencies of the highest occurrence notes of song music variance differ significantly or whether two song samples may be regarded as the same variance. If the calculated value of F-Test is compared with the table value for the degree of freedom for a sample song having larger variance and the degree of freedom for sample song having smaller variance at 5% and 1% level of significance, then the hypothesis is accepted. Hence, it may be regarded that two song frequency patterns have the same variance and their fundamental frequency patterns are similar.

Keywords Computational Musicology · Indian Classical Music · Behavioral Science · Music Similarity · Hypothesis · Sample · F-Test · Degree of freedom

1 Introduction

Music is one of the best tools for entertainment and Computational Musicology is the emerging field of the combination of both, Computer Science and Behavioral Science. The paper represents a method to find the music similarity through the fundamental frequencies using a statistical tool F-Test. F-Test states the null hypothesis and it is the ratio between the degree of freedom having a larger variance sample and the degree of freedom having a smaller variance sample. The F-Test statistical tool is the experimental basis of the proposed work.

S. Chakrabarty (✉)
Department of MCA, Techno India, Salt Lake, Kolkata, West Bengal, India

Md. R. Islam
Department of CSE, SMIT, Majhitar, East Sikkim, Rangpo, India

H. K. D. Sarma
Department of IT, SMIT, Majhitar, East Sikkim, Rangpo, India

S. Dhar et al. (eds.), *Advances in Communication, Devices and Networking*,
Lecture Notes in Electrical Engineering 776,
https://doi.org/10.1007/978-981-16-2911-2_16

F-Test is defined as

$$\text{F-Test} = \frac{\text{S1}^2}{\text{S2}^2} \tag{1}$$

where

$$\text{S1}^2 = \frac{\text{Total Sum of the Square of the Frequency Distances}}{\text{N1} - 1} \tag{2}$$

$$\text{S2}^2 = \frac{\text{Total Sum of the Square of the Frequency Distances}}{\text{N2} - 1} \tag{3}$$

It should be noted that S1^2 is always the larger estimate of variance, i.e. $\text{S1}^2 > \text{S}^2$.

Here, (N1 – 1) = Degree of Freedom having larger variance for the song sample and (N2 – 1) = Degree of Freedom having smaller variance for the song sample. The primary objective of the paper that whether the two given song inputs are similar at a certain percentage level or not. To map the similarity between songs, F-Test is the statistical basis of this work.

2 Related Work

Various researchers have worked in Indian Classical Music (ICM) though there are a lot of unexplored areas. In an effort, the authors have introduced a Quality Music model used for ICM using Genetic Algorithm [1]. Petri nets is a modeling tool and the two papers [2, 3] are used as music pattern recognition and music pattern analysis for Indian Classical Music where object-oriented methodology is the basis for analysis of musical patterns. One paper introduces the mechanism that efficiently chooses the most fitted parent rhythms of a set of rhythm chromosomes for creating offspring rhythm using the Genetic Algorithm Optimization technique in the context awareness pervasive music rhythm learning education Pervasive Education for Computational Musicology [4]. Another paper [5] has introduced the concept of hummed query, and user query depends on the feedback of user that improves retrieval performance through Genetic Algorithm. An approach that generates the song similarity percentage by the pitch values through Correlation Coefficient is proposed in the paper [6]. The paper [7] represents a method to generate a personalized song list using user's listening habits, and age factor of users for performing online. Again one paper represents the matching similarity between songs by their pitch values through coefficient of variance [8]. The paper [9] proposes one time slot-based music recommender using Neural Network, and the paper [10] represents another time slot-based Music Recommendation System using raga-time database. The paper [11] introduced an intelligent method that identifies the music rhythm density and music rhythm complexity of any particular music rhythm through software automatically.

Paper [12] describes a lot of music survey research and applications in the field of Musicology. Another paper represents that a song of a particular raga can be represented through Unified Modeling Language [13]. Another paper presents a method of Automatic Raga Recognition [14]. Another very interesting paper proposed an approach that finds out the complexity of any particular song based on some statistical measures [15]. Paper [16] introduces the music editor software that generates different versatile music which is the combination of both Vocal and Instrumental with one particular Tempo based on the Object-Oriented Concept Aggregation. The paper represents to create different music rhythms depending on the Memetic Algorithm [17]. The paper [18] proposed the method that computes the music similarity based on Fundamental Frequencies by Coefficient of Concurrent Deviation.

3 Proposed Work

The paper represents a method to find the music similarity through the fundamental frequencies using a statistical tool F-Test. The F-Test statistical tool is the experimental basis of the proposed work. The workflows of the proposed work are given as follows:

Step 1: Take two songs.

Step 2: Play the songs using WaveSurfer software and generate .f0 file that consists of all the pitches that are used in the song and the pitch data are saved in the ".f0" format. The .f0 file consists of a large number of pitches of monotonic songs and after that it should be converted into ".csv" format.

Step 3: Accept all those pitch values of the two songs from their corresponding .f0 format file.

Step 4: Then the number of occurrences has been calculated of all the fundamental frequencies of each song.

Step 5: Fix seven frequencies which have highest occurrence, respectively, from the list of frequencies of the .fo file of each of the songs.

Step 6: Calculate total frequency by the following formula:

Total frequency = Frequency x Occurrences (1).

Step 7: Compute Mean $= \frac{1}{N} \sum$ Total Frequency (2).

Here, N = 7.

Step 8: Calculate Frequency Distance = Total Frequency – Mean (3).

Step 9: Calculate Square of the Frequency Distance.

Step 10: Calculate the total sum of the squares of the frequency distances.

Step 11: Calculate

$$S1^2 = \frac{\text{Total Sum of the Square of the Frequency Distances}}{N1 - 1} \tag{4}$$

$$S2^2 = \frac{\text{Total Sum of the Square of the Frequency Distances}}{N2 - 1} \tag{5}$$

It should be remembered that $S1^2$ have always the larger estimate of variance whereas $S2^2$ have always the smaller estimate of variance, i.e. $S1^2 > S2^2$.

Here, (N1 – 1) = Degree of Freedom having larger variance for the song sample, and (N2 – 1) = Degree of Freedom having smaller variance for the song sample.

Step 12: Compute the value of

$$\mathrm{F-Test} = \frac{S1^2}{S2^2} \tag{6}$$

Step 13: If the computed value of F-Test is compared with the standard table value for the degree of freedom for the sample song having larger variance and the degree of freedom for the sample song having smaller variance at 5% and 1% level of significance, then the hypothesis is accepted.

4 Result Set Analysis

To establish the proposed work, a lot of Hindi movie songs have been considered, and their similarity through the proposed algorithm is found. In the actual experiment, the work has been considered for more than 100 test cases but for page limitation the paper only considers three test cases among four songs, and from these four songs one song is taken as the basis and the F-Test values of other three songs are found as follows:

Test Case 1:

Song1: "Beeti Na Bitayee Raina"; Movie Name: Parichay (1973).

Compared with

Song2: "Laaga Chunri Mein Daag"; Movie Name: Dil Hi To Hai (1963).

Table 1 depicts the measuring value of F-Test between Song1 and Song2. The calculated value of F-Test is less than the Standard value of variance at both 5% level and 1% level of significance for $v_1 = 7$ and $v_2 = 7$. The Standard F-Test value at 5% level is 3.7870 and at 1% level is 6.9928 [19]. Therefore the hypothesis is accepted. Hence, it may be concluded that the two song frequency patterns have the same variance, and their fundamental frequency patterns are similar.

Test Case 2:

Song1: "Beeti Na Bitayee Raina"; Movie Name: Parichay (1973).

Compared with

Song3: "Sun Ri Pawan Pawan Purvaiya"; Movie Name: Anuraag (1972).

Table 2 depicts the measuring value of F-Test between Song1 and Song3. The calcu-

Table 1 Measuring F-Test values between Song1 and Song2

Frequency	Occurrences	Total Frequency (F1)	$\sum$ F1	Mean of F1	Difference	(Difference)^2	Σ (Difference)^2	S1	F-Test
65	1689	109785	498052	7115029	38634.71	1492640817	2362504161	393750694	2.875925
87	892	77604			6453.71	41550372.8			
98	707	69286			−1864.29	3475577.2			
73	698	50954			−20196.29	407890130			
110	691	76010			4859.71	23616781.3			
86	648	55728			−15422.29	237847029			
97	605	58685			−12465.29	155383455			
82	855	70110	411827	5383243	11277.57	127183585	6794384234	1132397372	
109	689	75101			16268.57	264666370			
73	564	41172			−17660.43	311890788			
219	559	122421			63583.57	4043506235			
110	429	47190			−11642.43	135546176			
81	345	27945			−30887.43	954033332			
83	336	27888			−30944.43	957557748			

Table 2 Measuring F-Test values between Song1 and Song3

Frequency	Occurrences	Total Frequency (Fl)	ΣF1	Mean of F1	Difference	(Difference)^2	Σ (Difference)^2	S1	F-Test
65	1689	109785	498052	71150.29	38634.71	1492640817	2352504161	393750694	1.011782
87	892	77604			6453.71	41650372.8			
98	707	69286			−1864.29	3475577.2			
73	698	50954			−20196.29	407890130			
110	691	76010			4859.71	23516781.3			
86	648	55728			−15422.29	237847029			
97	605	58685			−12465.29	155383455			
123	539	66297	314265	44895	7464.57	55719805.3	2334992737	389165455	
69	447	30843			−27989.43	783408192			
124	435	53940			−4892.43	23935871.3			
138	246	33948			−24884.43	619234856			
155	226	35030			−23802.43	566555674			
278	161	44758			−14074.43	198089580			
311	159	49449			−9383.43	88048758.6			

lated value of F-Test is less than the Standard value of variance at both 5% level and 1% level of significance for $v_1 = 7$ and $v_2 = 7$. The Standard F-Test value at 5% level is 3.7870 and at 1% level is 6.9928 [23]. Therefore the hypothesis is accepted. Hence, it may be concluded that the two song frequency patterns have the same variance, and their fundamental frequency patterns are similar.

Test Case 3:

Song1: "Beeti Na Bitayee Raina"; Movie Name: Parichay (1973).

Compared with

Song4: "Tere Mere Sapne Ab Ek Rang Hain"; Movie Name: Guide (1965).

Table 3 depicts the measuring value of F-Test between Song1 and Song4. The calculated value of F-Test is greater than the Standard value of variance at both 5% level and 1% level of significance for $v_1 = 7$ and $v_2 = 7$. The Standard F-Test value at 5% level is 3.7870 and at 1% level is 6.9928 [19]. Therefore the hypothesis is rejected. Hence, it may be concluded that the two song frequency patterns don't have the same variance, and their fundamental frequency patterns are dissimilar.

5 Comparison Analysis

The comparative study on the basis work with other two already published works is depicted in Table 4. There are five grading features like Area of Work, Sample Size, Strength of the Work, Weakness of the work, and Accuracy level used for the Comparison Analysis.

6 Conclusion

The objective of the F-Test in the work is to find whether the two independent samples of the sets of fundamental frequency estimates of two song variances differ significantly or not. F-Test is one Statistical Measure which is the most appropriate method to calculate the similarities of the two different pitch samples from two given songs. Only for that reason F-Test is the standard factor of our proposed work. If the computed value of F-Test is compared with the standard table value [19] for the degree of freedom for sample song having larger variance and the degree of freedom for sample song having smaller variance at 5% and 1% level of significance, then the hypothesis is accepted. Hence, it can be concluded that the two song frequency patterns have the same variance, and their fundamental frequency patterns are similar and therefore they are very similar in respect of their frequency paradigm.

Table 3 Measuring F-Test values between Song1 and Song4

Frequency	Occurrences	Total Frequency (Fl)	ΣF1	Mean of F1	Difference	(Difference)^2	Σ (Difference)^2	S1	F-Test
65	1689	109785	498052	71150.29	38634.71	1492640817	2362504161	393750594	7.280119
87	892	77604			6453.71	41650372.76			
98	707	69286			−1864.29	3475577.204			
73	698	50954			−20195.29	407890129.8			
110	691	76010			4859.71	23616781.28			
86	648	55728			−15422.29	237847028.8			
97	605	58685			−12465.29	155383454.8			
73	2398	175054	391675	55953.57	107315.57	11516631564	15208489530	2534748255	
74	811	60014			1181.57	1396107.665			
77	514	39578			−19254.43	370733074.6			
72	430	30960			−27872.43	776872354.1			
123	373	45879			−12953.43	167791348.8			
98	286	28028			−30804.43	948912907.6			
92	229	21068			−37764.43	1426152173			

Table 4 Comparative study of the work with other two already published works

Features	Song similarity using F-test (Basis Work)	Song similarity using Pearson's correlation coefficient [6]	Song similarity using coefficient of concurrent deviation [18]
Area of work	Statistics	Statistics	Statistics
Sample size	7	7	16
Strength	Display only similar	Display a certain % similar	Display a certain % similar
Weakness	Based on Hypothesis	Based on measurement	Based on measurement
Accuracy	Medium	Medium	High

References

1. Chakrabarty S, De D (2012) Quality measure model of music Rhythm using genetic algorithm. In: Proceedings of international conference on radar, communication and computing (ICRCC), IEEE, pp 125–130
2. Roy S, Chakrabarty S, Bhakta P, De D (2013) Modelling high performing music computing using Petri Nets. In: Proceedings of international conference on control, instrumentation, energy and communication (CIEC), IEEE, pp 757–761
3. Roy S, Chakrabarty S, De D (2014) A framework of musical pattern recognition using Petri Nets. In: Proceedings of emerging trends in computing and communication (ETCC), Springer-Link Digital Library, pp 245–252
4. Chakrabarty S, Roy S, De D (2014) Pervasive diary in music Rhythm education: a context-aware learning tool using genetic algorithm. In: Proceedings of advanced computing, networking and informatics, Springer, pp 669–677
5. Rho S, Han B-J, Hwang E, Kim M (2008) MUSEMBLE: a novel music retrieval system with automatic voice query transcription and reformulation. J Syst Softw, Elsevier, pp 1065–1080
6. Sudipta C, Rahul IMd, Kumar DSH (2018) An approach towards the modeling of pattern similarity in music using statistical measures. In: Proceedings of the 5th international conference on parallel, distributed and grid computing (PDGC), IEEE, pp 436–441
7. Sudipta C, Sangeeta B, Rahul IMd, Kumar DSH (2018) Context aware song recommendation system. In: Proceedings of the 3rd national conference on communication, Cloud, and Big Data (CCB), Springer, pp 157–165
8. Sudipta C, Ruhul IMd, Debashis D (2017) Modelling of song pattern similarity using coefficient of variance. Int J Comput Sc Inf Secur, 388–394
9. Samarjit R, udipta C, Debashis D (2017) Time-Based raga recommendation and information retrieval of musical patterns in Indian classical music using neural network. IAES Int J Artif Intell (IJ-AI), 33–48
10. Sudipta C, Samarjit R, Debashis D (2016) Handbook of research on intelligent analysis of multimedia information, Chapter 12: Time-slot based intelligent music recommender in Indian music, IGI Global, pp 319–351
11. Sudipta C, Gobinda K, Rahul IMd, Debashis D (2017) Reckoning of music rhythm density and complexity through mathematical measures. In: Proceedings of the advanced computational and communication paradigm, LNEE Lecture Note, Springer, pp 387–394
12. Sudipta C, Samarjit R, Debashis D (2015) A foremost survey on State-Of-The-Art computational music research. In: Proceedings of the recent trends in computations and mathematical analysis in engineering and sciences. International science congress association, pp 16–25
13. Sudipta C, Roy S, Debashis D (2015) Behavioural modelling of Ragas of indian classical music using unified modelling language. In: Proceedings of the 2nd international conference on perception and machine intelligence, ACM, pp 151–160

14. Chakrabarty S, Samarjit R, Debashis D (2014) Automatic raga recognition using fundamental frequency range of extracted musical notes. In: the Proceedings of eight international multi-conference on image and signal processing (ICISP 2014), Elsevier, pp 337–345
15. Chakrabarty S, Samarjit R, Kumar DSH (2020) Measuring song complexity by statistical techniques. In: The Proceedings of second International Communication, Devices and Computing, Springer, pp 687–695
16. Chakrabarty S, Samarjit R, Kumar DSH (2020) Intelligent music abstraction tool for improvising the quality of music composition. Inte J Adv Trends Compur Sci Eng, Warse, pp 3641–3648
17. Sudipta C, Anushka B, Rahul IMd, Kumar DSH (2019) Algorithmic improvisation of music Rhythm. In: the Proceedings of international conference on communication, devices and networking, Springer, pp 323–335
18. Sudipta C, Sangeeta B, Rahul IMd, Kumar DSH (2019) Music similarity mapping through fundamental frequencies by coefficient of concurrent deviation. In: the Proceedings of international conference on computing, power and communication technologies, IEEE, pp 910–915
19. http://www.socr.ucla.edu/Applets.dir/F_Table.html The Link of F Distribution Tables

Hand Gesture Recognition Using Convex Hull-Based Approach

Kaustubh Wani and S. Ramya

Abstract Gesture recognition is a tool that can be used to control any common device by effortless hand gestures. The goal behind gesture recognition is to minimize the gap between the digital world and the physical world. This wireless interaction creates this algorithm much friendlier to the user. This paper tells the technological characteristics of gesture-controlled user interface (GCUI), and also recognizes its trend and application. It is observed that GCUI now offers practical opportunities for application-specific areas, especially for people who are not comfortable with input devices which are commonly used. This project implements an advanced image processing application to recognize the gestures and process them in real time for better and reliable results. To recognize the gesture the ratio of the percentage of area not covered by hand in the convex hull is found. The optimized code is easily integrated with a Raspberry-Pi processor or microcontroller for a fully functional robot.

Keywords Gesture recognition · Image recognition · Data analysis · GCUI · Image processing · Hand posture · Contour · Convex hull

1 Introduction

Humans use their five senses to interact with the physical world. However, since ancient times gesture has been used to communicate, even before the origination of language. In this generation of machines taking control of all our activities from small to big, interaction with these machines has become very essential. This paper will be focusing on gesture recognition and its implementation.

Gesture recognition is a major accomplishment in the technology and robotics field, which points to interpret user's gestures. Gesture technology is applied in several areas like robots with social assistive, augmented reality, identification of sign language, detection of facial emotion, virtual keyboard or mouse, controlling

K. Wani · S. Ramya (✉)
Electronics and Communication, MIT, MAHE, Manipal, India
e-mail: ramya.lokesh@manipal.edu

S. Dhar et al. (eds.), *Advances in Communication, Devices and Networking*,
Lecture Notes in Electrical Engineering 776,
https://doi.org/10.1007/978-981-16-2911-2_17

remotely, etc. Gestures can be analyzed and captured through a data sensor glove or through a camera. The embedded systems designed to reduce the cost and size of the device can be optimized to increase the reliability and performance of specific control functions. The application of these types of gesture recognition can be seen in military, surveillance, firefighting, space research, virtual gaming or places where human intervention is not possible.

This project will be introducing today's advanced image processing application to recognize the gestures and process them in real time for better and reliable results. This method uses python coding language to create the gesture recognition algorithm. This can be further integrated with a Raspberry-Pi processor or microcontroller for a fully functional robot.

2 Literature Survey

The emergence of gesture recognition has been significant in recent years. The main goal of gesture recognition was to convey information and identify specific human gestures or control devices. To help how gesture recognition works and what they are, a study of how other researchers view gestures is useful. Many engineering researchers have come up with different techniques to recognize gestures. Here are some key information and issues regarding hand gesture recognition presented by few researchers.

Vaishnav and Tiwari [1] worked on identifying gestures using accelerometer and transmit the data using RF transmitter and received using RF receiver, which is then decoded by HT12D. The decoded information is used by AT89C51 microcontroller to move the robot in different directions. The downside of an accelerometer is it is important that it requires a stable power supply and the current supply is constant to reduce signal noise generated by the accelerometer. The downside of an accelerometer with low impedance is that the sensor has a range and time constant that are internally fixed, potentially limiting its usability in some applications.

Mojeebi and Tulo [2] used an accelerometer and RF transmitter. As radiofrequency was used for wireless transmission, the range gets limited to around 50–80 m. This issue can be resolved by using a GSM module for wireless transmission. Almost all over the world, the GSM infrastructure is installed. GSM will not only provide wireless connectivity but also quite a large range. A monitoring on-board camera can be placed on a robot to surveillance faraway places. All they need is a wireless camera, which will help in broadcast and a receiver module, which will provide streaming live services.

Chanda et al. [3] used an accelerometer along with Bluetooth module HC-05 and Arduino for data transmission and moving the robot. The built-in device is easy to carry and cheap. Some additional cameras and sensors can be added to make it more productive. The limitation has been reduced to a great extent regarding hardware being associated with a system.

Kaura et al. [4] used C++ with OpenCV to recognize hand gestures and Arduino Duemilanove to remote the robot in a different direction. Data transmission is done by Wi-Fi Shield (WiFly). This is an alternative way for robot control. The more natural way of controlling devices makes robots more efficient and easy through gesture recognition. This provides two techniques for input through gestures: the direction of hand palm-based gesture control and finger count-based gesture control.

Matnani [5] worked on the literature review of glove-based and accelerometer-based gesture control. She used the same accelerometer-based technology on the glove to recognize hand gestures. The cons of using this system are noise from the wiring cable affects the results and reduces resolution; we should use cables with low noise, so it comes out to be more costly with the additional instrumentation requirement of special cables with low noise; the signal with high impedance nature makes it more vulnerable to noise, i.e., sensitive to dirt connectors; requires more expertise to operate the sensitive cable flex and system.

It can be concluded that using an accelerometer to recognize hand gestures in today's date has many downsides. Instead, image processing can be done to simplify the work and more reliable results are obtained as the camera can be easily accessible.

3 Methodology

In this paper, the basic methodology used is to define a simple region which can be called the region of interest (ROI) [6]. Whenever a gesture is shown inside this ROI, we recognize it and display the result since each gesture has a specific area of its own.

Distinguishing different gestures based on the area is easy. Therefore, by empirical testing, different thresholds of area ratio are found. This area ratio can be defined as a percentage of area not covered by hand in the convex hull. Figure 1 depicts the gestures which can be recognized through this algorithm and also many other gestures can be added for future expansion.

The sequence of steps or actions involved in this algorithm is given in the flowchart in Fig. 2.

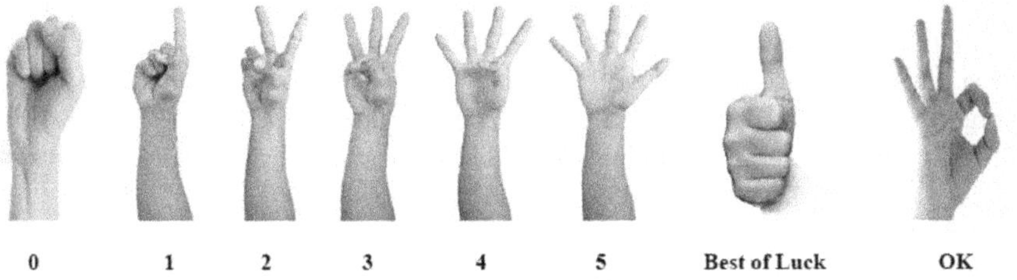

Fig. 1 Gestures used in this algorithm

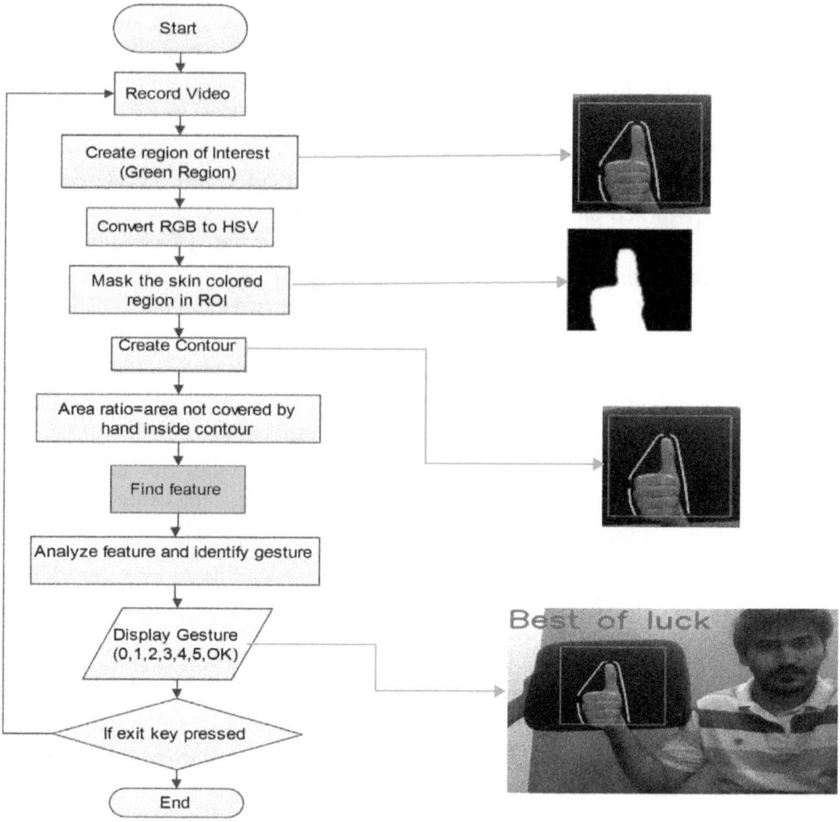

Fig. 2 Gesture recognition algorithm's flowchart

This is an algorithm for gesture recognition using **OpenCV-Python** [7]. It counts the number of fingers that are displayed in front of the camera. It recognizes gestures like the number of fingers pointed (0,1,2,3,4,5) and even special gestures like Best of Luck (Thumbs up), Ok sign.

The algorithm in Fig. 2 goes as follows:

- It starts with capturing the video from the front camera of a laptop and defines a variable **ROI** which is the region of interest (Green Frame).
- Everything that is captured in the frame is converted from red, green and blue colors, that is RGB, to hue, saturation and value colors, that is HSV [8]. To identify the hand and distinguish it with surrounding, it defines the upper and lower range of skin color in HSV, i.e., upper_skin and lower_skin.
- Next, it creates a mask that is anything under skin color is considered white and anything other than that is considered black.

- Then it finds the **contours**, i.e., the outline or perimeter formed by the region of interest. After finding the contour formed by the hand inside the region of interest, then it draws an outline around the hand which is called **convex hull** [6].
- To recognize the gesture it finds the ratio of the percentage of area not covered by hand in the convex hull.

Each gesture will have a unique area ratio. This is the basis on which it will differentiate different hand gestures.

- After this, it finds the gaps between the fingers (defects) in the convex cell, which are the regions not covered by the hand inside the convex hull. Since all the angles between the fingers are less than 90°, it ignores the defects making an angle greater than 90°. It will count rest of the gaps.
- Based on the area ratio and the number of gaps, it can be concluded that fingers pointed by the operator will be one plus number of gaps (1 + number of gaps) as shown in Fig. 3.
- Finally, it displays the frame and mask. Recognized gestures can be seen on the frame window.

Open Source Computer Vision Library, also known as OpenCV, is an open source software library for computer vision and machine learning. Its objective was to provide computer vision applications a familiar infrastructure and to push the use of machine insight into commercial products. Being a licensed product (BSD), OpenCV makes it easy for businesses to modify and utilize the code.

Contour is a simple curve joining all the continuous points (which are along the boundary), having similar intensity or color. It is a tool useful for object detection and shape analysis and recognition. It uses binary images for better accuracy. Therefore,

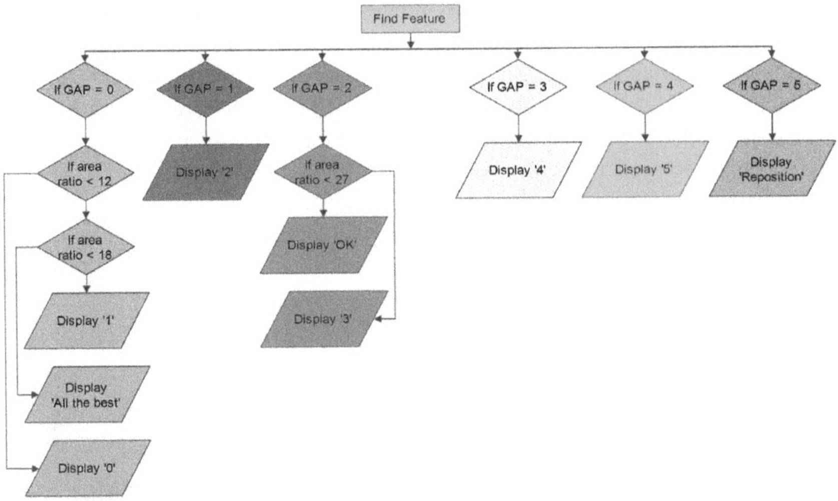

Fig. 3 Feature flowchart

it applies canny edge detection or threshold before finding contour. Since OpenCV 3.2, findContours() returns a modified image as the first of three return parameters instead of a modified source image. In OpenCV, contour finding is like identifying a white object from a black background. So remember, the background should be black, and the object to be found should be white.

In geometry, the convex closure or convex hull or convex envelope is the smallest convex set of a shape that contains it. Definition of a convex hull may be either as the intersection of all convex sets containing a given subset equivalently as the set of all convex combinations of points in the subset, or a subset of a Euclidean space. For a subset of the plane, which is bounded, the convex hull can be imagined as a shape surrounded by an elastic band stretched around the subset [6].

Red, Green and Blue (RGB) Color Model

The RGB color model constructs an additive color model in which the combination of red, green and blue light is added together in various ways to produce a broad variety of colors. The origin of the model name comes from the initials of the three additive primary colors—red, green and blue.

Hue, Saturation and Value (HSV) Color Model

HSV (hue, saturation and value), also known as hue, saturation and brightness (HSB), is often used by painters/artists due to its more natural and original to think about a color in terms of hue and saturation than in terms of additive or subtractive color components. HSV is found to be a transformation of an RGB color space, and its colorimeter and components are relative to the RGB color space from which it was originally derived.

Conversion of RGB to HSV is done because the object, which needs to be identified, is in a busy background and can be done efficiently by defining a threshold to make the original image into a binary image with the color it needs to detect [8]. The RGB component of the digital image corresponds to the amount of light that falls on the object and therefore differentiation between the object's components is difficult. The HSV color space abstracts color, which is hue by differentiating it from saturation and pseudo-illumination, making the components differentiable.

Conversion formula of RGB to HSV:

1. Divide r, g, b by 255
2. Calculate k_max, which is a maximum value of R,G,B; k_min, which is minimum value of R,G,B; and Δ which is

$$\Delta = k_{max} - k_{min}$$

3. **Compute Hue**:

- if k_max and k_min is equal to 0, then h = 0
- if k_max is equal to r, then compute $H = (60 * ((g - b)/\Delta) + 360) \% 360$
- if k_max is equal to g, then compute $H = (60 * ((b - r)/\Delta) + 120) \% 360$
- if k_max is equal to b, then compute $H = (60 * ((r - g)/\Delta) + 240) \% 360$

4. **Compute Saturation**:
   - if k_max = 0, then s = 0
   - if k_max is not equal to 0, then compute S = (Δ/k_max) * 100
5. **Compute Value**:
   - V = k_max * 100.

3.1 Specifications

For the algorithm it is written in Python language and Spyder (Anaconda) software which works efficiently and proficiently in a multi-language editor with a class/function browser, automatic code completion and code analysis tools. The author used an HP laptop with Intel Core i5-7200U (7-Gen) with clockspeed of 2.5 GHz and Intel HD-620 graphics. Its web camera is 0.9 megapixels and can record video in 720P at 30 fps. For hardware implementation, use an Arduino microcontroller or Raspberry-Pi processor.

4 Results and Discussion

Gesture recognition with this algorithm works well in all conditions and it is better than the traditional accelerometer as it has a fixed range in which hand can be moved. Also, it cannot recognize different finger gestures. This algorithm holds good under well lighted area. In addition, test this with different color backgrounds and different people, as listed in Table 1. The result holds true for different people with different hand sizes and skin colors. For best results, the hand should be kept at a distance of 50–60 cm from the camera. Gestures like numbers 0,1,2,3,4,5 can be appropriately recognized. Also, it identifies special gestures like OK signal and BEST OF LUCK signal as well.

The experiments are also carried out with different backgrounds. The test results show that the probability of error in the output increases if the user is sitting in front of an off-white background, i.e., color matching skin tone as listed in Table 2. Gesture error occurred due to off-white background and number 2 was misinterpreted as number 1. This happens because during the conversion of RGB color in the frame to HSV the algorithm recognizes the off-white (skin tone) color as part of the user's hand which in turn causes contour region to change making incorrect decisions.

Thus, gesture recognition through the convex hull is easy to handle, less complex, and can be implemented on any PC with a camera rather than using machine learning or artificial intelligence as mentioned in the literature survey [5]. This method is tested with different backgrounds such as black, purple, blue, green and red and found to be accurate. In total, this method has been tested with five backgrounds using 13

Table 1 Results of hand gesture recognition

Original Image	Contour	Mask	Result
'Best Of Luck' Gesture			
			Best of Luck
'OK' Gesture			
			OK
'0' Gesture			
	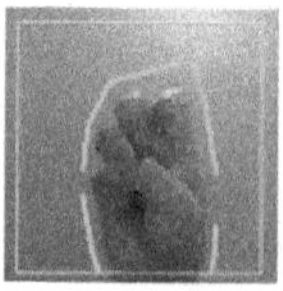		0
'1' Gesture			
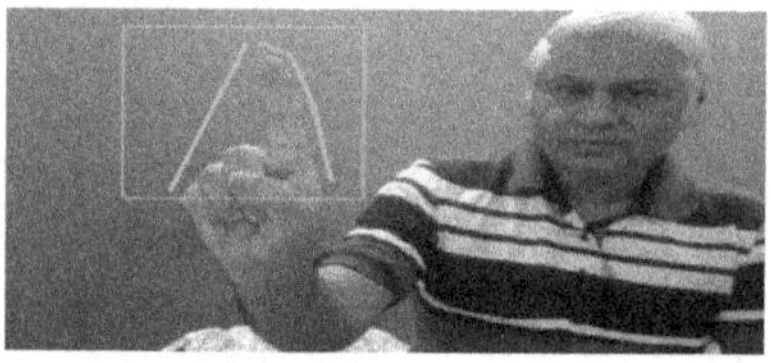			1
'2' Gesture			
			2

(continued)

Table 1 (continued)

'3' Gesture			
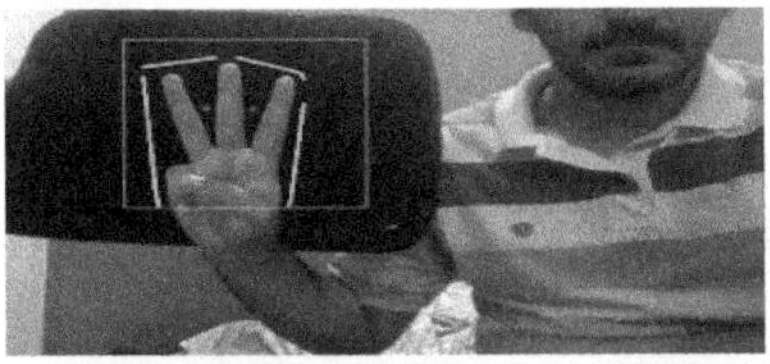	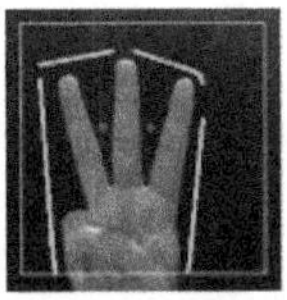		**3**
'4' Gesture			
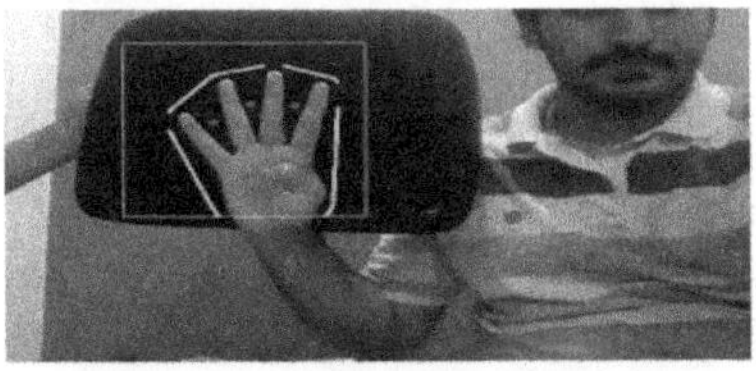	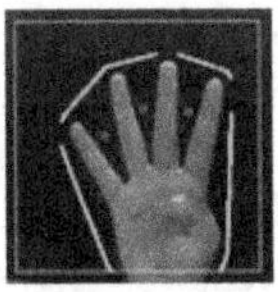		**4**
			**4** (Same gesture can be shown in different manner)
'5' Gesture			
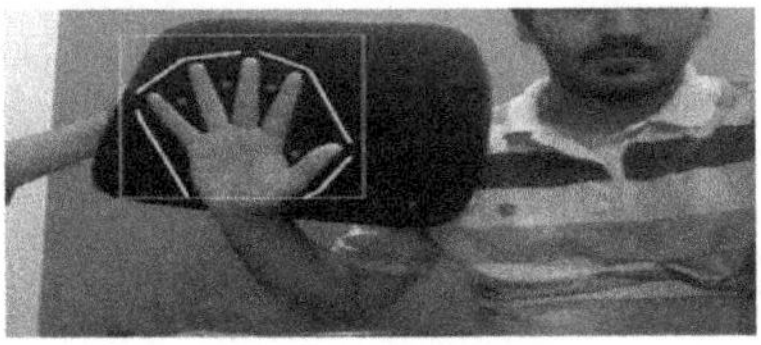			**5**

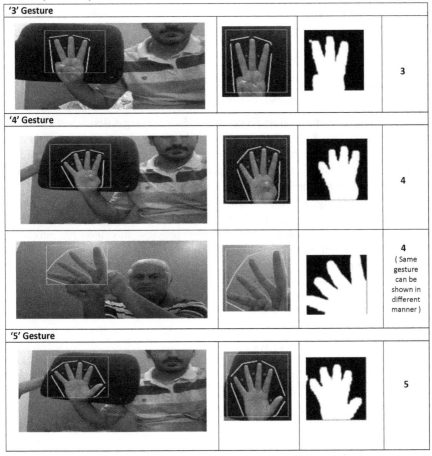

Table 2 Test results with off-white background

Gesture : Error occurred due to off-white background, number 2 was mis-interpreted as number 1 and number 3 as 4.			
	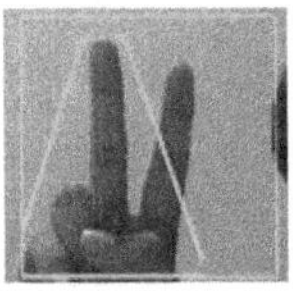		**1**

gestures with three different test subjects (people) and found to be accurate in all the cases. Misclassification is observed in the case of the background being similar to skin color.

5 Conclusion

The algorithm works in real-time processing and identifying different hand gestures. All the gestures like 1,2,3,4,5,0 Ok, Best of Luck are recognized correctly. The algorithm can always be expanded to identify more simple and complex gestures. The lighting condition of surrounding affects the results, so good lighting conditions are a must. The background color should not be in any shade of skin color as it may create noise and change the results. This gesture recognition technology can be used for different purposes like sign language recognition, immersive game technology, virtual controllers and remote control. For further progress, this algorithm can also be integrated with Arduino and Raspberry-Pi for moving the robot in different directions based on gestures.

References

1. Vaishnav P, Tiwari S (2015) Accelerometer based hand gesture controlled robot. Int J Sci Res 4(3)
2. Mojeebi T, Tulo SK (2016) Accelerometer gesture controlled robot using Arduino. Int J Eng Technol 3:38–41
3. Chanda P, Mukherjee P, Modak S, Nath A (2016) Gesture controlled robot using Arduino and android. Int J 6(6)
4. Kaura HK, Honrao V, Patil S, Shetty P (2013) Gesture controlled robot using image processing. Int J Adv Res Artif Intell 2(5)
5. Matnani P (2015) Glove based and accelerometer based gesture control. Int J Techn Res Appl 3(6):216–221
6. Avaible: https://docs.opencv.org/2.4/modules/imgproc/doc/structural_analysis_and_shape_descriptors.html?highlight=convexhull#convexhull
7. Avaible: https://docs.opencv.org/3.0beta/doc/py_tutorials/py_imgproc/py_contours/py_contours_more_functions/py_contours_more_functions.html
8. Available: https://www.peko-step.com/en/tool/hsvrgb_en.html

Wearable and Tactile E-skin for Large-Area Robots

Samta Sapra and Subhas Chandra Mukhopadhyay

Abstract The use of tactile sensing systems for ubiquitous robotic and telehealth monitoring will be one of the revolutionary changes in this century. Currently, robots are being used in almost all industries, but the growing need is to have robots that have a sense of touch and feel so that they can coexist with humans at the workplace and home. E-skin can promise that functionality by which they can feel the touch the same as humans do. The proposed E-skin uses the pressure-sensitive matrix (row–column) for measuring the change in capacitance using polydimethyl siloxane (PDMS) as substrate and copper electrodes. This system can be viewed as a large area E-skin with the capability of measuring both static and dynamic forces. It helps in telehealth, pain management, remote temperature monitoring, and robotic-assisted surgeries. This wearable E-skin shows the effectiveness of technology in evolving the human–robot (H-R) co-working and touch in the present era.

1 Introduction

Wearable is a sensing technology-based device that is worn on the human body. Wearable devices are also known as simple wearables, wearable technology, and wearable gadgets. Wearables are small enough to wear and include powerful sensor technologies that can collect and deliver information related to surroundings and vital signs of the body of the wearer. The combined collaboration, computation, and contextual awareness to enhance personal productivity. All these functions free up the individual mind and allow them to focus more on creative activities. Countries are struggling with an aged, diseased population and outbreaks. These are the reasons for the rising costs of the healthcare system. To overcome the above challenges and costs on the health system, wearables seem to be promising which can improve the quality of patient care by reducing the cost of healthcare by home monitoring of vital signs

S. Sapra · S. C. Mukhopadhyay (✉)
School of Engineering, Macquarie University, Sydney, NSW 2109, Australia
e-mail: Subhas.Mukhopadhyay@mq.edu.au

S. Dhar et al. (eds.), *Advances in Communication, Devices and Networking*,
Lecture Notes in Electrical Engineering 776,
https://doi.org/10.1007/978-981-16-2911-2_18

through the use of wearable [1]. Wearables can also be useful in the early detection and intervention of diseases by providing point-of-care wearable systems to monitor vitals continuously. Wearables allow people to convert biological, personal, and environmental data into valuable statistics and to convert these parameters into holistic decisions and goal-directed actions. For example, if doctors have access to statistics from wearables of patients and environmental data from smart sensors, then it becomes easy to diagnose and prevents earlier diseases or outbreaks in a community. Authorized access to this important data can also be filtered for use in businesses, retail, and entertainment industries for insight into consumer behaviors. Different research groups cited various characteristics of wearable devices depending on the wearability, ease of use, appealing design, functionality, and cost of the product. Wearable devices can be characterized by various criteria which include flexibility while in operation and portable, negligible operational delays, controlled by users, comfort, optical transparency, integration with existing technologies, and low cost. Gender, culture, image, and fashion play a significant role in the characteristics of wearables [2]. Wearables must be water-resistant/repellent so that they do not get affected by sweat. Wearables must be sensitive to measuring parameters as they rely on sensing various parameters by measuring and improving data. A wearable system may consist of sensors or sensor arrays, inertial sensors, including accelerometers gyroscope for user movement sensing and tracking, microcontroller, interface electronics, bus interface, A/D converters, multiplexers, filters, IoT-based wireless transmission systems, etc. There can be various other sensing components like cardiotachometers to monitor heart rates, and electromyography sensors for monitoring muscle status changes, bio-impedance sensors to measure blood flow resistance. Depending on applications these wearable systems may include various machine learning algorithms in software, the part which enhances the overall computation-intensive capabilities of these systems. These algorithms help from predicting the past, present, and future states to decision-making. Wearables are quite attractive to users due to their compactness and low cost and advancements day by day. Wearable technology is evolving as per the needs of humans. There are already many available systems like wearable gait analysis sensors, wearable gestural interface devices, drug infusion devices, heart rate monitoring, and health monitoring devices [3–5]. The ongoing demand for wearables has poised many research activities in this field [6]. Wearables are becoming context-aware technologies by adapting various data analysis techniques. With the pace at which wearables are advancing it can be inferred that future wearables will be even more user-friendly and convenient. Further advancements may lead them to become non-invasive [7] and eventually invisible [8, 9]. Despite being so much research being conducted in the field of wearables, there are few challenges for the technological adoption of wearable devices. Battery life is the main limiting factor as we must have very thin and small wearables due to which we are left with even smaller space for batteries and hence limits the longer usability. Wearables must satisfy a few goals for adoption in the daily life of people and are more connected with life. Few other challenges involved are free access availability of physiological data after use of wearables which need to be addressed in the future.

Our lab focuses on diverse fields and interdisciplinary areas. Successful research works have been conducted on the smart city, wearable flexible electronic [10–12], IoT-enabled devices for agricultural and water quality monitoring system [13–15], point of care devices for healthcare [16], wearable flexible electronic devices for motion sensing and other parameters [17], tactile-based pressure sensors [4], large area E-skin for robotics and prosthetics. Wearable pressure sensing tactile array along with ultrasonics can be used in the localization of tumors from soft tissues and artery localization in the wrist. Tactile sensing is complementary to an action-related feedback system and is different from all other senses due to its distributive nature rather than centralized nature.

Touch is one of the very important senses in humans. E-skin can be seen as a wearable for robots which enhances their capabilities Tactile pressure sensors have widespread applications and the use of tactile sensing systems for telehealth will be one of the important changes in this century. Despite so much research work done on real-time applications with robotic systems, one of the ongoing issues is that if we want to make robots work like and alongside humans then they must have E-skin which can feel touch the same as humans. This includes an Internet of Things (IoT)-based wearable, flexible tactile sensing system with high resolution and high sensitivity that can be used as E-skin on all body sites of robots. The proposed E-skin uses the pressure-sensitive matrix (row–column) for measuring the change in capacitance using PDMS as substrate and copper electrodes. This system can be viewed as a large area E-skin with the capability of measuring both static and dynamic forces. It helps in telehealth, pain management, remote temperature monitoring, and robotic-assisted surgeries.

2 Materials and Methods

This section includes materials used and fabrication steps involved. Polydimethyl siloxane (PDMS) was chosen as a substrate due to its dynamic flexibility by adjusting the elastomer ratio and curing agent. PDMS from Sylgard was used as an elastomer and curing agent in the ratio 10:1 [4]. A casting knife was used to keep the thickness of PDMS to 1 mm. Desiccation was done for 2 h to degas and air bubble removal. For developing substrate of sensor patch curing of the sample was done at 80 °C for 8 h. Copper tape of 4 mm was put on the cured PDMS by applying a very thin layer of PDMS at the places for electrodes and then the whole pad curing was done again for 8 h at 80 °C. Various shapes and geometry of electrodes like triangular, diamond shape, and rectangular shapes are simulated and were tested, but the rectangular shape of electrodes showed the ratio of maximum capacitance variation and has good performance than other shapes; therefore parallel rectangular electrodes were used for designing our sensor [11]. 5 * 5 Sensor matrix was developed in which the top and bottom layers with electrodes are separated by an insulator layer of PDMS forming a capacitor between two layers (Fig. 1).

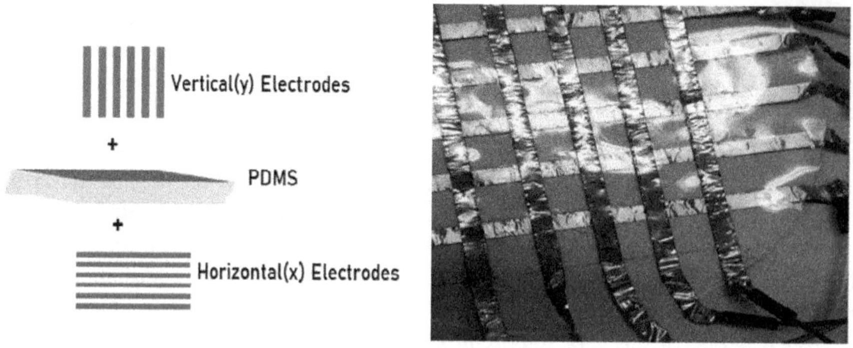

Fig. 1 Final sensor

3 Working Principle

The working principle of the sensor is proportional to the overlap area. The capacitance of a capacitive device can be written as:

$$C = (\epsilon_o * \epsilon_r * A/d)$$

where

C is the capacitance of the sensor,
ϵ_r is the relative permittivity,
$\epsilon_o = 8.85$ is the $\times\ 10^{-12}\ F{\cdot}m^{-1}$ is vacuum permittivity,
A is the overlap area of each capacitor formed,
d is the effective spacing between the X and Y electrodes.

The exertion of force on the capacitive sensor causes a change in the distance between the X and Y electrodes and the overlap area. A change in the value of d or A causes a change in the capacitance. The perpendicular force causes a change in the *d*, while the effective area between the plates is changed by the tangential force.

Therefore, these sensors can differentiate the normal and tangential forces. But the change in capacitance does not reflect much the type of applied force.

$$\Delta C = f(\Delta L, \Delta W, \Delta D)$$

The capacitive sensors transform the contact force signal to the output signal in two steps. In the first step, it converts contact/touch force into the change in capacitance and then measures the change in capacitance to an electrical output digital signal. There are two types of capacitive systems. The first type is self-capacitance, and the other is mutual capacitance. Self-capacitance is the load capacitance with respect to the ground that is where the finger (object) touches the sensor and causes an increase in the stray capacitance to the ground. The mutual capacitance is capacitive load when an object changes the capacitance between two electrodes (Fig. 2).

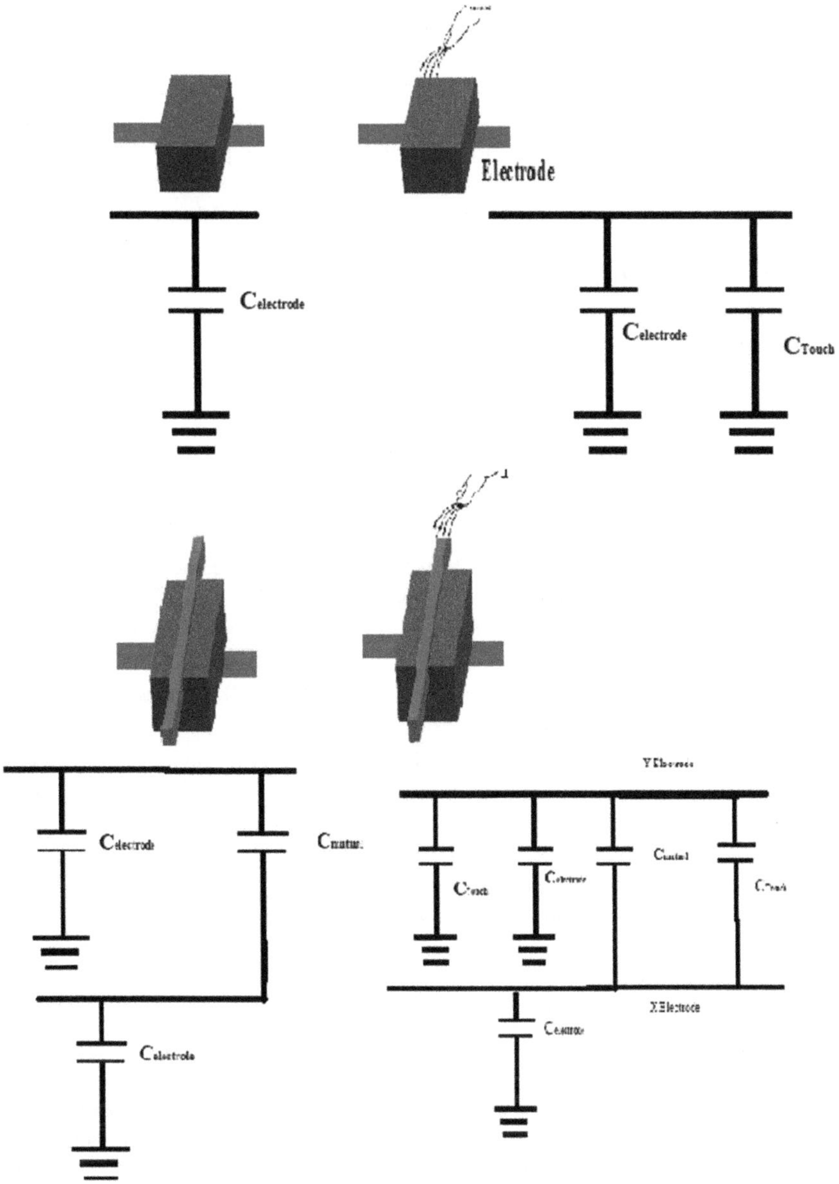

Fig. 2 The Self and Mutual capacitance touch sensor showing capacitance between X and Y electrodes

Various wiring arrangements were taken into consideration before the fabrication of the sensor. It was realized that using two wired approaches for each sensor causes 2(p*q) wires in a (p*q) sensor array. This approach can lead to many data wires given a lesser area sensor array. To decrease the number further, another approach was considered to use a shared return wire. By using this approach (p + q) +1 data wire will be needed for (p*q) array. Lastly, by using the row–column approach, only (p + q) data wires would be needed for array size of (p*q). This approach not only reduces data wires but also interface electronics, transmission cost plus the overall cost of the product [12]. Due to its advantages, the matrix arrangement is widely used in keypads or touch panels and can be used in large-area sensor arrays like E-skin applications.

4 Results and Discussions

This section includes the method, process, and instrument used for calculating the change in capacitance. The results were measured using the HIOKI IM3536 LCR meter by connecting sensor electrodes to Kelvin probes. The LCR meter was connected to the laptop using a USB cable. The contact parameters were saved in Microsoft Excel by using a data acquisition algorithm. The instrument was used in AC mode and the frequency was swept from 10 Hz to 10,000 Hz. First, the value of capacitance was measured at each electrode without any load. Then different forces from 1 N to 5 N were applied from area 1 to area 25 one by one, and the value of capacitance at each electrode was calculated with static force and the observed sensor response showed that maximum change of capacitance at those crossing or adjacent electrodes where force was applied. The exertion of force on the capacitive sensor causes a change in the distance between the X and Y electrodes and the overlap area. Figure 3a shows the observed maximum change in capacitance at electrodes a and g with the force of 5 N at area 1. Figure 3b shows the observed maximum change in capacitance at electrodes a and g for 5 N force. Figure 3c shows the location of unknown coordinates of object. This is achieved by observing change of capacitance at those co-ordinates of crossing electrodes as compared to all other regions. As observed where object is present that electrodes show maximum change in magnitude as compared to others. Here in this case object was identified at location where b, h (2*3) electrodes cross. It is to be noted that horizontal electrodes are named a to e from top to bottom whereas vertical electrodes are named f to j from left to right. Tactile sensing in E-skin can be seen in supplementary video results [18].

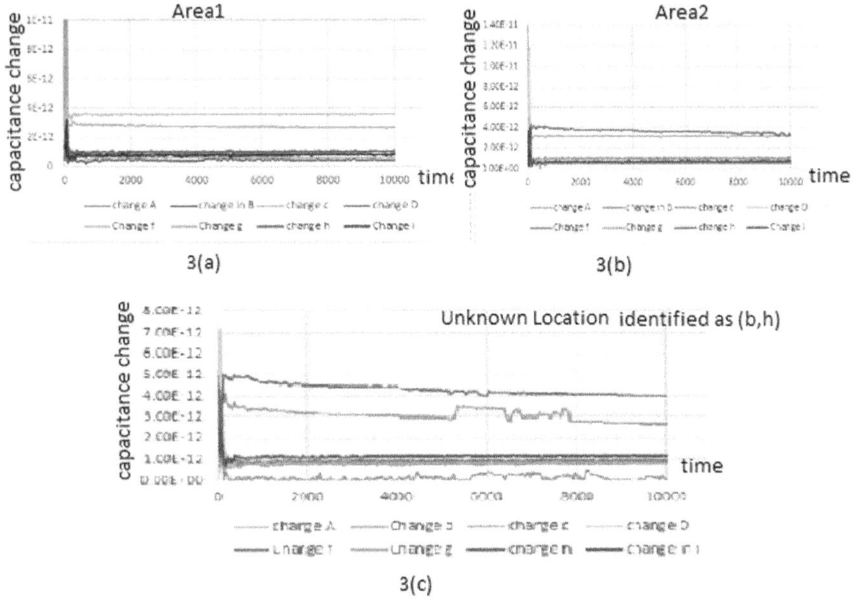

Fig. 3 **a** Change in capacitance when an object 5 N was placed at area1, **b** Change in capacitance when an object 5 N was placed at area 2, **c** the object 5 N was randomly placed in an unknown location (b, h identified location)

5 Conclusion

This novel, flexible sensor array of rows and columns can successfully measure localized pressure, at a location and from each sensing element. At each location where the top and bottom layer electrodes cross each other a capacitor is formed. Undoubtedly, increasing the number of electrodes allows more resolution but it may interfere with the sensitivity and efficiency of touch detection and it will also make the system more complex. So, there is a trade-off between resolution and efficiency by choosing the aspect ratio of sensor design and electrodes or sensing area. Touch sensitivity is a very important milestone in the development of flexible E-skin due to advantages like low cost of production and easy fabrication. There are still some gaps and advancements are needed for the implementation of this technology to be used for commercial purposes due to some irregularities in a change in capacitances and ghost value points due to some self-capacitance. Resolution improvement and flexibility issues must be involved. It can be made into completely energy-autonomous systems using body movement and mechanical energy to make it self-sufficient. Sensor resolution and efficiency improvement are the areas that need to be looked at for future work in E-skin. Some other features can also be added depending on applications like temperature and humidity sensing, touch-less sensing by using and experimenting with different materials. Therefore, various gaps can be filled in the future by incorporating some other parameters.

References

1. Zheng YL, Ding XR, Poon CCY, Lo BPL, Zhang H, Zhou XL, Yang GZ, Zhao N, Zhang YT (2014) Unobtrusive sensing and wearable devices for health informatics. IEEE Trans Biomed Eng 61(5):1538–1554
2. Adapa A, Nah FFH, Hall RH, Siau K, Smith SN (2018) Factors influencing the adoption of smart wearable devices. Int J Human-Comput Interact 34(5):399–409
3. Lee J, Kim D, Ryoo HY, Shin BS (2016) Sustainable wearables: wearable technology for enhancing the quality of human life. Sustainability 8(5):466
4. Sapra S, Nag A, Han T, Gooneratne CP (2019) Localisation of thin-film resistive sensors for force sensing applications. In: 2019 13th international conference on sensing technology (ICST) (pp 1–6). IEEE
5. Xu Y, Hu X, Kundu S, Nag A, Afsarimanesh N, Sapra S, Mukhopadhyay SC, Han T (2019) Silicon-based sensors for biomedical applications: a review. Sensors 19(13):2908
6. Akhter F, Khadivizand S, Siddiquei HR, Alahi MEE, Mukhopadhyay S (2019) IoT enabled intelligent sensor node for smart city: pedestrian counting and ambient monitoring. Sensors 19(15):3374
7. Nag A, Feng S, Mukhopadhyay S, Kosel J (2018) Development of printed sensors for shoe sensing applications. In: 2018 12th international symposium on medical information and communication technology (ISMICT) (pp 1–6). IEEE
8. Alahi MEE, Pereira-Ishak N, Mukhopadhyay SC, Burkitt L (2018) An internet-of-things enabled smart sensing system for nitrate monitoring. IEEE Internet Things J 5(6):4409–4417
9. Afsarimanesh N, Mukhopadhyay SC, Kruger M (2017) Molecularly imprinted polymer-based electrochemical biosensor for bone loss detection. IEEE Trans Biomed Eng 65(6):1264–1271
10. Han T, Nag A, Simorangkir RB, Afsarimanesh N, Liu H, Mukhopadhyay SC, Xu Y, Zhadobov M, Sauleau R (2019) Multifunctional flexible sensor based on laser-induced graphene. Sensors 19(16):3477
11. Ruan JY, Chao PCP, Chen WD (2010) November. A multi-touch interface circuit for a large-sized capacitive touch panel. In: Sensors, 2010 IEEE (pp 309–314). IEEE
12. https://pressureprofile.com/about/tactile-sensing
13. Nag A, Mukhopadhyay SC, Kosel J (2016) Flexible carbon nanotube nanocomposite sensor for multiple physiological parameter monitoring. Sens Actuators A 251:148–155
14. Perera C, Zaslavsky A, Christen P, Georgakopoulos D (2013) Context-aware computing for the internet of things: a survey. IEEE Commun Surv Tutorials 16(1):414–454
15. Han T, Nag A, Afsarimanesh N, Akhter F, Liu H, Sapra S, Mukhopadhyay S, Xu Y (2019) Gold/polyimide-based resistive strain sensors. Electronics 8(5):565
16. Dahiya RS, Valle M (2013) Tactile sensing technologies. In: Robotic tactile sensing (pp 79–136). Springer, Dordrecht
17. Nag A, Simorangkir RB, Valentin E, Björninen T, Ukkonen L, Hashmi RM, Mukhopadhyay SC (2018) A transparent strain sensor based on PDMS-embedded conductive fabric for wearable sensing applications. IEEE Access 6:71020–71027
18. https://www.youtube.com/watch?v=Hj4_VeUkub4

Multi-constellation GNSS Performance Study Under Indian Forest Canopy

Sukabya Dan, Atanu Santra, Somnath Mahato, Sumit Dey, Chaitali Koley, and Anindya Bose

Abstract GPS-based forest survey is being increasingly popular over the conventional methods; signals from other operational GNSS constellations are also being used for the purpose. The forest environment is different from the open-sky and urban environments as the type, height, density of the trees and the season-dependent foliage properties degrade the GPS/GNSS signals passing through the foliage within a dense multipath environment. This manuscript presents the results of a study on the solution quality of standalone and hybrid constellation GNSS (GPS, GLONASS, Galileo) operation within an Indian forest predominantly containing evergreen or dry-session deciduous Sal (*Shorea Robusta*) trees of the Dipterocarpaceae family, abundantly found all over India. The results show that in the single-frequency, static operation, single constellation GNSS solution quality degrades in comparison to open-sky operation with intense fluctuations in received signal strength reaching the GNSS antenna. Multi-constellation, hybrid GNSS operation improves the situation and GPS + GLONASS + Galileo provides better than 1.5 m horizontal (2DRMS) precision. The results of the study would be useful for the survey, especially for the forest survey community.

Keywords GNSS · Forest canopy · Precision · Elevation · C/N_0

1 Introduction

With the advancement of satellite-based navigation systems, the Global Navigation Satellite System (GNSS) is extensively being used in forest surveys. This has become more popular over the conventional handheld digital range finders and total stations because of the associated advantages of accuracy and cost-effectiveness. Survey

S. Dan (✉) · C. Koley
Department of ECE, National Institute of Technology Mizoram, Chaltlang, Aizawl 796012, Mizoram, India
e-mail: sukabya_dan@rediffmail.com

S. Dan · A. Santra · S. Mahato · S. Dey · A. Bose
Department of Physics, The University of Burdwan, Burdwan 713104, India

S. Dhar et al. (eds.), *Advances in Communication, Devices and Networking*, Lecture Notes in Electrical Engineering 776,
https://doi.org/10.1007/978-981-16-2911-2_19

departments/agencies generally use GPS in forest surveying for mapping the forest boundaries, estimating forest areas, mapping of forest resources and spatial managements, and other GIS-based applications [1–4]. Apart from GPS, other constellations like GLONASS, Galileo, BeiDou also provide a position solution through which a geodetic survey can be completed in the forests. Regional navigation systems, QZSS, and NavIC provide service over a limited service area of the globe. Within forest areas, GNSS-based positioning is associated with various issues related to signal quality and strength due to the dense presence of the trees and foliage surrounding the operation area. GNSS satellite signals are affected while passing through the foliage, and the dense trunks of the trees around the antenna create a multipath environment. As a result, the signal availability, quality, and position solution accuracies are degraded in comparison to the open sky conditions. Different GNSS constellations use varying data multiplexing methods, different signal coding techniques and they transmit signals in different frequencies; therefore their signal attributes are different. Hence, under the forest canopy, the signal performance of different satellite systems would be different. Types, height profile, the density of trees present in the forest area, and the associated foliage characteristic would also affect the signals passing through the forest canopy. The relative location of the individual satellite of different constellations in the sky is also different from a particular location in terms of the elevation and azimuth angle from the observation point. The signals will pass through different foliage thicknesses before reaching the antennas that change with time. So, the GNSS satellite signal quality in the forest environment is different from the open sky conditions. Because of the growing use of GPS and/or GNSS, various types of studies have been carried out on the use of GNSS signals in forest areas.

The performance of GPS Precise Point Positioning (PPP) under forest canopies of coniferous trees was studied by Næsset and Gjevestad. They found the increased accuracy with the decreased forest stand density, and the expected positioning accuracy correlates significantly with the stand basal area and number of trees [5]. Rodriguez-Perez et al. compared the accuracy and precision of different GPS receivers and compared the relative advantages of those receivers for position solutions under the forest canopy with careful data-acquisition protocols [6]. Kaartinen et al. investigated the positioning accuracy of the GNSS receiver integrated with the Inertial Measurement Unit (IMU) devices under forest canopies in varying forest conditions. They used GLONASS along with GPS for the purpose and found that using standalone GNSS, the required accuracy level of 1 m cannot be achieved, but an accuracy of even less than 1 m can be achieved while using real-time or post-processed differential GNSS [7]. Li et al. reported that hybrid GNSS (GPS, GLONASS, BeiDou, and Galileo) operation significantly increases the number of observed satellites and better satellite geometry at a survey site resulting in improved accuracy, continuity, convergence time, and reliability of positioning [8]. Sigrist and Hermy found that under the forest canopy PDOP is not a good indicator of obtained position solution precision and the signal attenuation increased with an increase of canopy density [9]. Mahato et al. represented the initial results of two different baseline RTK forest experiments with a compact, low-cost receiver. It was observed that meter-level precision has been achieved in a short baseline RTK forest environment experiment. For longer baseline distance, RTK forest experiment, GPS + GAL + QZSS mode was better and reliable than GPS + GAL + GLO + QZSS mode [10].

24.56% of the Indian landmass has a forest cover of different densities [11] and forest surveys are frequent for resource management. Many reports on GNSS-based efforts from India could not be found although India is located in a region of very high GNSS signal visibility, and therefore has the potential of using multi-GNSS for forest survey purposes. So, sustained and systematic efforts on GNSS-based forest surveys are required to access the suitability and related advantages in the Indian forest environments. This paper presents such an effort on GNSS-based surveys made within an Indian medium dense forest environment predominantly consisting of Sal (*Shorea Robusta*) trees. Standalone, static GNSS (GPS, GLONASS, and Galileo) solution accuracies in single and hybrid-constellation modes under forest canopy have been studied and compared. The procedure and analysis are presented in the subsequent sections.

2 Experimental Setup

The experiment is performed using a geodetic GNSS receiver, JAVAD DELTA G3T [12] in standalone, static operation mode as shown in Fig. 1. To access the open-sky performances, the experiment is first carried out at the rooftop of GNSS Laboratory Burdwan (GLB), Department of Physics, The University of Burdwan, India (23.2545 °N, 87.8467 °E, −11.402 m). National Marine Electronics Association (NMEA) data from the receiver is recorded for 20 h each @1 Hz for standalone GPS, GLONASS, Galileo operation.

A subsequent experiment using a similar setup is carried out at a static point located within a forest area (Shibpur forest (23.6061°N, 87.42833°E) near Durgapur,

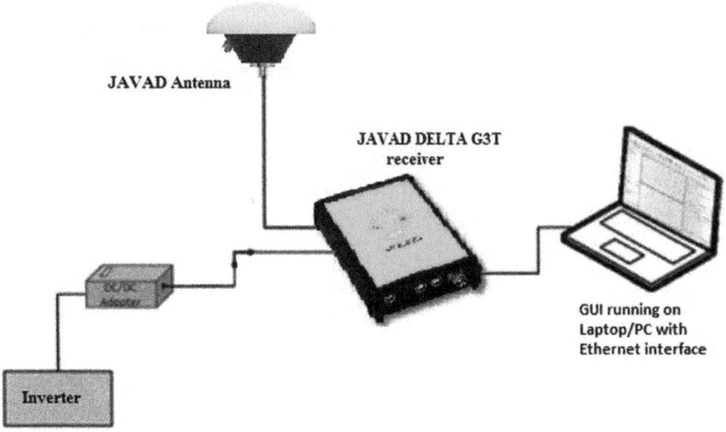

Fig. 1 Experimental setup for data logging

Fig. 2 Snapshot of the survey location

India approximately 60 km away from GLB) for studying the GNSS performance under the forest canopy. In this case, due to time constraints, NMEA data is recorded for a 30 min span @1 Hz using each constellation or a combination thereof. The forest is predominantly covered with Sal (*Shorea Robusta*) trees of 15–20 m height as measured using transit theodolite and is located with 4–5 m inter-tree distance. Sal is abundantly found all over India, is evergreen in wetter areas and dry-session deciduous variety in dry areas, and has leaves of size 10–25 cm long and 5–15 cm broad. In the survey location, the trees have sufficient leaves as shown in the snapshot of the location in Fig. 2.

The recorded NMEA data is post-processed to calculate the precision parameters of the solution and the signal strength variations. The standard statistical terms that are used to analyze accuracy and precision are: Distance Root Mean Square (DRMS) and Circular Error Probable (CEP), for horizontal (2d) solutions, and Spherical Error Probable (SEP), and Mean Radial Spherical Error (MRSE) for three-dimensional (3d) solutions [13–15]. To study the signal strength variations obtained in terms of C/N_0 (dB-Hz), the elevation angle of all used satellites are grouped for every 1° elevation angle, and the corresponding coefficient of variation (CV) of the signal strength is calculated using the following formula:

$$CV(\%) = \frac{\text{Standard Deviation}}{\text{Mean}} \times 100\% \tag{1}$$

3 Result and Discussion

3.1 Standalone Single-Frequency Operation

This section compares the precision parameters and signal strengths obtained under the open sky conditions and forest canopy. The precision parameters under open sky conditions for each standalone constellation are shown in Table 1 and are taken as the reference for subsequent studies. The precision parameters of the position solution for the standalone single-frequency operation of each GPS, GLONASS, Galileo under the forest canopy are presented in Table 2. It is to be noted that the average coordinate values during the data recording period are taken as reference coordinate for calculating the precision parameters [12–14].

From Table 1, it can be seen that the best two-dimensional (2d) precision, i.e., 2DRMS and CEP is obtained by standalone GPS followed by Galileo and GLONASS. In the case of 3d precision, the best result is obtained for Galileo with comparable values obtained from GPS. Figure 3 shows the plot of the coefficient of variation (%) of the signal strength with respect to the elevation angle of all the used satellites taken together. It is observed that the CV (%) decreases with increasing elevation angle and the same pattern is followed by all three constellations. This indicates that the variation of the signal strength is more for lower elevation angles due to multipath and atmospheric effects. The least variation is observed for Galileo E1 C/A coded signal followed by GLONASS L1 C/A coded and GPS L1 band C/A coded signals. Up to 70° elevation angles, the CV (%) values remain within 2–8, and beyond that,

Table 1 The precision of the position solution for standalone single-frequency constellation under open sky condition (data logging duration: 20 h session each @1 Hz)

Constellation used	Maximum Variation p2p (m)			2DRMS (m)	CEP (m)	SEP (m)	MRSE (m)
	Lat. (m)	Long. (m)	Alti. (m)				
GPS	5.00	2.84	10.74	2.90	1.14	2.10	2.65
GLONASS	13.81	16.32	145.77	6.94	2.89	7.36	10.16
Galileo	5.41	6.11	23.31	3.05	1.26	2.07	2.48

Table 2 The precision of position solution for standalone single-frequency constellation under the forest canopy (data logging duration: 30 min session each @1 Hz)

Constellation used	Maximum Variation p2p (m)			2DRMS (m)	CEP (m)	SEP (m)	MRSE (m)
	Lat. (m)	Long. (m)	Alti. (m)				
GPS	5.93	5.24	20.83	3.29	1.35	3.09	4.15
GLONASS	16.05	12.82	149.66	7.99	3.33	19.09	32.02
Galileo	3.73	8.97	67.45	3.39	1.35	7.01	11.53

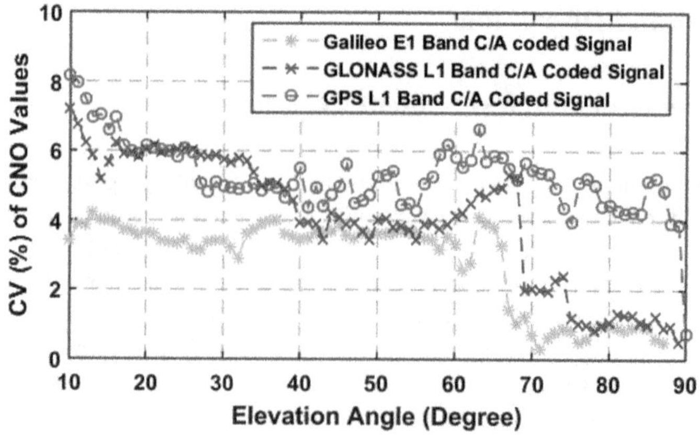

Fig. 3 Coefficient of variation (%) of signal strength variation for different standalone constellations under open sky condition

the values remain below 2 for GLONASS and Galileo and between 2 and 6 for GPS. This observation is consistent with the results shown by Dutta and Bose [16].

Similar results for the forest data shown in Table 2 indicate that the 2d precision parameters are almost similar for GPS and Galileo, whereas GLONASS provides the worst 2d precision among the three systems. In the case of 3d precision, GPS provides a better result than that for Galileo. In the case of 2d solutions, the precision degrades and degrades significantly in the case of 3d solutions.

GNSS signal strength variation in terms of coefficient of variation (CV, %) with respect to the elevation angle within the forest environment is plotted in Fig. 4. It shows a similar variation pattern with increasing elevation angle. But under the forest

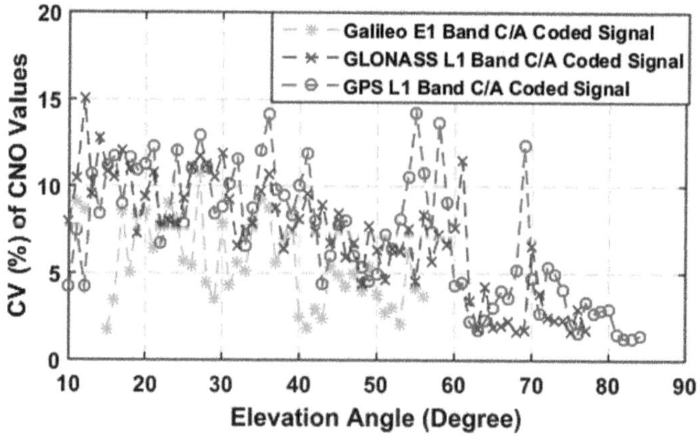

Fig. 4 Coefficient of variation (%) of the signal strength variation for different standalone constellations under the forest canopy

Table 3 The precision of the position solution for hybrid single-frequency operation under the forest canopy (data logging duration: 30 min each @1 Hz)

Constellation used	Maximum Variation p2p (m)			2DRMS (m)	CEP (m)	SEP (m)	MRSE (m)
	Lat. (m)	Long. (m)	Alti. (m)				
GPS + GLONASS	4.76	3.32	14.75	2.59	1.06	2.31	3.03
GPS + Galileo	3.84	3.22	6.93	2.07	0.86	1.37	1.61
GPS + Galileo + GLONASS	2.21	3.06	8.67	1.28	0.53	1.08	1.38

canopy much higher and more frequent fluctuations (observed in terms of CV) are observed up to a value of 15 for 10°–70° elevation angles because the satellite signals get affected due to the foliage, and multipath is created due to the presence of dense trees at the survey location. The sudden and strong fluctuations may lead to loss of lock for the receivers affecting seamless operation.

3.2 *Hybrid Single-Frequency Operation*

In this section, the precision parameters for single-frequency hybrid operation (GPS + GLONASS, GPS + Galileo, and GPS + GLONASS + Galileo) under the forest environment are presented. Hybrid constellations provide better performance than that of individual standalone operations in the forest environment. The hybrid operation provides a higher number of usable satellites, better satellite geometry, and improvement in precision compared to single constellation operation as shown in Table 3. GPS + GLONASS + Galileo provides the best 2d and 3d precision over the other combinations, and GPS + Galileo and GPS + GLONASS operations come sequentially in terms of precision.

4 Conclusion and Future Work

This paper presents the preliminary results on the use of different GNSS constellations for forest surveys in India. The performance of GNSS degrades while operating under the forest canopy in comparison to the open-sky operation due to the foliage with intense amplitude fluctuations and the multipath environment. Hybrid GNSS operation improves the situation and Galileo supports conventional GPS in comparison to GLONASS. Future work would include studies within other forest environments, the use of low-cost, compact GNSS receivers, dual-frequency measurements, and the multi-GNSS real-time kinematic (RTK) method for the purpose. The results of the current study would help the forest survey community.

Acknowledgments The authors acknowledge the Defense Research and Development Organization (DRDO), New Delhi (Project Code: ERIP/ER/DG-MSS/990516601/M/01/1658) and University Grants Commission (UGC) through CAS-II program for the financial support to carry out the study.

References

1. Tachiki Y, Yoshimura T, Hasegawa H, Mita T, Sakai T, Nakamura F (2005) Effects of polyline simplification of dynamic GPS data under forest canopy on area and perimeter estimations. J Forest Res 10(6):419–427
2. McDonald TP, Carter EA, Taylor SE (2011) Using the global positioning system to map disturbance patterns of forest harvesting machine. Can J For Res 32(2):310–319
3. Wing MG, Bettinger P (2003) GIS: An updated primer on a powerful management tool. J Forest 101(4):4–8
4. Wing MG, Kellogg LD (2004) Digital data collection and analysis techniques for forestry applications. 12th international conference on Geoinformatics, University of Gävle, Sweden, pp 77–83)
5. Næsset E, Gjevestad JG (2008) Performance of GPS precise point positioning under conifer forest canopies. Photogramm Eng Remote Sens 74(5):661–668
6. Rodríguez Pérez JR, Álvarez Taboada MF, Sanz Ablanedo E, Gavela E (2006) Comparison of GPS receiver accuracy and precision in forest environments: practical recommendations regarding methods and receiver selection. GNSS Processing and applications, Munich, Germany
7. Kaartinen H, Hyyppä J, Vastaranta M, Kukko A, Jaakkola A, Yu X, Pyörälä J, Liang X, Liu J, Wang Y, Kaijaluoto R (2015) Accuracy of kinematic positioning using global satellite navigation systems under forest canopies. Forests 6(9):3218–3236
8. Li X, Ge M, Dai X, Ren X, Fritsche M, Wickert J, Schuh H (2015) Accuracy and reliability of multi-GNSS real-time precise positioning: GPS, GLONASS, BeiDou, and Galileo. J Geodesy 89(6):607–635
9. Bettinger P, Merry KL (2012) Influence of the juxtaposition of trees on consumer-grade GPS position quality. Math Comput Forestry Natural-Resour Sci 4(2):81–91
10. Mahato S, Shaw G, Santra A, Dan S, Kundu S, Bose A (2020) Low Cost GNSS Receiver RTK performance in forest environment. 2020 URSI regional conference on radio science, Varanasi
11. Ministry of Environment, Forest and Climate Change. [Online]. https://pib.gov.in/PressReleasePage.aspx?PRID=1597987. Accessed 10th Jan, 2020
12. Javad Delta G3T Data Sheet. [Online]. http://download.javad.com/sheets/DeltaG3T_Datasheet.pdf. Accessed 18th Sept, 2019
13. GPS position Accuracy measures, Novatel, APN-029 Rev. 1, 03 December 2003.[Online]. https://www.novatel.com/assets/Documents/Bulletins/apn029.pdf. Accessed: 09 Oct. 2018
14. Deakin RE, Kildea DG (1999) A note on standard deviation and RMS. Australian Surv 44(1):74–79
15. Santra A, Mahato S, Dan S, Bose A (2019) Precision of satellite based navigation position solution: a review using NavIC data. J Inf Optim Sci 40(8):1683–1691
16. Dutta D, Bose A (2019) Studies on Variation of GNSS Signal Strengths from India. Indian J Radio Space Phys 68:64–71

Effect of Intrinsic Base Resistance on Rise Time of Transistor Laser

R. Ranjith and K. Kaviyarasi

Abstract Rise time of a transistor laser is analyzed by giving an input pulse to the base terminal in a common emitter configuration. The effect of rise time of transistor laser parameters is analyzed by varying the current gain, collector-emitter voltage and spontaneous lifetime of transistor laser using HSPICE circuit simulation software. It is found that the rise time of transistor laser increases for increasing the current gain, collector-emitter voltage and spontaneous lifetime of transistor laser. It is observed that the switching speed of a transistor laser decreases for increasing the current gain, collector-emitter voltage and spontaneous lifetime of transistor laser.

Keywords Light-emitting transistor (LET) · Transistor laser · Rise time analysis · Intrinsic base resistance of transistor laser

1 Introduction

In the year 1947, the bipolar junction transistor (BJT) was invented by Bardeen and Brattain. A transistor consists of three terminals with carrier injection, base recombination and carrier collection. It is also considered as a two-port device mainly used for signal amplification and switching. Carriers emitted by the emitter terminal reach the collector through the base region allowing nonradiative recombination in the base region. Light-emitting transistor was invented by Feng et al. [1]. Transistor laser uses direct bandgap material with an active region in the base region that allows radiative recombination (light emission). Transistor laser combines both transistor and laser and considered as a replacement for laser diode in the near future. Transistor laser consists of one input port and two output ports (electrical and optical). The collector terminal acts as an electrical output and the active region in the base acts as an optical output port.

Continuous-wave operation of transistor laser at 980 nm was demonstrated [1–5] and the transistor laser at a longer wavelength (1.3 μm) was demonstrated [6] in both

R. Ranjith (✉) · K. Kaviyarasi
Electronics and Communication Engineering, Government College of Engineering Bargur, Bargur, Tamilnadu, India

S. Dhar et al. (eds.), *Advances in Communication, Devices and Networking*, Lecture Notes in Electrical Engineering 776,
https://doi.org/10.1007/978-981-16-2911-2_20

CE and CB configurations. The coupled-rate equation model of transistor laser was developed and various parameters of transistor laser were studied [7, 8]. The circuit model of transistor laser is developed and analyzed [9–12], including three terminals with associated currents and voltages.

The carriers emitted from the emitter terminal reach the base region. In the base region, part of the carriers is captured by the quantum well allowing photon emission, and the remaining carriers will be collected by the reverse-biased collector terminal allowing collector current. The relaxation time of the carriers in the base region is highly reduced by the reverse-biased collector region, thus allows high-speed operation compared with the laser diode. With both electrical and optical outputs, the transistor laser finds enormous applications compared with the laser diode. Transistor laser offers higher bandwidth and better distortion characteristics [13].

In this paper, the circuit model of the transistor laser shown in Fig. 1 is analyzed with HSPICE circuit simulation software. DC analysis is done by applying DC current at the input base terminal and large signal analysis is done by applying a pulse current at the input base terminal. The effect on rise time of transistor laser is analyzed by varying the intrinsic base resistance, current gain and base recombination lifetime.

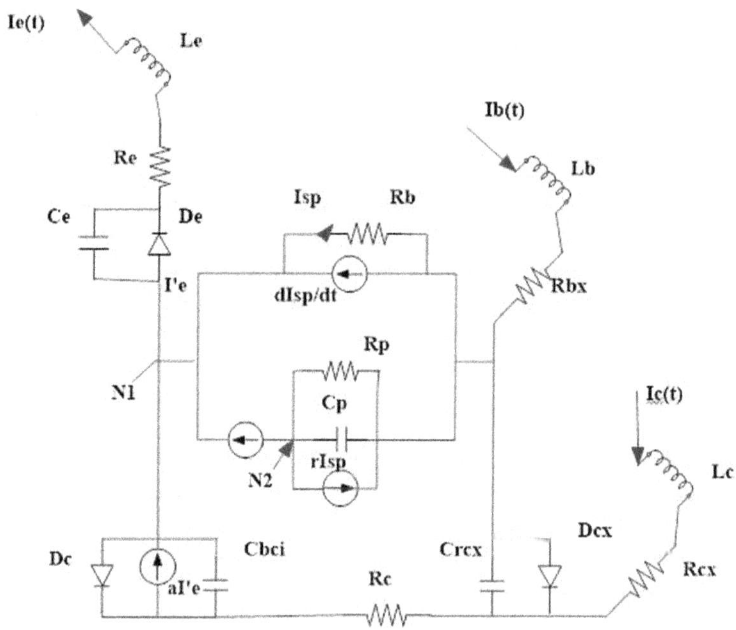

Fig. 1 Equivalent circuit model of transistor laser (TL) [12]

2 Equivalent Electrical Circuit of Transistor Laser

The circuit model of the transistor laser shown in the above figure models both electrical and optical processes of the transistor laser [12]. Resonance circuit found in the above figure models the active region of the transistor laser that includes both stimulated and spontaneous recombination. Photons stored in the resonator cavity will be approximated by the photon capacitor (C_p), and the charge stored by the capacitor is considered as photons stored in the resonant cavity. The number of photons generated by measuring the voltage across the photon capacitor is calculated using Eq. (1).

$$qN_{ph} = \frac{C_p}{q} \times qV_p \tag{1}$$

The parameters of the circuit model are tabulated in Table 1.

Table 1 Circuit parameters

Symbol	Definition	Unit	Values
I_B	Base recombination current	mA	0 to 50
R_p	Photon resistance	Ohm (Ω)	1.6
R_E	Extrinsic resistance (emitter)	Ohm (Ω)	3.6
R_{bx}	Extrinsic resistance (base)	Ohm (Ω)	0.9
R_C	Extrinsic resistance (collector)	Ohm (Ω)	0.5
R_{cx}	Extrinsic resistance (collector)	Ohm (Ω)	1.2
R_p	Photon resistance	Ohm (Ω)	1.6
C_p	Photon capacitance	pF	1.7
C_E	Emitter capacitance	pF	1.2
C_{bcx}	Extrinsic base collector depletion capacitance	pF	15
C_{bci}	Intrinsic base collector depletion capacitance	pF	1.5
R_B	Intrinsic base QW resistance	Ohm (Ω)	0.7
L_E	Parasitic inductance (emitter)	pH	0.5
L_C	Parasitic inductance (collector)	pH	21
L_B	Parasitic inductance (base)	pH	43
A	Current gain		0.2 to 0.8
ϒ	Gama		1.3%
τ_B	Spontaneous lifetime	Ps	37, 50 and 70

3 Simulation Results

3.1 Large Signal Analysis

Large signal analysis is done by giving a square wave signal as an input to the base terminal. The bias point is chosen such that the TL works in cutoff and saturation regions. In the above figure, the output voltage across the tank circuit (photon voltage) is observed by giving an input current at the base terminal of the TL. From the above figure, the parameters such as rise time and fall time are calculated. The effect of current gain, base resistance, base recombination lifetime and collector voltage on rise time of TL is analyzed (Fig. 2).

3.2 Rise Time Analysis

The rise time of the transistor laser is analyzed by varying the alpha (current gain) for different base resistance (r_b), base recombination time (t_b) and collector voltage (V_{CE}) (Fig. 3).

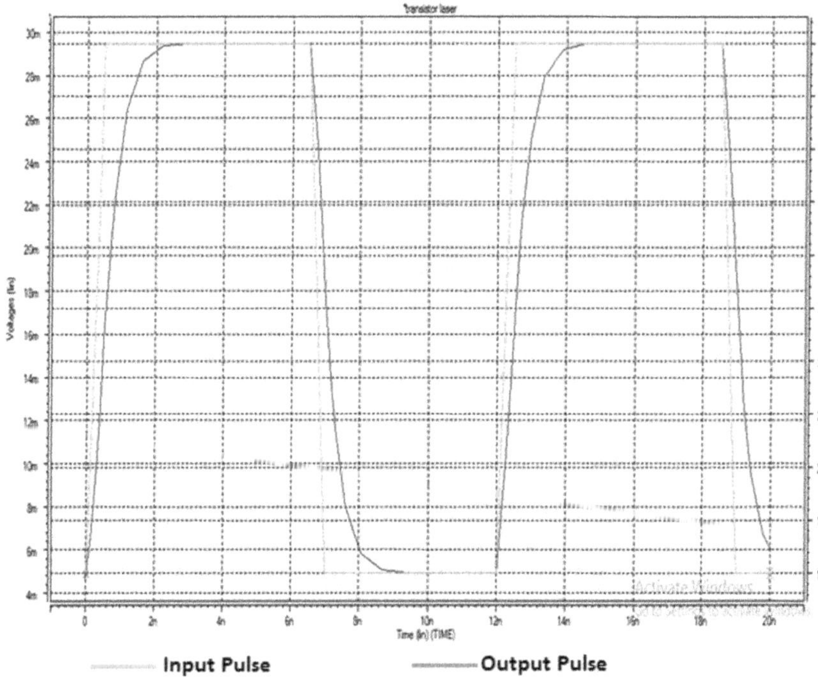

Fig. 2 Large signal analysis of transistor laser

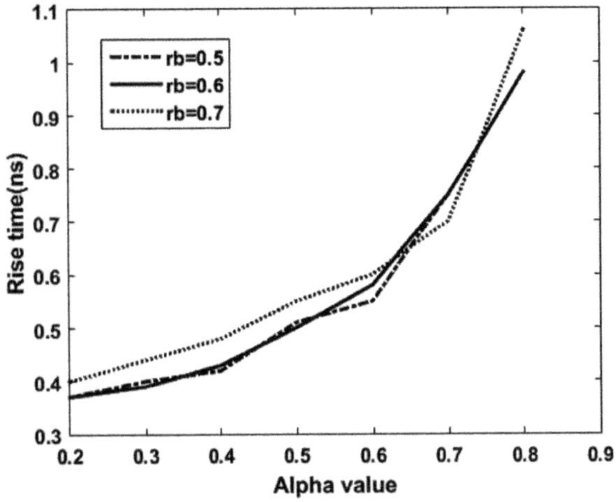

Fig. 3 Rise time versus current gain for different values of r_b with fixed $V_{CE} = 3v$ and $\tau_b = 37$ ps

From the above figure, it is found that the rise time of the TL increases for increasing the alpha value for different base resistance (r_b). It is also found that the low base resistance provides a lower rise time (Fig. 4).

From the above figure, it is found that the rise time of the TL increases for increasing the alpha value (current gain) for different base recombination lifetime (t_b). It is also found that the lower recombination lifetime provides lower rise time.

From the above figure, it is found that the rise time of the TL increases for increasing the alpha value (current Gain) for different collector voltage (V_{CE}). It is also found that the lower collector voltage provides lower rise time (Fig. 5).

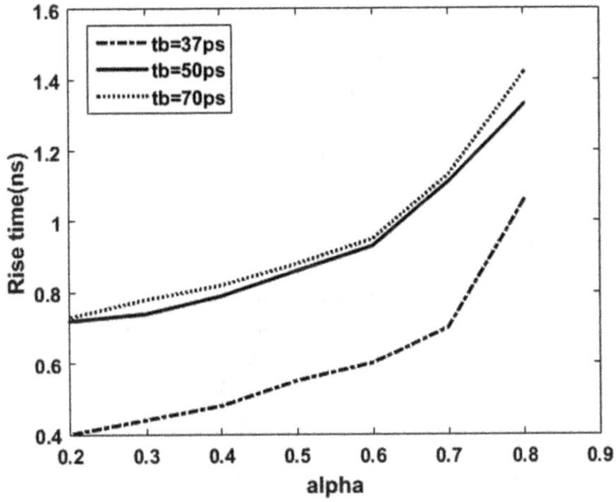

Fig. 4 Rise time versus current gain for different τ_b with constant $r_b = 0.7$ and $V_{CE} = 3v$

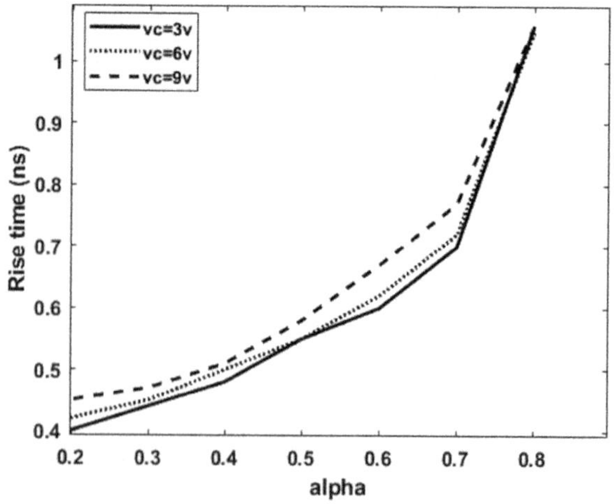

Fig. 5 Rise time versus current gain for different V_{CE} with constant $\tau_b = 37$ ps and $r_b = 0.7$

4 Conclusion

DC analysis and large signal analysis of transistor laser are simulated and analyzed. The effect of current gain, intrinsic base resistance and base recombination lifetime and collector voltage on rise time of TL is analyzed. It is found that the low current gain provides a low rise time, and thus provides a high data rate. For low current gain, lower resistance, base recombination lifetime and collector voltage provide a low rise time value.

References

1. Holonyak N, Feng M (2012) The transistor laser. IEEE Spectrum IEEE
2. Feng M, Holonyak N, Walter G, Chan R (2005) Room temperature continuous wave operation of a Heterojunction Bipolar transistor laser. Appl Phys Lett 87(13):131103–31103-3. Print
3. Walter G, Holonyak N (2018) Tunneling modulation of transistor laser: theory and experiment. IEEE J Quantom Electron 54(2)
4. Holonyak M-KN, Feng M (2012) Surface emission Vrtical cavity transistor laser. IEEE Photon Technol Lett 24
5. Feng M, Holonyak N (2011) The transistor laser. 2011 conference on laser and electro optics Europe and 12th European Quantum Electronics conference
6. Sato N, Shirao M, Sato T, Yukinari M, Nishiyama N, Amemiya T, Arai S (2013) Design and characterization of AlGaInAs/InP Buried Heterostructure transistor lasers emitting at 1.3 μm wavelength. IEEE J Sel Topics Quant Electron 19
7. Zhang L, Leburton J-P (2009) Modeling of the transient characteristics of Heterojunction Bipolar transistor lasers. IEEE J Quantum Electron 45:359–366
8. Faraji B, Shi W, Pulfrey DL, Chrostowski L (2009) Analytical modeling of the Transistor laser. IEEE J Sel Top Quantum Electron 15(3):594–603
9. Feng M, Liu M, Wang C, Holonyak N, Vertical cavity transistor laser for on-chip OICs Department of Electrical and Computer Engineering, University of Illinois at Urbana-Champaign, 1406 W. Green St., Urbana, IL 61801 USA Email: mfeng@illinois.edu

10. Feng M, Fellow, IEEE, Nick Holonyak, Jr., Fellow, IEEE, Richard Chan, Student Member, IEEE, Adam James, Student Member, IEEE, and Gabriel Walter" High-Speed (1 GHz) Electrical and Optical Adding, Mixing, and Processing of Square-Wave Signals With a Transistor Laser
11. Feng M, Holonyak N (2013) Transistor laser for electronic-photonic integrated Circuits. 2013 IEEE compound semiconductor Integrated Circuit symposium (CSICS)
12. Then HW, Feng M, Holonyak N, Microwave circuit model of the three-port transistor laser. J Appl Phys
13. Ranjith R, Piramasubramanian S, Ganesh Madhan M (2017) Distortion analysis of 1.3 μm AlGaInAs/InP transistor laser. In: Bhattacharya I, Chakrabarti S, Reehal H, Lakshminarayanan V (eds) advances in optical science and engineering. Springer Proceedings in Physics, vol 194. Springer, Singapore

A Reconfigurable Bandpass-Bandstop Microwave Filter Using PIN Diode for Wireless Applications

Hashinur Islam, Tanushree Bose, and Saumya Das

Abstract This paper presents a compact reconfigurable bandpass-bandstop microwave filter composed of a dual concentric U-shaped parasitic resonator. SMP1320-079LF PIN diodes from Skyworks are used to allow switching between bandpass and bandstop response. The proposed reconfigurable filter achieved four different switchable modes (two bandpass and two bandstop) using two PIN diodes with acceptable losses and rejection levels. In addition, the filter performs a flat group delay within the passband of approximately less than 0.5 ns in all switching conditions that are implemented using the CST-MWS simulator. The primary benefit of this microwave filtering system is the efficient ability to switch between two different operating modes by controlling the center frequency and bandwidth simultaneously. For multiple wireless communication systems, the proposed filter may be a good choice.

Keywords Reconfigurable filter · Switchable filter · Bandpass filter · Bandstop filter · U-shaped resonator · PIN diode

1 Introduction

Microwave filters are of immense significance when dealing with RF communication systems. New advances in this field have led to the need for compact, lightweight, multipurpose, and effective designs. Under this paradigm, electronically switchable filters are gaining significant interest from researchers and developers [1, 2]. Microwave filters are designed to keep both structure compactness and ease in mind throughout their switching between alternative operating modes. Microwave filters can be designed to switch between different geometrical shapes with different resonant frequencies using commercially available PIN diodes, resulting in reconfigurable structures [3]. Nowadays, modern wireless communication systems need

H. Islam (✉) · T. Bose · S. Das
Department of Electronics and Communication Engineering, Sikkim Manipal Institute of Technology, Sikkim Manipal University, Sikkim, India

S. Dhar et al. (eds.), *Advances in Communication, Devices and Networking*, Lecture Notes in Electrical Engineering 776,
https://doi.org/10.1007/978-981-16-2911-2_21

multiband, multifunctional, and switchable capabilities. In such conditions, a single filter cannot fulfill the criteria for all operating bands, while the use of several filters occupies a wider surface area with higher design costs [4]. In order to solve this problem, reconfigurable fillers [5] are needed.

For any communication device, bandpass and bandstop filters are two of the most widely used elements. Bandpass filters are used to pass a specific frequency band, while bandstop filters are used to eliminate certain frequency bands [6]. Thus, to miniaturize all wireless communication systems and obtain the facility in one platform, a switchable bandpass to bandstop filter is a promising solution to this problem, moving toward a smart filtering system. This type of smart filter allows a single filter structure with built-in electronic components [7, 8] to pick various frequency bands and make the device adaptable and flexible as well. In the broader sense, several concerns should be considered when selecting an exact topology to reconfigure the microwave filter structure.

Recently, a switchable microstrip filter with four operating modes has been published in [9]. To achieve switchability, four PIN diodes are used here with a filter dimension of 22.5 mm × 24.3 mm. Article [10] reported a bandpass filter to a single/double-notch bandnotch filter where five PIN diodes are used for switching the proposed filter with the dimension of 26 mm × 15 mm. A reconfigurable bandpass filter with a size of 21.8 mm × 22 mm is introduced in [11], where the fractional bandwidth varies from 118% to 15%, by controlling the biasing voltages of two PIN diodes, keeping a notch at 2.4 GHz. So, if the device needs a variable center frequency, it will not accept it.

[12] presents a reconfigurable bandpass to bandstop switchable filter using two PIN diodes with the dimension of the filter 55 mm × 60 mm. A switchable bandstop to bandpass reconfigurable filter based on PIN diodes has been reported in [13]. Here, the discreet tunability of the proposed filter is achieved by using two PIN diodes with a relatively larger filter size of 55 mm × 60 mm. However, due to the limited size, the large dimension might not be suitable for compact wireless devices.

In this work, a reconfigurable bandpass-bandstop (BP-BS) microwave filter is designed using a dual concentric U-shaped parasitic resonator with a reduced dimension of 20 mm × 20 mm. The BP-BS transformation response is achieved using electronic switching, i.e., by adjusting two PIN diodes. The proposed BP-BS microwave filter achieved four separate switchable modes with simultaneous control of center frequency and bandwidth. The benefit of this filter is its compact size, low insertion loss, good rejection, and high attenuation. The proposed reconfigurable BP-BS filter is simulated on FR-4 substrate having $\varepsilon_r = 4.3$, $\tan\delta = 0.025$, and a thickness of 0.8 mm.

2 Filter Design and Configuration

The proposed reconfigurable bandpass-bandstop (BP-BS) filter designed using a dual concentric U-shaped parasitic resonator fed by a 50Ω feed line is shown in Fig. 1. In this work, the filter is simulated using cost-effective FR-4 substrate having $\varepsilon_r =$

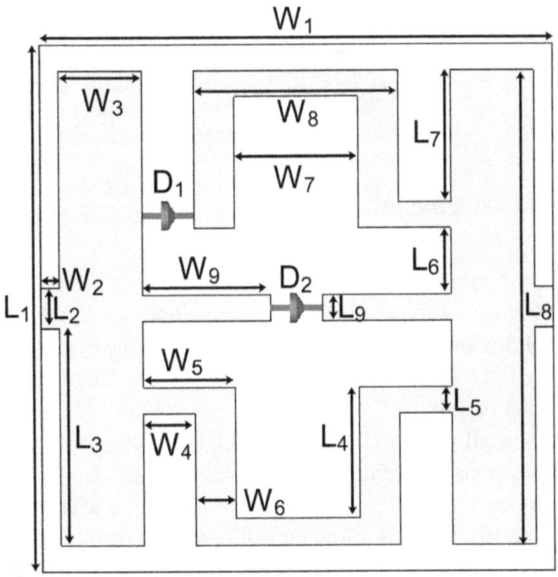

Fig. 1 Geometry of the proposed reconfigurable BP-BS filter

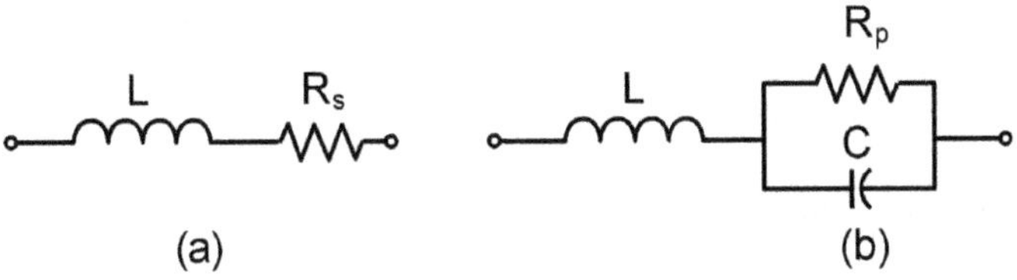

Fig. 2 Equivalent circuit models for SMP1320-079LF PIN diodes **a** ON condition, **b** OFF condition

4.3 and loss tangent tanδ = 0.025 with a substrate thickness of 0.8 mm. Two PIN diodes (SMP1320-079LF) from Skyworks are incorporated to produce a switchable bandpass to bandstop filter design. As shown in Fig. 1, PIN diodes are inserted into the strip of the proposed BP-BS filter.

Metal strips with dimensions 1 mm × 2 mm are used within the main slot for applying DC voltage to PIN diodes. The PIN diodes are modeled as a series RL circuit for ON condition and a combination of parallel series RLC circuits for OFF condition, shown in Fig. 2. As a result of searching for appropriate PIN diodes, SMP1320-079LF was selected with high-frequency switching, low inductance series, and low forward resistance [14]. A supply of 0.7 V is applied to metal strips for biasing PIN diodes. These PIN diodes have ON state resistance and inductance R_s = 0.75 Ω and L = 0.7 nH, respectively. The PIN diode with OFF state having C = 0.23 pF and R_p = 0.4 MΩ shunt capacitance and reverse resistance, respectively, with L = 0.16 μH is shown in Fig. 2. The total dimension of the proposed BP-BS reconfigurable filter is 20 mm × 20 mm, and other physical parameters are specified in Table 1.

Table 1 Dimensions of proposed reconfigurable BP-BS filter

Parameter	W_1	W_2	W_3	W_4	W_5	W_6	W_7	W_8	W_9	L_1	L_2	L_3	L_4	L_5	L_6	L_7	L_8	L_9
Value (mm)	20	0.75	3.25	2	3.6	1.6	4.8	8	5	20	1.5	8.25	5	1	2.5	5	18	1

3 Results and Discussion

The proposed reconfigurable BP-BS filter simulations have been carried out using the CST-MWS simulator for different diode switching conditions. In order to reconfigure the filter, two PIN diodes have been considered, which have achieved four possible switching modes. Four different modes, as well as the frequency bands of these modes, are included in Table 2.

In Mode 1, when all diodes (D_1 and D_2) are simultaneously in OFF condition, a bandstop filter response is achieved with a bandwidth of –10 dB ranging 1.32–2.24 GHz with a rejection level greater than 30 dB and an average loss of 0.67 dB. Figure 3a shows the simulated S-parameter of bandstop responses for Mode 1. This bandstop filter mode is designed to suppress signals ranging 1.32–2.24 GHz, which covers multiple communication signals, i.e., 1.575 GHz (GPS), 1.8 GHz (DCS), 1.85 GHz (GSM), 1.9 GHz (DECT/PHS), and 2.1 GHz (UMTS).

In Mode 2, when the diode D_1 is in OFF condition and D_2 is in ON condition, the bandstop filter response is achieved with a –10 dB bandwidth ranging 1.9–2.4 GHz with a rejection level greater than 25 dB and an average loss of 0.77 dB. This bandstop filter mode is mainly designed to suppress signals from 2.1 GHz (UMTS) and 2.3 GHz (WCS) communication signal, as shown in Fig. 3b.

Table 2 Switching modes of proposed reconfigurable BP-BS filter

Modes	D_1	D_2	Function	Frequency Range (GHz)	Operating Frequency (GHz)	Bandwidth (%)	Applications
Mode 1	OFF	OFF	BSF	1.32–2.24	1.575, 1.8, 1.85, 1.9, 2.1	58.41, 51.11, 49.73, 48.42, 43.81 (10 dB FBW)	GPS, DCS, GSM, DECT/PHS, UMTS
Mode 2	OFF	ON	BSF	1.9–2.4	2.1, 2.3	23.81, 21.74 (10 dB FBW)	UMTS, WCS
Mode 3	ON	OFF	BPF	1.27–2.1	1.575, 1.8, 1.85, 1.9	52.70, 46.11, 44.86, 43.68 (3dB FBW)	GPS, DCS, GSM, DECT/PHS
Mode 4	ON	ON	BPF	2.02–2.7	2.1, 2.3, 2.4, 2.5	32.38, 29.57, 28.33, 27.2 (3 dB FBW)	UTMS, WCS, WLAN/ISM, WiMAX

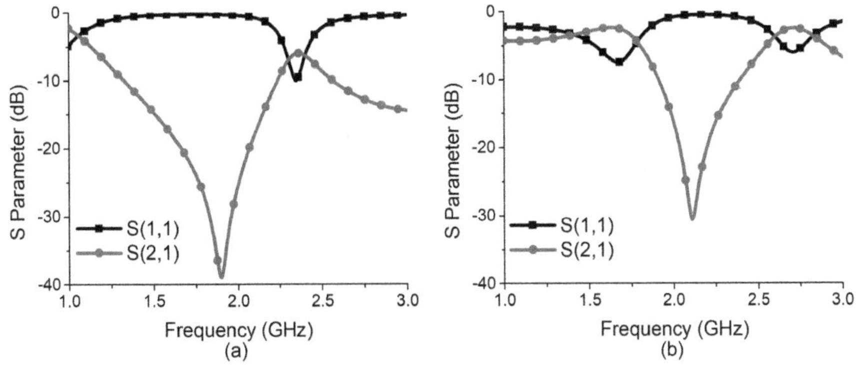

Fig. 3 Simulated S-parameter response for **a** Mode 1, **b** Mode 2

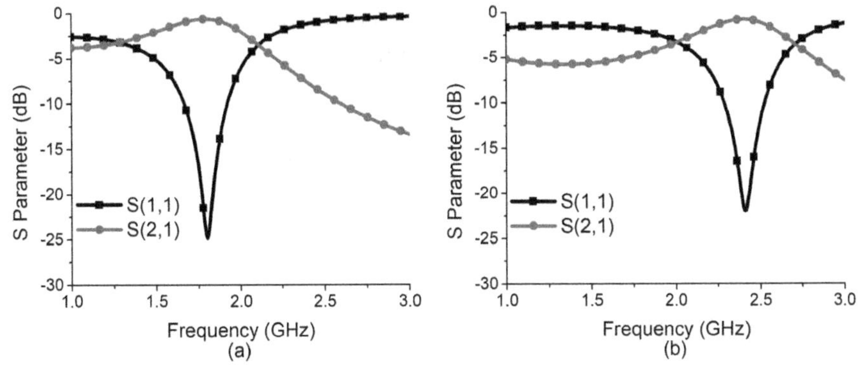

Fig. 4 Simulated S-parameter response for **a** Mode 3, **b** Mode 4

In Mode 3, when diode D_1 is in ON condition and diode D_2 is in OFF condition, a bandpass filter response has a passband zone of –3 dB ranging 1.27–2.1 GHz with an average insertion loss of around 0.7 dB while a return loss significantly lower than –20 dB is achieved. The simulated S-parameter of bandpass response for Mode 3 is presented in Fig. 4a. This bandpass filter is designed to pass signals ranging 1.27–2.1 GHz, which covers four communication signals, i.e., 1.575 GHz (GPS), 1.8 GHz (DCS), 1.85 GHz (GSM), and 1.9 GHz (DECT/PHS).

Finally, in Mode 4, when both diodes (D_1 and D_2) are simultaneously in ON condition, a bandpass filter response is achieved having –3 dB passband zone ranging 2.02–2.7 GHz with an average insertion loss of around 0.78 dB while having a return loss below –20 dB. Figure 4b shows the simulated S-parameter response of the bandpass filter for Mode 4. This bandpass filter is designed to pass signals ranging 2.02–2.7 GHz, which covers four communication signals, i.e., 2.1 GHz (UTMS), 2.3 GHz (WCS), 2.4 GHz (WLAN/ISM), and 2.5 GHz (WiMAX).

The phase properties of the filter have to be evaluated in order to estimate its transmission distortion. The group delay (GD) is calculated and displayed in Fig. 5a

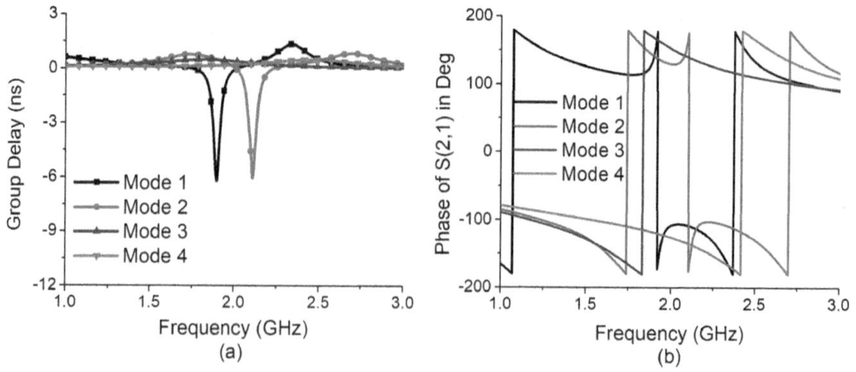

Fig. 5 **a** Group delay response, **b** Phase response of the filter

Table 3 Compares some other reconfigurable BP-BS filter with the current works

References	Dimension (mm^2)	Number of PIN diodes	Characteristics	Frequency tuning	Bandwidth tuning
[9]	22.5 × 24.3	4	BPF, BSF, ASF	Yes	Yes
[10]	26 × 15	5	BPF to BSF	Yes	Yes
[11]	21.8 × 22	1	BPF to BSF	No	Yes
[12]	55 × 60	2	BPF to BSF	Yes	Yes
[13]	55 × 60	2	BPF to BSF	Yes	Yes
This work	20 × 20	2	BPF to BSF	Yes	Yes

BPF: Bandpass Filter; BSF: Bandstop Filter; ASF: All-stop filter

to achieve these objectives. The group delay characteristics of the reconfigurable BP-BS filter are calculated and compared in Fig. 5a for different diode switching conditions. From the flat group delay response in the passband, it can be inferred that the transmission through the proposed filter is less distorted. For all the switching conditions, the observed group delay is less than 0.5 ns on the passband. Figure 5b displays the phase response of the reconfigurable BP-BS filter.

A detailed comparison of the reconfigurable BP-BS filter with other works around the globe is presented in Table 3. The comparison shows that the proposed reconfigurable BP-BS filter is compact in size and also needs only two PIN diodes to make it switchable. Both the attributes together make the proposed filter a better choice for researchers.

4 Conclusion

A compact reconfigurable BP-BS microwave filter is presented in this article. The filter structure is composed of a dual concentric U-shaped parasitic resonator. The proposed reconfigurable BP-BS filter is simulated on a low-cost FR-4 substrate having $\varepsilon_r = 4.3$, $\tan\delta = 0.025$, and a thickness of 0.8 mm, which is commercially available. Two PIN diode (SMP1320-079LF) cells are used to detect the transition by monitoring the forward bias (ON condition) and reverse bias (OFF condition). The proposed reconfigurable BP-BS filter provides a bandstop filter response with a rejection level greater than 25 dB and an average loss of 0.72 dB, while the bandpass filter response achieves an average insertion loss of around 0.75 dB and a return loss below –20dB. In addition, at passband, the estimated group delay is less than 0.5 ns for all switching conditions. The advantage of this reconfigurable BP-BS filter is the simultaneous control of center frequency and bandwidth, which is a significant tool for fulfilling modern device requirements.

Acknowledgements The authors would like to thank Sikkim Manipal University, Sikkim, India, for providing TMA Pai University Research Fund (Ref. No. 118/SMU/REG/UOO/ 104/2019) for this work.

References

1. Islam H, Sarkar S, Bose T, Das S (2020) Design of a novel reconfigurable Microstrip Bandpass filter with electronic switching. In: Advances in communication, devices and networking, vol 662, pp 191–197. Springer
2. Islam H, Dhar S, Das S, Bose T, Rai B (2020) Design of F-shaped reconfigurable Bandstop filter for interference resistant UWB communication. In: Advances in communication, devices and networking, vol. 662, pp 335–340. Springer
3. Islam H, Das S, Bose T, Ali T (2020) Diode based reconfigurable microwave filters for cognitive radio applications: a review. IEEE Access, 1
4. Matthaei GL, Young L, Jones EMT (1980) Microwave filters, impedance-matching networks, and coupling structures. Artech House
5. Karim MF, Liu AQ, Alphones A, Yu A (2007) A reconfigurable micromachined switching filter using periodic structures. IEEE Trans Microwave Theory Techn 55(6):1154–1162
6. Hong JSG, Lancaster MJ (2004) Microstrip filters for RF/microwave applications, vol 167. Wiley
7. Hsieh LH, Chang K (2002) Slow-wave bandpass filters using ring or stepped-impedance hairpin resonators. IEEE Trans Microwave Theory Techn 50(7):1795–1800
8. Tu WH (2010) Switchable microstrip bandpass filters with reconfigurable on-state frequency responses. IEEE Microwave Wirel Compon Lett 20(5):259–261
9. Xu J (2016) A microstrip switchable filter with four operating modes. IEEE Microwave Wirel Compon Lett 26(2):101–103
10. Mohamed HA, El-Shaarawy HB, Abdallah EA, El-Hennawy HM (2015) Frequency-reconfigurable microstrip filter with dual-mode resonators using rf pin diodes and dgs. Int J Microwave Wirel Technol 7(6):661–669
11. Rabbi K, Athukorala L, Panagamuwa C, Vardaxoglou JC, Budimir D (2013) Highly linear microstrip wideband bandpass filter with switchable notched band for wireless applications. Microwave Opt Technol Lett 55(6):1331–1335

12. Arain S, Abbassi MAB, Nikolaou S, Vryonides P (2015) A Reconfigurable Bandpass to Bandstop filter using pin diodes based on the square ring resonator. In: Progress in electromagnetics research symposium proceedings, pp 1415–1418. Electromagnetics Academy
13. Nachouane H, Najid A, Tribak A, Riouch F (2017) A switchable bandstop-to-bandpass reconfigurable filter for cognitive radio applications. Int J Microwave Wirel Technol 9(4):765
14. SMP1320 Series: Low Resistance, Low Capacitance, Plastic Packaged PIN Diodes . https://www.skyworksinc.com/-/media/SkyWorks/Documents/Products/101-200/SMP1320_Series_200047S.pdf. Accessed 15 Oct 2020

A Miniaturized Four-Port Annular MIMO Antenna for Ultra-Wideband Frequency Range Application

Jai Mangal and Mansi Parmar

Abstract This paper contemplates the potential outcomes to develop an adaptable, lightweight, miniaturized and precisely vigorous four-port MIMO antenna for the applications over WiMAX (2.5, 3.5 GHz) and unlicensed ultra-wideband frequency range of 3.1–10.6 GHz. The proposed antenna achieves a reflection coefficient <−10 dB in the frequency range of 2.5–11 GHz. The dimensions of the proposed antenna are 24 mm × 24 mm × 1.6 mm. The patch antenna is fabricated over a substrate material called FR-4. This paper introduces a concept of annular patches, diagonal arrangement of circles and partial grounds, which are helpful in achieving the wide range of operating frequencies with miniaturized dimensions. The proposed antenna achieves a maximum peak gain of −0.1 dBi at 8.7 GHz and minimum radiation efficiency of 88% in the entire operating band. The innovative antenna permits coordination with portable UWB high-throughput application devices.

Keywords Annular patch · MIMO antenna · Miniaturized · Partial ground · Ultra-wideband antenna · Unlicensed frequency · WiMAX

1 Introduction

In today's world, with the increasing number of subscribers and load on mobile communication channels where the higher speed of internet is needed at every instant of time, the use of MIMO antennas becomes very important. To attain the higher throughput and connectivity at every instant even while the subscriber is continuously moving, it is very important to come up with a miniaturized multiple-input multiple-output antenna. An MIMO antenna takes multiple inputs and provides multiple

J. Mangal (✉) · M. Parmar
ECE, VIT Bhopal University, Bhopal, India
e-mail: jai.mangal2018@vitbhopal.ac.in

M. Parmar
e-mail: mansi.parmar2018@vitbhopal.ac.in

S. Dhar et al. (eds.), *Advances in Communication, Devices and Networking*, Lecture Notes in Electrical Engineering 776,
https://doi.org/10.1007/978-981-16-2911-2_22

outputs. Whenever any signal reaches toward the antenna, it always takes the signal with high SNR and provides a maximum throughput to the communication link.

The antenna is said to be the backbone of communication. The MIMO antenna can be taken as the backbone of future generation communication. The major advantage of the MIMO antenna is that it provides data bandwidth. In today's world, we connect at least four devices with a home Wi-Fi. The MIMO antenna will provide parallel and equal bandwidth to all four devices. The MIMO antennas manage to provide increased data rates by using spatial multiplexing and diversity.

The beauty of MIMO is that it has every option to suit all scenarios. In modern wireless standards like IEEE 802.11n, 3GPP LTE and mobile WiMAX systems, MIMO antennas were used widely for communication. The MIMO antennas were also used in 3GPP UMTS, HSPA+ , LTE. It has become the modern-day trend and necessity of the communication system to use MIMO antennas. Further improvement in the MIMO antennas can be incorporated by making them miniaturized and increase their operating bandwidth over a wide frequency.

2 Related Works

The antenna design and dimensions are observed from many papers [1–10]. In [1], the authors proposed a compact ultra-wideband MIMO antenna with WLAN band-rejected operation for mobile devices. The dimensions of the proposed antenna are 55 mm × 86.5 mm × 0.8 mm. The antenna is fabricated over the substrate known as FR4 epoxy. It is provided with two input ports. It resonates at the resonating frequency of 6.1 GHz. The antenna attains the reflection coefficient of −35 dB. The −10 dB bandwidth of the proposed design is from 1.92 to 10.6 GHz. The antenna attains the peak gain of 4.96 dBi. The peak gain frequency comes out to be 5 GHz.

In [2], the authors proposed a compact ultra-wideband MIMO antenna with a half slot structure. The dimensions of the proposed antenna are 23 mm × 18 mm × 0.8 mm. The antenna is fabricated over the substrate known as F4b-2. It is provided with two inputs. It attains the −10 dB bandwidth ranging from 3 to 12.4 GHz. The antenna resonates at the resonating frequency of 11.1 GHz. It attains the reflection coefficient of −35 dB. It attains the peak gain of 4 dBi. The peak gain frequency comes out to be 11 GHz.

In [3], the authors proposed a compact ultra-wideband MIMO antenna using QSCA for high isolation. The dimensions of the proposed antenna are 60 mm × 41 mm × 1 mm. The antenna is fabricated over the substrate called FR4 epoxy. It is provided with four inputs. It covers the −10 dB bandwidth ranging from 3.1 to 10.6 GHz. The antenna resonates at the resonating frequency of 5.6 GHz. It attains the reflection coefficient of −35 dB.

In [4], the authors proposed a compact CPW-fed ultra-wideband MIMO antenna using hexagonal ring monopole antenna elements. The dimensions of the proposed antenna are 47 mm × 47 mm × 0.8 mm. The antenna is fabricated over the substrate called FR4 epoxy. It is provided with four inputs. It attains the −10 dB bandwidth ranging from 2.3 to 11.2 GHz. The antenna resonates at the resonating frequency of 4.6 GHz. It attains the reflection coefficient of −45 dB.

In [5], the authors proposed a compact MIMO slot antenna for ultra-wideband applications. The dimensions of the proposed antenna are 42 mm × 25 mm × 1.6 mm. The antenna is fabricated over the substrate called FR4 epoxy. It is provided with four input ports. It attains the −10 dB bandwidth ranging from 3.2 to 12 GHz. The antenna operates at the resonating frequency of 3.5 GHz. It attains the reflection coefficient of −20 dB. It achieves the peak gain of 6.2 dBi. The antenna peak gain frequency comes out to be 11 GHz.

In [6], the authors proposed a compact MIMO antenna for portable devices in ultra-wideband applications. The dimensions of the proposed antenna are 26 mm × 40 mm × 0.8 mm. The antenna is fabricated over the substrate called Roger. It is provided with two inputs. It attains the −10 dB bandwidth ranging from 3.1 to 10.6 GHz. The antenna operates at the resonating frequency of 9.5 GHz. It attains the reflection coefficient of −25 dB. It achieves the peak gain of 6 dBi. The peak gain frequency comes out to be 11 GHz.

In [7], the authors proposed an MIMO antenna with a built-in circular-shaped isolator for sub-6 GHz 5G applications. The dimensions of the proposed antenna are 50 mm × 45 mm × 0.787 mm. The antenna is fabricated over the substrate called FR4 epoxy. It is provided with four inputs. It attains the −10 dB bandwidth ranging from 2.7 to 10.9 GHz. It operates at resonating frequency of 8.3 GHz. The antenna attains the reflection coefficient of −42 dB.

In [8], the authors proposed a compact printed MIMO antenna for ultra-wideband applications. The dimensions of the proposed antenna are 32 mm × 32 mm × 0.8 mm. The antenna is fabricated over the surface called FR4 epoxy. It is provided with two input ports. It attains the −10 dB bandwidth ranging from 2.9 to 12 GHz. It operates at a resonating frequency of 6.8 GHz. The antenna attains the reflection coefficient of −28 dB. It attains the peak gain of 3.5 dBi. The antenna peak gain frequency comes out to be 10 GHz.

In [9], the authors proposed a compact 4 × 4 MIMO antenna for ultra-wideband applications. The dimensions of the proposed antenna are 35 mm × 35 mm × 0.8 mm. The antenna is provided with four inputs. Its patch is made of a funnel-like structure. It attains the −10 dB bandwidth ranging from 3.2 to 12 GHz. The antenna operates at the resonating frequency of 3.4 GHz. It attains the reflection coefficient of −18 dB. It achieves the peak gain of 4.8 dBi. The antenna peak gain frequency comes out to be 12 GHz.

In [10], the authors proposed an ultra-wideband MIMO diversity antenna with a tree-like structure to enhance wideband isolation. The dimensions of the proposed antenna are 35 mm × 40 mm × 0.8 mm. The antenna is provided with two input ports. It is fabricated over the substrate called Taconic ORCER RF 35. The antenna attains the −10 dB bandwidth ranging from 3.1 to 10.6 GHz. It operates at the resonating frequency of 4.3 GHz. It attains the reflection coefficient of −20 dB. It achieves the peak gain of 6.2 dBi. The antenna peak gain frequency comes out to be 10.6 GHz.

The aim of this study is to design and implement a four-port MIMO antenna using HFSS in the WiMAX and unlicensed ultra-wideband frequency range for portable device applications. The proposed antenna achieves a reflection coefficient of < -10 dB in the frequency range of 2.5–11 GHz by using the annular patch structure over the substrate. The proposed UWB antenna exhibits stable wideband characteristics. Hence, it can be used in many UWB applications, like WBAN, WLAN, WiMAX, cognitive radio spectrum sensing, wireless sensor networks and many more.

3 Proposed Work

This section discusses the design of the proposed four-port MIMO antenna for ultra-wideband frequency applications with the combination of circular patches arranged in a diagonal manner. The dimensions of the proposed antenna design are 24 mm $\times$ 24 mm $\times$ 1.6 mm. In the top view, as shown in Fig. 1, the structure consists of a diagonally placed annular patch using copper material by placing six circular patches. Each circle is 1.5 mm in radius and identical in nature. The dimensions of the microstrip-feeded transmission line are 5 mm $\times$ 2.5 mm $\times$ 0.035 mm.

All the four antennas that were made on the upper patch were identical in nature but aligned in a different manner. The concept of partial ground is used in this paper. To achieve the partial ground five rectangular slots were positioned in the ground plane. This concept of partial ground is introduced to get the reflection coefficient below -10 dB over a wide range of frequencies. The proposed UWB MIMO antenna top view is shown in Fig. 1. The proposed UWB MIMO antenna bottom view is shown in Fig. 2.

The simulated reflection coefficient curve of the proposed antenna is shown in Fig. 3.

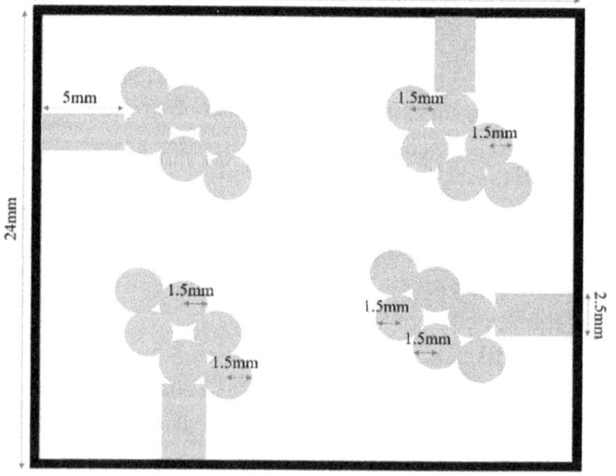

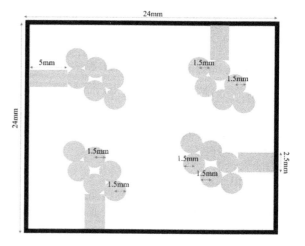

Fig. 1 Top view of the proposed antenna

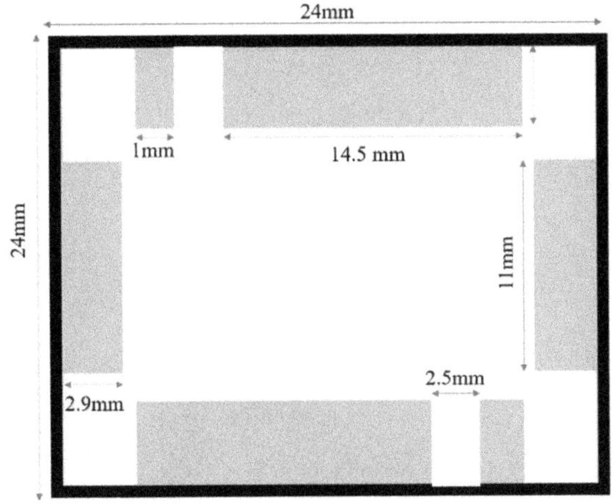

Fig. 2 Bottom view of the proposed antenna

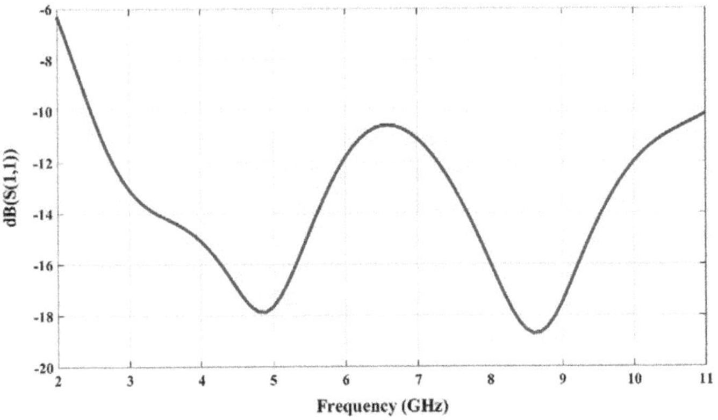

Fig. 3 Simulated reflection coefficient of the MIMO antenna

From Fig. 3 it can be clearly seen that the antenna attains the reflection coefficient below −10 dB over an unlicensed UWB frequency range, i.e., from 2.5 to 11 GHz.

The simulated peak gain of the antenna is shown in Fig. 4.

From Fig. 4 it can be noted that the maximum peak gain was found to be −0.19 dBi at 8.7 GHz. The 3D radiation pattern at 8.7 GHz is shown in Fig. 5.

Fig. 5 also confirms that the gain at 8.7 GHz is −0.19 dBi.

Figure 6 shows the 2D radiation pattern of the proposed antenna at a frequency of 8.7 GHz. It shows the gain for both the phases, i.e., for $\phi = 0°$ and $\phi = 90°$.

The current distribution of the proposed antenna at the frequency of 8.7 GHz is shown in Fig. 7.

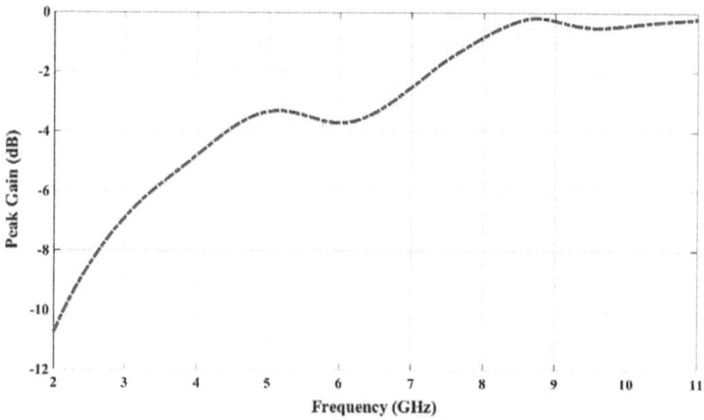

Fig. 4 Simulated peak gain of the MIMO antenna

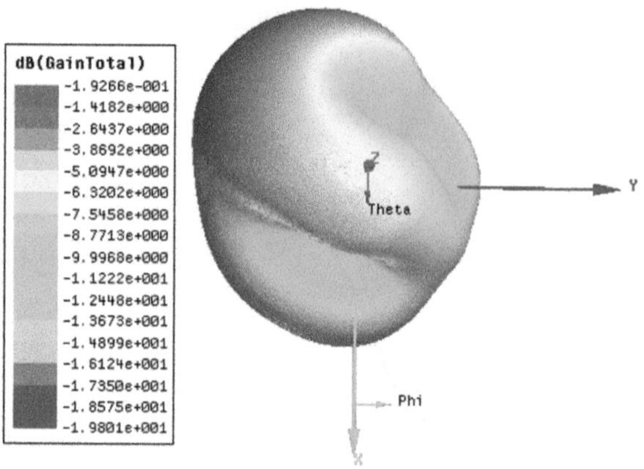

Fig. 5 3D radiation pattern at 8.7 GHz

The radiation efficiency over the operating range is shown in Fig. 8. Figure 8 depicts that a minimum radiation efficiency of 88% is found in the entire operating band. Also, at lower frequencies the radiation efficiency is high and it is decreasing as frequency increases.

The fabricated UWB MIMO antenna top view is shown in Fig. 9.

The fabricated UWB MIMO antenna bottom view is shown in Fig. 10.

The measured reflection coefficient curve of the fabricated antenna is shown in Fig. 11.

The summary of antenna design and parameters is shown in Table 1.

A summary table that shows the literature work and the proposed antenna is shown in Table 2.

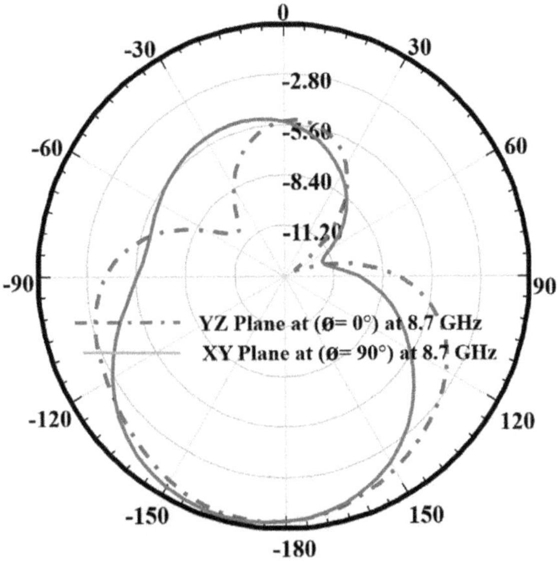

Fig. 6 2D radiation pattern at 8.7 GHz

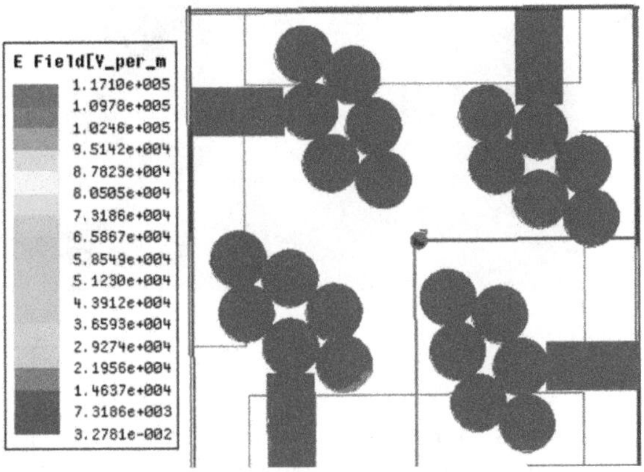

Fig. 7 Surface current distribution on the proposed antenna at 8.7 GHz

4 Conclusion

This paper discusses 24 mm × 24 mm × 1.6 mm size four-port ultra-wideband MIMO antenna performance characteristics for unlicensed frequency range applications. This antenna has achieved the reflection coefficient of −10 dB in the unlicensed UWB 2.5–11 GHz frequency range. The design is able to achieve the maximum

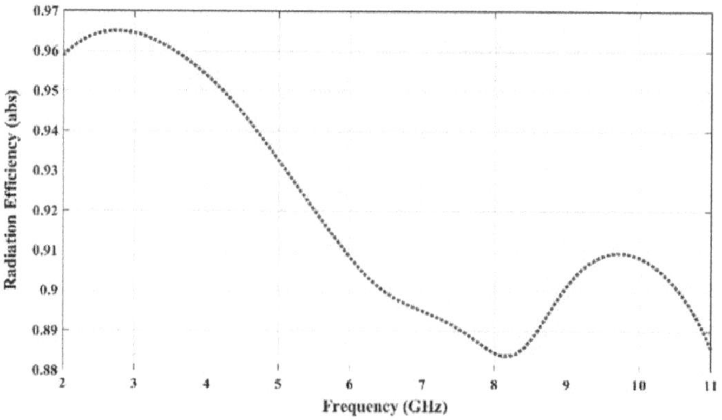

Fig. 8 Radiation efficiency over the UWB

Fig. 9 Top view of the fabricated antenna

peak gain of –0.19 dBi at 8.7 GHz and minimum radiation efficiency of 88% in the entire operating band. Because of its miniaturized size and best performance this antenna is best suitable for the unlicensed ultra-wideband frequency range.

Fig. 10 Bottom view of the fabricated antenna

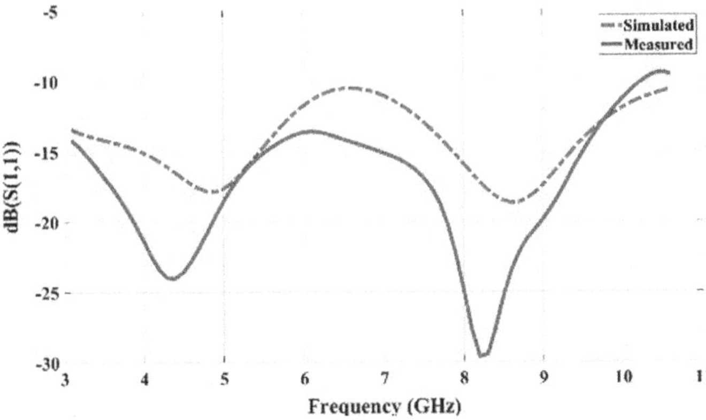

Fig. 11 Measured reflection coefficient of the MIMO antenna

Table 1 Antenna parameter table

Desired parameter	Calculated value
Dimensions of substrate	24 mm × 24 mm × 1.6 mm
Thickness of patch	0.035 mm
Substrate material	FR4 epoxy
Patch material	Copper
Bandwidth	2.5–11 GHz
Peak gain	−0.19 dB
Peak gain frequency	8.7 GHz
Polarization	Linear

Table 2 Literature review summary table

References	Dimensions (mm^3)	B.W. (GHz)	No. of input ports
[1]	55 × 86.5 × 0.8	1.92–10.6	2
[2]	23 × 18 × 0.8	3–12.4	2
[3]	60 × 41 × 1	3.1–10.6	4
[4]	47 × 47 × 1	2.3–11.2	4
[5]	42 × 25 × 1.6	3.2–12	4
[6]	26 × 40 × 0.8	3.1–10.6	2
[7]	50 × 40 × 0.787	2.7–10.9	4
[8]	32 × 32 × 0.8	2.9–12	2
[9]	35 × 35 × 0.8	3.2–12	4
[10]	35 × 40 × 0.8	3.2–10.6	2
Proposed	24 × 24 × 1.6	2.5–11	4

References

1. Lee JM, Kim KB, Ryu HK, Woo JM (2012) A compact ultrawideband MIMO antenna with WLAN band-rejected operation for mobile devices. IEEE Antennas Wirel Propag Lett 11:990–993
2. Tao J, Feng Q (2016) Compact ultrawideband MIMO antenna with half-slot structure. IEEE Antennas Wirel Propag Lett 16:792–795
3. Liu XL, Wang ZD, Yin YZ, Ren J, Wu JJ (2014) A compact ultrawideband MIMO antenna using QSCA for high isolation. IEEE Antennas Wirel Propag Lett 13:1497–1500
4. Mathur R, Dwari S (2018) Compact CPW-fed ultrawideband MIMO antenna using hexagonal ring monopole antenna elements. AEU-Int J Electron Commun 93:1–6
5. Srivastava G, Mohan A (2015) Compact MIMO slot antenna for UWB applications. IEEE Antennas Wirel Propag Lett 15:1057–1060
6. Liu L, Cheung SW, Yuk TI (2013) Compact MIMO antenna for portable devices in UWB applications. IEEE Trans Antennas Propag 61(8):4257–4264
7. Femina Beegum S, Mishra SK (2016) Compact WLAN band-notched printed ultrawideband MIMO antenna with polarization diversity. Progr Electromagn Res 61:149–159
8. Ren J, Hu W, Yin Y, Fan R (2014) Compact printed MIMO antenna for UWB applications. IEEE Antennas Wirel Propag Lett 13:1517–1520
9. Wani Z, Kumar D (2016) A compact 4 × 4 MIMO antenna for UWB applications. Microw Opt Technol Lett 58(6):1433–1436
10. Zhang S, Ying Z, Xiong J, He S (2009) Ultrawideband MIMO/Diversity antennas with a tree-like structure to enhance wideband isolation. IEEE Antennas Wirel Propag Lett 8:1279–1282

TM Mode Analysis in Optical Waveguide Study

Anup Kumar Thander and Sucharita Bhattacharyya

Abstract An accurate and efficient polarized mode solver in the form of a higher (fourth) order compact finite difference method (FDM) along with the successive over-relaxation technique is presented here for transverse magnetic or TM mode analysis of propagating electromagnetic waves through rib-structured GaAs and GeSi waveguides. For analysis of performances of such waveguides, the effective index concept has been identified as a key parameter. Accordingly, modal/effective index, normalized index and modal birefringence parameters are introduced in the present study and their obtained values show close agreement with other published data. Most significantly, the scheme successfully identified the optimized values of used waveguide structure parameters and explained their importance in the efficient propagation of transverse electromagnetic waves which might play a crucial role in the fabrication of such waveguides.

Keywords Polarized mode solver · Higher (fourth) order compact · Rib-structured waveguide · Finite difference method · Successive over-relaxation technique

1 Introduction

In the case of communication and signals processing, the optical beam is chosen as a carrier for which waveguide is used as a basic guiding component. To realize the transmission properties of various waveguides appropriately, their geometric shapes and corresponding refractive index profiles are studied in detail [1–12]. Specially structured rib waveguides have drawn much attention in recent days, particularly, to mention, in the field of photonics and opto-electronics because of their functional variations at the output, though their fabrications are easier. It may be noted that even

A. K. Thander · S. Bhattacharyya (✉)
Department of Applied Science and Humanities, Guru Nanak Institute of Technology, Kolkata 700114, India
e-mail: sucharita.bhattacharyya@gnit.ac.in

A. K. Thander
e-mail: anup.thander@gnit.ac.in

S. Dhar et al. (eds.), *Advances in Communication, Devices and Networking*, Lecture Notes in Electrical Engineering 776,
https://doi.org/10.1007/978-981-16-2911-2_23

for the case of waveguides with the simplest geometries, a single analytical approach is not sufficient to optimize the designs as per the requirements. As a result, a wide class of numerical and statistical methods has been developed to study them using their modal analysis.

Essentially, almost all waveguides support two modes of independent polarizations: the transverse electric (TE) and the transverse magnetic (TM) field components. The developed scheme is already being used successfully for TE mode analysis [1–4]. Inspired by the result it is now used here to determine the polarized electric field solutions of the resulting wave equation of Helmholtz in the TM mode. In this work, GeSi and GaAs rib-structured waveguides are used where normalized index and modal index concepts have been introduced [3]. With the change of materials, how the structure parameter optimization affects the transmission characteristics of the propagating waves is clearly shown here. After a brief introduction, the theoretical concept and numerical schemes are discussed in Sect. 2. Analysis of the results is done in Sect. 3 considering the variation in the parameters of the used rib structures. The results of other researchers agree closely with the present work. In Sect. 4, an overall conclusion is given.

2 Theoretical Concept and Numerical Simulations

Figure 1 shows the structure of the used rib waveguide. The inner rib height and rib width used here are represented by h_r and w, respectively. The rib outer region thickness/rib depth is d, the height of the rib is h and λ is the optical wavelength in the free space [4]. The core, guiding and substrate regions used in the structure have refractive indices of n_C, n_G and n_S, respectively, at the chosen λ value. Inside the rib,

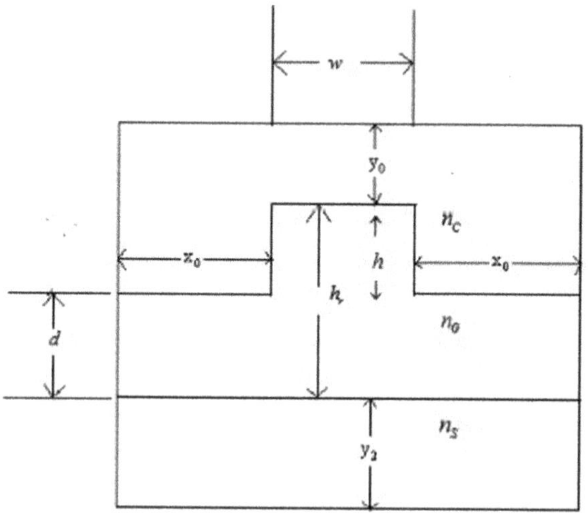

Fig. 1 Rib waveguide structure [1]

harmonic wave propagation is considered along the z-direction. Then by following the usual procedure [1–5], Helmholtz wave equations are obtained here as the eigen solutions in the form as follows:

$$\left[\nabla_T^2 + k^2\right]\vec{F} = \beta^2\,\vec{F} \tag{1}$$

Here $\vec{E}$ or $\vec{H}$ can replace $\vec{F}$ to consider either the electric field or the magnetic field.

For modal analysis, Eq. (1) is solved numerically by compact fourth-order FDM [1–4, 12], where the used solution domain (rectangular in shape) is decomposed into uniform discrete nine-point stencils with grid points, where h_x and h_x are considered as the h_y length and x length, respectively (Fig. 2). Here, the internal grid points of the decomposed rib waveguide solution domains are positioned at the center of the cells where the refractive indices are permitted to change at the borders of the cells only (Fig. 2). The refractive indices of adjacent cell borders are related and are obtained following the work of Stern [5]. Finite difference mesh points used here are scanned to minimize the bandwidths of the coefficient matrices of the algebraic eigenvalue system (corresponding to governing wave equation) by fully utilizing geometrical symmetry.

It is observed that continuity of E_z across all horizontal and vertical cell boundaries (see Fig. 2) implies continuity of $\frac{\partial E_y}{\partial y}$, $\frac{\partial E_y}{\partial x}$ across the same dielectric interfaces. Hence quasi-TM modes are the eigen solutions of the scalar Helmholtz equation

$$\left[\nabla_T^2 + k^2\right]E_y = \beta^2 E_y \tag{2}$$

with E_y maintaining continuity across the interfaces parallel to the $y - z$ plane but remains discontinuous over all interfaces parallel to the $x - z$ plane, whereas

Fig. 2 Nine-point stencil of the field domain [1]

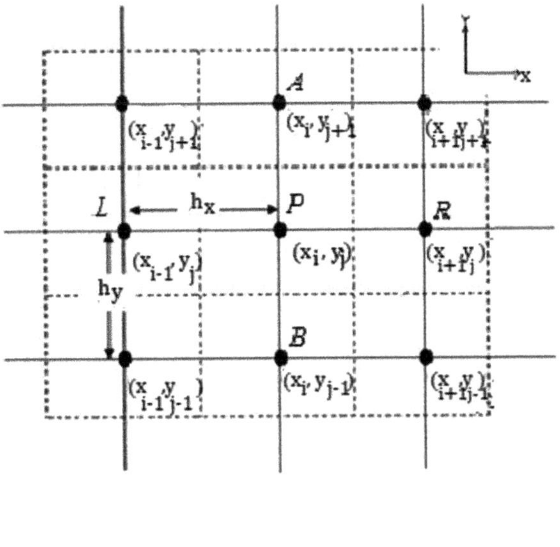

E_x maintains continuity over the interfaces parallel to its $x - z$ plane but remains discontinuous across the interfaces parallel to the $y - z$ plane.

Following the scheme developed [1–3], a system of algebraic equations with a nine-diagonal coefficient matrix is formed as

$$\sum_{\substack{k=-1,0,1\\ l=-1,0,1}} W_{i,j}^{k,l} E_{i+k,j+l} = 0 \tag{3}$$

where

$$W_{i,j}^{1,1} = a_{i,j}\left[\frac{2k_{i+1,j+1}^2}{k_{i+1,j+1}^2 + k_{i+1,j}^2}\right]$$

$$W_{i,j}^{1,-1} = b_{i,j}\left[\frac{2k_{i+1,j-1}^2}{k_{i+1,j-1}^2 + k_{i+1,j}^2}\right]$$

$$W_{i,j}^{-1,1} = c_{i,j}\left[\frac{2k_{i-1,j+1}^2}{k_{i-1,j}^2 + k_{i-1,j+1}^2}\right]$$

$$W_{i,j}^{-1,-1} = d_{i,j}\left[\frac{2k_{i-1,j-1}^2}{k_{i-1,j}^2 + k_{i-1,j-1}^2}\right]$$

$$W_{i,j}^{-1,0} = c_{i,j}\left[\frac{k_{i-1,j+1}^2 - k_{i-1,j}^2}{k_{i-1,j+1}^2 + k_{i-1,j}^2}\right] + f_{i,j}\left[\frac{2k_{i-1,j}^2}{k_{i,j}^2 + k_{i-1,j}^2}\right] + d_{i,j}\left[\frac{k_{i-1,j-1}^2 - k_{i-1,j}^2}{k_{i-1,j-1}^2 + k_{i-1,j}^2}\right]$$

$$W_{i,j}^{1,0} = a_{i,j}\left[\frac{k_{i+1,j+1}^2 - k_{i+1,j}^2}{k_{i+1,j+1}^2 + k_{i+1,j}^2}\right] + q_{i,j}\left[\frac{2k_{i+1,j}^2}{k_{i,j}^2 + k_{i+1,j}^2}\right] + b_{i,j}\left[\frac{k_{i+1,j-1}^2 - k_{i+1,j}^2}{k_{i+1,j-1}^2 + k_{i+1,j}^2}\right]$$

$$W_{i,j}^{0,-1} = h_{i,j}\left[\frac{2k_{i,j-1}^2}{k_{i,j}^2 + k_{i,j-1}^2}\right]$$

$$W_{i,j}^{0,1} = g_{i,j}\left[\frac{2k_{i,j+1}^2}{k_{i,j}^2 + k_{i,j+1}^2}\right]$$

$$W_{i,j}^{0,0} = q_{i,j}\left[\frac{k_{i+1,i}^2 - k_{i,i}^2}{k_{i+1,i}^2 + k_{i,i}^2}\right] + f_{i,j}\left[\frac{k_{i-1,i}^2 - k_{i,i}^2}{k_{i-1,i}^2 + k_{i,i}^2}\right] + g_{i,j}\left[\frac{k_{i,i+1}^2 - k_{i,i}^2}{k_{i,i+1}^2 + k_{i,i}^2}\right] + h_{i,j}\left[\frac{k_{i,i-1}^2 - k_{i,i}^2}{k_{i,i-1}^2 + k_{i,i}^2}\right] - m_{i,j}$$

Here, it may be noted that the finite difference scheme for the TE mode of an electric field [2, 3] has been developed similarly.

The finite difference scheme represented by Eq. (3) approximating Eq. (2) is giving the eigenvalue equation in the algebraic form as

$$A_{TM}E = \beta_{TM}^2 E$$

in which A_{TM} is the non-symmetric and real band matrix, β_{TM}^2 is the eigenvalue related to propagation. Obviously, here normalized eigenvector E represents the field profile in TM mode. Zero-field condition in the outer border of the solution domain is considered in the work.

3 Results and Discussion

The developed scheme including the Rayleigh quotient solution formula is applied here to estimate the modal/effective index (n_{eff}) and the normalized index (b). The structure parameters of the GeSi and GaAs waveguides for structures 1–3 (S1–S3) as considered are shown in Table 1 [1–4].

It is found from Table 2 that the obtained values of the normalized index and effective refractive index (modal) in both TE and TM modes for electric field agree reasonably well with the corresponding results of Stern [5] for the structures under consideration. At $\lambda = 1.55$, the values of b are found the highest for structure 1 and decrease for structures 2 and 3, which implies lesser vertical spread for the horizontal as well as vertical polarization modes TE and TM of electric and magnetic fields' propagation for structure 1, strongly guiding one, compared to the structures 2 and 3 (S2 and S3).

Table 1 Optical and geometrical structure parameters of GaAs and GeSi rib waveguides [2]

Structure	n_G		n_S		n_C		h_r	h	d	w	x_0	y_0	y_2	λ
	GaAs	GeSi	GaAs	GeSi	GaAs	GeSi								
S1	3.44	3.6	3.34	3.5	1.0	1.0	1.3	1.1	0.2	2	3.00	0.525	5.025	1.55
S2	3.44	3.6	3.36	3.5	1.0	1.0	1.0	0.1	0.9	3	3.05	0.525	5.025	1.55
S3	3.44	3.6	3.435	3.5	1.0	1.0	6.0	2.5	3.5	4	4.34	0.550	7.550	1.55

Table 2 Comparison of the effective index (n_{eff}) and normalized index (b) values for E field in structures 1–3 [6]

	Stern [5]				Present work			
	TE mode		TM mode		TE mode [1, 2]		TM mode	
Structure	n_{eff}	b	n_{eff}	b	n_{eff}	b	n_{eff}	b
S1	3.3869266	0.4655	3.3867447	0.46377	3.3902787	0.4991	3.38824	0.4788
S2	3.3953942	0.4395	3.39055	0.3792	3.3958327	0.4449	0.389573	0.3669
S3	3.4366674	0.3333	3.4266	0.3274	3.4367243	0.3446	3.43672	0.3441

The basic requirement of the opto-electronics fabrication [9] industry is to develop such waveguide structures that can produce efficient output in a cost-effective way. Accordingly, the behavior of the waveguides in the transmission of the propagating EM waves [6–8] needs to be optimized, both in terms of structure parameters and of the properties of the materials. Hence, to design the waveguide devices, various factors are studied to choose their optimized structure parameters obviously, based on the simulation results. So in this analysis, the variation in normalized index and effective index [1–6] for GaAs and GeSi in the TM mode with outer rib thickness d for a particular rib structure S2 are studied, and their results are shown in Table 3. It may be mentioned that similar study can be extended to other two structures and structure parameters like rib width w.

Analysis of these results is done to identify the optimized values of waveguide structure parameters for different materials. The range of d is chosen here from 0.6 to 1 μm and the variation of n_{eff} and b with d shows that both increase with increasing rib depth for GeSi and GaAs but shows material dependence with their different values, as expected at a particular wavelength ($\lambda = 1.55$ μm). For $d \leq 0.8$ μm, in each case, the b value for GeSi is larger than that for GaAs, but for $d \geq 0.9$ μm, the situation becomes opposite where a larger spread in the vertical direction (as b is smaller) is estimated for GeSi than GaAs. It shows that the gallium-arsenide (GaAs) has a better confinement capability at this particular wavelength for TM mode also, like TE mode [4]. The results thus confirm that GaAs material works in a better way than germanium-silicon (GeSi) material for the confinement of transmitting waves at the specified wavelength for both TE and TM polarizations. This behavior is related to the fact that the electron mobility of silicon material [3, 6, 7] is intrinsically lower

Table 3 n_{eff} and b variation with rib depth d

	GaAs		GeSi	
d	n_{eff}	b	n_{eff}	b
0.6	3.37851	0.2293	3.52326	0.2301
0.7	3.38222	0.2755	3.52801	0.2773
0.8	3.38585	0.3206	3.53258	0.3228
0.9	**3.38957**	**0.3669**	**3.53675**	**0.3642**
1.0	3.39294	0.4089	3.54093	0.4059

Table 4 Birefringence (modal) comparison with d, the rib depth for S2 structure in E-field

Rib depth/Inner rib height (d) in μm	Δn_{eff} in GaAs	Δn_{eff} in GeSi
0.6	−0.014044202	−0.011919097
0.7	−0.001051407	0.001144337
0.8	0.005054042	0.006965338
0.9	0.00626154	0.008086595
1.0	0.005558052	0.006865843

than GaAs material. Also, the direct energy band gap of GaAs material may be related to the recombination of electrons and holes more efficiently.

The phenomenon of modal birefringence [11] is also studied in this work (Table 4) for the E-field propagation to have an insight into the effect of polarization on transmitting waves. For a chosen λ and the refractive index profile (inside the materials) of a rib waveguide, d, the rib depth is again considered to adjust as a critical parameter to realize the condition of birefringence.

For TE and TM modes in S2 structure, positive birefringence is obtained for $d \geq 0.8\ \mu m$ in the case of GaAs and for $d \geq 0.7\ \mu m$ in GeSi, which gives an idea about the change in polarization characteristics of the chosen materials for higher d closer to its optimized value of $0.9\ \mu m$ (see [2, 3, 6]).

4 Conclusions

In this research work, the authors tried to look at certain aspects of optical waveguide communications within the framework of electromagnetic field theory where various semiconductor optical waveguides are used as fundamental components for the analysis of corresponding TM modes.

For that purpose, the HOC type FDM is considered along with the successive over-relaxation (SOR) iteration scheme. The normalized index, as well as the effective index values, of a typical rib waveguide structure for material systems of GeSi and GaAs is obtained as output. The results show good agreement with other available published works. Also, the present method signifies the role of parameter optimization on the propagating waves through rib structures successfully. The information can be taken into consideration at the time of fabrication of such waveguides which are normally used in miniaturization and integration of the communication systems in a cost-effective way.

Here it may be noted that the authors have already started extending their works for two-rib structure in addition to the slab waveguide case whose results are expected to be reported very soon.

Acknowledgements The authors are really grateful to the DST, Govt. of India for the research fund (grant no. SR/FST/COLLEGE-/2018/466) required to carry out this work, and to the Guru Nanak Institute of Technology, Kolkata authority, for providing the infrastructure.

References

1. Thander AK, Bhattacharyya S (2014) Study of optical wave guide using HOC scheme Appl Math Sci 8(79):3931–3938
2. Thander AK, Bhattacharyya S (2015) Optical wave guide analysis using higher order compact FDM in combination with Newton Krylov subspace methods. In: 2015 IEEE international conference on research in computational intelligence and communication networks, pp 66–71

3. Thander AK, Bhattacharyya S (2016) Optical Confinement study of different semi conductor rib wave guides using higher order compact finite difference method. Optik 127:2116–2120
4. Thander AK, Bhattacharyya S (2017) Study of optical modal index for semi conductor rib wave guides using higher order compact finite difference method. Optik 131:775–784
5. Stern MS (1988) Semivectorial polarised finite difference method for optical waveguides with arbitrary index profiles. IEE Proc-Optoelectron 135(1):56–63
6. Bhattacharyya S, Thander AK (2019) Study of H-field using higher order compact (HOC) finite difference method in Semiconductor rib waveguide structure. J Opt 48:1–12
7. Robertson MJ, Ritchie S, Dayan P (1985) Semiconductor waveguides: analysis of optical propagation in single rib structures and directional couplers. IEE Proc J 132:336–342
8. Soref RA, Schmidtchen J, Petermann K. Large Single-Mode rib waveguides in GeSi-Si and Si-on-Sio_2. IEEE J Quant Electron 27(8)
9. Huang W, Hermann Hauss A (1991) A simple variational approach to optical rib waveguides. J Lightwave Technol 9(1):56–61
10. Hoffman JD (1992) Numerical methods for engineers and scientists. Marcel Dekker, Inc., New York
11. Rahman BMA (2005) Birefringence compensation of Silica waveguides. IEEE Photonics Technol Lett 17(6):1205–1207
12. Bhattacharyya S, Thander AK (2018) Newton–Krylov subspace method to study structure parameter optimization in rib waveguide communication. In: International conference on industry interactive innovations in science, engineering and technology. Lecture notes in networks and systems, vol 11, pp 229–238

Enhanced Performance of Microstrip Antenna with Meta-material: A Review

Neetu Agrawal

Abstract Meta-material is an artificial or unreal material with extraordinary electromagnetic properties (negative permeability and negative permittivity) that are not present in the existing materials. Likewise, meta-material structures are typically placed close to the patch, inserted in the substrate, etched from the ground plane or installed as a substrate layer to maximize bandwidth and gain, and also miniaturized in the size of traditional patch antennas. In modern wireless commutation systems, the requirements of broad bandwidth, high gain and low-profile antenna are strongly considered in recent research analysis. Nowadays, the patch antenna is widely used in wireless technology except their lightweight, low-cost, and easy to fabricate. It has few limitations like narrow bandwidth, low gain and reduce power managing capacity. Meta-material structured antennas have raised huge attention in solving these issues. Therefore, this paper aims to assess and examine a recent research study on the meta-material developments for the performance enhancement of rectangular patch antenna.

Keywords Meta-material · Patch antenna · Split-Ring Resonator (SRR) · Complementary Split-Ring Resonator (CSRR)

1 Introduction

In modern world communication, antenna plays an important role and is required everywhere in any wireless communication system. An antenna is an electronic device that converts electromagnetic energy into electric or magnetic energy or vice versa. An antenna is a means of radiating and receiving radio waves. In Fig. 1, antennas are used in a transmitter as well as in a receiver.

Many antennas are available in a simple configuration, such as wire antenna, a dipole antenna, a monopole antenna, a helical antenna, loop antenna and microstrip

N. Agrawal (✉)
Electronics and Communication Department, GLA University, Mathura, UP, India
e-mail: neetuagrawal.ec@gla.ac.in

S. Dhar et al. (eds.), *Advances in Communication, Devices and Networking*,
Lecture Notes in Electrical Engineering 776,
https://doi.org/10.1007/978-981-16-2911-2_24

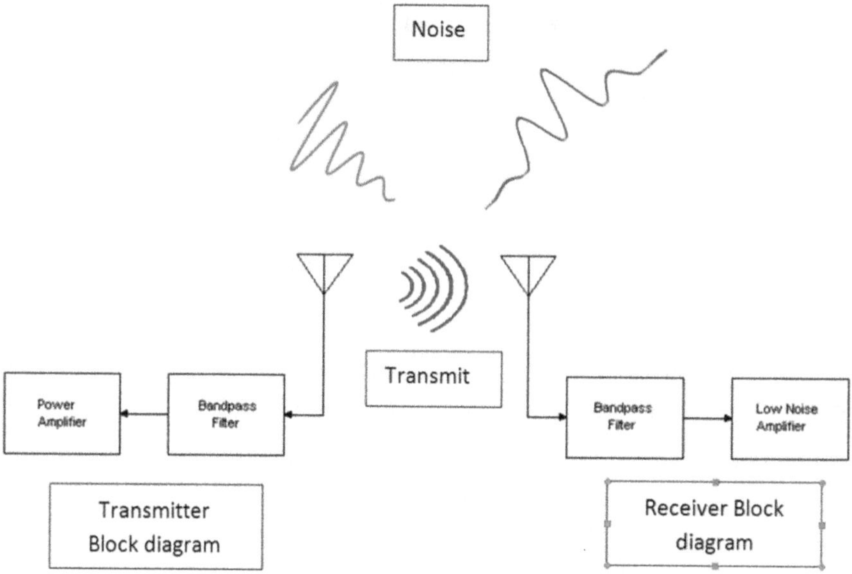

Fig. 1 Transmitter and receiver block diagram

antenna (MSA). These different antennas are required for different–different applications, like monopole and MSA used in cell phones. Two decades back, generally, loop and helical antennas were used in cell phones but people were not satisfied with that, so microstrip antenna and printed monopole antenna came into the picture. Nowadays, microstrip antennas are widely used in many electronic gadgets (mobile phone, pager etc.) and also in satellite and defense communication. To design any antenna in a practical aspect for different applications, the performance of the antenna is needed. The fundamental parameters that determine the performance of an antenna are directivity, bandwidth, radiation pattern, gain and many more.

This paper discusses many literatures that have studied the enhancement or improvement of these fundamental parameters of antennas and their performance. These parameters can be improved by using meta-material-based microstrip antenna Here, we present a lot of literature surveys on meta-material-based microstrip antenna

The rest of the paper is organized as follows: Sect. 2 presents a brief overview of microstrip antenna, advantages and disadvantages, and essential equations are required for designing rectangular microstrip antenna. Section 3 presents a discussion on meta-material and its properties and the desire equations needed to determine the value of ε and μ. Section 4 presents the roles of meta-material-based microstrip antenna and its comparative study. Finally, Sect. 5 presents the conclusion.

2 Microstrip Antenna

A microstrip antenna has an intermediate layer between a radiating patch and the ground plane called a dielectric substrate. On the top side of the substrate etch a patch. The common shapes of the patch are shown in Fig. 2. The radiating patch acts as a metallic plate made of material such as copper or gold. Radiation takes place in a microstrip antenna from the fringing fields. For achieving excellent performances, like higher bandwidth, higher efficiency and best radiation, the substrate must be wide and the dielectric constant should be low and suitable. But during this case, the size of the patch antenna increases, which is not desirable in modern days. If we try to reduce the size, a substrate with high dielectric constants must be used which affects other antenna parameters, like narrow bandwidth and less efficiency. The dielectric constants of a substrate used in a microstrip antenna should be not less than $\in_r = 2.2$ and not exceed $\in_r = 12$ usually. Figure 3 shows rectangular a microstrip antenna (RMSA) which is very simple and quickly designed.

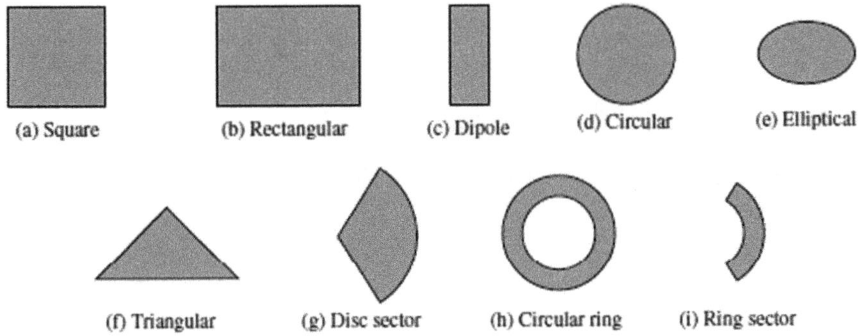

Fig. 2 Common shapes of microstrip patch antenna

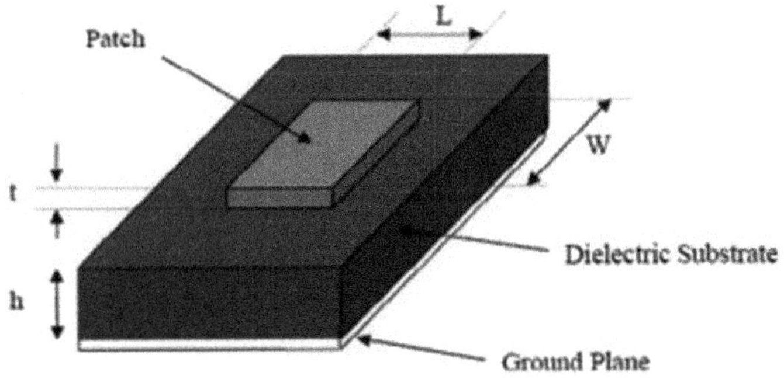

Fig. 3 Rectangular microstrip patch antenna

2.1 Essential Equations for Design RMSA

For any given frequency and dielectric constant $\in_r$, the design of RMSA is possible by the use of the equation as mentioned below.

The necessary steps are:

Step 1: To calculate width (W) of RMSA

$$W = \frac{c}{2f_{0\sqrt{\frac{(\in_r+1)}{2}}}} \tag{1}$$

where c = light velocity in free-space

f_0 = resonance frequency

$\in_r$ = dielectric constant/relative permittivity of substrate

W = width of patch.

Step 2: To calculate effective dielectric constant

$$\in_{\text{eff}} = \frac{(\in_r + 1)}{2} + \frac{(\in_r - 1)}{2}\left(1 + 12\frac{h}{W}\right)^{-1/2} \tag{2}$$

where $\in_{\text{eff}}$ = effective dielectric constant

h = substrate height.

Step 3: To calculate the actual length of the patch (L)

$$L = L_{\text{eff}} - 2\Delta L \tag{3}$$

where

$$L = \frac{c}{2f_{0\sqrt{\in_{\text{eff}}}}} \tag{4}$$

Step 4: To calculate length extension

$$\frac{\Delta L}{h} = 0.412\frac{(\in_{\text{eff}} + 0.3)\left(\frac{W}{h} + 0.264\right)}{(\in_{\text{eff}} - 0.258)\left(\frac{W}{h} + 0.8\right)} \tag{5}$$

where w/h = width to height ratio

ΔL = extended length of each side of the patch.

2.2 Advantages/Disadvantages of Microstrip Antenna

There are major advantages and disadvantages of microstrip patch antennas which are listed as follows:

- Weight is light, less volume
- Low profile or compact size
- Low fabrication cost
- Both linear and circular polarizations are possible
- Realization of dual- and triple-frequency operations.

Microstrip patch antennas suffer from radiation losses which increase by using a thick and low dielectric constant substrate. The substrate should be thick for larger bandwidth, but at the same time degrades the other one like in the size that becomes large. It is undesirable in modern days. Some disadvantages of microstrip antenna are:

- Narrow bandwidth
- Low efficiency and directivity
- Low gain
- Low power-handling capacity.

3 Meta-material

The meta-material is first recommended by a Russian researcher Victor Vesalago, in 1968. Meta-materials are unreal or unnatural materials with illustrative EM properties that do not occur in natural materials. It shows extraordinary electromagnetic properties, both negative permittivity and negative permeability. Permittivity and permeability of a material describe the electromagnetic properties in an electric field and the magnetic field, respectively. The material has been classified by Buriak et al. on the basis of the value of permittivity and permeability (Fig. 4) [13].

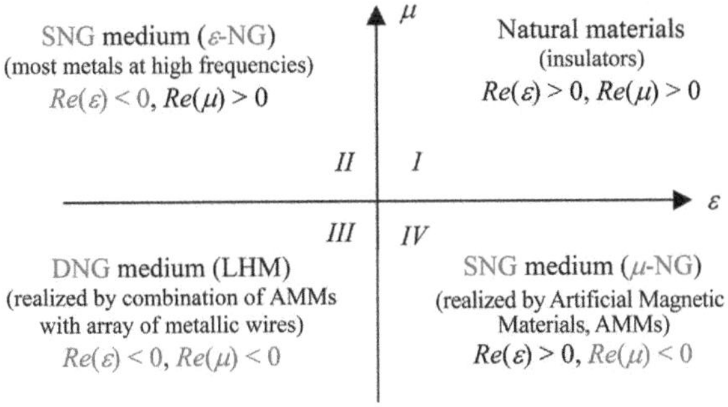

Fig. 4 Classification of material on the basis of ε and μ

I Quad shows natural material that has positive values of ε and μ both. It is called double positive (DPS) or right-handed material (RHM) and III Quad has negative ε and μ, so it is called dual-handed negative (DNG) or left-handed material.

3.1 Unit Cell Design

In the conceptual reduction coupling system, the CSRR systems are ideal for modifying and redistributing electrical and magnetic quantities. Double CSRRs that cut from the two sides of the substrate are the alternative explanation. This concept redistributes and eliminates the reciprocal coupling of the surface current between the two patches. The CSRR is implemented to act at the optimal frequency in order to apply this technique. The operating frequency is set to be 28 GHz. By using series-slotted gaps that are continuously repeated on a microstrip transmission line, a broadband single-negative meta-material (SNG), negative μ or negative ε medium can be produced. The electromagnetic characteristics have the same SNG behavior in both split-ring resonators (SRRs) and CSRRs. In fact, SRR has a negative permeability, although there is a negative permittivity for CSRR. The mathematical concept and SRRs and CSRRs analysis show that they act as an analogous LC circuit. The design of the CSRR and its fundamental circuit equivalent model is also seen in Fig. 5.

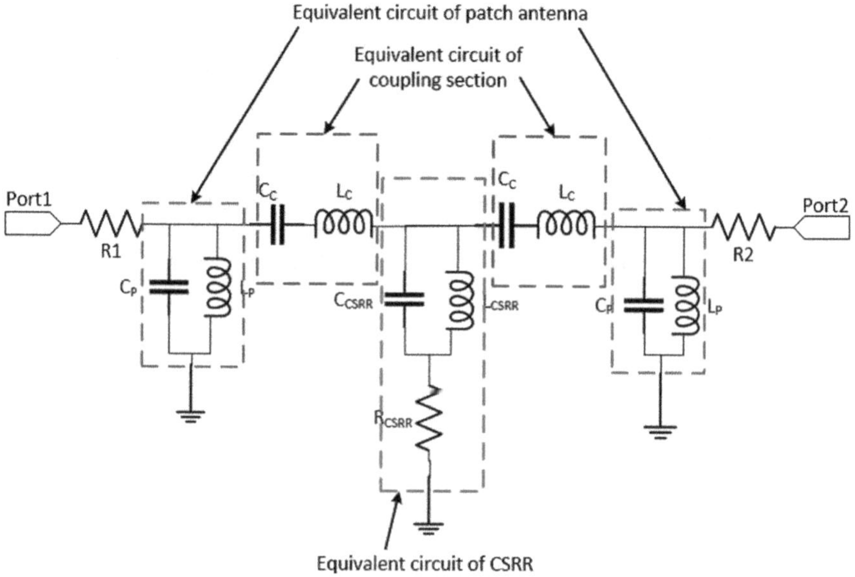

Fig. 5 Equivalent circuit of CSRR

3.2 *Equation Required*

Desire equations are used to determine the permittivity (ε_r) and permeability (μ_r) of a material using the NRW method.

$$\mu_r = \frac{2(1 - V_2)}{\omega d(1 + V_2)} \tag{6}$$

$$\varepsilon_r = \mu_r +^{2S} \frac{11Ci}{\omega d} \tag{7}$$

$$V_1 = S_{11} + S_{21} \tag{8}$$

$$V_2 = S_{21} - S_{11} \tag{9}$$

is the radian frequency, d is the substrate height, $C = 3 \times 10^8$ m/s,

$_1$ = maximum voltage, and $_2$ = minimum voltage.

4 Meta-material Roles in Microstrip Patch Antenna

By adding the meta-material in the microstrip patch antenna as a substrate and superstrate, the following parameters can be improved:

- Improve gain and efficiency
- Increase directivity
- Enhancing the strength of main lobe level
- Reduce the size of the antenna

A lot of research has been done by researchers which validates whether the above performance parameters of antenna are achieved. Some literatures are reviewed in this paper, which are as follows.

Ravirajsinh a Raulji et al. presented a paper on microstrip patch antenna using meta-material. Meta-material is very helpful to design a broadband microstrip antenna. Its structure is used to improve the gain, bandwidth and size of the antenna. The gain is enhanced by 1.5–7 dB and size is reduced by 50% [1]. Fei Wang et al. presented a paper on compact UWB MIMO antenna meta-material-inspired isolator [2]. Here, with the use of meta-material in ultrawideband MIMO antenna various performance parameters are improved, like directivity, isolation, and compact size. The authors also developed a dual-band notch.

Kioumars Pedram et al. have presented a paper on the meta-material-based fractal antenna. In this paper, the fractal structure with a composite right–left transmission line is proposed [3].

An emerging trend in high gain antennas for wireless communication is presented in [4]. In this paper, the performance parameters of an antenna, like gain and bandwidth, have been enhanced, and also strengthened the main lobe level by avoiding side lobes due to the ultrarefraction phenomenon.

Vu Van Yem and Nguyen Ngoc LanIn had presented a paper "Gain and Bandwidth Enhancement of Array Antenna Using a Novel Metamaterial Structure" [5]. In this paper, with the help of a novel meta-material structure, both the bandwidth and gain parameters of the antenna have been improved. Also, the array size of the antenna is reduced by enhancing the slow-wave factor.

T. Sathiyapriya, V. Gurunathan and A. Shafeek had presented a paper "DESIGN OF METAMATERIAL INSPIRED HIGH GAIN PATCH ANTENNA" [6]. This paper shows meta-material-loaded patch antenna operating at X band. Improving gain and directivity is the major objective of this paper.

Akash Agrawal, Narendra Kumar Garg had presented "REDUCTION IN THE SIZE OF RECTANGULAR MICROSTRIP PATCH ANTENNA" [7]. In this paper, by adding meta-material, the design of small size and highly efficient microstrip patch antenna is found at an operating frequency of 1.548 GHz. It is the main objective of this paper because in the modern age of nanotechnology, size is an important factor for any device.

Shengyuan Luo, Yingsong Li, Tao Jiang1, Beiming Li had presented a paper "A multi-layers frequency selective surface (FSS) with meta-material" [8]. In this paper, the mutual coupling is reduced in an MIMO antenna array by the realization of FSS elements with the periodic split-ring resonators (SSRs). The technique introduced in this paper can provide a 2.5 dB gain enhancement and 20 dB coupling reduction in two elements.

Wang et al. had presented a paper "Compact UWB MIMO Antenna with Metamaterial-inspired Isolator". In this paper, a dual-band CSRR to improve the dual band-notch effect (-2.5 dB at 3.55 GHz and -3 dB at 5.80 GHz). At the same time, design a novel SRR as an isolator to reduce the coupling in both bands (from 3 to 4 GHz and from 8 GHz to 11 GHz). This meta-material-inspired isolator can achieve high isolation ($|S_{21}|$ is under -19 dB) and minimization in both bands only using a *single* unit cell. Based on these novel metamaterials, the dual band-notch UWB MIMO antenna with a small size (13.5 mm $\times$ 34 mm) is developed. The experimental results show that the proposed UWB MIMO antenna has excellent performance such as compact structure, good directivity and high isolation. This meta-material-inspired UWB MIMO antenna can be used in miniaturized mobile devices.

Table 1 shows the performance summary of different meta-material-based microstrip antennas. A comparative study has been done among different meta-material-based microstrip antennas in terms of gain, bandwidth, directivity, efficiency, operating frequency, size, reflection coefficient, isolation and so on.

As seen from Table 1, meta-material-inspired patch antenna gives 100% efficiency by Ref. [9] but other approaches are less efficient. But in terms of gain, the meta-material used as a superstrate antenna [10] is better than the others.

Table 1 Performance comparison among different meta-material-based antenna

S. no.	References	Methodology	Total number of parameters	Approximated value
1	[10]	Meta-material structure with SRR and wire element	2	Gain: improve from 7 to 15 dB
				Size: 50% reduced
2	[5]	Dual-band complementary SRR	5	Size: 13.5 mm × 34 mm
				Bandwidth (BW): 3–12 GHz
				Isolation: >19 dB
				Peak gain: 4 dBi
				Radiation efficiency > 60%
3	[6]	Compact dual-band CRLH-TL-based antenna	3	Resonance freq.: 4.4, 6.1 GHz
				Gain: 3.8, 6.2 dB
				Size: 20 mm × 20 mm
4	[5]	CSRR-loaded patch antenna with 2 × 2 hexagonal array	3	Resonance frequency: 2.4 and 5 GHz
				Gain: 6.64 and 10.1 dB
				Size: 20 × 25 × 0.6 mm^3
5	[4]	Novel meta-material structure	6	BW: 1.1 GHz
				Efficiency: 87%
				Gain: 11.2 dBi
				Impedance matching: good
				Size: 115 × 118 mm^2
				Central frequency: 8.15 GHz
6	[11]	Multilayer FSS with periodic SRR	3	Reflection coefficient: −28 dB
				Isolation: −25 dB
				BW: Better than without using FSS
7	[12]	DNG meta-material superstrate	4	Center frequency: 5.7 GHz
				Isolation: −24 dB
				Gain: increase by 1.68 dBi
				Efficiency: >85%

(continued)

Table 1 (continued)

S. no.	References	Methodology	Total number of parameters	Approximated value
8	[10]	Meta-material used as a superstrate	2	Gain: increase from 7.1 to 12.21 dB
				BW: Increase from 71.5 to 145 MHz
9	[13]	Meta-material-inspired patch antenna	6	Gain: 7.9 dB
				Efficiency: 100%
				Reflection coefficient: –24 dB
				Operating band: X band
				Resonance frequency: 11.8 GHz
				Application: satellite communication
10	[14]	Meta-material structure	4	Operating frequency: 1.548 GHz
				Return loss: −42 dB
				Efficiency: 69.88%
				Directivity: 6.725 dBi

5 Conclusion

In this research paper, microstrip antenna, its advantages and disadvantages, how to overcome the limitations and how meta-material is helpful to improve the performance of the existing antennas are studied. In this regard, many literature works have been presented in this paper. As seen from Table 1 many performance parameters are improved using meta-material structure rather than without it. In conclusion, a lot of research has been done to enhance the performance of antenna parameters by using various techniques. Antenna using meta-material is one of them. Using meta-material properties to enhance various performance parameters, like gain, size, directivity, radiated power, isolation etc., has been discussed in this paper.

References

1. Jayakrishnan VM, Liya ML (2020) A survey of electromagnetic waves based metamaterials and applications in various domains. In: 2020 third international conference on smart systems and inventive technology (ICSSIT), Tirunelveli, India, pp 662–666. https://doi.org/10.1109/icssit48917.2020.9214163
2. Mahmood SN, Ishak AJ, Saeidi T, Alsariera H, Alani S, Ismail A, Soh AC (2020) Recent advances in wearable antenna technologies: a review. Progr Electromagn Res B 89:1–27. https://doi.org/10.2528/pierb20071803

3. Wu CTM, Chen PY (2020) Low-profile metamaterial-based adaptative beamforming techniques. In: Modern printed circuit antennas. IntechOpen. https://doi.org/10.5772/intechopen.90012
4. Hariyadi T, Rodiah N, Pantjawati AB (2019) The effect of split ring resonator (SRR) metamaterials on the bandwidth of circular microstrip patch antennas. J Phys: Conf Ser 1387(1):012095. IOP Publishing
5. Wang F, Duan Z, Li S, Wang Z-L, Gong Y-B (2018) Compact UWB MIMO antenna with metamaterial-inspired isolator. Progr Electromagn Res C 84:61–74. https://doi.org/10.2528/pierc18030201
6. Pedram K, Nourinia J, Ghobadi C, Pouyanfar N, Karamirad M (2020) Compact and miniaturized metamaterial-based microstrip fractal antenna with reconfigurable qualification. AEU-Int J Electron Commun 114:152959. https://doi.org/10.1016/j.aeue.2019.152959
7. Pedram K et al (2020) Compact and miniaturized metamaterial-based microstrip fractal antenna with reconfigurable qualification. AEU-Int J Electron Commun 114(2020):152959 https://doi.org/10.1016/j.aeue.2019.152959
8. Luo S et al (2019) FSS and meta-material based low mutual coupling MIMO antenna array. In: 2019 IEEE international symposium on antennas and propagation and USNC-URSI radio science meeting. IEEE. https://doi.org/10.1109/APUSNCURSINRSM.2019.8888621
9. Saxena P, Kothari A (2017) Mathematical modeling of n-sided polygon metamaterial split ring resonators for 5.8 GHz ISM band applications. Wireless Pers Commun 96:5959–5971 (2017). https://doi.org/10.1007/s11277-017-4457-z
10. Kaushal A, Bharadwaj G (2019) A comprehensive survey on design approaches for metamaterial based microstrip patch antenna. In: 2019 international conference on computing, communication, and intelligent systems (ICCCIS), Greater Noida, India, pp 478–483. https://doi.org/10.1109/icccis48478.2019.8974496
11. Setia V, Sharma KK, Kishen Koul S (2019) Triple-band metamaterial inspired microstrip antenna using split ring resonators for WLAN/WiMAX applications. In: 2019 IEEE Indian conference on antennas and propogation (InCAP), Ahmedabad, India, pp 1–4. https://doi.org/10.1109/incap47789.2019.9134602
12. Arayeshnia A, Bayat A, Keshtkar-Bagheri M, Jarchi S (2019) Miniaturized low-profile antenna based on uniplanar quasi-composite right/left-handed metamaterial. Int J RF Microw Comput Aided Eng 29(10). https://doi.org/10.1002/mmce.21888
13. Kaur H, Sharma A (2017) Microstrip patch antennas using metamaterials: a review. Int J Electron Electr Comput Sys 6(6):130–133
14. Stutzman WL, Thiele GA (2012) Antenna theory and design. Wiley

Detection of Chemical Warfare Agents Using Kretschmann–Raether Configuration-Based Surface Plasmon Resonance (SPR) Biosensor

Jitendra Singh Tamang, Saket Kumar Jha, Rudra Sankar Dhar, and Somenath Chatterjee

Abstract The manuscript aims at detecting different and harmful chemical warfare agents which are highly toxic and health hazard. The detection is possible by designing a biosensor keeping in mind the concept of surface plasmon resonance (SPR) phenomenon. The SPR biosensors are too beneficial for defense-related applications; therefore, detection of such chemical warfare agents has to be accurate which is very essential. Designing such biosensors can be achieved by considering the characteristic transfer matrix (CTM) method and ensuing it in the MATLAB environment. Furthermore, the sensitivity and detection accuracy values are calculated and tabulated, respectively.

Keywords Surface plasmon resonance · Chemical warfare agents · Kretschmann–Raether configuration · Sensitivity · Figure-of-merit

1 Introduction

Defense-related warfare agents are very widely used nowadays. Chemical materials play a very vital role in causing injury and harm in the war fields. Even a few chemical agents can be easily dissolved and mixed in the solutions which makes it difficult in detecting them. Therefore, it becomes very critical to detect and absorb them. Surface plasmon resonance (SPR) biosensors can be designed in order to solve this problem to some extent. SPR, obeying the optical occurrence of light and following the Drude model [1], becomes highly efficient to identify and sense these warfare agents.

J. S. Tamang (✉) · S. K. Jha · S. Chatterjee (✉)
Department of Electronics and Communication Engineering, Sikkim Manipal Institute of Technology (SMIT), Sikkim Manipal University (SMU), Majitar, Sikkim 737136, India

J. S. Tamang · R. S. Dhar (✉)
Department of Electronics and Communication Engineering, National Institute of Technology, Aizwal, Mizoram 796012, India
e-mail: rdhar@uwaterloo.ca

S. Dhar et al. (eds.), *Advances in Communication, Devices and Networking*, Lecture Notes in Electrical Engineering 776,
https://doi.org/10.1007/978-981-16-2911-2_25

Literature surveys reveal that chemical and biological interactions among the biomolecules can lead to health hazards. Even a few toxic and poisons can be considered as warfare agents [2]. Efforts have been made to significantly detect numerous chemical agents in industries, and thus, designing different types of materials is very interesting. Toxic and unstable chemicals and various organic elements can be detected even if they are mixed with water. Several studies and reviews suggest that the detection of chemical toxic agents can be vital in industrial applications [3]. The industrial action of the rapid spread of warfare agents if exposed can be treated effectively by several antioxidants. These antioxidants can efficiently act to protect against the agents and show promising results in treating the toxic and warfare agents [4]. In order to detect and characterize the warfare agents, several computational and experimental methods are used in accordance with the water absorption. Agents like sarin, soman, mustard gas and several agents are chosen, and their performance is analyzed and experimentally synthesized. Even porous materials are also considered in such analysis [5]. Absorption and removal of health hazardous chemicals are very essential. Various materials and methods have been selected and implemented to detect these agents. In some methods, catalysts are too considered as it increases the efficiency and speed of detection. Among many catalysts, metal–organic frameworks (MOFs) materials are chosen in order to speed up the catalysis and hydrolysis process and detection of warfare agents [6]. The commonly known warfare agents like sarin, mustard gas can be absorbed and decontaminated by the metal–organic frameworks (MOFs). Few MOFs having better surface areas and volumes can be highly efficient in influencing the reactive properties of such warfare agents [7]. The metal–organic frameworks (MOFs) can be modified by adding zirconium metal for better efficiency and accurate detection of warfare agents. This method assists in detecting gaseous substances by using a bimetallic structure and compared with another catalyst made of aluminum [8]. Another method to further modify the metal–organic frameworks (MOFs) is by considering titanium metal as well [9]. Another approach by which the agents can be detected efficiently is by designing a convenient and handy Raman spectrometer. This equipment can be used for defense applications to measure and analyze the chemical properties of the warfare agents within the wavelength of 790–1200 nm. Few agents were detected in 795 nm and 1064 nm based on their responsive, sharp curves. Even the comparison of concentrations between different warfare agents proved that the wavelength of 1064 nm seemed much better than the 790 nm excitation for the detection process [10, 11]. Another chemical warfare agent detection method involves other organic compounds as hosts which are based on supramolecular interfaces. Sensors designed based on such interactions proved to be highly selective and sensible for the detection mechanism [12, 13].

In this article, the effort has been made to design an effective SPR biosensor model to detect different chemical warfare agents based on two different parameters, viz., sensitivity and figure-of-merit (F.O.M.).

2 Modeling Methods

Chemical agents are detected by considering one of the basic SPR configurations, i.e., Kretschmann–Raether configuration. In such configuration, a three-layer plasmonic structure, as shown in Fig. 1, is taken which consists of SF10 (Prism), metal nanostructure (gold) and sensing medium (warfare agents). The refractive indices of each layer in the configuration are as follows:

***SF10 (npr)*—1.72148**

***Gold, Metal (ng)*—0.15557 + 3.6024*i*.**

***Sensing mediums (chemical warfare agents)*:**

(i) ***Chlorine (nchl)*—1.000773**
(ii) ***Chloroform (nchlf)*—1.4417**
(iii) ***Chloropicrin (nchlp)*—1.461**
(iv) ***Cyanogen chloride (ncycl)*—1.3668**
(v) ***Hydrogen cynaide (nhycn)*—1.2675**

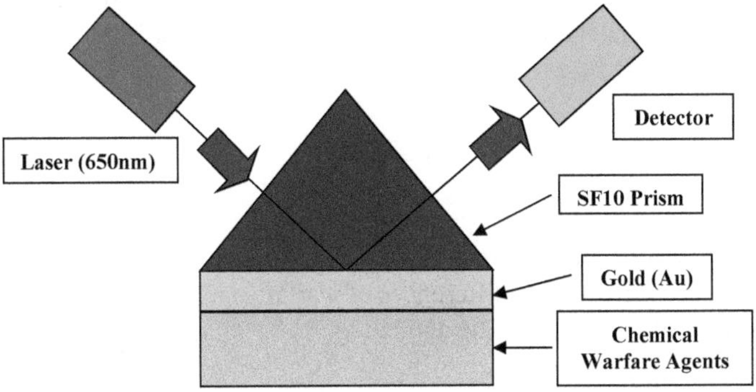

Fig. 1 Three-layer SPR configuration (SF10||Metal||Chemical warfare agent)

Fig. 2 Comparison of two SPR curves indicating sensitivity and F.O.M. factors

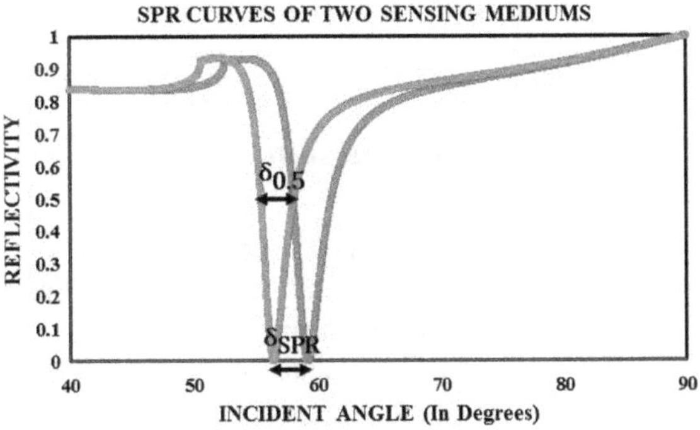

2.1 Three-Layer Mathematical Model

The analysis of the three-layer model can further proceed by expressing the SPR curve obtained in the detector mathematically. The SPR curve provides reflectivity, R-value (equation below) with respect to the incident angle.

$$R = |r^p_{\text{coefficient}}|^2 = \left| \frac{r^p(\text{npr} + \text{ng}) + r^p(\text{npr} + \text{ncwa})e^{2\text{iop}_{\text{coeff.}}\text{tg}_1}}{1 + r^p(\text{npr} + \text{ng})\ r^p(\text{npr} + \text{ncwa})e^{2\text{iop}_{\text{coeff}}\text{tg}_1}} \right| \quad (1)$$

where r^p is the reflectivity coefficient for the p-polarization wave.

npr + ng are the refractive indices of SF10 Prism and metal (Au).

npr + ncwa are the refractive indices of SF10 Prism and chemical warfare agents.

$\text{op}_{\text{coeff.}}$ is the optical coefficient of the mediums.

tg is the thickness of the metal (Au) layer.

Furthermore, while comparing the SPR curves of the two warfare agents simultaneously, two other parameters, viz., sensitivity (S_n) and figure-of-merit (F.O.M.) play a very vital role in understanding the efficiency of the sensor in terms of detection. These two factors can be expressed as follows:

$$\text{S} = \frac{\delta_{\text{SPR}}}{\delta_{\text{newa}}} \quad (2)$$

where δ_{SPR} is the SPR angle differences.

δ_{newa} is the refractive index differences of warfare agents.

Figure-of-merit (F.O.M.) can be defined as

$$\text{F.O.M} = \frac{S}{\delta_{0.5}} \quad (3)$$

where $\delta_{0.5}$ is the half-width of the two SPR curves obtained.

3 Results and Discussion

The two SPR curves as shown in Figs. 3 and 4 are plotted with respect to two different types of warfare agents. Chlorine and chloroform both are considered to be toxic and pungent chemical agents and thus become very essential to detect them. The sensitivity and F.O.M. values are calculated and tabulated in Table 1.

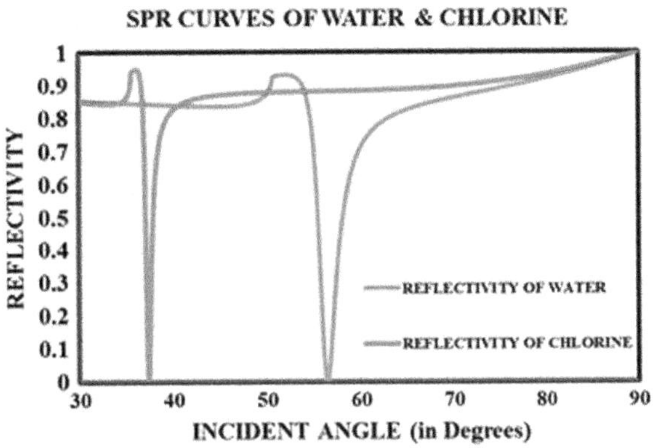

Fig. 3 Reflectivity curves of water and chlorine

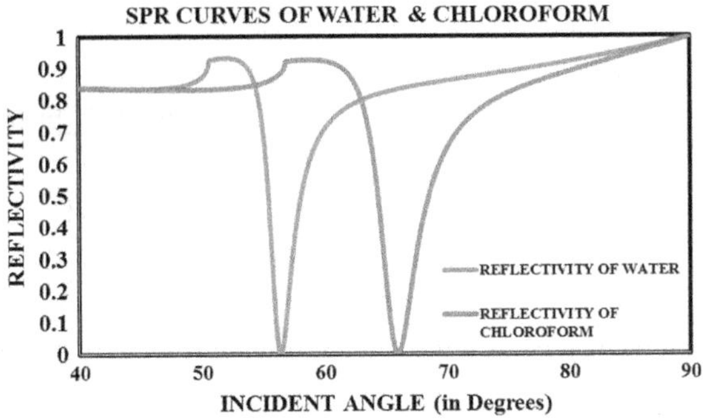

Fig. 4 Reflectivity curves of water and chloroform

Table 1 Sensitivity and F.O.M. values of different chemical warfare agents

S. no.	Chemical warfare agent	Sensitivity (S)	Figure-of-merit (F.O.M.)
1	Chlorine	57.80	3.143
2	Chloroform	85.99	9.68
3	Chloropicrin	89	8.26
4	Cyanogen chloride	92	34.5
5	Hydrogen cyanide	69.44	16.57

Chloropicrin and cyanogen chloride are poisonous and choking agents that hinder the blood circulation and respiratory process; therefore, the detection requires to be precise and accurate whose values are tabulated in Table 1.

This organic compound is also considered to be a very toxic and harmful agent because of its extremely high toxicity concentration level. The detection values have been tabulated in Table 1.

Table 1 depicts the maximum and minimum values of sensitivity and F.O.M. values of the designed SPR sensor. Based on the above parametric values, cyanogen chloride

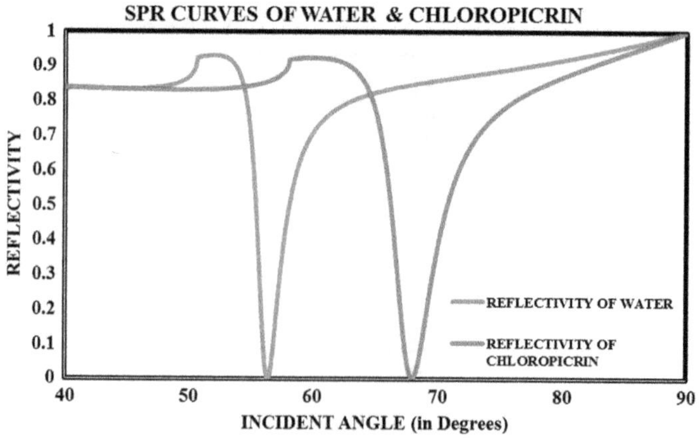

Fig. 5 Reflectivity curves of water and chloropicrin

Fig. 6 Reflectivity curves of water and cyanogen chloride

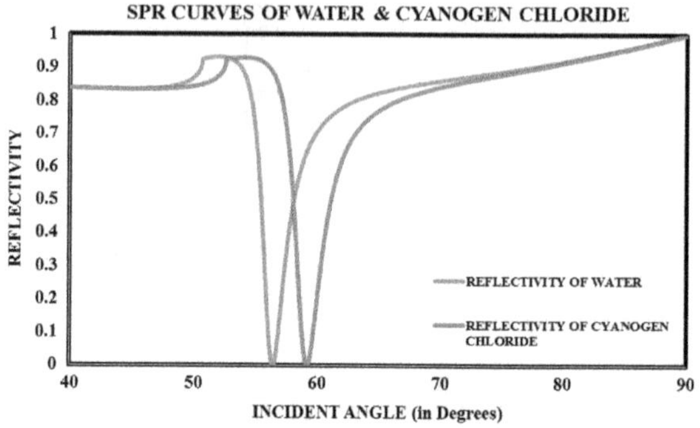

Fig. 7 Reflectivity curves of water and hydrogen cyanide

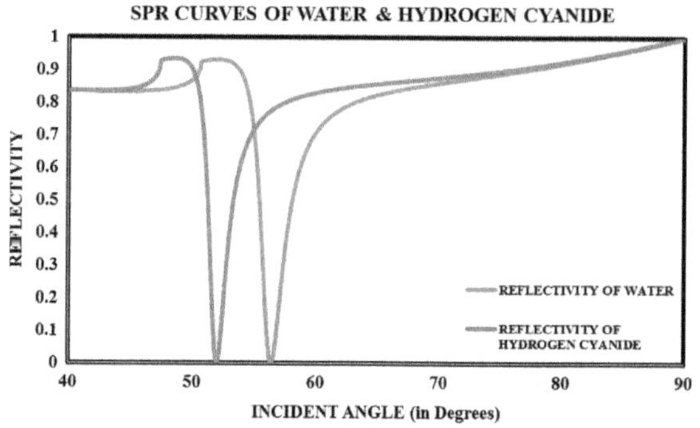

can be detected very easily and accurately as compared to other agents. Though some of the warfare agents show a better sensitivity, detection of such agents cannot be accurate.

4 Conclusion

SPR biosensor based on the basic Kretschmann configuration makes an effort to detect and sense the most hazardous warfare agents. However, only a few agents can be detected accurately. The sensing property of the designed sensor is extremely useful for detecting and analyzing agents, viz., cyanogen chloride, chloropicrin and chloroform even if they are soluble with water or any aqueous medium. The comparison of the most toxic agents based on the reflectivity, sensitivity and detection property reveals the performance of the sensor designed. Furthermore, these agents can be identified, analyzed and detected based on their chemical properties and design parameters of the sensor.

References

1. Paul D (1900) ZurElektronentheorie der metalle. Ann Phys 306(3):566–613
2. Schwenk M (2018) Chemical warfare agents. Classes and targets. Toxicol Lett 293:253–263
3. Scott BN, Mendonca ML, Howarth AJ, Islamoglu T, Hupp JT, Farha OK, Snurr RQ (2017) Metal–organic frameworks for the removal of toxic industrial chemicals and chemical warfare agents. Chem Soc Rev 46(11):3357–3385
4. McElroy CS, Day BJ (2016) Antioxidants as potential medical countermeasures for chemical warfare agents and toxic industrial chemicals. Biochem Pharmacol 100:1–11
5. Matito-Martos I, Moghadam PZ, Li A, Colombo V, Navarro JAR, Calero S, Fairen-Jimenez D (2018) Discovery of an optimal porous crystalline material for the capture of chemical warfare agents. Chem Mater 30(14):4571–4579
6. Liu Y, Howarth AJ, Vermeulen NA, Moon S-Y, Hupp JT, Farha OK (2017) Catalytic degradation of chemical warfare agents and their simulants by metal-organic frameworks. Coord Chem Rev 346:101–111
7. Son FA, Wasson MC, Islamoglu T, Chen Z, Gong X, Hanna SL, Lyu J et al (2020) Uncovering the role of metal–organic framework topology on the capture and reactivity of chemical warfare agents. Chem Mater 32(11):4609–4617
8. Chen H, Snurr RQ (2020) Insights into catalytic gas-phase hydrolysis of organophosphate chemical warfare agents by MOF-supported bimetallic Metal-Oxo clusters. ACS Appl Mater Interf 12(13):14631–14640
9. Soares CV, Leitão AA, Maurin G (2019) Computational evaluation of the chemical warfare agents capture performances of robust MOFs. Microporous Mesoporous Mater 280:97–104
10. Kondo T, Hashimoto R, Ohrui Y, Sekioka R, Nogami T, Muta F, Seto Y (2018) Analysis of chemical warfare agents by portable Raman spectrometer with both 785 nm and 1064 nm excitation. Forensic Sci Int 291:23–38
11. Kawai NT, Spencer KM (2004) Raman spectroscopy for homeland defense applications. Raman Technol Today's Spectroscopists 54–58 (2004)
12. Puglisi R, Pappalardo A, Gulino A, Sfrazzetto GT (2019) Multitopic supramolecular detection of chemical warfare agents by fluorescent sensors. ACS Omega 4(4):7550–7555
13. Diauudin FN, Rashid JIA, Knight VF, Yunus WMZW, Ong KK, Kasim NAM, Halim NA, Noor SAM (2019) A review of current advances in the detection of organophosphorus chemical warfare agents based biosensor approaches. Sens Bio-Sensing Res 26:100305

Machine Learning Capability in the Detection of Malicious Agents

Anurag Sharma, Puja Archana Das, Muhammad Fazal Ijaz, and Abu ul Hassan S. Rana

Abstract The variety and volume of cyber-attacks have exponentially increased over the years. This calls for a strong security defense mechanism against the attacks. This paper discusses the advancements made in the field of cyber-security using various machine learning techniques. We review some of the common machine learning techniques used in cyber-security and also discuss the issues related to cyber-security. Overall, we focus on exploring the idea of a combination of deep learning, machine learning and human supervision.

Keywords Machine learning · Intrusion · Malicious code · Neural network · Accuracy

1 Introduction

Hacking in cyber-security is advancing daily in this world. There must be some higher security protection to stop these attacks. Several kinds of attacks are done by the attacker; some are D-dosing, man-in-the-middle, information escape, SQL injection, remote to local. They use such techniques to illegally enter restricted networks, websites or personal data from your device [1]. The attackers from within or outside are finding innovative techniques to crack the information, money or any sensitive

A. Sharma · P. A. Das
School of Computer Engineering, Kalinga Institute of Industrial Technology (KIIT) Deemed to be University, Bhubaneswar, Odisha, India
e-mail: 1829006@kiit.ac.in

P. A. Das
e-mail: 1829080@kiit.ac.in

M. F. Ijaz (✉) · A. H. S. Rana
Department of Intelligent Mechatronics Engineering, Sejong University, Seoul 05006, South Korea
e-mail: fazal@sejong.ac.kr

A. H. S. Rana
e-mail: rana@sejong.ac.kr

S. Dhar et al. (eds.), *Advances in Communication, Devices and Networking*, Lecture Notes in Electrical Engineering 776,
https://doi.org/10.1007/978-981-16-2911-2_26

information. The innovative ideas and new methods developed promise to stop or try to reduce new methods created or developed by the hackers. Cyber-security can be stated as a method or technique to defend against various cyber-attacks done by hackers and shield sensitive data from attackers. Cyber-security within the year 2016 had multiple advances updates in machine learning techniques like auto-cars, linguistic communication process, medical field, and virtual AI [2]. These need to be used to find various databases related to the matter of various intrusion detection. Thus implement it using machine learning to update and make the security better against the intrusion. First, we need to input these into the machine learning (ML) model. This model gets practiced by the dataset model and then is known as the trained model. Once we input the dataset, next we use the machine learning formula on the dataset sample. [3]. ML formula plays a crucial part in increasing security for intrusion detection systems [3]. Machine learning algorithms are separated into two groups: unsupervised learning and supervised learning. They are distinguished based on data (i.e. input) they are settled for.

Unsupervised learning refers to the algorithms of training information that are unlabeled, with the job of deducing the classes all by itself. Supervised learning refers to the algorithms of training information that are labeled and acknowledge what differentiates the labels. The labeled information is extremely rare and the chore of the labeled data is itself exceptionally exhausting and we may not be able to sight if labels really exist.

2 Common Machine Learning Techniques Used in Cyber-Security

Regression

In regression, values of the dependent attributes are approximated on the basis of values of the independent attributes through studying the currently existing data connected to previous events. This understanding is also used to manage the new events. Regression is used to solve fraud detection in cyber-security. When a model is understood on the basis of the past database proceedings by observing the current attributes, it determines fraudulent transactions. We can learn decision tree, support vector machine, linear regression, random forest, polynomial regression and some more regression models from machine learning. Venkatesh Jaganathan used multiple regression techniques for prognosticating the effect of cyber-attacks. The all-inclusive common vulnerability scoring system (CVSS) level is taken to be a co-related feature while two non-co-related features as Y1 (vulnerabilities count) and Y2 (mean traffic). For privacy identification in a smart environment, Daria Lavrova suggested a multiple regression model, which helped to uncover the known and unknown attacks.

Classification

Classification is one of the broadly used supervisory machine learning tasks. The use of the following machine learning tools is possible due to the accessibility of a huge collection of labeled data. In cyber-security, classifications are made on the basis of ML which discriminates the provided email messages as spam or that are not used in spam detection. The spam messages are separated from non-spam messages by the spam filter models. Classifications made based on deep learning frameworks which involve recurrent neural networks (RNN), convolutional neural networks (CNN), restricted Boltzmann machines (RBM) or long short-term memory (LSTMs) cells for attribute selection through multi-layer and non-sparse neural network tend to be quite effective in handling complicated tasks with the availability of a huge collection of the past dataset. The machine learning techniques used for classification involve naïve Bayes, logistic regression, K-nearest neighbors, decision tree, support vector machine, random forest classification.

Clustering

It is indispensable to have data with the label as regression and classification in supervised learning models. But clustering is an unsupervised learning method that retrieves general patterns from raw data even though it is unlabeled. A set of indistinguishable events establishes a cluster as they share common attributes that define a specific behavioral pattern. Clustering, in cyber-security, is used for the analysis of malware, forensic analysis, anomaly detection, etc. Self-organizing maps (SOMs) based on neural networks may be useful for cluster analysis. In cyber-security, some of the ML clustering techniques used are K-means, K-Medoids, DBSCAN, Gaussian mixture model and agglomerative clustering.

3 Issues in Cyber-Security

Machine learning algorithms have an important part in four different areas, which are intrusion detection system, malware analysis, Andriod malware detection and spam detection (Fig. 1).

Intrusion Detection

If there is any exploitation of the information by malacious software or violation of company policy, intrusion detection is used. Intrusion detection can be done in many ways. It is mainly classified into two types based on signature (signature-based) and anomaly (anomaly-based) intrusion. All packages that are received are first cross-checked with the signatures present for similarities with a known malicious threat. This is signature-based intrusion detection. Monitoring of the network traffic is done by an established normality baseline in anomaly-based intrusion detection. Biswas [4] displayed machine learning-based ways that are very useful in making a better intrusion detection system. Combinations of feature selection techniques gave them

Fig. 1 Issues in cyber-security

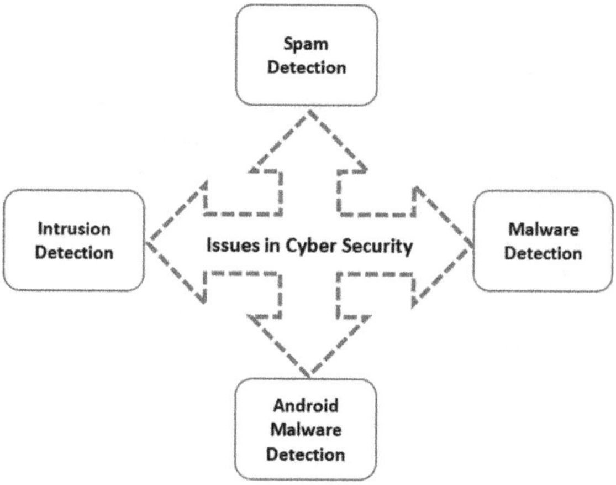

great results. Vinaya Kumar [5] put forward a scale hybrid IDS AlertNet system which helps in analyzing networks and activities done by the host. We used deep neural networks (DNNs) to create the model. Deep belief networks for intrusion detection are proposed by Zahangir Alom [6]. We have used the features of the trained set of two-layer restricted Boltzmann machine (RBM). Shone et al. [7] gave us a DL model for intrusion detection systems operation in networks using features of machine learning and deep learning.

Malware Detection

Malware is a short form of malicious software and is one of the types of cyber-threats software in the cyber-world. It is usually used for unauthorized attacks on organizations, like filching information or getting control over the entry or deal damage to personal data of the organizer and so on. The term coined as malware is usually given for programs which are malicious in nature, like virus, bugs, bots, adware, rootkits, Trojan horses, worms, spyware, ransomware, Keylogger, backdoor. Most of the malware can be subdivided into a number of families. For example, we can classify ransomware into Jisut family, Pletor family, Simplocker family, Charger family, Koler family, RansomBO family, Svpeng family, etc. The programs which are malicious in nature can be transported concealed in a secure file and operating systems. There can be many examples, like executable and linkable files or UNIX ELF, Windows PE files (portable executables with .exe, dll, efi). Malware programs can also be document-based and kept hidden inside doc files, pdfs and rtf files. Extensions and plug-ins for famous software platforms can also have malware in the form of extensions; for example, extension for web browsers and frameworks.

Uppal et al. [8] used ngram method to put forward a classification and detection system for malware. Chowdhury et al. [9] showed a neural network-based method for malware detection. Kalash et al. [10] proposed classifying malware using CNN. They applied CNN classification to them after they converted their codes of 25 families of malware binaries to grayscale images.

Android Malware Detection

Android is exceedingly attacked by mobile malware makers as it is one of the most extensively used mobile platforms. With an alarming increase in the volume and variant of Android malware, it has become exceptionally difficult to detect and classify the types of mobile malware. Researchers have made a large number of attempts toward mobile malware detection. Arp et al. [11], Varsha et al. [12] and Sharma and Dash [13] extracted static features from Android apps and they attained satisfactory results by using machine algorithms, like decision tree, SVM, K-NN, random forest, naïve Bayes to attain satisfactory results.

Spam Detection

Spam email comes in various flavors. Many are just exasperating messages aiming to draw attention to a cause or spread wrong information. Some of them are phishing emails with the intent of attracting the receiver into clicking on a malicious link or downloading malware. Spam detection is a supervised machine learning problem. This means you must develop your machine learning model with a set of samples of spam and ham messages and let it find the pertinent patterns that separate two discrete categories [19–23].

4 Real-Life Case Scenario of Cyber-Security Risk Analysis Using Machine Learning

A real-time scenario is highlighted in this section. The main goal is to see the ability of machine learning classifiers to differentiate the different types of responses given by the classifiers for the input malicious code [14]. We used four types of machine learning algorithms to classify the malicious codes, namely naïve Bayes (NB), neural network, radial basis function and support vector machine (SVM). From four different organizations, we took a combined dataset [15, 16]. The incidents that happened in the organization were collected by a centralized hub, and then the summed up data were used for the research with a goal of analyzing the results given by the classifiers in differentiating between the various accidents that took place and learning that how different data are taken from the different organizations can help in improving the accuracy of classification [17, 18].

From the given dataset table in Table 1, we have four different organizations, and the number of events occurring whose summation is in total 1900 was used to find the behavior of malware in different classifiers. First, we calculated the precision analysis in Table 2 which shows the precision of different classifiers and how well they react to the malware. Accordingly, the rows were designed where SVM has the highest recovery precision. Table 3 shows the recall analysis which is the correct malware detection divided by the number of malware that should have returned by using different classifiers. Table 4 is the F-score analysis (the higher the F-score, the more the precision and recall) of the different classifiers and shows that for different

Table 1 Data samples distribution among four different organizations

Organization name	Number of events
Organization 1	400
Organization 2	550
Organization 3	600
Organization 4	350
Total	***1900***

Table 2 Precision analysis of classifiers in identifying the different types of responses based on malware

	Naïve Bayes	Neural network	Radial basis function	Support vector machine
None	0.0	0.0	0.0	0.0
Recovered	0.0	0.0	0.0	0.76
Segregated	0.96	0.94	0.88	0.9
Dropped	0.98	0.93	0.94	0.92
Undefined	0.92	0.91	0.95	0.93
Blocked	0.0	0.0	0.0	0.64

Table 3 Recall analysis of classifiers in identifying the different types of response based on malware

	Naïve Bayes	Neural network	Radial basis function	Support vector machine
None	0.0	0.0	0.0	0.0
Recovered	0.0	0.0	0.0	0.78
Segregated	0.92	0.91	0.91	0.87
Dropped	0.95	0.94	0.93	0.93
Undefined	0.93	0.92	0.94	0.95
Blocked	0.0	0.0	0.0	0.68

Table 4 F-score analysis of classifiers in identifying the different types of responses based on malware

	Naïve Bayes	Neural network	Radial basis function	Support vector machine
None	0.0	0.0	0.0	0.0
Recovered	0.0	0.0	0.0	0.77
Segregated	0.94	0.93	0.89	0.89
Dropped	0.96	0.93	0.92	0.92
Undefined	0.92	0.91	0.95	0.94
Blocked	0.0	0.0	0.0	0.67

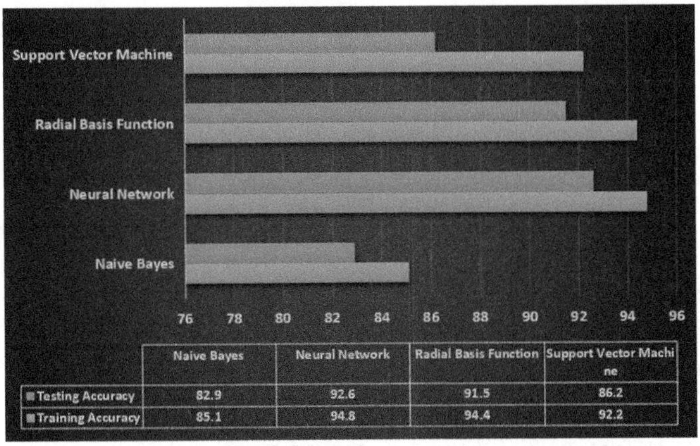

Fig. 2 Classification accuracy rate analysis using four classifiers

functions different classifiers are better, i.e., the F-score for different functions varies for different classifiers. Figure 2 shows the training and testing accuracy of different machine learning algorithm, where the neural network has the highest training and testing accuracy; naïve Bayes has the lowest testing accuracy and SVM has a drastic decrease for the unknown dataset (resting set) whereas it works much better for the trained dataset. The radial basis function is very similar to a neural network and also shows great results in both the training and testing sets.

5 Conclusion

In order to resolve various types of cyber-security problems, machine learning techniques are extensively used. The current advancements are made in the area of deep learning and machine learning and provide encouraging solutions for cyber-security threats. But it is equivalently crucial to recognize the correct algorithm acceptable for the required application. To achieve high detection rates and to keep the solution hard against malware attacks, a multi-layered proposal is required. While solving a cyber-security problem, it is important to select the right model. In this paper, for cyber-security problems, the authors investigated state-of-the-art mechanisms. The desired results for cyber-security can be achieved by the amalgamation of machine learning techniques and human supervision.

References

1. Hatcher WG, Yu W (2018) A survey of deep learning: platforms, applications and emerging research trends. IEEE Access 6. https://doi.org/10.1109/ACCESS.2018.2830661
2. Mishra S, Tripathy HK, Mallick PK, Bhoi AK, Barsocchi P (2020) EAGA-MLP—an enhanced and adaptive hybrid classification model for diabetes diagnosis. Sensors 20(14):4036
3. Mallick PK, Mishra S, Chae GS (2020) Digital media news categorization using Bernoulli document model for web content convergence. Pers Ubiquit Comput. https://doi.org/10.1007/s00779-020-01461-9
4. Mishra S, Mallick PK, Jena L, Chae GS (2020) Optimization of skewed data using sampling-based preprocessing approach. Front Public Health 8:274. https://doi.org/10.3389/fpubh.2020.00274
5. Vinayakumar R, Alazab M (Senior Member, IEEE), Soman KP, Poornachandran P, AlNemrat A, Venkatraman AN (2019) Deep learning approach for intelligent intrusion detection system. IEEE Access 7. https://doi.org/10.1109/ACCESS.2019.2895334
6. Zahangir Alom M, Bontupalli VR, Taha TM (2015) Intrusion detection using deep belief networks. 978-1-4673-7565-8/15/$31.00 ©2015 IEEE
7. Shone N, Phai VD, Ngoc TN, Shi Q (2018) A deep learning approach to network intrusion detection. IEEE Trans Emerg Top Comput Intell 41–50, February 2018
8. Uppal D, Jain V, Sinha R, Mehra V. Malware detection and classification based on extraction of API sequences. 978-1-4799-3080-7/14/$31.00_c 2014 IEEE
9. Chowdhury M, Rahman A, Islam R (2017) Protecting data from malware threats using machine learning technique. In: 2017 12th IEEE conference on industrial electronics and applications (ICIEA)
10. Kalash M, Rochan M, Mohammed N, Bruce NDB, Wang Y, Iqbal F (2018) Malware classification with deep convolutional neural networks. 978-1-5386-3662-6/18/$31.00 ©2018 IEEE
11. Arp D, Spreitzenbarth M, Hubner M, Gascon H, Rieck K (2014) Drebin: efficient and explainable detection of android malware in your pocket. In: Proceedings of 20th annual network. distributed system security symposium (NDSS), San Diego, CA, USA, February 2014, pp 1–15
12. Varsha MV, Vinod P, Dhanya KA (2017) Identification of malicious Android app using manifest and opcode features. J Comput Virol Hacking Tech 13(2):125–138
13. Sharma A, Dash SK (2014) Mining API calls and permissions for Android malware detection. In: Cryptology and network security. Springer International, Cham, Switzerland, pp 191–205
14. Mishra M, Mishra S, Mishra BK, Choudhury P (2017) Analysis of power aware protocols and standards for critical E-health applications. In: Internet of Things and big data technologies for next generation healthcare. Springer, Cham, pp 281–305
15. Mishra S, Mahanty C, Dash S, Mishra BK (2019) Implementation of BFS-NB hybrid model in intrusion detection system. In: recent developments in machine learning and data analytics. Springer, Singapore, pp 167–175
16. Mishra S, Thakkar H, Chakrabarty A, Kimtani D (2012) Dynamic cluster based data aggregation in WSN (FDDA). Int J Electron Commun Comput Technol (IJECCT) 2(5):227–230
17. Mishra S, Mallick PK, Tripathy HK, Bhoi AK, González-Briones A (2020) Performance evaluation of a proposed machine learning model for chronic disease datasets using an integrated attribute evaluator and an improved decision tree classifier. Appl Sci 10(22):8137
18. Mishra S, Tripathy HK, Mishra BK (2018) Implementation of biologically motivated optimisation approach for tumour categorisation. Int J Comput Aided Eng Technol 10(3):244–256
19. Bhoi AK, Sherpa KS (2014) QRS complex detection and analysis of cardiovascular abnormalities: a review. Int J Bioautom 18(3):181–194
20. Bhoi AK, Sherpa KS, Khandelwal B (2018) Arrhythmia and ischemia classification and clustering using QRS-ST-T (QT) analysis of electrocardiogram. Clust Comput 21(1):1033–1044
21. Bhoi AK, Sherpa KS, Khandelwal B (2018) Ischemia and Arrhythmia classification using time-frequency domain features of QRS complex. Procedia Comput Sci 132:606–613

22. Bhoi AK, Sherpa KS (2016) Statistical analysis of QRS-complex to evaluate the QR versus RS interval alteration during ischemia. J Med Imaging Health Inform 6(1):210–214
23. Bhoi AK (2017) Classification and clustering of Parkinson's and healthy control gait dynamics using LDA and K-means. Int J Bioautom 21(1)

Deep Learning Approach for Object Features Detection

Ambik Mitra, Debasis Mohanty, Muhammad Fazal Ijaz, and Abu ul Hassan S. Rana

Abstract Object detection and identification system basically find real-world objects in a digital image or any video. In order to detect an object in any video or digital image, few components are necessary, which are a model database, a feature detector, a hypothesizer and a hypothesizer verifier. This paper shows several techniques that are used for object detection, such as localization of an object, categorization of an object, extraction of features, showing information and so on, in images and videos. The document also stresses the idea of a possible solution for multi-class object detection. This paper basically relates to the researchers who are working on this particular domain. Object detection is a field distributing into a lot of enterprises, along with its uses extended from personal security to a healthy working environment. This field has its applications in various categories of image processing, such as image retrieval, security, observation, computerized vehicle systems and machine investigation. The outcomes of this particular system are inestimable on the basis of future use cases for this particular field.

Keywords Object detection · Multi-class · Enterprises · Picture retrieval and hypothesizer

---

A. Mitra
School of Electronics and Computer Engineering, Kalinga Institute of Industrial Technology (KIIT) Deemed To Be University, Bhubaneswar, Odisha, India
e-mail: 1830007@kiit.ac.in

D. Mohanty
iNurture Education Solutions Private Limited, Bengaluru, Karnataka, India
e-mail: debasis.m@inurture.co.in

M. F. Ijaz (✉) · A. H. S. Rana
Department of Intelligent Mechatronics Engineering, Sejong University, Seoul 05006, South Korea
e-mail: fazal@sejong.ac.kr

A. H. S. Rana
e-mail: rana@sejong.ac.kr

S. Dhar et al. (eds.), *Advances in Communication, Devices and Networking*,
Lecture Notes in Electrical Engineering 776,
https://doi.org/10.1007/978-981-16-2911-2_27

1 Introduction

The document explains the whole object detection and identification system. The python library ImageAI plays a major role in this particular domain along with some needful libraries, like TensorFlow, SciPy, Pillow, Matplotlib, H5py, Keras, OpenCV and NumPy. To build this kind of system, refer to the object detection model file (RetinaNet). This is a domain in deep learning. The method of deep learning used in this system is computer vision as well as image processing [1]. This is separated into computer vision and image processing. Computer vision means computers or machines help in better analysis of the input digital images or videos, due to the automated tasks that can be done by the human visual system. Although computer vision uses many ways and one of them is image processing [2], through image processing, the images can be enhanced by tuning the specific parameters and features of the images. So, a subset of computer vision is image processing, which means in some way it comes under computer vision. Here, an input image is provided with various transformations, and then we get the resultant output image. The transformations applied are sharpness, smoothness, stretching etc. [3]. In this document, we will be making a system that would be performing dense shape detection in live imagery using the application of computer vision and image processing using the model file RetinaNet. The project will be done through python programming using the python libraries such as ImageAI, TensorFlow, SciPy, Pillow, Matplotlib, H5py, Keras, OpenCV and NumPy.

2 Literature Survey

Karasulu and Korukoglu [4] discussed that on video processing, the issue of tracking moving objects over time was challenging. They discussed the different object detection and tracking algorithms and their disadvantages. Wu et al. [5] had proposed an actual time and accurate object detection model named C4 which detected objects on the basis of their contour information with the use of a cascade classifier and the CENTRIST visual descriptor. Kumar et al. [6] had introduced a localized mechanism and a gradient directional masking-based object detection algorithm for stationary background video surveillance systems. Stalder et al. [7] introduced dynamic objectness in a discriminative tracking framework to periodically rediscover the tracked objects based on their motion. They used KLT point detection and tracking with forward–backward verification for providing robustness in tracking the object. Dennis and Leibe [8] and Bastian Leibe proposed a hybrid framework for moving multi-person detection. It could only work on some small region of interest. Jiménez et al. [9] had developed a web-based solution with the use of box-count computation technique and skeleton generation algorithm in WebGL to compute and analyze the 3D fractal dimension. This solution was analogous to desktop computer-based fractal image analysis. Fustes et al. [10] found and located the marine spills using a Sentinazos tool which they developed. They covered the maximum areas using radar scenes. They had to be prepared in less time, because of the temporary requirements. They used grid technologies, for example, map, through the use of which they could enhance the algorithm and reduce the computational time. Frame differentiating-based object detection algorithm detected the object with high accuracy [11]. The computational requirement for this algorithm is from low to moderate. It is the easiest method. It will support only static background and not dynamic background. Mohammadian et al. [12] presented the combination of differential method and active blob method for detecting and tracking the object in an optimized way. Nema and Sazena [13] introduced an efficient method to find dynamic objects using a canny edge operator and some morphological processes. Several ML-based works have been carried out in the healthcare domain in regards to classify and identification of abnormal events or features in biomedical signals [15–19].

In this work, they first converted the frame image into an edge frame, then the frame difference between the two successive input frames gave the position of the mobile object.

3 Proposed Model and Description

Figure 1 shows the processes undergoing in the object detection and identification system. The python libraries are applied to these use cases to make out an object from any image or any video.

- Object recognition is defined as a group of similar computer vision tasks involving the identification of objects in digital images.
- Classification of images is to predict the class of one object in an image.
- Localization of objects means identifying the location of more than one object in an image.
- Object detection is the union of image classification and object localization. After combining the above-mentioned two tasks (points 2 and 3), it localizes and classifies more than one object in an image.

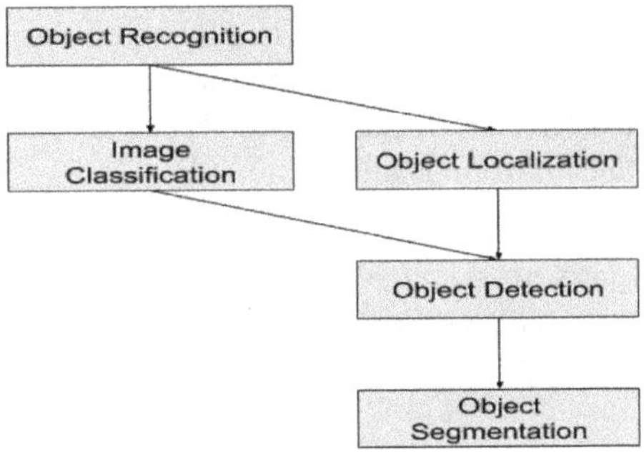

Fig. 1 The whole object detection and identification system

- Object segmentation is defined as the process to divide an object into a group of smaller fixed-size objects for optimizing storage and resources for large objects. In this process, the objects are labeled with their respective confidence number or score. The confidence score is described as the level of agreement between multiple contributors.

The python libraries used to perform these tasks are:

ImageAI—ImageAI consists of powerful classes and functions for performing object detection and also for extraction [20]. ImageAI allows you to do all the tasks along with deep learning algorithms, for example, RetinaNet, YOLOv3 and TinyYOLOv3.

TensorFlow—It is Google's publicly available machine learning framework. It is used to perform dataflow programming across a range of tasks. It finds objects using RetinaNet-50 and RetinaNet-101 feature aspirators skilled on the COCO Dataset for 4 million replications.

NumPy—NumPy is a library of python, which supports large, multi-dimensional arrays and matrices, with large groups of advanced mathematical functions for operating over these arrays [21].

Pillow—Pillow is a python imaging library (PIL). PIL provides different standard procedures to manipulate images. It accepts the image file extensions, like PNG, JPEG, PPM, GIF, TIFF and BMP.

SciPy—Modules for performing many optimizations, linearity in algebra, integration, interpolation, special function, FFT, image and signal processing are contained in SciPy. SciPy abstracts mainly on NumPy array object and is the part of the stack in NumPy, which consists of tools like Matplotlib, pandas, SymPy, etc. This NumPy stack and other applications like MATLAB, Octave, and Scilab have similar uses. The SciPy stack is another name given to NumPy stack.

OpenCV—OpenCV, a large open-source python library, is used for computer vision, machine learning, and processing of images, and currently, it acts the main role in actual-time operation which is a necessity in a system nowadays. This allows an individual to process images and videos for identifying objects, faces, and also the handwriting of a person [22, 23].

Matplotlib—Matplotlib is a Python library used to plot graphs. It also provides an API that is object-oriented for the purpose of using general-purpose graphical user interface (GUI) toolkits, for example, Tkinter, wxPython, Qt, or GTK + to lodge plots into applications.

H5py—This software includes both advanced as well as a subordinate interface for HDF5 library of Python. The subordinate interface covers the HDF5 API and the advanced component uses established NumPy concepts that give support to access HDF5 files, datasets and groups.

Keras—Keras is referred to as a publicly available neural network library in relation to Python. It has the ability to run over TensorFlow, Microsoft Cognitive Toolkit and PlaidML. It enables quick processing along with deep neural networks. It is friendly to the user, built-up and expandable.

Along with this, we need a model file for working on the detection process and giving proper labels to the object with a bounded box in an image.

3.1 RetinaNet

RetinaNet is the best one-stage object detection model that works with compact and smaller objects. It is a part of convolution neural networks (CNNs). So, it is a prevalent object detection structure used with aeriform and satellite imagery [14]. RetinaNet is formed by presenting two improvements on top of the existing single-stage object detection structures.

Feature Pyramid Networks (FPN)—It is a feature pyramid built upon image pyramids. This shows that an image can be classified into lower resolution and smaller size images. Hand-engineered characteristics are taken from each layer of the pyramid for detecting objects. So, the pyramid becomes scale constant. This particular process is reckoned and memory exhaustive.

Focal Loss—Focal loss is an improvement over cross-entropy (CE) loss, which manages the problem of class imbalance with the help of single-stage object detection models, which has major face foreground and background class disparity problem because of compact sampled, anchor boxes, which possibly are object locations.

Table 1 shows the juxtaposition of how the RetinaNet is better in object detection compared to other methods. Here, AP is the average precision for detecting the object, which is better compared to other methods. **Average precision** is defined as the average of the precision scores of the retrieved documents.

Table 1 RetinaNet versus other methods

Methods used for object detection	Average precision (AP)
Two-stage methods	
Faster R-CNN w FPN	36.1
Faster R-CNN+++	34.6
Faster R-CNN w TDM	36.9
Faster R-CNN by G-RMI	34.6
One-stage methods	
SSD513	31.1
RetinaNet	39.2
DSSD513	33
YOLOv2	21.5

In this experiment, we are using the COCO dataset which is used where the images and videos are extracted for image processing. The images are taken and the RetinaNet is used for detecting dense shapes.

3.1.1 COCO Dataset

Common objects in context (COCO) is a huge dataset used to detect, segment and caption any object. During the image processing, each object is tagged with a number named the confidence number. Those objects will be only identified whose confidence number is greater than the threshold, which is taken as 0.5 here.

4 Results and Analysis

In this experiment, we focused on dense shape detection in live imagery. These are the results we had achieved.

Figure 2 shows the graph of the variation in average precision (AP) value of RetinaNet. Therefore in general, we could say that the average precision (AP) is better than any other method in a convolution neural network (CNN). This is the reason why RetinaNet was preferred over other methods. It was easier to access compared to other methods in CNN.

Figure 3 shows the variation in inference time values of two types of RetinaNet, which are RetinaNet-101 and RetinaNet-50. Inference time is the time to find the object from an image. We could observe from the above graph that both the variations vary in inference time and average precision (AP).

Figure 4 shows the image after the detection process had undergone. The objects had been identified and labeled with proper boxes with the confidence score written

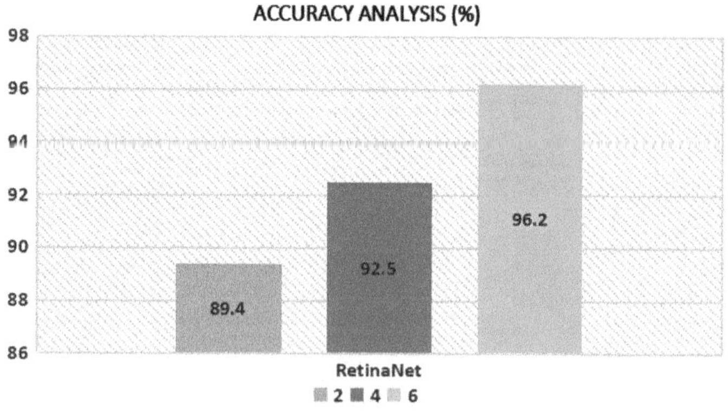

Fig. 2 Graph showing accuracy

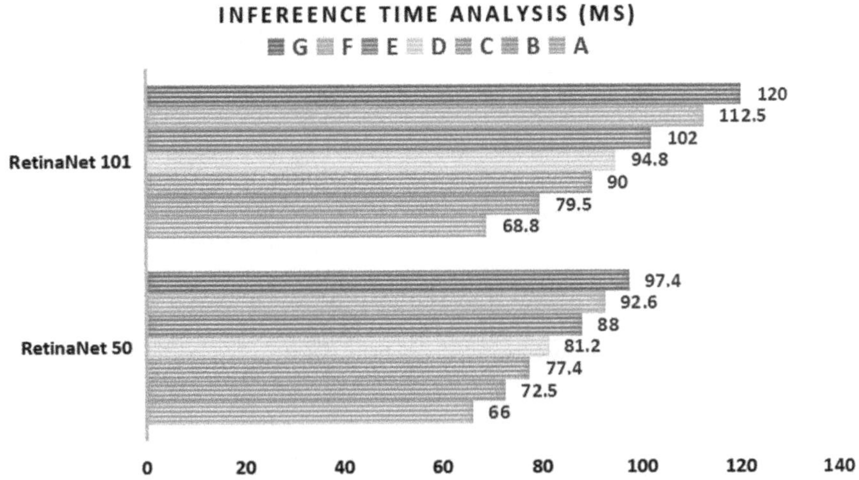

Fig. 3 Two variations of RetinaNet

Fig. 4 Image after detection

at every object. Those objects had been identified whose confidence number or score is > 0.5.

Table 2 shows the objects identified from the live imagery using the RetinaNet. The objects have been marked with their confidence number. The values are very close and do not vary much.

Table 2 Objects identified with their confidence number

Object	Confidence score
Man	0.746
Man	0.765
Mobile	0.694
Woman	0.678
Bag	0.634
Bottle	0.669
Man	0.578
Table	0.545
Woman	0.583
Glass	0.567
Chair	0.589
Watch	0.586
Man	0.542

5 Conclusion

An outcome we derived was that the COCO dataset detects objects from unknown scenes. The whole process took 5–7 s. RetinaNet detects all the identifiable objects in particular scene. All the python libraries were used in this method (named RetinaNet).

The graphs were plotted from the confidence scores of each detected object in an image from the COCO dataset and we could finally see the localized objects in the live imagery.

References

1. Mishra S, Tripathy HK, Mallick PK, Bhoi AK, Barsocchi P (2020) EAGA-MLP—an enhanced and adaptive hybrid classification model for diabetes diagnosis. Sensors 20(14):4036
2. Mallick PK, Mishra S, Chae GS (2020) Digital media news categorization using Bernoulli document model for web content convergence. Pers Ubiquit Comput. https://doi.org/10.1007/s00779 020 01461 9
3. Mishra S, Mallick PK, Jena L, Chae GS (2020) Optimization of skewed data using sampling-based preprocessing approach. Front Public Health 8:274. https://doi.org/10.3389/fpubh.2020.00274
4. Karasulu B, Korukoglu S (2013) Moving object detection and tracking in videos. In: Karasulu B, Korukoglu S (eds) Performance evaluation software: moving object detection and tracking in videos, Springer, pp 7–30. https://doi.org/10.1007/978-1-4614-6534-8_2
5. Wu J, Liu N, Geyer C, Rehg JM (2013) C4: a real-time object detection framework. IEEE Trans Image Process: Publication IEEE Signal Proces Soc 22(10):4096–4107. https://doi.org/10.1109/TIP.2013.2270111
6. Kumar P, Khan MI, Kumar A, GuptaS, Hasan DMH, Kim J-M (2012) An efficient real time moving object detection method for video surveillance system. Int J Signal Process Image Process Pattern Recogn 5(3), September 2012

7. Stalder S, Grabner H, Van Gool L (2013) Dynamic objectness for adaptive tracking. In: Lee KM, Matsushita Y, Rehg JM, Hu Z (eds) Computer vision – ACCV 2012. ACCV 2012. Lecture notes in computer science, vol 7726, Springer, Berlin, Heidelberg. https://doi.org/10.1007/978-3-642-37431-9_4
8. Mitzel D, Leibe B (2011) Real-time multi-person tracking with detector assisted structure propagation. In: IEEE international conference on computer vision workshops (ICCV workshops), pp 974–981
9. Jiménez J, López AM, Cruz J et al (2014) A web platform for the interactive visualization and analysis of the 3D fractal dimension of MRI data. J Bio Med Inform. Elsevier
10. Fustes D, Cantorna D, Dafonte C, Arcay B, Iglesias A, Manteiga M (2013) A cloud-integrated web platform for marine monitoring using GIS and remote sensing. Application to oil spill detection through SAR images. J Fut Gen Comput Syst. Elsevier
11. Lee J-Y, Yu W (2011) Visual tracking by partition-based histogram backprojection and maximum support criteria. In: 2011 IEEE international conference on robotics and biomimetics, pp 2860–2865. https://doi.org/10.1109/ROBIO.2011.6181739
12. Mohammadian H, Esfandiarijahromi E, Esmailani L (2012) Optimization real-time tracking of moving objects based on differential and active blob. In: International conference on intelligent systems (ICIS'2012), Penang, Malaysia, 19–20 May 2012
13. Nema R, Sazena AK (2013) Modified approach for object detection in video sequences. Am Int J Res Sci Technol Eng Math 122–126. ISSN:2328–3491
14. Mishra S, Mahanty C, Dash S, Mishra BK (2019) Implementation of BFS-NB hybrid model in intrusion detection system. In: Recent developments in machine learning and data analytics. Springer, Singapore, pp 167–175
15. Bhoi AK, Sherpa KS (2014) QRS complex detection and analysis of cardiovascular abnormalities: a review. Int J Bioautom 18(3):181–194
16. Bhoi AK, Sherpa KS, Khandelwal B (2018) Arrhythmia and ischemia classification and clustering using QRS-ST-T (QT) analysis of electrocardiogram. Clust Comput 21(1):1033–1044
17. Bhoi AK, Sherpa KS, Khandelwal B (2018) Ischemia and Arrhythmia classification using time-frequency domain features of QRS complex. Procedia Comput Sci 132:606–613
18. Bhoi AK, Sherpa KS (2016) Statistical analysis of QRS-complex to evaluate the QR versus RS interval alteration during ischemia. J Med Imaging Health Inform 6(1):210–214
19. Bhoi AK (2017) Classification and clustering of parkinson's and healthy control gait dynamics using LDA and K-means. Int J Bioautom 21(1)
20. Mishra S, Mallick PK, Tripathy HK, Bhoi AK, González-Briones A (2020) Performance evaluation of a proposed machine learning model for chronic disease datasets using an integrated attribute evaluator and an improved decision tree classifier. Appl Sci 10(22):8137
21. Jena L, Patra B, Nayak S, Mishra S, Tripathy S (2019) Risk prediction of kidney disease using machine learning strategies. Intelligent and cloud computing. Springer, Singapore, pp 485–494
22. Ray C, Tripathy HK, Mishra S (2019) Assessment of autistic disorder using machine learning approach. In: International conference on intelligent computing and communication. Springer, Singapore, pp 209–219
23. Mishra S, Koner D, Jena L, Ranjan P (2019) Leaves shape categorization using convolution neural network model. In: Intelligent and cloud computing. Springer, Singapore, pp 375–383

Student Behavioral Analysis Using Computer Vision

Anushka Sharma, Debasis Mohanty, Muhammad Fazal Ijaz, and Abu ul Hassan S. Rana

Abstract Monitoring students through the camera in the classroom has now become a conventional approach and is losing its effectiveness. A student's academic achievements depend on various factors. One of the main factors among these is student behavior and attention levels. For a long time, researchers and educators have heavily relied on teacher observation and student self-reporting to gauge student learning behaviors in class. The objective of this research work is to develop an automated model to enable informative-based functionalities to supervisors that helps the faculties to gather and summarize student behaviors in the classroom based on their facial expressions and other activities as a kind of information gathering to facilitate in making decisions. This paper uses some popular machine learning classifiers like Naive Bayes, neural network and KNN algorithm to train the model and used for pattern classification purposes. It was observed that an error rate of 0.2189 was minimum with diagonal seating arrangement while row-wise seating gave a high RMSE value of 0.3042. The highest classification accuracy rate of 92.5% was generated with the decision tree classifier while a relatively low accuracy of 84.6% was the outcome with Naive Bayes. The developed model can be effectively used to monitor the entire teaching session in the classroom or virtual classes, which helps faculties decide whether and when students are paying attention or not. This will result in an improvised learning session and will be more effective.

A. Sharma
School of Computer Engineering, Kalinga Institute of Industrial Technology (KIIT), Deemed To Be University, Bhubaneswar, Odisha, India
e-mail: 1829130@kiit.ac.in

D. Mohanty
iNurture Education Solutions Private Limited, Bengaluru, India
e-mail: debasis.m@inurture.co.in

M. F. Ijaz (✉) · A. H. S. Rana
Department of Intelligent Mechatronics Engineering, Sejong University, Seoul 05006, South Korea
e-mail: fazal@sejong.ac.kr

A. H. S. Rana
e-mail: rana@sejong.ac.kr

S. Dhar et al. (eds.), *Advances in Communication, Devices and Networking*,
Lecture Notes in Electrical Engineering 776,
https://doi.org/10.1007/978-981-16-2911-2_28

Keywords Student behavior · Teacher observation · Student self-reporting · Learning session

1 Introduction

Numerous criteria impact a student's academic performance. They can be the quality of teaching, study hours, learning environment and one of the most important things is student attention levels [1, 2]. Student behavior can be tracked by teachers by asking them questions time to time, observing them and conducting surveys. However, an accurate observation is not possible for a large crowd of students where teachers cannot keep a check on the behavior and activities of each and every student, and cameras fitted in classrooms are also not much effective and accurate as they are over the head and not that much precise [3]. So it is the need for the hour to propose a model that will help teachers in conducting a fair and precise observation and then take appropriate steps for a more productive learning session.

This paper deals with the presentation of an intelligent student behavior monitoring model using machine learning algorithms. The first section deals with the introductory part of research. The second section highlights the literature survey and related research works undertaken in the domain. The third section deals with the proposed methodology and its explanation. Result analysis forms the fourth section along with the visual representation of the results in graphs and tables. Finally, the last section concludes the research study.

2 Literature Survey

This section is about the state of art techniques, i.e. the already existing work in this field. The Easy with Eve acts as an ATS that provides learning aggregation task, which is helpful to preliminary level students. At first, the model was prepared as an automatic facial expression determination agent and gesture analysis but eventually, it was useful for capturing the expression of face during class sessions. The work discussed in [4] shows the engagement degree of students is directly proportional to critical thinking as well as grades fetched for any subject. The interest and curiosity of students also depend on the teachers [5]. The student's learning attitude will be somehow being in sync with the teacher's behavior and positive attitude. This will not only help teachers in their mental well-being but will also increase the interest and focus of students and will prove to be propitious. Additionally, Blatchford et al. [6] suggested that the perimeter of a classroom also impacts the attention level of students in a class. However, Raca et al. [7] highlighted that around 65% of students use to be attentive in class during lectures. Face detection and recognition of facial expression seem to be the most efficacious methods of classroom management to date. Studies show that cameras in classrooms are not much functional due to many factors such as the classroom brightness level, which depends on the windows are closed or open or the lights are on or off. These factors affect image quality and can produce distorted images. A research in [8] proved that an uncalibrated image further degrades the face identification accuracy rate. Camera calibration is among the vital criteria in face detection algorithms to work properly as low-resolution images are a backdrop for this system [9]. Face recognition is also a very crucial methodology for effective management in classroom. A study developed in [10] presented a model, which utilized a cascade CNN model to detect face and applied a ResNet1 01 layers convolution neural network for face recognition. Analysis of eye movement is an important metric to measure the attention level of students. The head motion and the gaze tracking levels were configured and analyzed to introduce a new mechanism [11]. Several such relevant works have been undertaken in this domain [12, 13, 15–20].

3 Proposed Methodology

An image is collected for analysis from the acquisition unit. Image is analyzed by calibration of the camera. The camera is used to retain the high resolution of images. Facial features are detected and relevant features are extracted for processing. Face tracking block helps in retrieving the features. Facial expressions and body language of students are tracked and recognized by comparing their real position and previous positions. All information details related to students' behavior and attention level in classrooms are stored into vectors. The mean average value of the attention level of every student is computed by the 'process data' module, which is later stored in the data store. Different machine learning algorithms like KNN, neural network and Naive Bayes were used for the face realization unit [12, 21]. They are responsible for the pattern matching with the images suggesting student attention level and are used to determine whether a particular student is attentive inside class during the lecture or not (Fig. 1).

4 Results Analysis

As many as 1200 students' data were used for the evaluation. Among them, 700 students' data were used as training set while remaining 500 were useful for testing purposes. The proposed model was evaluated using student detail information, which include ID of students, movement position of students, row and column number where they are seated. Performance metrics like accuracy rate and RMSE value was used for the validation [22, 23]. Since the data are unbalanced hence F-score metric was also used for the purpose [24]. The error rate was computed with respect to the seating position of students. Column-wise, row-wise and diagonal seating arrangement was considered for the demonstration using testing data. It was observed that an error rate

Table 1 Error rate analysis based on seating pattern

Seating arrangement	RMSE
Column-wise seated	0.2368
Row-wise seated	0.3042
Diagonally seated	0.2189

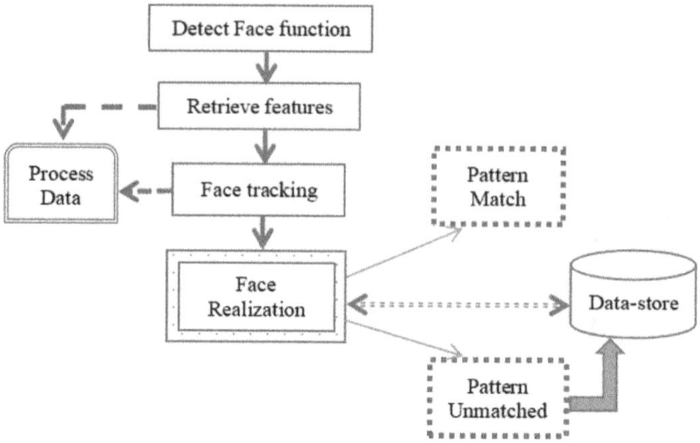

Fig. 1 Proposed model for student behavior detection

of 0.2189 was minimum with diagonal seating arrangement while row-wise seating gave a high RMSE value of 0.3042. Table 1 highlights the evaluation process.

Different machine learning classifiers were used in pattern evaluation when integrated with face realization unit. The highest classification accuracy rate of 92.5% was generated with decision tree classifier while a relatively low accuracy of 84.6% was the outcome with Naive Bayes. Figure 2 illustrates the classification accuracy comparison analysis.

5 Conclusion

This work presented an intelligent model to monitor classroom events. A research was undertaken on computer vision approach to be applicable for classroom scenarios. They were integrated in a workflow to showcase the areas of behavior of students as well as teachers during the classroom session and also to demonstrate the areas of interest and disinterest by the students as preliminary results of its prototype. We believe that this agent has the capability of transforming the classroom into a more discern environment, by providing guidance and feedback to both teachers and students by showing them where they are lagging behind and their faults and behavior so that they both can give their 100% on their part and increase the productivity of the learning session. The automated attention monitoring model can be successfully

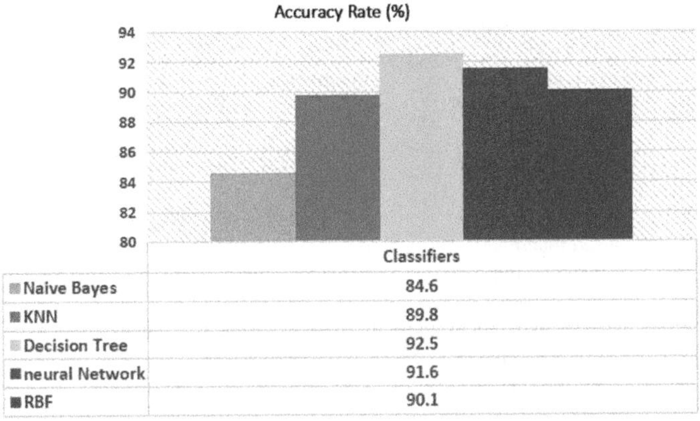

Fig. 2 Classification accuracy analysis using computational methods

used as an effective agent in automatic analysis in teaching-learning paradigm thereby guiding as an approach in gauging behavioral aspects of a student in the classroom. The objective of this work is to provide visual feedback to the supervisors about the average level of students' attention thus facilitating students' counseling of their behavior in class.

References

1. Grasha AF, 2nd (ed) (2002) Teaching with style: a practical guide to enhancing learning by understanding teaching and learning styles. Alliance Publishers, San Barnadino, CA, USA, pp 167–174
2. Richardson M, Abraham C (2013) Modeling antecedents of university students' study behavior and grade point average. J Appl Soc Psychol 43:626–637
3. Ning HK, Downing K (2011) The interrelationship between student learning experience and study behaviour. High Educ Res Dev 30:765–778
4. Carini RM, Kuh GD, Klein SP (2006) Student engagement and student learning: testing the linkages. Res High Educ 47(1):1–32
5. Hagenauer G, Tina H, Volet SE (2015) Teacher emotions in the classroom: associations with students' engagement, classroom discipline and the interpersonal teacher-student relationship. Eur J Psychol Educ 30(4):385–403
6. Blatchford P, Bassett P, Brown P (2011) Examining the effect of class size on classroom engagement and teacher-pupil interaction: differences in relation to pupil prior attainment and primary versus secondary schools. Learn Instr 21(6):715–730
7. Raca M, Kidzinski L, Dillenbourg P (2015) Translating head motion into attentiontowards processing of student's body-language. In: Proceedings of the 8th international conference on educational data mining. No EPFL-CONF-207803. APA
8. Zhao W et al (2003) Face recognition: a literature survey. ACM Comput Surv (CSUR) 35(4):399–458
9. Neves AJR, Trifan A, Cunha B (2014) Self-calibration of colormetric parameters in vision systems for autonomous soccer robots. In: Behnke S, Veloso M, Visser A, Xiong R (eds)

RoboCup 2013. LNCS (LNAI). Springer, Heidelberg, vol 8371, pp 183–194. https://doi.org/10.1007/978-3-662-44468-9 17
10. Fu R et al (2017) University classroom attendance based on deep learning. In: 2017 10th international conference on intelligent computation technology and automation (ICICTA). IEEE
11. Zhang K et al (2016) Joint face detection and alignment using multitask cascaded convolutional networks. IEEE Sig Process Lett 23(10):1499–1503
12. Mishra S, Mallick PK, Tripathy HK, Bhoi AK, González-Briones A (2020) Performance evaluation of a proposed machine learning model for chronic disease datasets using an integrated attribute evaluator and an improved decision tree classifier. Appl Sci 10(22):8137
13. Mishra S, Tripathy HK, Mallick PK, Bhoi AK, Barsocchi P (2020) EAGA-MLP—an enhanced and adaptive hybrid classification model for diabetes diagnosis. Sensors 20(14):4036
14. Mallick PK, Mishra S, Chae GS (2020) Digital media news categorization using Bernoulli document model for web content convergence. Pers Ubiquit Comput. https://doi.org/10.1007/s00779-020-01461-9
15. Mishra S, Mallick PK, Jena L, Chae GS (2020) Optimization of skewed data using sampling-based preprocessing approach. Front Public Health 8:274. https://doi.org/10.3389/fpubh.2020.00274
16. Bhoi AK, Sherpa KS (2014) QRS complex detection and analysis of cardiovascular abnormalities: a review. Int J Bioautom 18(3):181–194
17. Bhoi AK, Sherpa KS, Khandelwal B (2018) Arrhythmia and ischemia classification and clustering using QRS-ST-T (QT) analysis of electrocardiogram. Clust Comput 21(1):1033–1044
18. Bhoi AK, Sherpa KS, Khandelwal B (2018) Ischemia and Arrhythmia classification using time-frequency domain features of QRS complex. Procedia computer science 132:606–613
19. Bhoi AK, Sherpa KS (2016) Statistical analysis of QRS-complex to evaluate the QR versus RS interval alteration during ischemia. J Med Imaging Health Inf 6(1):210–214
20. Bhoi AK (2017) Classification and clustering of Parkinson's and healthy control gait dynamics using LDA and K-means. Int J Bioautom 21(1)
21. Mishra S, Mallick PK, Tripathy HK, Bhoi AK, González-Briones A (2020) Performance evaluation of a proposed machine learning model for chronic disease datasets using an integrated attribute evaluator and an improved decision tree classifier. Appl Sci 10(22):8137
22. Jena L, Patra B, Nayak S, Mishra S, Tripathy S (2019) Risk prediction of kidney disease using machine learning strategies. Intelligent and cloud computing. Springer, Singapore, pp 485–494
23. Ray C, Tripathy HK, Mishra S (2019) Assessment of Autistic disorder using machine learning approach. In: International conference on intelligent computing and communication. Springer, Singapore, pp 209–219
24. Mishra S, Koner D, Jena L, Ranjan P (2019) Leaves shape categorization using convolution neural network model. In: Intelligent and cloud computing. Springer, Singapore, pp 375–383

Mus-Emo: An Automated Facial Emotion-Based Music Recommendation System Using Convolutional Neural Network

Shubham Mittal, Anand Ranjan, Bijoyeta Roy, and Vaibhav Rathore

Abstract Facial expressions are considered to be the best way to identify human emotions. With the rise of digital music, an automated music recommendation system is beneficial for users to select suitable pieces of music from a large music repository. The framework presented over here is used to generate music according to the emotional state of mind of the person, which is being recognized through facial expression. First, the emotion of the user is being recognized by capturing the facial expression using Viola–Jones algorithm by identifying the Haar features. The emotions are then classified into five different classes of emotions using a convolutional neural network (CNN). Finally, the emotion is being sent to server and matched with the music label to be played. Experimental findings show that the proposed music recommendation based on emotion achieves an average accuracy of 73.05%.

Keywords Emotion · Convolutional neural network · Music recommendation · Facial expression · Viola–Jones

1 Introduction

Music plays a great role to mould and uplift the human state of mind. Music recommendation system in content and collaborative filtering techniques is based on historical preferences of user's choice of music [1]. It does not consider the emotional state or facial expression of the listener while recommending the music [1, 2]. The main purpose of music recommendation engine is to explore a certain pattern from the previous data of user's preference of music and suggesting music as per the need and requirement of the user. Facial expression helps in reading human mind and emotions. A system that can recommend music by understanding emotions from the facial expression of humans will have a significant influence on daily life of people. Manual browsing and selecting songs from the playlist to enhance the mood are

S. Mittal · A. Ranjan · B. Roy (✉) · V. Rathore
Sikkim Manipal Institute of Technology , SMU, Gangtok, Sikkim, India
e-mail: bijoyeta.r@smit.smu.edu.in

S. Dhar et al. (eds.), *Advances in Communication, Devices and Networking*,
Lecture Notes in Electrical Engineering 776,
https://doi.org/10.1007/978-981-16-2911-2_29

labor-intensive tasks and might not always land up in selecting appropriate song that can uplift the emotional state of mind. This motivated us to build a recommendation engine, which not only considers user's previous music choice history but also plays music based on the mood and emotion of the user that is captured through the facial expressions.

In this paper, we have proposed a model of emotion recognition and how it is connected to the application of playing music. This model is divided into three major parts, which is being explained in Sect. 3. First, the emotion recognition model is used to recognize the emotions via facial expressions of the user that is sad, happy, angry, neutral and surprise using convolutional neural network (CNN). The real-time image is being processed by the Viola–Jones algorithm, which is widely used for face detection and then the image is sent to convolutional neural network model. Second, emotion recognized is then sent into the server to verify the matching emotion file and play the suitable music.

In the emotion recognition model, to capture the real-time image of the user, we have used Viola–Jones algorithm, which is being used to detect the human face. This algorithm is used purposefully to detect the face of the user using Haar features and then the Adaboost feature of the algorithm is applied to eliminate the redundant features. Following this, the processing of the image in the CNN model is done to detect the emotion of the user, and generate the music from the server. Our model can be incorporated as part of the Spotify music player and snapchat stories. In Spotify music player, for example, if the user does not want to type the emotion of the music he wants to listen, he or she can simply use the emotion recognition option to play the music easily. This process is faster and less time-consuming. Similarly, in Snapchat stories, the user may want to play music in the background according to the emotional state he or she is in to make it more attractive. So far combining the emotion recognition and music recommendation together is not being extensively explored to be used properly in an application and to be used by the common people for listening to music. Rigorous training with large number of epochs is done but addition of more information and training the convolutional neural network model on new data is necessary to increase the accuracy for identifying the emotions is needed to produce more accurate results.

In this paper, in Sect. 2, we have discussed previous related works in this area followed by Sect. 3 where we have described the proposed framework for the emotion-based music generation model. Then in Sect. 4, the experimental results obtained by analyzing the results are shown, which is followed by a conclusion in Sect. 5.

2 Background Study

Due to the popularity of digital music in human life, music information retrieval has become important of research. In the last decade, the area of machine learning associated with automatic music generation has gain attention and was extensively explored by various authors. Out of the various recommendation search engines, the most commonly used were content-based recommendation and collaborative filtering techniques [3]. However, these traditional techniques do not take human emotion into consideration.

Shlok Gilda et al. [1] in their paper "Smart Music Player Integrating Facial Emotion Recognition and Music Mood Recommendation" used CNN for emotion recognition and received good results. Aurobind V. Iyer et al. [2] have proposed a system in which the face detection is done through Viola–Jones algorithm, and the emotion classification is done by using Fisherfaces Classifier. A facial recognition process based on Fisher Linear Discriminant was suggested and the proposed method achieved lower error rates as compared with Eigenface concept [4]. Hyeon-Jung Lee and Kwang-Seok Hong in their study [5] have developed an application in which they classified seven emotions using the deep learning technique of CNN plus positive and negative emotions using graphs and percentages. A transition network was developed based on the mental transition state of humans describing emotions such as sad, angry, happy, fear, surprise, etc. Transitions taking place between any two states are computed from the test dataset and represented it with some probabilistic values [6]. Man-Kwan Shan et al. [7] proposed a generic framework where the music emotion model was constructed from film music using music affinity graph, and their experimental results achieved an average accuracy rate of 85%. The important modules associated were extraction of features, detection of emotions and association discovery. In feature extraction process, chord tempo and rhythm were extracted. Viola–Jones face detection algorithm was deployed to detect face from the input image and then KNN classifier is used for the evaluation of facial and emotion detection [8]. Database is created by extracting the features of the face and this database is further used for evaluating the facial expressions and their associated emotions by using different standard algorithms.

3 Proposed Framework

a. Overview of the Recommendation system

The proposed framework consists of two models. First is the emotion detection model that captures the facial image of the user along with its expression and detects the emotion whether it is happy, sad, neutral, angry or surprise. The second part of the proposed work is the music recommendation model. Figure 1 depicts how appropriate emotion is captured from the facial expression.

In the emotion recognition model, the image of the user is captured using the webcam and the detection of face is done using Viola–Jones face detector algorithm. After the face of the user is detected, it is converted into 48 × 48 Gy scale image, which is then passed into the convolutional neural network trained model consisting of 11 hidden layers. The first layer and second layers from which the image is passed consist of 32 features of size 3 × 3 each. The output of the layers is then passed

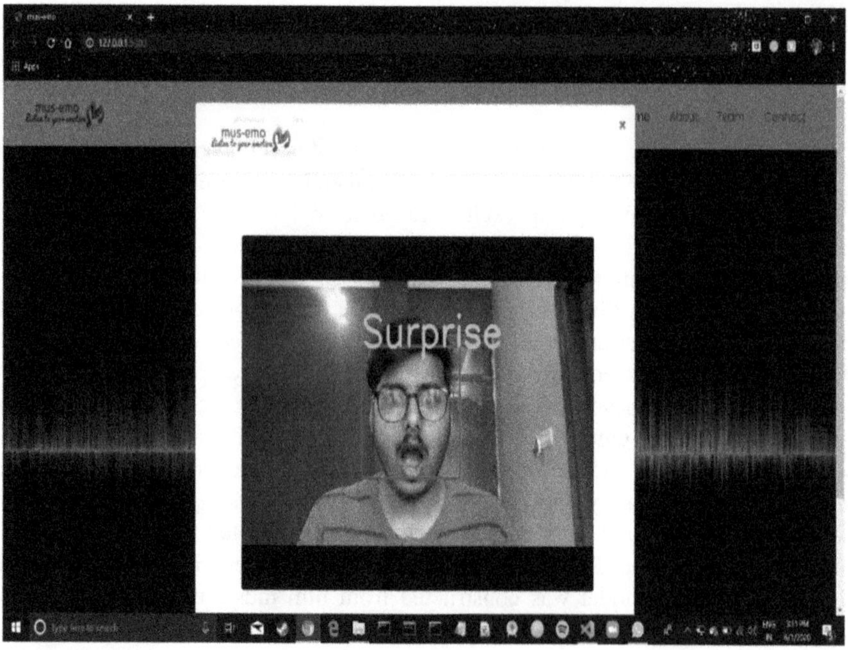

Fig. 1 Showing Facial Expression of being surprise

onto the third layer and fourth layer that consist of 64 features of size 3 × 3 each. Similarly, the fifth and sixth layers consist of 128 features of size 3 × 3 each and seventh and eighth layers consist of 256 features of size 3 × 3 each. Following the hidden layers, after every second layer, there is a max-pooling layer of pool size 2 × 2 and stride size of 2 × 2. At the end, there are two fully connected layers or dense layer having five neurons each neuron depicting the number of emotions. This is described diagrammatically in Fig. 4.

In music generating model, the user's emotion detected by the emotion recognition model is utilized to detect the music whose emotion has matching to it. Then the music is played according to the matching emotion captured to enhance the mood of the user. It helps in making the work easy and faster for the user to play the music based on his emotion manually. The pictorial description of this model is described in Fig. 6.

b. **Viola–Jones algorithm**

This algorithm is used to capture the image and process it for face detection and the feature elimination of redundant features. This algorithm works upon how two different parts of the face have different shades in the grayscale image [9]. This algorithm has four different stages. In the first stage, the input image is converted into an integral image. The value of the pixel (X, Y) is in an integral image represents the cumulative sum of all the pixel values above and the left of (X, Y). The same is depicted in Figs. 2 and 3.

Second, Haar features are used to detect the face in the image and analyze a particular sub-feature using characteristics consisting of two rectangles or more [10]. Then comes Ada-boosting, which is used to eliminate the redundant features. A strong classifier is constructed by a linear combination of the weak classifier.

$$\mathbf{F(n)} = \alpha_1 \mathbf{f_1(n)} + \alpha_2 \mathbf{f_2(n)} + \nu + \alpha_n \mathbf{f_n(n)}$$

F(n) is a strong classifier and $\alpha_1, \alpha_2, \ldots, \alpha_n$ are weights applied and f(n) are weak classifier. At last comes the cascading in which cascade of multiple classifiers can be used to increase system efficiency.

c. **Emotion Recognition Model**

The convolutional neural network model has been used to classify images according to their respective emotions. The model used batch training with the size of 32 images at a time to reduce the computational time of the system. Consequently, all images were passed through the CNN model using one forward pass method. After one

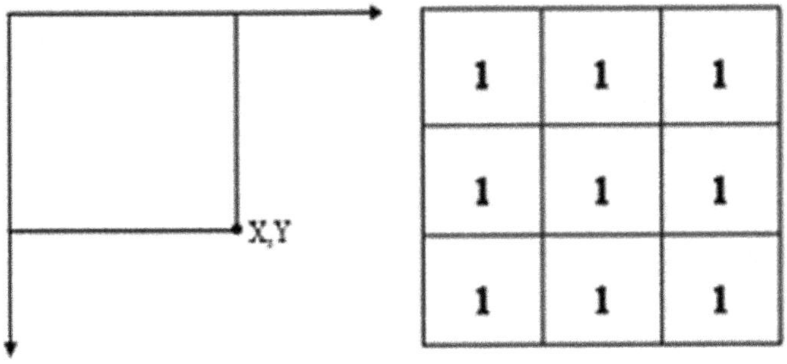

Fig. 2 Input image

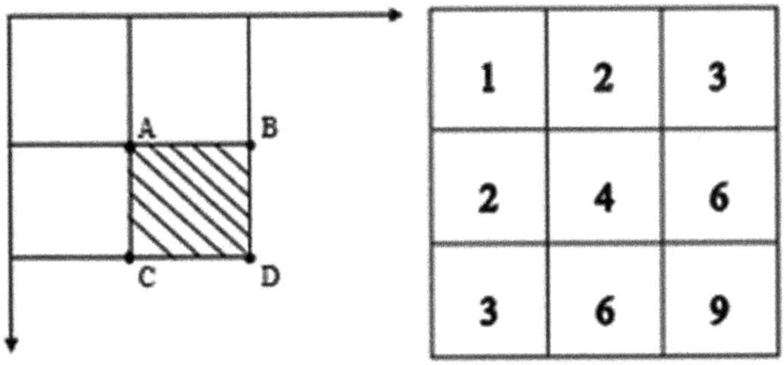

Fig. 3 Integral image

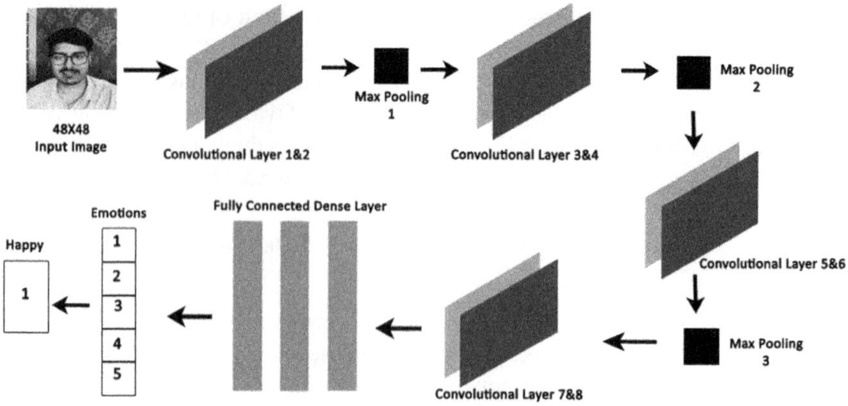

Fig. 4 CNN model for emotion recognition

forward pass, the categorical cross-entropy loss function is computed. To optimize weights, Adam has been used. 1 epoch took around 23 min and 45 s to run on a machine having 2.5 GHz Intel Core i5 7th Gen processor. The CNN model has 11 hidden layers with 8 convolutional layers followed by 3 max-pooling layers, which were alternated and also 3 fully connected layers having 5 neurons depicting 5 classes of emotions (Fig. 4).

Pseudocode for Model Creation:

1. Load the image to be processed
2. Resize the image into 48 × 48 dimension
3. Convert it into GRAY scale image from RGB
4. Stan CNN model to train for emotion recognition
5. model= sequential()
6. model.add
   - 6.1 CONV2D()
   - 6.2 Activation('relu')
   - 6.3 BatchNormalization()
   - 6.4 CONV2D()
   - 6.5 Activation('relu')
   - 6.6 BatchNormalization()
   - 6.7 MaxPooling2D()
7. Compile model and save model
8. Return weights

d. **Music Recommendation Model**

We have put forth a web application called "mus-emo" in this paper. This is a music recommendation app, which suggests songs to the users based on their emotions. In

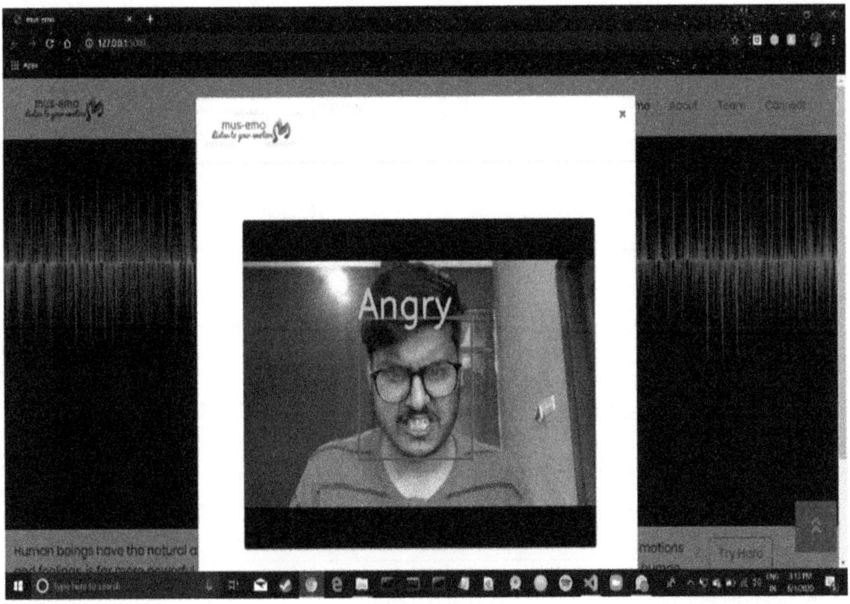

Fig. 5 Angry emotion recognized

this web application, first, an option is given to the user whether he/she wants to have a recommendation or not (Fig. 5).

If the user chooses "NO" option then the user is presented with five different emotion options, from which the user can select anyone. After the selection of the emotion, user will be presented with the list of the songs accordingly. Now in the other case, when the user selects the "YES" option. Then in that case, the camera activity starts and using the Viola–Jones algorithm, the face detection is done, and the live image of the user is captured after that the captured image is sent to the server where deep learning's CNN technique is used to predict the emotion of the user and according to the predicted emotion, the songs' playlist will be fetched and presented before the user. In the web application, songs are suggested to the user in such a way so that it can lift up their mood. For instance, if the user's current emotion is sad then the user will be presented with the happy songs, playlist for uplifting their mood. Figure 6 explains the entire system architecture in greater depth.

4 Results and Discussions

Dataset Information

The dataset used consists of 24,256 training and 3,006 validation 48×48 grayscale images, which are classified into five categories of emotions. Music files mostly

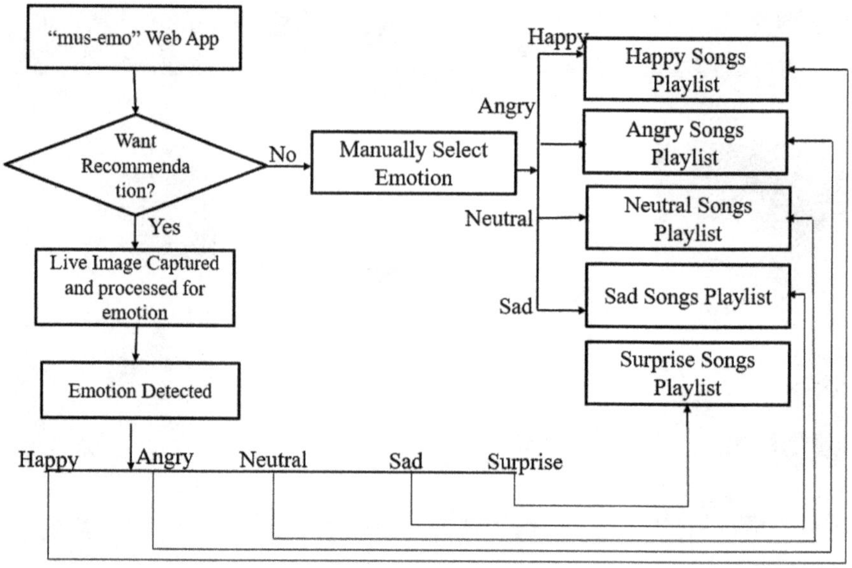

Fig. 6 Architecture of music recommendation model

contain songs released in albums of Beatles, Elvis Presley and some more renowned musicians.

The accuracy obtained by the model is 73.05%. It took 100 epochs to obtain the accuracy.

Figure 7 shows the loss that is the number of errors the model made while identifying and recognizing the emotion of the user. The orange graph line represents the testing loss and the corresponding blue line shows the training loss.

In Fig. 8, the accuracy of the model is shown. Accuracy graphs show that the model has comparable performance on both training and testing datasets. From the graph, it has been analyzed there is little overfitting, which is caused due to the insufficient data present in the dataset used.

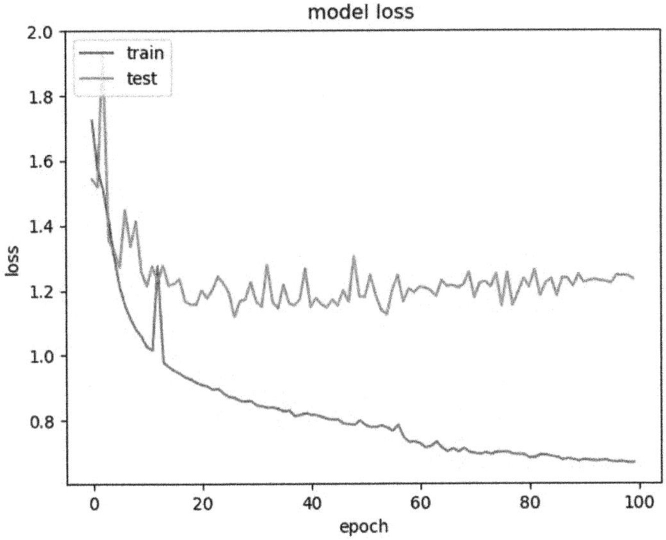

Fig. 7 Loss function

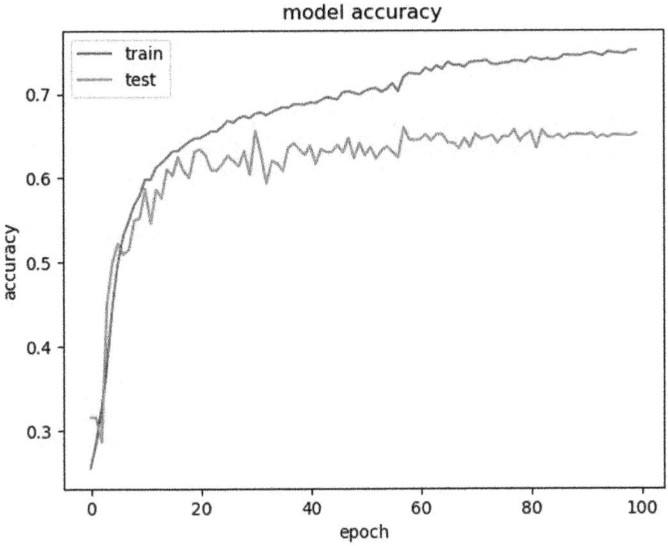

Fig. 8 Accuracy graph

5 Conclusion

In this paper, Mus-Emo framework is being presented, which is able to play music according to the facial emotion recognized. It is able to do so by first detecting the face in the real-time image captured using webcam or front mobile camera and then processing it using CNN model to detect the emotional status of the user via facial expressions. Then the label of the emotion assigned is being passed into the server to match up with the music file. Finally, the music is played according to the emotion detected. This process is faster and easy way to play relevant music similar to the emotional state of the user. The accuracy achieved by this model is 73.05%. The proposed model can be used in the corporate world of music as a feature to reduce time for selection of music or can be integrated into the applications, which deals with the images like Instagram, Snapchat, Twitter and many more to play music in the background of the image automatically according to the emotion depicted in the image.

References

1. Gilda S, Zafar H, Soni C, Waghurdekar K (2017) Smart music player integrating facial emotion recognition and music mood recommendation. In: 2017 international conference on wireless communications, signal processing and networking (WiSPNET), Chennai, pp 154–158. https://doi.org/10.1109/WiSPNET.2017.8299738
2. Iyer AV, Pasad V, Sankhe SR, Prajapati K (2017) Emotion based mood enhancing music recommendation. In: 2017 2nd IEEE international conference on recent trends in electronics, information & communication technology (RTEICT), Bangalore, pp 1573–1577. https://doi.org/10.1109/RTEICT.2017.8256863
3. Isinkaye F, Folajimi Y, Ojokoh B (2015) Recommendation systems: principles, methods and evaluation. Egypt Inf J 16(3):261–273
4. Belhumeur P, Hespanha J, Kriegman D (1997) Eigenfaces versus Fisherfaces: recognition using class specific linear projection. IEEE Trans Pattern Anal Mach Intell 19(7):711–720
5. Lee H, Hong K (2017) A study on emotion recognition method and its application using face image. In: 2017 international conference on information and communication technology convergence (ICTC), Jeju, 2017, pp 370–372
6. Xiang H, Ren F, Kuroiwa S, Jiang P (2005) An experimentation on creating a mental state transition network. In: Proceedings of the 2005 IEEE international conference on information acquisition, pp 432–436
7. Shan MK, Kuo FF, Chiang MF, Lee SY (2009) Emotion-based music recommendation by affinity discovery from film music. Expert Syst Appl 36(4):7666–7674
8. Reney D, Tripaathi N (2015) An efficient method to face and emotion detection. In: Fifth international conference on communication systems and network technologies
9. Pantic M, Rothkrantz JM (2004) Facial action recognition for facial expression analysis from static face images. IEEE Trans Syst Man Arid Cybern Part B 34(3):1449–1461
10. Viola P, Jones MJ (2004) Robust real-time object detection. Int J Comput Vis 57(2):137–154
11. Madhok R, Goel S, Garg S (2018) SentiMozart: music generation based on emotions. ICAART
12. Matre GN, Shah SK (2013) Facial expression detection. In: 2013 IEEE international conference on computational intelligence and computing research, Enathi, 2013, pp 1–3

Secure-M2FBalancer: A Secure Mist to Fog Computing-Based Distributed Load Balancing Framework for Smart City Application

Subhranshu Sekhar Tripathy, Rabindra K. Barik, and Diptendu Sinha Roy

Abstract Nowadays, urban communities are planned to change to a smart city. As per late investigations, the utilization of information from contributors and physical items in numerous urban communities plays a key component in the change toward a smart city. The 'smart city' standard is described by omnipresent computing resources for the watching and basic control of such city's structure, medical services, environment, transportation, and utilities. Mist computing is considered as registering model, which performs IoT applications at the edge of the physical devices. To keep up the quality of service (QoS), it is great to utilize context-aware computing as well as cloud computing simultaneously. In this article, the author implements an optimization strategy applying a dynamic resource allocation method based on genetic algorithm and reinforcement learning in combination with a load balancing procedure in a secured environment. The proposed Secure-M2FBalancer model directs the traffic in the network unremittingly, gathers the data about each server load, moves the incoming query, and distributes them among available servers equally utilizing dynamic resource allocation strategy. The proposed model comprises four layers, i.e. IoT layer, mist layer, fog layer, and cloud layer. Finally, the results demonstrate that the solutions enhance QoS in the mist-assisted cloud environment concerning maximization of resource utilization and minimizing the makespan with advanced security features. Therefore, Secure-M2FBalancer is an effective method to utilize the resources efficiently by securely ensuring uninterrupted service.

Keywords Fog computing · Cloud computing · Mist computing · Load balancing · Security · Smart cities

---

S. S. Tripathy (✉) · D. S. Roy
Department of Computer Science and Engineering, National Institute of Technology Meghalaya, Shillong, India

D. S. Roy
e-mail: diptendu.sr@nitm.ac.in

R. K. Barik
School of Computer Applications, KIIT Deemed To Be University, Bhubaneswar, India

S. Dhar et al. (eds.), *Advances in Communication, Devices and Networking*, Lecture Notes in Electrical Engineering 776,
https://doi.org/10.1007/978-981-16-2911-2_30

1 Introduction

With the help of the Internet of Things (IoT), the interconnected set of devices, physical machines, and sensors to communicate data over the network without the interference of humans [1]. Every day, a day-to-day existence utilization of IoT is available, for example, medical care applications. To get an advantage from IoT gadgets, it is fundamental to interface with fog computing or mist computing. The utilization of IoT objects plays an imperative part in the medical care segment. It incorporates remote health check-ups, old-age care, and incurable diseases [2].

The utilization of sensors builds the exhibition of the medical care framework and limits the expense of the administration this prompts expands the effectiveness. For the most part, different types of sensors are utilized for various applications in the clinical division. Heartbeat sensors are utilized for the beats of the body a rundown of genuine cases, for example, cardiovascular failure, low and hypertension, and respiratory issues can discover [3]. Temperature sensors are utilized to distinguish the temperature of the body [4]. Pulse (BP) sensors are mindful to quantify hypertension [5]. Heartbeat oximetry sensors are used to discover the level of oxygen in the blood [6].

The principal goals of fog computing are to limit latency, utilization of energy, and bandwidth efficiency [7]. In a fog environment, the utilization of cloud is to bring the service close to the clients, however, for a smaller scale. The assignment with more computational complexity, however, less delay sensitivity might be sent to the cloud whereas assignment with lower computational complexity might be sent to the closer fog server. Cloud-fog computing architecture is composed of a four-layer architecture that is an IoT layer, mist layer, fog layer, and cloud layer. The fog layer is partitioned into various areas; thusly, two possibilities may emerge one, server of a region may be over-burden, and second, there are possibilities of availability of ideal resources in different locales. For the calculation of incoming requests, it needs various resources dependent on the sorts of calculation, the load balancing mechanism will distribute resources. Even though they are having requirements, for example, different kinds of computational resources are available in the computational condition. If a fog server is becoming busy, at that point, the quality of service of the framework will not be maintained due to which results in a higher delay.

The detail of the present paper is explained as follows: Sect. 2 presents the related works and presents a detailed overview from the perspective of cloud, fog, mist, and IoT computing in a cooperative environment and security. Section 3 explains the proposed model, for load balancing and task scheduling with advanced security features. Section 4 highlights the experimental setup and analyses the simulation-related result followed by a discussion. Finally, Sect. 5 ends with concluding remarks with future insight.

2 Related Work

2.1 Smart City

The smart city concept is being developed for enhancing the economy, portability, ecosystem, people, and standards of living and management of cities. The smart city aimed for providing basic and necessary facilities to its citizens. They all are having equipped with various objects for seamless solutions to their problems. Nowadays, cities across the world collect a massive amount of data that are associated with all the requirements of living organisms stayed in the city. These collections of a huge amount of data will lead to the creation of useful and significant content that will help the technology to solve the complicated problems faced by the citizens of the smart city [8].

A vast analysis had been done how the use of a huge amount of data in smart city and related issues in specific fields of transportation, public healthcare management, energy distribution, and sustainability [9]. Figure 1 illustrates the various components associated with the smart city.

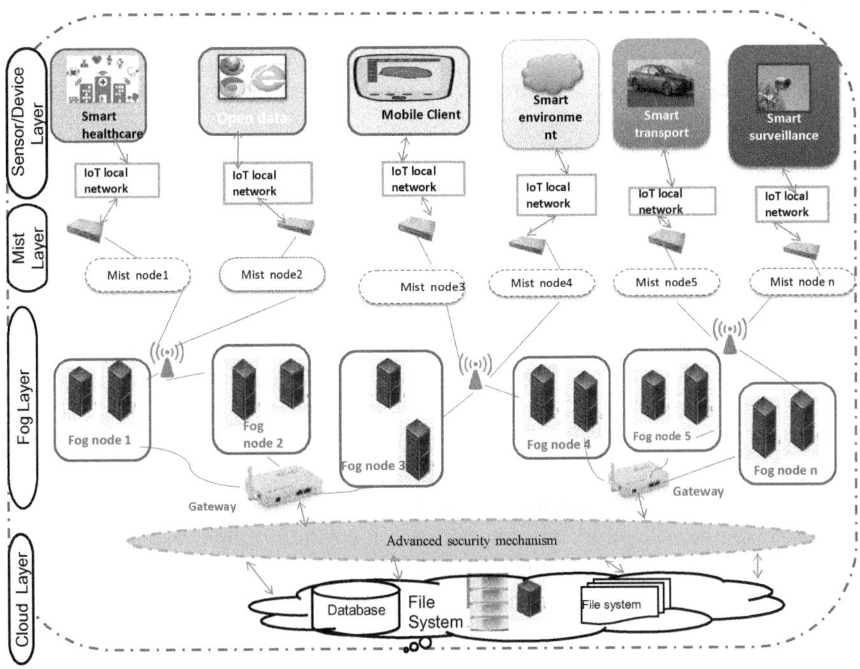

Fig. 1 General structure of smart city

2.2 Cloud, Fog, and Mist Computing

Cloud computing is characterized as the arrangement of inter-linked thousands of PCs, having various quantities of applications and records facilitated on a cloud and all the hubs are intricately associated through the web. It additionally executes the idea of parallel and distributed computing. Fog computing is a networking architecture that utilizes IoT gadgets for doing computation, storage, and communication in a predetermined region [10, 11]. Fog computing is a cooperative framework where IoT applications are introduced. Various challenges faced by the computing environment such as architecture, interfacing, and programming offloading of computation, resource allocation, load balancing, the security of data, and storage of data are discussed in detail in ref. [12]. During the process of transmission of data between the IoT-enabled devices and fog servers, there will be an impact of processing delay and transmission delay [13].

Mist computing gives the facility of, the arrangement of computing facility on the edge of the genuine gadgets, of the organization. The microchips or microcontrollers connected to the physical items will figure the entire information detected by the objects. In the wake of figuring this information, they will move this information to the organization. Mist computing environment requires less cloud storage and transmission power for long-term examination information. For example, the mist computing environment is working close to the edge, physical gadgets; it gives low power gateway that expanded throughput with reduced latency at the edge of the gadgets.

2.3 Load Balancing Strategy

The strategies for distributing incoming requests among all the servers similarly are known as load balancing. The objective of load balancing in fog computing is to affirm that there is no over-burden. The merits of using load balancing techniques are (1) waiting for time reduction, (2) minimization of response time, (3) efficient resource utilization, (4) improves the throughput, and (5) enhance overall performance and efficiency of the system. The Secure-M2FBalancer is a traffic checking program among server and client demands. The fundamental objective of the balancer is to increase resource utilization. It is critical to distribute the available resources between various servers available in a viable way. The advantages of using dynamic resource allocation in a computing environment are lessens the area overhead and no software or hardware overhead [14]. The load balancing strategy is actualized to scale up the capacity to expand the server limit by moving the overloaded tasks among accessible underloaded servers. For every server, it is up to date the information continuously based on the sequence of connections [15]. The load balancer is having the list of free servers, and the approaching query will be sent similarly to everybody in sequence [16]. It is similar to the round-robin in assigning the task in a time-shared manner but it varies in the sense that the job is having the highest number of requests will be assigned a higher weight [17]. Dynamic round-robin is a method of assigning the weight to the servers depending on the server load and capacity. As per the server load, the load balancer moves assignment to the compatible server [18].

2.4 Security

With the development of cloud computing, some related works have tended to some security issues in the cloud, for example, the security for the cloud system, area protection in the portable cloud, security in cloud storage. The pattern is to have this immense information put away in an untrusted outsider cloud framework in an encoded structure because of the privacy and affectability of the datasets. The major difficulty incorporate is that data are situated in separate places and encrypted with various keys. To ensure the protection of data, all calculations, intermediate outcomes created during the process must be secure. To improve the effectiveness, calculation ought to be finished by the cloud server to diminish the calculation/transmission cost of the data providers.

3 Proposed Security Model

Here, it portrays a mist to cloud-based load balancing algorithm and optimization strategy in a cooperative fog computing environment. In this paper, the authors proposed a mist computing-based load balancing algorithm for a four-layer-based framework with advanced security features. It is implementing an IoT-fog-based healthcare management system, which is shown in Fig. 2. Among these four layers, the role of the IoT layer is to receive pulses from the patients. The mist layer is responsible for allocation and migration to a suitable server. The third layer is the fog layer, which is used to decide the status of the server and migrate the job from an overloaded server to an underloaded server. The fourth layer that is cloud layer is responsible for the storage of secured forwarding of data to and from the fog layer.

The incoming request from different regions will be assigned with priorities by the Secure-M2FBalancer, and it will assign the optimized scheduling algorithm for processing the incoming task. In consequence, the Secure-M2FBalancer gets the least execution time and minimum waiting time. The primary goal behind the cooperative architecture is to attain a low response time. The fog layer is having a Secure-M2FBalancer module that is responsible to determine the status of the server such as underloaded, overloaded, and normal. The allocation cost and response time are also minimized so that it enhances the QoS in the proposed environment. The Secure-M2FBalancer is capable to pick the specific server and can handle the incoming request. Cloud security model and algorithms are explained in Sections I and II, respectively.

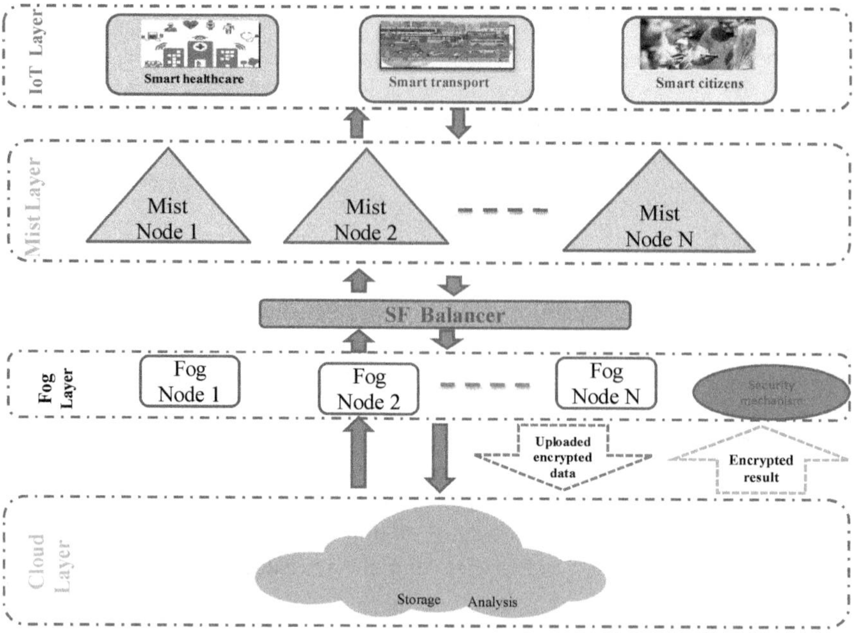

Fig. 2 Proposed Secure-M2FBalancer model

i. Cloud security model
Step 1. Client- >Request register key- >Cloud authentication system(CAS)
Step 2. TTP-based CAS- > one series of key(Id0)- > Client
Step 3. Client request cloud server with the register key
Step 4. Cloud server confirms the register key in the CAS
Step 5. CAS sends the acknowledgment to the cloud server
Step 6. Cloud server sends the requested data to the client authentication encryption key:Id1(Id1 = h(Id0)) transmission encryption material (P = Id1 + Id0 {Data}) to Cloud Server.$C \rightarrow S$: Idn + 1 = h(Idn),P = Idn + 1 + Idn{Data} (Initial Value: 0)
Step 7. Cloud server receives the encryption material and uses originalIdn, Hash Function to obtain Idn + 1 confirmation comparison Step 6 Idn + 1 weather Is equal to Step 7 Idn + 1; if equal, then it uses Kn to decipher the material. S: use Kn. Create Idn + 1 (Idn + 1 = h (Idn)). If Idn + 1 = Idn + 1 $\Rightarrow$Identify Cthen use Kn decryption K{Data} S-c: Idn + 1 = h (Idn),P = Idn + 1 + Idn{Data}

ii. Resource Migration-based algorithm
Input:Task, resource, the information in the SIT Output:Achieve a normalized system with low response time and high performance
Start Step 1: for all the tasks Do: Create a Tasklist table consisting of two columns [State: available, under loaded and normalize mist server, Action: selecting the compatible mist server to execute task]initialized to zero;
Step 2: Initiate the Task assignment Algorithm;
Step 3: The user interacts with mist server and updates to state-action pair in the task list table;
Step 4: The user selects the compatible mist server based on low response time by looking to Tasklist table;
Step 5: Updating the Tasklist table;
Step 6: Modify the values in the task list table;
Step 7: Transfer the overloaded task to the compatible under loaded resources;
Step 8: Broker update information in SIT by updating the load and status for each mist server

4 Result and Analysis

To authenticate the viability of Secure-M2FBalancer, all the trials have been simulated on MATLAB. The proposed load-balancing algorithm is being contrasted and recently utilized calculations in mist registering for least association, round-robin, and weighted round-robin. The simulation results show that the proposed technique gives an eminent improvement for makespan and resource utilization. Dynamic autonomous tasks with variable lengths are considered to approve the viability of the proposed algorithm. The proposed security algorithm provides a secure environment for the transmission of data in the cooperative mist-assisted framework (Fig. 3).

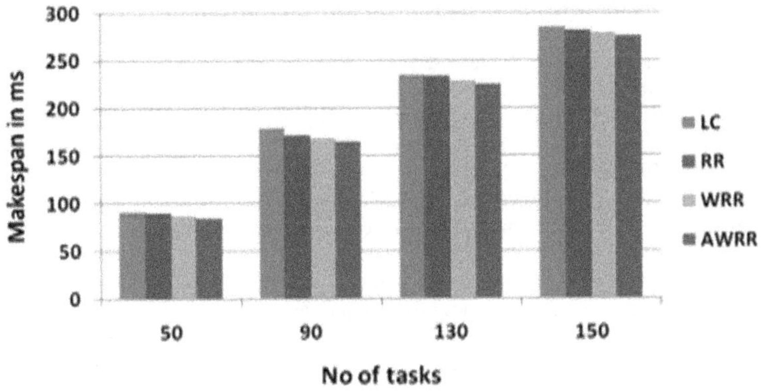

Fig. 3 Makespan for AWRR with the compared algorithms

5 Conclusion

In this paper, it actualizes a dynamic resource assignment technique utilizing a supported learning and genetic algorithm in a mix with a load balancing method. The proposed Secure-M2FBalancer directs the traffic in the organization ceaselessly, gathers the data about a load of every server, move the on-request demand, and scatter them among accessible server consistently. The investigations are led, and the recreation results show that the proposed algorithm improved the considered QoS boundaries of the mist assisted cloud computing environment regarding the average resource utilization and makespan. Subsequently, Secure-M2FBalancer is a compelling method to build up a high usage of resources by guaranteeing continuous service. The implementation of the cloud security model provides security for secure transactions in formations in a cooperative framework. In future work, we are planning to implement other security algorithms in a cooperative mist-assisted framework.

References

1. Ghobaei-Arani M, Souri A, Rahmanian AA (2019) Resource management approaches in fog computing: a comprehensive review. J Grid Comput 18(2):1–42
2. Shang C, Chang C-Y, Liu J, Zhao S, Roy DS (2020) FIID: feature-based implicit irregularity detection using unsupervised learning from IoT data for homecare of elderly. IEEE Int Things J 7(2):10884–10896
3. Islam SR, Kwak D, Kabir MH, Hossain M, Kwak KS (2015) The internet of things for health care: a comprehensive survey. IEEE Access 3(2):678–708
4. Zenko J, Kos M, Kramberger I (2016) Pulse rate variability and blood oxidation content identification using miniature wearable wrist device. In: International conference on systems, signals and image processing (IWSSIP), pp 1–4
5. Narczyk P, Siwiec K, Pleskacz WA (2016) Precision human body temperature measurement based on thermistor sensor. In: IEEE 19th international symposium on design and diagnostics of electronic circuits & systems (DDECS), pp 1–5
6. Aqueveque P, Gutiérrez C, Rodríguez FS, Pino EJ, Morales AS, Wiechmann EP (2017) Monitoring physiological variables of mining workers at high altitude. In: IEEE transactions on industry applications, vol 53(2), 2628–2634
7. Gubbi SV, Amrutur B (2015) Adaptive pulse width control and sampling for low power pulse oximetry. In: IEEE transactions on biomedical circuits and systems, vol 9(2), pp 272–283
8. Hussain M, Beg MM (2019) Fog computing for internet of things (IoT)-aided smart grid architectures. Big Data Cogn Comput 3(1):12–23
9. Reddy KHK, Luhach AK, Pradhan B, Dash JK, Roy DS (2020) A genetic algorithm for energy efficient fog layer resource management in context-aware smart cities. Sustain Cities Soc 63:102428
10. Clarke A, Steele R (2011) How personal fitness data can be re-used by smart cities. In: IEEE seventh international conference on intelligent sensors, sensor networks and information processing, pp 395–400
11. Roy DS, Behera RK, Reddy KHK, Buyya R (2019) A context-aware Fog enabled scheme for real-time cross-vertical IoT applications. IEEE Internet Things J 6(2):2400–2412

12. Reddy KHK, Behera RK, Chakrabarty A, Roy DS (2020) A service delay minimization scheme for QoS-Constrained. Context-Aware Unified IoT Appl IEEE Internet Things J 7(10):10527–10534
13. Yuxuan J, Huang Z, Tsang DHK (2018) Challenges and solutions in Fog computing orchestration. IEEE Netw 32(3):122–129
14. Liu Y, Fieldsend JE, Min G (2017) A framework of fog comput-ing: architecture, challenges, and optimization. IEEE Access 5:25445–25454
15. Mishra K, Majhi S (2020) A state-of-art on cloud load balancing algorithms. Int J Comput Digit Syst 9(2):201–220
16. Barik RK, Dubey H, Samaddar AB, Gupta RD, Ray PK (2016) FogGIS: fog computing for geospatial big data analytics. In: IEEE international conference on electrical, computer and electronics engineering (UPCON), pp 613–618
17. Youm DH, Yadav R (2016) Load balancing strategy using round robin algorithm. Asia-pacific J Converg Res Interchang 2(3):1–10
18. Phi NX, Hung TC (2017) Load balancing algorithm to improve response time on cloud computing. Int J Cloud Comput: Serv Archit 7(6):1–12

Intelligent Node Placement for Improving Traffic Engineering in Hybrid SDN

Mir Wajahat Hussain and Diptendu Sinha Roy

Abstract Software-defined network (SDN) has become quite promising with separating control plane from the data plane but the considerable capital requirements limit the full deployment of SDN, and hence a desirable way out is to move to a hybrid SDN (h-SDN). Traffic engineering (TE) has become quite a concern, which optimizes the network performance. A critical issue in the network is how to lower the maximum link utilization, which directly affects the network performance. This issue is handled with the proper placement of SDN nodes add, thus, ensuring flexibility in the network. Existing works have only focused on the greedy placement of SDN nodes, which is oblivious of the traffic details and data emanating to/from the nodes. This paper proposes a novel intelligent node placement (INP) scheme whereby SDN nodes get placed based on both the degrees of a node and traffic characteristics of the network. Experiments performed in MATLAB show that the MLU is reduced by an average of about 5–19% as compared with the state of art protocols in various real-time topologies.

Keywords Software-defined network · Link utilization · Traffic engineering · Openflow

1 Introduction

Software-defined networking (SDN) has evolved as a prominent network paradigm to address several concerns, which were intrinsic in the traditional networking paradigm [1]. SDN leaves out the control plane from the forwarding plane and hence allows both of them to be evolved independently [2]. The communication between control and data plane is allowed through a standard interface called OpenFlow [3]. SDN

M. W. Hussain (✉) · D. Sinha Roy
Department of Computer Science and Engineering, National Institute of Technology, Meghalaya, Shillong, Meghalaya, India
e-mail: mir.wajahat.hussain@nitm.ac.in

D. Sinha Roy
e-mail: diptendu.sr@nitm.ac.in

S. Dhar et al. (eds.), *Advances in Communication, Devices and Networking*,
Lecture Notes in Electrical Engineering 776,
https://doi.org/10.1007/978-981-16-2911-2_31

devices forward data based on the flow concept as contrary to destination-based forwarding in traditional devices. The flow-based forwarding ensures flexibility in the network by splitting the flows via various paths based on its QoS. However, due to the high capitals required for replacing the existing network architecture with SDN, a full deployment is once in a blue moon. Thus, a possible way out is to utilize both the centralized and decentralized paradigms (h-SDN) to manage and configure a network [4, 5].

TE has been a core concern, which ensures minimizing both overall cost and MLU in a network and is of utmost importance in h-SDN owing to the running of multiple paradigms [5]. TE plays a pivotal role in optimizing the network performance by noting the pattern for transmitting data over the links for effective routing. Routing in traditional devices utilizes Open Shortest Path First (OSPF) protocol, which routes data through the shortest paths. Routing traffic along the shortest paths might not utilize the network effectively as some paths get utilized more as compared with paths with higher cost. Since SDN has the possibility of splitting the flows across any outgoing links via shortest/non-shortest paths, thus, placement of SDN is crucial to address the concerns for efficient TE. As pointed out earlier, the transition of legacy to a full SDN deployment is not going to happen soon, so careful placement of SDN nodes is the need for the hour to utilize the links effectively during the routing in the network.

Previous research attempts for the placement of SDN nodes have placed the nodes in a greedy-based approach (place nodes where the numbers of links are more) or some works have considered placement of SDN based on the traffic size [6]. Also, since SDN nodes arbitrarily split the outgoing flows so effective planning must be devised not leading to forwarding loops or black holes in the network [4]. Furthermore, some works have increased the number of outgoing paths after carefully planning the migration sequence of the SDN nodes but these works did not consider all the paths for routing the network owing to QoS violation and unwanted high bandwidth requirements [6].

In this paper, the placement of SDN nodes is done in a h-SDN. The placement of SDN is done by considering traffic requirements of the node as well as traffic from other intermediate nodes, links which are highly/under-utilized and thus ensuring loop-free paths in the network. Experiments performed in the paper suggest that the MLU reduces by an average of about 5–19% when considering several real-time topologies.

The rest of the paper is organized as follows: Section 2 deals with the related works. The proposed INP is presented in Section 3. Flowchart for the INP scheme is described in Section 4. Section 5 presents the experimental details, results and discussion. Section 6 presents the further avenues of the research and, thus, concludes the paper.

2 Related Work

Several research attempts have focused on the incremental deployment of SDN nodes in h-SDN. Some of them are discussed as;

Aggarwal et al. [7] proposed a full polynomial-time approximate algorithm (FPTAS) in a partial deployment. The placement of SDN nodes was guided by the network topology, and the placement was determined in advance. Besides, the authors used random and greedy solutions to place the SDN nodes. Marcel et al. [6] proposed a migration scheduling algorithm where key nodes are to be migrated to improve the TE in the partial deployment. The key nodes considered for migration were based on the greedy mechanism (the place where alternative paths are more). The placement of SDN nodes was not based on the traffic size. Sun et al. [8] proposed a multipath load balancing in the context of h-SDN. The authors propose a routing algorithm based on Dijkstra-Repeat algorithm where multiple paths can be given to flows, which have to traverse the same destination thus ensuring load balancing. This arbitrary splitting by the SDN nodes does not guarantee loopless paths owing to running multiple paradigms in the h-SDN. Fakhteh et al. [9] clustered the network based on the concept of closeness-centrality (average graph distance to other nodes) and then addressed the migration problem in a h-SDN. This work had addressed several issues including improvements in the network control ability (NCA), minimizing the cost of upgradation for improving NCA, increasing flexibility and ensuring link fault recovery algorithm. Chen et al. [10] formulated an optimization problem and hence an incremental strategy of HybridScore is proposed to upgrade nodes to SDN. HybridScore utilizes the real network pattern and topology for upgradation of legacy to SDN nodes in the network. In each iteration of the network, only a single node was upgraded to SDN pertaining to several constraints like SDN flow table size, VLAN size of traditional switches and the link capacity. Guo et al. [11] formulated an optimization problem for the incremental deployment of SDN nodes in the network. The authors propose a heuristic-based genetic algorithm to seek a migration sequence of legacy-enabled routers to SDN routers for improving TE. Vissicchio et al. [12] defined several types of h-SDN models viz. topology-based, service-based, class-based and integrated-based. These models were classified based on the network services offered and the usage scenario but, however, this work did not focus on improving the link utilization in h-SDN.

All of the above works have primarily focused on the placement of SDN nodes based on the greedy sequence. The placement of SDN nodes based on the greedy mechanism does not guarantee loopless paths and hence exacerbates the costly bandwidth in the network. The proposed INP scheme is promising on account of placement based on the incoming traffic matrix, ensuring loopless paths and the link utilization is kept steady, which ensures QoS is met. The subsequent subsection presents the routing scenario of OSPF in traditional networks and the greedy-based mechanism in an incremental SDN deployment.

2.1 OSPF Routing in Traditional Networks

Consider a network comprised of four nodes as described in Fig. 1. Each node is linked with a server, which is implicit. Assume the cost of all the paths is same. Each row represents the traffic emanating from the respective node starting as A, B, C and D. Let the utilization of the network is represented by a 4 × 4 matrix shown as X. Shortest paths for node A, B, C and D are (A-B, A-C, A-B-D), (B-A, B-C, B-D), (C-A, C-B, C-D) and (D-B-A, D-B, D-C), respectively.

$$X = \begin{pmatrix} \mathrm{A} & \mathrm{B} & \mathrm{C} & \mathrm{D} \\ - & 2 & 4 & 6 \\ 2 & - & 6 & 4 \\ 4 & 2 & - & 2 \\ 4 & 1 & 2 & - \end{pmatrix}$$

As pointed out earlier, all the cost moving to the next node is same so based on the shortest path routing, the overall link utilization is described in Fig. 1. From the figure, we can see some links are highly utilized (A-B is 14, B-D is 15), and some are lightly utilized (A-C is 8, B-C is 8, C-D is 4). The traffic utilization matrix suggests that both nodes A and B transfer the same amount of data 12, which is higher than both C and D (8 and 7). To find the links that are highly utilized, compute

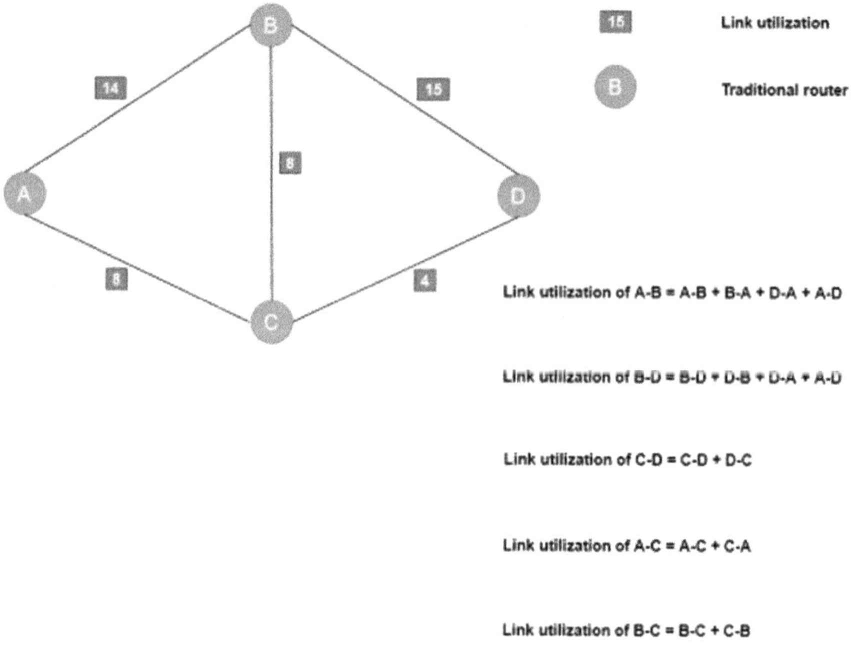

Fig. 1 OSPF routing in traditional network

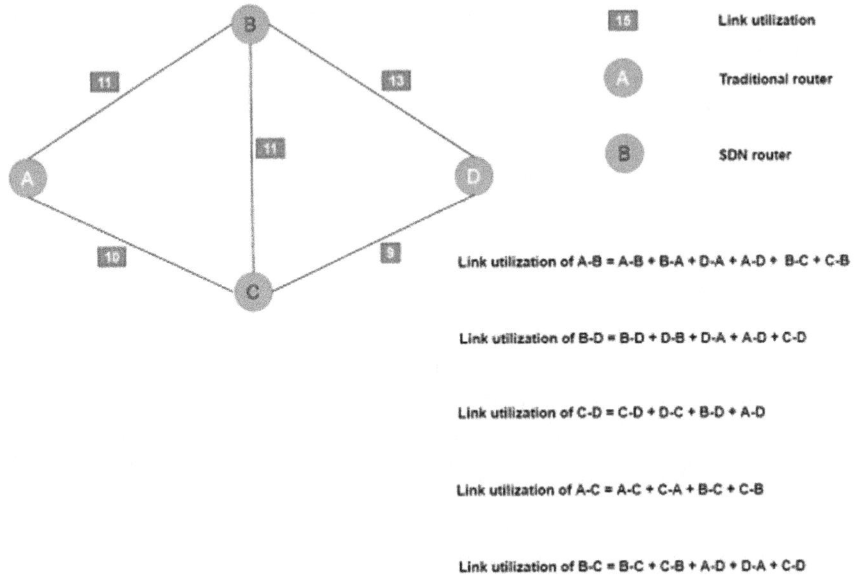

Fig. 2 Greedy-based placement of SDN nodes in h-SDN

the average of all the links. After calculation of the average, calculate the standard deviation [3]. Links whose utilization comes beyond a level greater/lesser with the standard deviation than the average link utilization are highly utilized/underutilized links, so in this case, links A-B, B-D are highly utilized and C-D is an under-utilized link and, hence, require proper SDN node placement to minimize this MLU. Routing data across highly utilized links deteriorate the performance and thus need arises to address such scenarios in the network.

2.2 *Greedy-Based Approach*

As per the greedy-based approach to SDN placement, should be placed at the nodes with the highest degree and are described in Fig. 2. Nodes placed at the highest degree ensure the flow splitting (multipath routing) across a number of links so all the links are uniformly utilized.

Based on this arrangement, SDN nodes should be placed at B and C. With this arrangement, data coming/emanating from nodes B and C can be split across any outgoing links. Data that traverses/emanate nodes B and C can be easily split across two and three paths (one path of the OSPF protocol with path length-1, other paths via the B-C link with path length-2, using A/D as the intermediate node with the largest path length-3). However, if the node B has to send data to node A, there are only two paths (shortest path, using B-C path as an intermediate) actually because if the other path B-D-C-A is used then there is no shortest path to A from D via D-C-A and, hence, loops might occur. Thus, SDN nodes should avoid sending data to legacy nodes where the shortest paths converge with the already used paths. In this example, the splitting ratio of SDN nodes is kept at 0.5. The overall utilization of the highly utilized links (A-B is 11, B-D is 13) and the lightly utilized links are (A-C is 10, B-C is 11 and C-D is 9).

3 Proposed INP Scheme

As pointed out earlier, most of the works have placed nodes based on the place where the outgoing numbers of paths are more. Placing nodes just based on the outgoing paths is not efficient because if the node is not highly utilized the TE is not going to improve. The proposed scheme of intelligent node placement is done by taking care of the traffic matrix, degree of a node and ensuring loopless paths after OSPF is run in a traditional network.

4 Flowchart Description

In h-SDN the controller obtains topology information either by layer two or layer three protocols [13, 14]. In network, nodes disseminate Link State Advertisement (LSA) between each other to gather the topology [14]. With the reception of LSA from the entire nodes, node fetches the global topology of the network. Each node runs the shortest path algorithm to route the data from the sender to the receiver. After the routing is completed based on the traffic matrix, link utilization of all the links is calculated. Based on the link utilization of all the links, the average of links is calculated. After completion of the average, standard deviation is computed and thus over-utilized links are noted as mentioned earlier. Now, a count is kept of the number of links that are highly utilized. If all the highly utilized links belong to a single node then a single SDN placement is to lower the link utilization. Otherwise, if the links belong to different nodes then all the nodes need to be replaced by SDN nodes as shown in Fig. 3. Since now a link is a relation between two nodes, so if a link is over-utilized then which nodes need to be replaced? First, priority is given to a node based on the number of highly utilized links it has. Second, priority is given to a node if the traffic emanating from the node is high. Finally, priority is given based on the number of degrees a node has with other nodes in the network. Figure 3 illustrates the steps to place SDN in an incremental fashion in h-SDN. The intelligent SDN placement algorithm keeps a focus to avoid loops in the network. Each replaced SDN node ensures that there is no violation of the shortest path routing by the legacy nodes. The SDN node does not forward packets to a node where there is a chance for the violation of the shortest path by the creation of loops [15].

5 Experiments and Evaluation

5.1 Simulation Environment

The simulation experiment of this paper is carried out in a real network topology obtained from SNDLIB [16]. SNDLIB has the topology information and the demand traffic information of the topology. Three different network topologies viz. Abilene, Atlanta and Geant are used. Table 1 lists all the network topology-related information used in the evaluation. The entire network is modeled based on the directed graph (V, E) where V represents the vertices of the topology and E represents the edges of a directed graph. The proposed INP algorithm is tested on MATLAB R2016b, and running on a system with Windows 10. Results presented herein are compared with the OSPF and Greedy-based approaches. Each experiment is run multiple times to note the average performance.

Table 1 SNDLIB topologies used

Topology name	Nodes	Links
Abilene	12	15
Atlanta	15	22
Geant	22	36

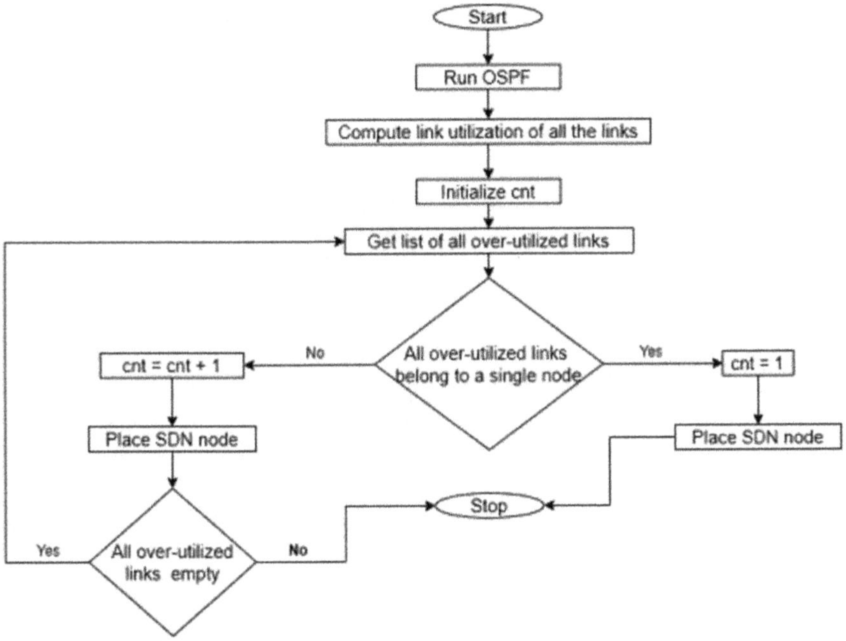

Fig. 3 Flowchart for the proposed INP scheme

Fig. 4 MLU graph of Abilene topology

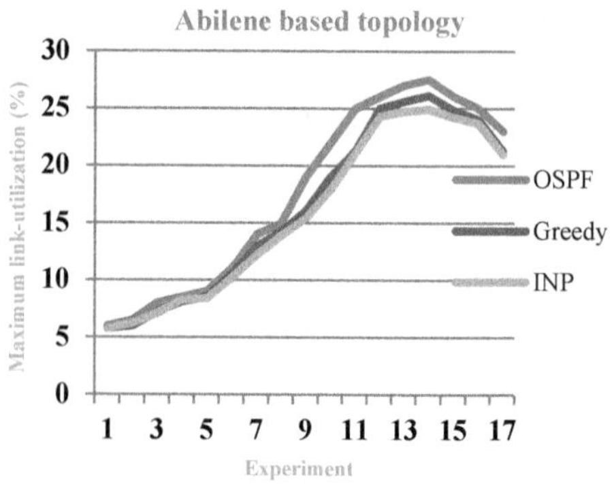

5.2 MLU

Figures 4, 5 and 6 show the cumulative distribution function versus the MLU of the Abilene, Atlanta and Geant-based topologies. For the Greedy-based approaches, the number of SDN nodes in Abilene, Atlanta and Geant is 3, 4 and 6, respectively. In

Fig. 5 MLU graph of Atlanta topology

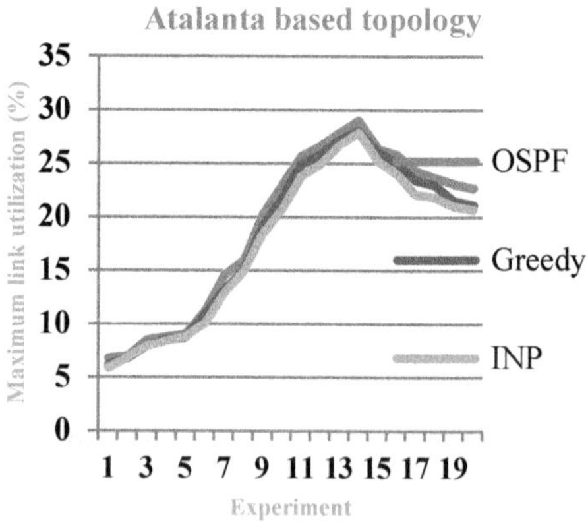

Fig. 6 MLU graph of Geant topology

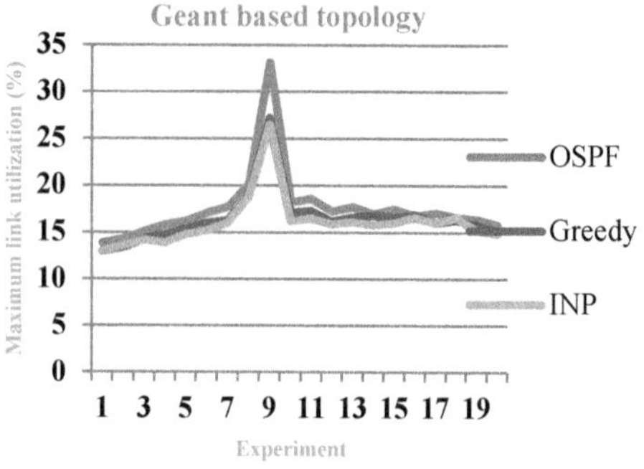

INP, the placement of SDN nodes is not anonymous but based on the traffic, link utilization and the degree of the nodes, hence, the number of SDN nodes required is quite less as compared with Greedy-based approaches, which are 2, 2, 4, in Abilene, Atlanta and Geant. Since OSPF as a routing algorithm is both oblivious of the inherent link loading and traffic characteristics so it performs worst in all cases of the topology under investigation. OSPF has the highest MLU on account of using only the shortest paths for routing. Greedy-based approaches place SDN nodes at strategic points so it performs better than OSPF. Owing to the high usage of SDN nodes in the network, in some cases, Greedy technique outperforms the proposed INP. INP performs best in almost all the cases of topology under investigation. INP shows a lower MLU as compared with both OSPF and Greedy approaches.

Compared with the OSPF routing, INP shows a reduction of 15–19% on average in the MLU of the topologies under investigation. INP also performs better in comparison to Greedy approaches, and improvement of about 5–7% is noted in various topologies.

6 Conclusion

In this paper, we have proposed INP for improving TE by proper placement of SDN nodes in the network. INP handles the node placement by analyzing traffic information and considering the degree of the node which placing SDN in the network. Experimental results show that the proposed INP scheme lowers the MLU by an average of about 5–19% in various real-time topologies. INP does not consider the traffic characteristics such as jitter, delay while re-routing the data to the destination which we intend to work in the future.

References

1. Hussain MW, Pradhan B, Gao XZ, Reddy KHK, Roy DS (2020) Clonal selection algorithm for energy minimization in software defined networks. Appl Soft Comput 96 106617
2. Kreutz D, Ramos F, Verissimo P, Rothenberg CE, Azodolmolky S, Uhlig S (2014) Software-defined networking: a comprehensive survey. arXiv preprint arXiv:1406.0440
3. Hussain MW, Reddy KHK, Roy DS (2019) Resource aware execution of speculated tasks in hadoop with SDN. Int J Adv Sci Technol 28(13): 72–84
4. Rathee S, Sinha Y, Haribabu K (2017) A survey: hybrid SDN. J Netw Comput Appl 100:35–55
5. Hussain MW, Roy DS (2021, May) Enabling indirect link discovery between SDN switches. In: Proceedings of the International Conference on Computing and Communication Systems: I3CS 2020, NEHU, Shillong, India, vol 170, p 471, Springer Nature
6. Caria M, Jukan A, Hoffmann M (2013) A performance study of network migration to SDN-enabled traffic engineering. In: 2013 IEEE global communications conference (GLOBECOM) 1391–1396
7. Agarwal S, Kodialam M, Lakshman TV (2013) Traffic engineering in software defined networks. In: 2013 Proceedings IEEE INFOCOM 2211–2219

8. Sun X, Jia Z, Zhao M, Zhang Z (2016) Multipath load balancing in SDN/OSPF hybrid network. In: IFIP international conference on network and parallel computing. Springer Cham. 93–100
9. Fakhteh AH, Sattari-Naeini V, Naji HR (2019) Increasing the network control ability and flexibility in incremental switch deployment for hybrid software-defined networks. In 2019 9th international conference on computer and knowledge engineering (ICCKE). 263–268 IEEE
10. Chen MH, Wang WM, Chung IH, Chou CF (2017) Incremental hybrid SDN deployment for enterprise networks. In: 2017 IEEE 15th International conference on dependable, autonomic and secure computing. 15th international conference on pervasive intelligence and computing. 3rd international conference on big data intelligence and computing and cyber science and technology congress (DASC/PiCom/DataCom/CyberSciTech). 1143–1149 IEEE
11. Guo Y, Wang Z, Yin X, Shi X, Wu J, Zhang H (2015) Incremental deployment for traffic engineering in hybrid SDN network. In: 2015 IEEE 34th international performance computing and communications conference (IPCCC). 1–8 IEEE
12. Vissicchio S, Vanbever L, Bonaventure O (2014) Opportunities and research challenges of hybrid software defined networks. ACM SIGCOMM Computer Communication Review 44(2):70–75
13. Hussain MW, Reddy KHK, Rodrigues JJ, Roy DS (2020) An indirect controller-legacy switch forwarding scheme for link discovery in hybrid SDN. Syst J IEEE
14. Hong DK, Ma Y, Banerjee S, Mao ZM (2016) Incremental deployment of SDN in hybrid enterprise and ISP networks. In: Proceedings of the symposium on SDN research. 1–7
15. Wang W, He W, Su J (2017) Enhancing the effectiveness of traffic engineering in hybrid SDN. In: 2017 IEEE international conference on communications (ICC). 1–6 IEEE
16. "Welcome to survivable fixed telecommunication network design "http://sndlib.zib.de/home.action.

A Comparison of Two Popular Deep Learning Methods for Nowcasting of Rainfall

Bishal Paudel, Nandita Sarkar, and Swastika Chaktraborty

Abstract Rainfall nowcasting is an essential technique for various public services as well as the decision-making process for disaster prevention. LSTM, NAR and NARX algorithms are used in this work for the prediction of a stochastic time series of rainfall taking experience from the effect of rainfall on radio signal degradation. LSTM works much better than NAR or NARX network while analyzing the error performance of each of the networks in this prediction problem.

Keywords Nowcasting · Radio signal degradation · Rainfall · LSTM · NAR · NARX

1 Introduction

Rainfall, one of the major stochastic environmental phenomena, affects human life in different ways. It has an adverse effect and significant impact on a number of activities ranging from crop production to life-threatening landslide events. Forecasting of rainfall has been done since long back in numerous literatures. As climate change within a very short interval of time is very natural, that is why long-term forecasting is not accurate with the actual happening. The alternative way, i.e. nowcasting, attributed to forecasting just before a short interval of actual happening is shown to prove a better matching with the actual happening.

Random forest algorithm, one of the elementary decision-making machine learning algorithms, is found to perform well for precipitation nowcasting, as it

B. Paudel · N. Sarkar · S. Chaktraborty (✉)
Sikkim Manipal Institute of Technology, Sikim Manipal University, Majitar, Sikkim 737132, India
e-mail: swastika.c@smit.smu.edu.in

B. Paudel
e-mail: bishal_201700240@smit.smu.edu.in

N. Sarkar
e-mail: nandita_201700342@smit.smu.edu.in

S. Dhar et al. (eds.), *Advances in Communication, Devices and Networking*,
Lecture Notes in Electrical Engineering 776,
https://doi.org/10.1007/978-981-16-2911-2_32

is able to capture the non-linear pattern of rainfall [1]. However, nowcasting accuracy has been relatively more in the neural network models as rainfall is completely chaotic in nature [2, 4–6]. With the advancement of neural networks, deep learning provides multiple models for time-series problems. Autoregressive networks have gained popularity nowadays in nowcasting problems because of their ability to learn chaotic time sequences. Autoregressive moving average model (ARMA) is based on the stochastic process. Autoregressive moving average and autoregressive integrated moving average model (ARIMA) eliminate the non-stationarity of the sequence [2, 6]. Non-linear autoregressive exogenous model (NARX) learns the non-linear relationships between the response series and the input series [3, 4].

Convolutional long short-term memory (ConvLSTM) gives distinct outcomes with very low errors. It processes satellite images or radio detection and ranging (RADAR) images and uses it in long short-term memory (LSTM), a recurrent neural network [7, 8]. Recent studies show LSTM networks perform best in precipitation nowcasting compared with all other methods. This model has memory cells and input, output and forget gates rather than traditional hidden layers. It handles noise and chaotic sequence with non-decaying error backpropagation [2, 5–10]. The network adapts itself to changing trends and possesses human-like expertise. In this work, nowcasting of precipitation has been done with the knowledge of signal attenuation or fading of high-frequency signals using the models NARX and LSTM.

2 Dataset

The dataset used for rainfall prediction contains received signal data in the presence of rain, which can be termed as attenuated signal data and the rain rate data. The signal is received from the GSAT-14 satellite at a beacon frequency 11.3 GHz, i.e. Ku-band with the antenna placed at Sikkim (27.18 °N and 88.50 °E) at an elevation angle 48°. The receiving system contains a spectrum analyzer, low noise block converter (LNB) and a computer to store and analyze the data. The precipitation value is obtained from an in-situ laser precipitation monitor (LPM) made by Theis Clima. All this arrangement is shown in Fig. 1. The rain rate dataset used here has per minute temporal resolution. We extracted rainfall events from May to July, 2020. We have used 18 rain events for training the dataset and 3 events for testing the dataset for the purpose of nowcasting. For prediction, past 3 hours signal attenuation data of temporal resolution 0.1 s are used for nowcasting 10, 30 and 60 min rainfall. Figure 2 shows recorded Ku-band signal attenuation during a rain event on 31 May 2020 and simultaneous variation of precipitation intensity with respect to signal attenuation.

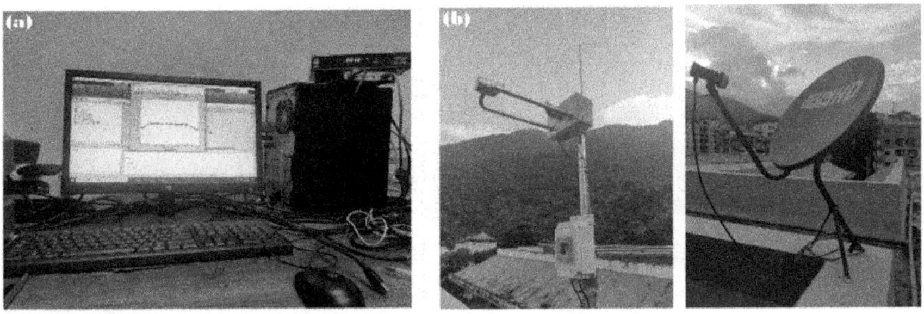

Fig. 1 **a** Indoor unit **b** Outdoor unit of LPM and receiving antenna

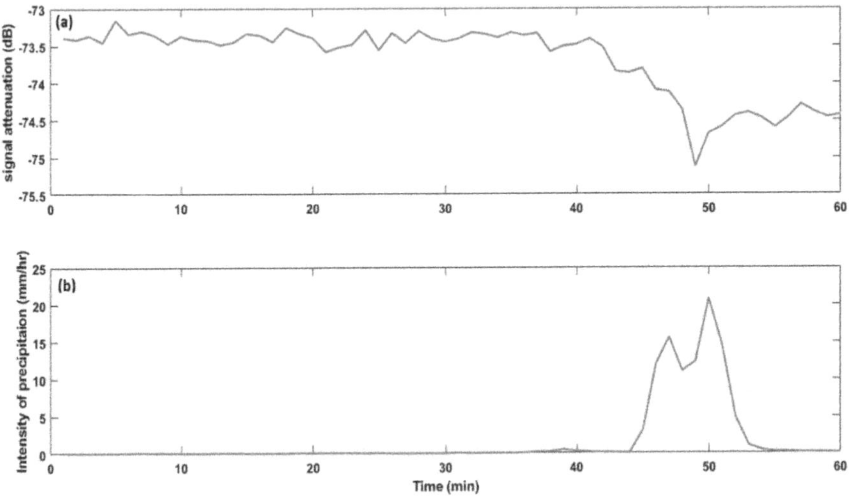

Fig. 2 **a** Recorded Ku-band signal attenuation during a rain event on 31 May 2020.**b** Simultaneous variation of precipitation intensity with respect to signal attenuation

3 Methods

3.1 Artificial Neural Network

Rainfall is essentially non-linear and chaotic in nature. Non-linear autoregressive network (NAR) is well suited for such time-series problem as stated in the literature. NAR network can learn non-linear relationships between the variables by evaluating the preceding trend and predict accordingly. A NAR network describes a discrete, non-linear autoregressive model when applied to time series forecasting. The training function assigns the weight and bias for the neurons. NAR network predicts a time series from the series of past values Y(t-1), Y(t-2), ..., Y(t-d) known as the feedback delays, with d as the time delay parameter.

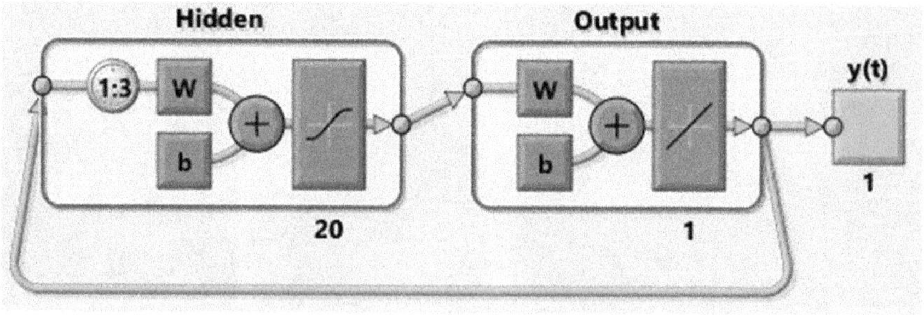

Fig. 3 Close loop NAR network

Training of the network is done in an open loop configuration with real target values as the response. Then the trained network is converted into a closed loop and the predicted values are fed as the response inputs to the network. The NAR network can be mathematically represented by the equation

$$Y_t = f(Y_{t-1}, Y_{t-2}, \ldots, Y_{t-d}) + \varepsilon_t$$

where, f is the non-linear function that the network implements to optimize the weight and bias of the neurons while training. ε_t is the error parameter and Y being the network response with d delay.

Non-linear autoregressive network with exogenous input (NARX) has a framework of dynamic recurrent neural network. It has feedback connections linking various layers of the neural network. NARX has similar structure as NAR except that it has an exogenous set of variables on which the response of the network is dependent. NARX can be mathematically represented by the equation

$$Y_t = f(Y_{t-1}, Y_{t-2}, \ldots, Y_{t-n}, X_{t-1}, X_{t-2}, \ldots, X_{t-m}) + \varepsilon_t$$

where, f is the training function, ε_t is the error parameter at time t, Y is the response series for n delay and X is the set of exogenous input with delay m. Figures 3 and 4 give a pictorial representation of NAR and NARX prediction network of our case.

3.2 *Long Short-Term Memory Network (LSTM)*

Long short-term memory (LSTM) is a type of recurrent neural network that learns long-term dependencies between time steps of a series. It has feedback connections, hence, has a memory of past states of the network. LSTM comprises of cells or memory states, and three regulators or gates—input, forget and output gate.

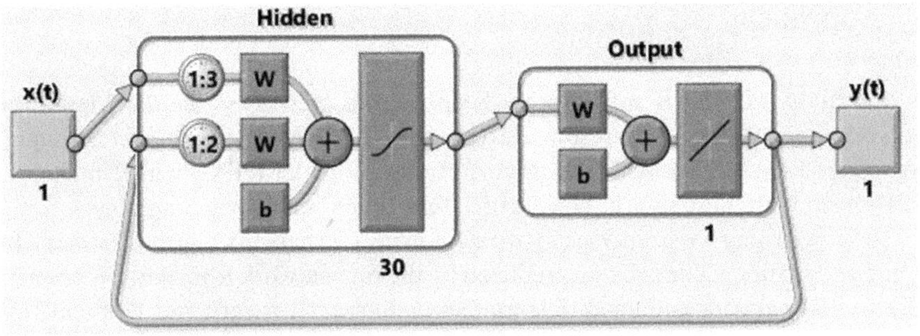

Fig. 4 Close loop NARX network

The input gate updates the cell state by providing new information. The forget gate resets the cell state and the output gate gives the output of the hidden states.

At a given time step *t*, the cell state is given by the equation

$$c_t = f_t \odot c_{t-1} + i_t \odot g_t$$

where $\odot$ denotes the Hadamard product, *i* denotes the input gate, *f* denotes the forget gate, *g* denotes cell candidate.

The hidden state at time step *t* is given by the equation

$$\mathbf{h}_t = o_t \odot \sigma_c(\mathbf{c}_t),$$

where o denotes the output gate, σ_c denotes the state activation function i.e. hyperbolic tangent function(tanh). Figure 5 shows the pictorial representation of LSTM network.

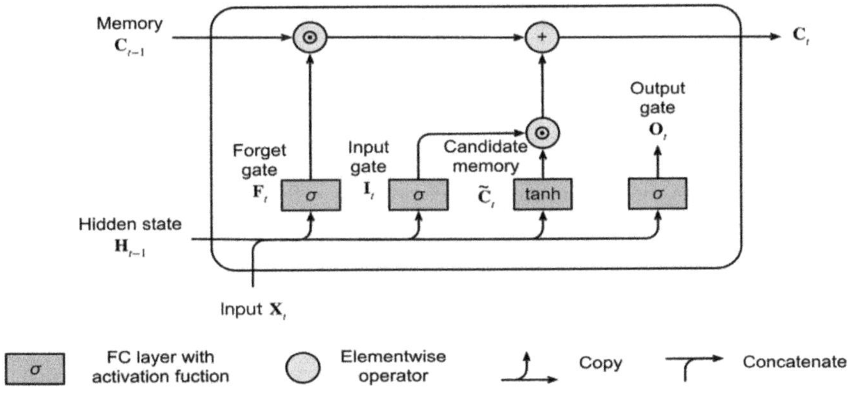

Fig. 5 Gates structure of LSTM [11]

4 Approach

In this study, MATLAB software has been used to preprocess the data, build the networks and to analyze the result. The experiments have been carried on the super-computer 'PARAM Shavak', a high-performing computer (HPC) with 2 multicore CPUs each with minimum 12 cores and 3 Tera-Flops peak computing power.

The nowcasting has been done in two steps using two networks. In the first step, the future attenuation value is being predicted by the first network from past experience. The second network is trained to learn the relationship between signal attenuation and rainfall rate. Hence, the predicted attenuation value is fed to the second network as input, and the response is predicted for the rainfall rate.

The error parameter root mean square error (RMSE) and mean average error (MAE) are used to analyze the results.

In the artificial neural network (ANN), the signal attenuation is predicted using the NAR network. The NAR network is trained with 20 hidden layers. The feedback delay is set to 3 with the training function as Bayesian regularization backpropagation. It updates the weights and biases of neurons to minimize the squared errors. The training, validation and testing are divided as 80%, 10% and 10% of the dataset. The network is trained in the open loop and the prediction is made using the closed loop network. The error autocorrelation is confined within the confidence limit with these parameters.

The predicted attenuation is fed to the input of the NARX network according to which the rainfall rate is predicted. The NARX network is trained with 30 hidden layers with input delay as 3 and feedback delay 2. Here the target series is rainfall, and the input series is the signal attenuation. It is also trained with Bayesian regularization backpropagation function in the open loop. The predicted attenuation is then fed to get the output rainfall. The network error autocorrelation and cross-correlation are confined within the confidence limit with these parameters.

Table 1 shows the specification of NAR and NARX networks used for this study.

The LSTM network also follows the similar fashion. The response of the first LSTM network is signal attenuation. The network has one sequence input (past values of attenuation) and one sequence output (future attenuation values) and 200

Table1 Parameters of NAR and NARX networks used

Specifications	NAR network	NARX network
Hidden layers	20	30
Feedback delay	3	2
Input delay	-	3
Training function	Bayesian regularization backpropagation	Bayesian regularization backpropagation
Output	Attenuation (10, 30 and 60-min prediction)	Rainfall (10, 30 and 60-min prediction)

Table 2 Parameters of LSTM networks used

Specifications	LSTM network 1	LSTM network 2
Epochs	200	400
Hidden units	200	200
Initial learning rate	0.03	0.03
Output	Attenuation (10, 30 and 60-min prediction)	Rainfall (10, 30 and 60-min prediction)

hidden units. The network is trained for 200 epochs with four layers, sequence input layer, LSTM layer, fully connected layer and regression layer with initial learning rate 0.03.

The second LSTM network also has one sequence input (attenuation values) and one output layer (rainfall values) with 200 hidden units. The network is trained for 400 epochs and initial learn rate 0.03 with the same four layers. We get the minimum network loss with these parameters.

Table 2 shows the specification of LSTM network used for this study.

The networks are trained for 10, 30 and 60 min lag time. The networks are updated at their lag times. Thus, for a 60 min prediction, the network is updated 6 times for lag time 10 min, and for 30 min the network is updated twice and for 60 min lag time there is no update. The network performs best for lag time 10 min that is visible from Fig. 6 a, b, and c is, therefore, opted for the study.

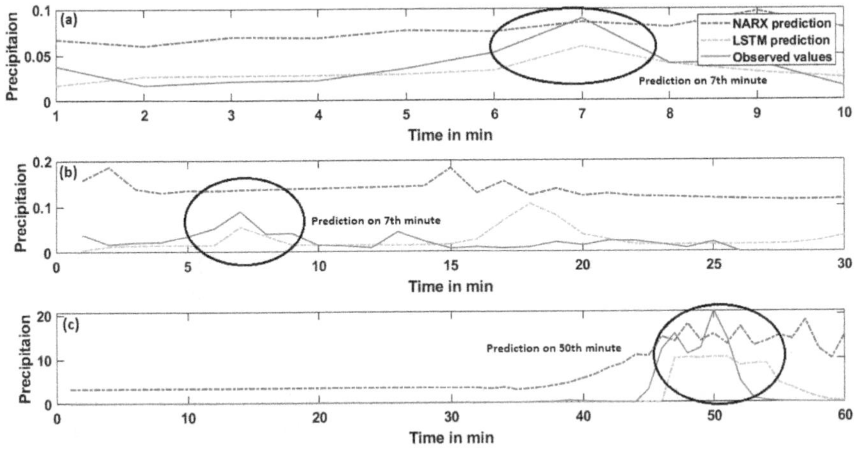

Fig. 6 **a** Precipitation prediction for lag time 10 min, **b** Precipitation prediction for lag time 30 min, **c** Precipitation prediction for lag time 60 min

Table 3 Comparison of performance of LSTM and NARX

Error parameter	NARX			LSTM		
Prediction time (min)	10	30	60	10	30	60
RMSE	0.042	0.115	6.606	0.022	0.031	4.615
MAE	0.039	0.112	5.331	0.016	0.022	2.909

5 Results

The nowcasting is performed in three prediction intervals, i.e. 10, 30 and 60 min. The error parameters used are root mean square error (RMSE) and mean average error (MAE). The comparison of the error parameters of both algorithms for the time slots is given in Table 2.

The predicted graphs for 10, 30 and 60 min are shown in Fig. 6. The good prediction points are encircled.

From Table 3, the RMSE for 10 min prediction time in the LSTM network is 0.022 and in NARX 0.042, i.e. the RMSE for LSTM is less than NARX network. Similarly, we can observe that the RMSE of LSTM for all the prediction time is less than NARX network. Also, MAE of LSTM network is better than NARX network.

However, the error parameter for both the network increases as the prediction interval increases. As we can observe, for NARX network, the RMSE for 10 min is 0.042, 30 min is 0.115 and 60 min is 6.6. This is because as the prediction time increases the anomaly of rainfall prediction increases. Hence, the network performance degrades for long prediction intervals.

Therefore, based on the error parameters, we can conclude that the LSTM network outperforms NARX network and gives a better prediction of precipitation values.

6 Summary/Conclusion

A comparison of two deep learning methods LSTM and NARX has been done in this work for the prediction of the complex time-series problem of rainfall looking at received electromagnetic signal degradation prior to rainfall. The error analysis has been done in terms of MSE and RMSE.

The RMSE and MAE for 30 min prediction time are 0.115 and 0.112, respectively, for NAR and NARX network and 0.031 and 0.022, respectively, for LSTM network as shown in Table 3. We can clearly observe that LSTM has a lower error value than NAR and NARX. Analyzing the error parameters, the performance of LSTM is quite better than NAR and NARX networks. Training with more volume of data may produce better results in future study.

References

1. Das S, Chakraborty R, Maitra (2017) A: a random forest algorithm for nowcasting of intense precipitation events. Adv Space Res Vol 60(6):1271–1282
2. Zeyi C, Pu F, Yin Y, Han B, Chen X (2018) Research on real-time local rainfall prediction based on MEMS sensors. J SensS 2018:1–9. https://doi.org/10.1155/2018/6184713
3. Benevides P, Catalao J, Nico G (2019) Neural network approach to forecast hourly intense rainfall using GNSS precipitable water vapor and meteorological sensors. Remote Sens vol 11(8) 966
4. Biswas S, Leniency M, Purkayastha B, Chakraborty M, Singh H, Bordoloi M (2016) Rainfall forecasting by relevant attributes using artificial neural networks - a comparative study. Int J Big Data Intell 3(2):111–121
5. Aswin S, Geetha P, Vinayakumar R (2018) Deep learning models for the prediction of rainfall. international conference on communication and signal processing (ICCSP), Chennai, 0657–0661
6. Poornima S, Pushpalatha M (2019) Prediction of rainfall using intensified LSTM based recurrent neural network with weighted linear units. Atmosphere 10 668
7. Chen L, Cao Y, Ma L, Zhang J (2020) A deep learning-based methodology for precipitation nowcasting with radar. Earth Space Sci 7
8. Kumar A, Islam T, Sekimoto Y, Mattmann C, Wilson B (2020) Convcast: an embedded convolutional LSTM based architecture for precipitation nowcasting using satellite data. PLoS ONE 15(3):e0230114
9. Zhang C-J, Zeng J, Wang H-Y, Ma L-M, Chu H (2019) Correction model for rainfall forecasts using the LSTM with multiple meteorological factors. Meteorol Appl 27 e1852
10. Hu C, Wu Q, Li H, Jian S, Li N, Lou Z (2018) Deep learning with a long short-term memory networks approach for rainfall-runoff simulation. Water 10(11):1543
11. https://d2l.ai/_images/lstm_2.svg

Comparison of IoT Application Layer Protocols on Soft Computing Paradigms: A Survey

Abhimanyu Sharma, Kiran Gautam, and Tawal Kumar Koirala

Abstract IoT is the term most used nowadays. "Internet of Things," people think it is a new emerging technology, but they cannot be more wrong. It has existed for more than a decade and still, no one can completely grasp its vast use in different fields. This occurs due to a lack of research work and improper implementation. The upcoming IoT market focuses on the things that can communicate intermediately with each other, and every major and minor company is now trying to implement this technology for the betterment of human life. The main goal of IoT is to connect the things they use to communicate with each other for data analysis and sharing. If all the things that people use in a default household could perform this task, then the work that the average human being does could be a lot easier. The content is presented as a survey on application layer protocol and the major role it plays in communication between two machines and how they use the protocol to communicate with each other. The uses of various soft computing paradigms that can be used are discussed in this paper.

Keywords IoT · MQTT · AMQP · CoAP · XMPP

1 Introduction

IoT is advancing with the new concept of computing and communication mediums are being created to cater to the ubiquitous nature of information processing and its delivery from one machine to another. In this, the user can access their data generated by the machine remotely from anywhere, anytime until there is a proper network

A. Sharma (✉) · K. Gautam · T. K. Koirala
Department of Computer Science and Engineering, Sikkim Manipal Institute of Technology, Sikkim Manipal University, Gangtok, India

K. Gautam
e-mail: kiran.g@smit.smu.edu.in

T. K. Koirala
e-mail: tawal.k@smit.smu.edu.in

S. Dhar et al. (eds.), *Advances in Communication, Devices and Networking*, Lecture Notes in Electrical Engineering 776,
https://doi.org/10.1007/978-981-16-2911-2_33

connection or the user is connected to the internet. Currently, due to smartphones that earlier were largely expensive, it has been possible for the masses to stay connected to the world of internet. The data service provider's tariffs are also affordable and can appeal to the masses. This may also be one of the major reasons why the IoT implementation caught its roots among the masses. But the most asked question is, why use IoT? Because IoT enables all kinds of devices to connect with each other and share information seamlessly, and a number of things that are connected to the internet are more than people present on earth for that reason people started using IoT [1]. Every device contains some kind of electrical circuitry that helps power the device, now visualize the system connected with a processor and you get the information about the state of the device and its surrounding environments.

Well earlier, it may just be an image but now this has been the truth. Now you may be having many similar devices in your house, and the entire device in your house can communicate with each other and give you the information about the state of the device. This can be done using IoT. After learning about IoT, the knowledge about "How can the device communicate with each other rather share information regarding its state"? To know this, an attempt has been made to first learn about the architecture of IoT. A commonly agreed architecture of the IoT comprises three layers: application layer, network layer, and perception layer [2].

In this paper, the focus will be on application layer protocol and the role it plays for machine to machine communication. A comparative study will also be performed for the protocols and conclude which protocol can be used regarding various parameters that will be beneficial for real-time use. The comparison of four protocols like MQTT, CoAP, AMQP, XMPP that can Restful and Web-Socket bases APIs taking various aspects in mind for their use.

2 Protocols for IoT

In computer networks, the use of OSI layer protocol transports the packets of data from one network to the other using layers of unique encapsulation and decapsulation. But for IoT, it can categorize it into three layers, session/communication layer, transport layer, and data link layer, as shown in Fig. 1.

In data link layer, the connectivity module will connect the IoT devices with other IoT-enabled devices and will send data via it. Module such as Wi-Fi, Bluetooth, LTE, ZigBee, etc. is used in this module.

The network layer is divided into two parts, routing and encapsulation. In routing, the use of RPL, CORPL, CARP, etc. as routing protocols provides the shortest path for routing the data from one IoT-enabled device to the other. In encapsulation, the data are encapsulated so that no one other than the client or the interconnected IoT device could access the data. Some of the encapsulation techniques used are 6LowPAN, 6TiSCH, Thread, etc.

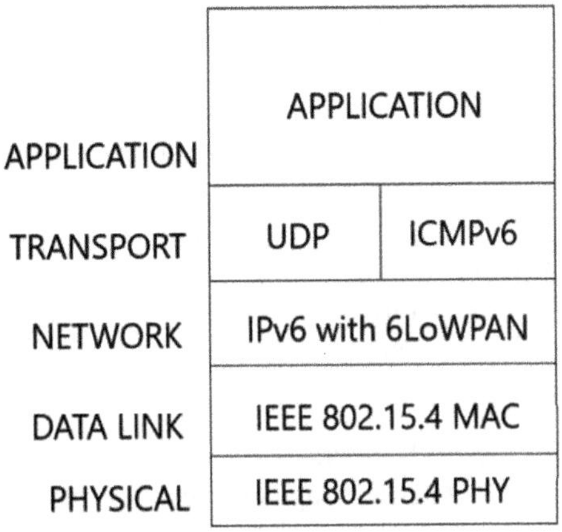

Fig. 1 Protocols of IoT

The application layer will be using the protocols to communicate with each other, protocols such as MQTT, AMQP, CoAP, XMPP, etc. In this paper, a study will be performed discussing about these protocols.

The layer that is most important in IoT is the physical layer, which consists of sensors, which collect the data that are processed by the above-mentioned three layers.

3 Related Work

In this paper, the transferring of message between two or more IoT devices, hence looking over some papers regarding the state of transferring the data and the area where IoT may have a major impact, and also looked upon various devices that can be used for making IoT application a reality.

References [1, 4, 5] in this paper, the authors write about the different application layer protocols and the challenge faced while creating IoT-enabled devices. The comparative study of the application layer was not conducted in this paper [2]. In this paper, the author has conducted an experimental model comparing the various application layer protocols and simulating a simple IoT network and by generating and exchanging environment data [3]. In this paper, the authors have discussed about the architectural implementations, gaps in IoT security as well the pros and cons of using certain application layer protocols that are currently being used in IoT's environment.

4 Application Layer Protocol

The application layer is the layer that moves/transfers the packets from one machine to the other. It resides on the top of the networking protocol. The application layer solves the challenges that occur during the transmission of data, challenges such as "communication between machines on a local network," "communication between machines through the internet," and "communication between server and stored data" [9]. The application layers use TCP and UDP to overcome such challenges.

The various application layer protocols that can be used in the implementation of IoT are:

1. XMPP (Extensible Messaging and Presence Protocol).
2. MQTT (Message Queue Telemetry Transport).
3. CoAP (Constrained Application Protocol).
4. AMQP (Advanced Message Queuing Protocol).
5. Web-Socket.
6. RESTFUL.

I XMPP

XMPP stands for Extensible Messaging Presence Protocol. Jabber was the original name for XMPP and was basically developed as a messaging protocol [11]. It uses an XML that is Extensible Mark-up language, a widely used messaging protocol. XMPP was an old protocol that was going to be of obsolete use but is compatibility with IoT regained its popularity back. It runs on publish/subscribe and also request/response messaging protocol. It was designed for real-time use as it uses a small message footprint and also has a low latency rate for message exchange. The main feature of this protocol is its addressing mechanism. It identifies the devices/nodes in the IoT network using the address known as Jabber ID (JID) [11].

Messages are usually transmitted over the TCP connection. Pooling mechanism is used to figure out the destination of the message. XMPP follows the client–server architecture. XMPP is an open protocol, which means anyone can have their own XMPP server in their own personal network without internet connectivity. However, additional overheads are created due to unnecessary tags and require XML parsing, which increases the need for additional computational ability, which in return increases power consumption [4].

II MQTT

Message Queuing Telemetry Transport is a publish/subscribe protocol used in machine for machine/IoT connectivity protocol. It was mainly designed for transferring lightweight messages. The importance of MQTT protocol is due to its simplicity and the no need for high CPU and memory usage (the reason is the lightweight protocol) [1]. MQTT works on the basis of publish/subscribe, that is, a client who is sending a particular message (e.g. temperature readings) to another client has to use a broker who can deliver the message but cannot store it. MQTT works in real time where the data are generated and delivered when the subscriber requests for it.

The broker can publish to one or more subscriber based on if the subscriber has requested/subscribed to one or more data/message from the broker (Figs. 2 and 3).

It has various ways to deliver the message and provide quality of service (QoS) [4]

i. Fire and forget: A message is sent once and no acknowledgment is required.
ii. Deliver at least once: A message is sent at least once and an acknowledgment is required.
iii. Deliver exactly once: A four-way handshake mechanism is used to ensure the message is delivered exactly one time.

MQTT runs on top of the TCP stack, although it is designed to have low overhead compared with other TCP-based application layer protocols [4].

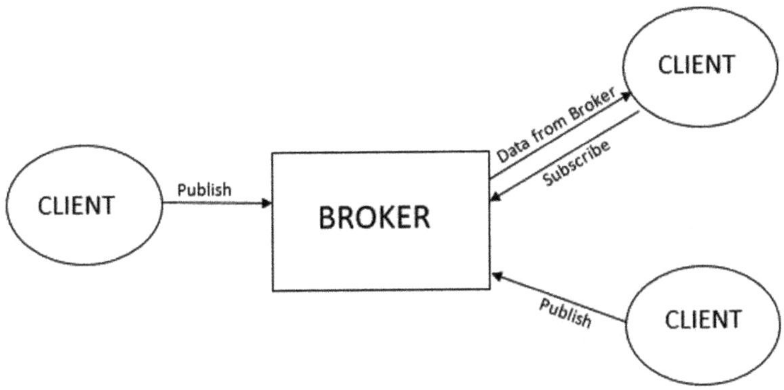

Fig. 2 MQTT Publish/subscribe communication many to one

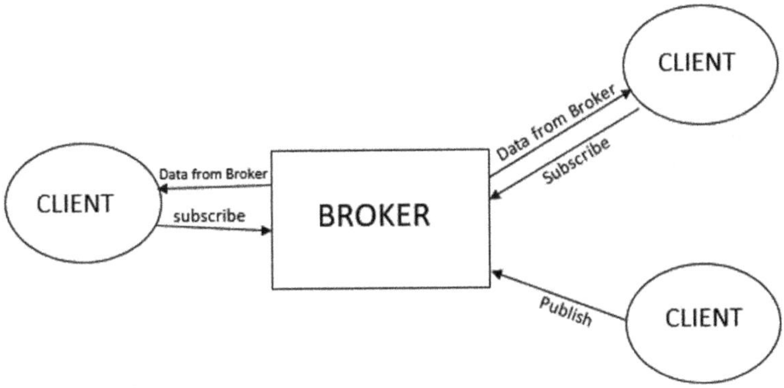

Fig. 3 MQTT publish/subscribe communication one too many

III CoAP

CoAP is a Constrain Application protocol that uses a request/response communication module for delivering its message from one machine to the other. It is like the client–server protocol. This protocol is only sufficient in a constrained environment such as constrained node with low capability in RAM or CPU and constrained network such as lower power using wireless personal area network (WPAN) [5]. It uses less overhead for message delivery, which increases its simplicity for effective delivery of messages. CoAP uses the simple interface such as HTTP, which helps message delivery faster and a lot simpler. As CoAP uses UDP its overall message delivery is simpler.

It uses HTTP commands such as POST, GET and Delete and provides resource-oriented interaction in client–server architecture [4]. TCP provides QoS (Quality of Service) to ensure reliability for message Delivery and has its own set of mechanism to ensure this:

i. Non-confirmable
ii. Confirmable
iii. Reset
iv. Acknowledgment

IV AMQP

The Advanced Message Queuing Protocol (AMQP) is an open standard for passing business messages between applications or organizations. It connects systems, feeds business processes with the information they need and reliably transmits onward the instructions that achieve their goals [12]. It uses the publish/subscribe communication model. It can be mainly implemented in large businesses/industries as it has the capability to store data and publish data when required, unlike MQTT, which just forwards the message to the subscriber who is subscribed to the message in the instance of time. The data storage feature of AMQP ensures reliability when there is some unwanted obstruction in the network (Figs. 4 and 5).

It has three important thins, i.e. Exchange, Message queue and Binding.

i. Exchange: It routes messages to respective queues that are received from publishers.
ii. Message queue: It keeps the message in the queues until received by the clients.
iii. Binding: It describes the state between the messages queue and the change.

5 Comparison Between Different Layers

In this paper, comparison of the four-application layer protocol based on their header size, power consumption, transport medium over a network, architecture, QoS, security, area of application for each protocol and their limitations.

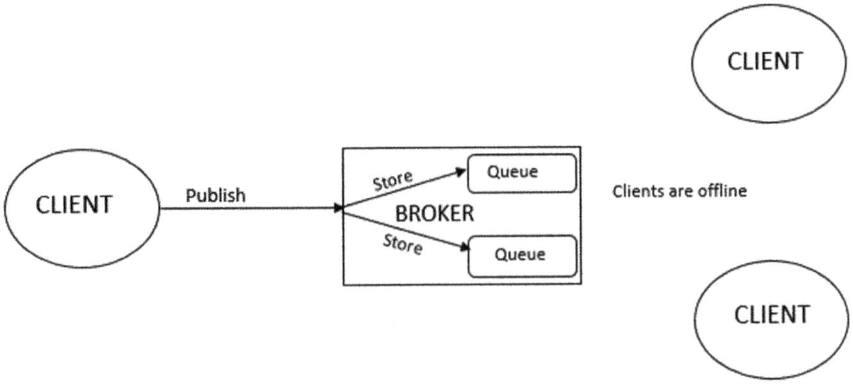

Fig. 4 AMQP Communication when clients are offline

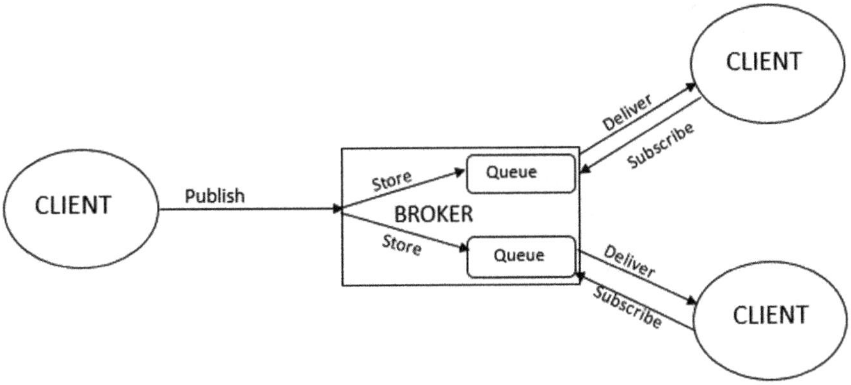

Fig. 5 AMQP communication when clients are online

CoAP: Constrain application protocol as discussed above is meant to work in a constrained environment like your PAN. Their architecture is based on synchronous that is request/response protocol. They do provide a certain aspect of QoS. They are mostly used with UDP as transport medium, which makes them unreliable when sending a message. It uses datagram transport layer security (DTLS), which is designed to prevent the tampering of data, eavesdropping and no forgery of data. It is used in other IoT-based application sensor networks and live data communication. Its main advantages are it provides enhanced reliability, multicasting, reduced bandwidth usage, content-type support, reduced latency, single parsing, good PDR. It has some limitations such as packet losses due to TCP retransmissions, its deployment and implementation is high cost, network robustness, and application deployment errors.

MQTT: Message Queuing Telemetry Transport is made for wide-area implementation using WAN. The architecture is based on asynchronous that is publish/subscribe

Table 1 Comparison between different application layer protocols

Protocol	QoS	Transport	Architecture	Security
MQTT	Yes	TCP	Publish/Subscribe	TSL/SSL
CoAP	Yes	UDP	Request/Response	DTLS
AMQP	Yes	TCP	Publish/Subscribe	TSL/SSL
XMPP	No	TCP	Publish/Subscribe Request/Response	TSL/SSL

protocol. They provide QoS, which makes them reliable. They transport the message using TCP, which makes their message delivery reliable. It uses SSL/TLS as its security. Some of the fields MQTT can be implemented in consist of Healthcare, Energy monitoring and management, Industrial application and agricultural, social networking, and other IoT-based applications. They provide some advantages over CoAP as they consume less energy, they have a much lesser header size is much smaller compared with CoAP, it is better in the management of internet traffic and keeps memory and CPU usage to optimal and many more. Its limitation is that it requires a moderate bandwidth for transporting the message.

AMQP: Advance Message Queuing Protocol is made for large organizations who want to share their IoT data with each other or with other organizations. It uses asynchronous that is publish/subscribe protocol. They do provide QoS and are reliable. It uses TCP as a transport medium, which makes it reliable. It uses SSL/TLS security to ensure the system remains secure. It is mostly applied in applications, which are based on control and server, which performs the analysis. It provides advantages such as it can connect across various technology organizations and time domains, and stores and forwards strategy for good reliability of the system. It has certain limitations as it is not suitable for constrained real-time applications, no support for automation discovery.

XMPP: Extensible Messaging and Presence Protocol.

Is it still a developing protocol in the field of IoT? It uses both publish/subscribe and request/response architecture for message delivery. But despite having low overhead, it provides no support for QoS. It uses SSL/TSL security to ensure the safety of its messages and runs on TCP as its transport medium. Their fields of applications are voice and video calls, chatting and message exchange applications, and mostly message delivery when playing online games. Its main advantage is that if an application is build using XML then it is easier to implement this protocol as it provides inter-operability between systems.

The table below gives us brief idea about the working of different application layer protocols (Table 1).

6 IoT Implementation Using Soft Computing Paradigms

IoT can be adapted to many fields such as medical, traffic, household, personal assistance system and the list goes on. The working of the wireless sensor nodes (WSN's) along with the software layer created new possibilities of implementing, communicating, and sharing. The data are shared in a raw form, which is most readable to humans. The extraction of these data is known as semantic data extraction, where information is collected from the sensory nodes and are stored and processed into some readable format. The data could be further classified using machine learning technique [22]. The method to separate the data accordingly is known as classification of data. In classification of data, the data are separated into predefined categories and the data extracted are placed into these categories. This system could further be improved by teaching the system to automatically place the data using supervised learning. In supervised learning, we teach the system how to categorize the data into their respective fields using some training sets. The training sets can be changed to make the system adaptive to changes in the data.

There are other methods of training the data such as unsupervised learning and deep reinforcement learning. In unsupervised learning, the trainer trains the system for a brief amount of time and let it learn by itself. It means that the system can adapt to the changes in the data without the intervention of external source. Humans can be considered a perfect example of unsupervised learning, where an infant is taught to walk but the infant learns to run by itself.

In reinforcement learning, we use rewards as a base to teach the machine, what has to be done to the data. When data are transferred from the destination to the source, the system is rewarded. The traveling salesman problem can be used as a perfect example of reinforcement learning.

Deep reinforcement learning uses an agent for message transfer. The agent works on the concept of exploring and learn. The agent gets the data and will deliver the data to the nodes weather its right or wrong. This is how the agent learns about the source where the data need to be delivered. If delivered correctly, the agent is rewarded by the system.

These methods can be used to optimize the message delivery of various application layer protocols. In the future, the implementation of these methodologies can be used for data delivery and optimization with machine learning and training the artificial neural network.

7 Future Work and Implementation

In this paper, the basic overview of IoT's application layer protocol has been discussed. The future implementation can be conducted using MQTT and CoAP for the constrained environment. Eclipse Mosquitto can be used for the implementation of MQTT service for constrained environment. MQTT-SN is mostly used for sensor network and designed to work over low power and lossy network can be implemented.

For CoAP, Erbium can be used for full-fledged REST engine and CoAP implementation. Erbium is based on Contiki Operating system, and for real-time execution, the software Cooja can be used. As IoT being a novel field and has rapid rise, there are various fields that can be explored from machine learning, artificial intelligence and blockchain technology can be used alongside the IoT devices, but the main aspect that should be known is that all IoT devices work in constrain environment, i.e. having limited resources to work with. The resources are memory, energy, security, network, etc. so focus can be kept on these aspects.

8 Conclusion

In this paper, the comparison of different application protocols was done and has looked briefly on how data can be transferred using a different application layer. Looking on some aspects regarding some of the constraint that may hold back the mass deployment of IoT tools in real-time, like cost, security, power requirement, and network constraints. Conclusion can be made that the application layer protocol can be used according to the environment that suits it and the needs of clients/users. Keeping this aspect in the deployment of IoT system, each application layer protocol is environment-centric and can be used to perform tasks suited for their own unique environment.

The discussion on various learning techniques was discussed in this paper, from which the use of various machine learning techniques can be implemented along with the current existing IoT technologies, and their optimized delivery of data is investigated.

References

1. Asim M (2017) A survey on application layer protocols for internet of things (IoT). Int J AdvRes Comput Sci 8(3)
2. Năstase L, Sandu IE, Popescu N (2017) An experimental evaluation of application layer protocols for the internet of things . Stud Inf Control 26(4):403–412
3. Narayanaswamy S, Kumar AV (2019) Application layer security authentication protocols for the internet of things: a survey. Adv Sci Technol Eng Syst J 4(1):317–328
4. Vazquez-Gallego F, Alonso-Zarate J (2015) Transaction on IoT and cloud computing 2015 a survey on application layer protocols for the internet of things
5. Yassein MB, Dua' Al-zoubi, Shatnawi MQ (2016) Application layer protocols for the internet of things: a survey
6. Sagar PJ, Manoj D (2019) "Proceedings of national conference on machine learning, 26th Mar 2019", "Optimization of application layer protocols for the IoT system"
7. Proceeding of second international conference on circuits, controls and communications 2017 IEEE a study on application layer protocols used in IoT

8. Mirjana M, Vladimir V, Nikola D, Vladimir M, Branko P (2014) Raspberry Pi as internet of things hardware: performances and constraints
9. https://blog.eduonix.com/internet-of-things/learn-application-layer-protocols-used-iot/. Accessed 25 Sept 2019
10. Toby J (2019)https://www.eclipse.org/community/eclipse_newsletter/2014/february/article2.phpcited. Accessed 28 Sept 2019
11. Alok B (2019) https://www.einfochips.com/blog/a-quick-guide-to-understanding-iot-application-messaging-protocols/cited. Accessed 28 Sept 2019
12. AMQP (2019) https://www.amqp.org/about/what. Accessed 25 Sept 2019
13. Tirupathi V, Sagar K (2019) A research on interoperability issues in internet of things at application layer. Int J Recent Technol Eng (IJRTE) 8(1S4):2277–3878
14. From CoAP https://coap.technology/ cited on 25 Sept 2019
15. From MQTT.org http://mqtt.org/. Accessed 25 Sept 2019
16. Samer J (2019) Communication protocols of an industrial internet of things environment: a comparative study. Futur Internet 11(66)
17. (2013) Comparison of protocols used in remote monitoring: DNP 3.0, IEC 870–5–101 and modbus. 27th International conference on advanced information networking and applications workshops. IEEE
18. One M2M, Technical report. Accessed 28 Sept 2019
19. Adrian M, Hakim C (2013) Designing internet of things
20. David H, Gonzalo S, Patrick G, Robert B, Jerome H (2017) IoT fundamentals: networking technologies, protocols, and use cases for the internet of things
21. Harwood T, IoT Standards and Protocols. https://www.postscapes.com/internet-of-things-protocols/. Accessed Oct 3
22. Wagle S (2016) International conference on internet of things and applications (IOTA). semantic data extraction over MQTT for IoT centric wireless sensor networks. Maharashtra institute of technology, Pune, India, Jan 22–24

Dual-Input DC-DC Cascaded Converters for Hybrid Renewable System

Rubi Kumari, Moumi Pandit, and K. S. Sherpa

Abstract A cascaded converter is proposed with dual input fed from two different renewable sources. The two sources are solar and wind energy. A solar PV panel has been used to receive energy from solar energy. Another prototype is built to provide wind energy. The two sources are connected in series and fed as input to the cascaded converter. Boost and buck-boost converters are cascaded to provide regulated output voltage with high voltage gain. The dual-feed cascaded converter is a noble technique to provide regulated output from renewable sources with increased voltage gain and less complexity.

Keywords Hybrid system · Cascaded converters · Solar energy · Wind energy

1 Introduction

One of the basic reasons behind the large emission of carbon dioxide is the burning of fossil fuels for the generation of electricity. This major cause of pollution has shifted the attention of researchers towards renewable energy. Renewable energy is the only alternative to generate energy without pollution. But there are few limitations associated with renewable sources, i.e., their efficiency is very low and their availability is not reliable. Also, the continuous increase in the cost of fossil fuels has diverted the attention of researchers toward the use of renewable energy for the generation of electric power.

Solar and wind energies are abundant forms of energy that can be used as the solution to generate electric power energy without affecting the environment. The combination of solar, hydel, wind and other energies can produce megawatts of electric power with zero effect on the environment [1]. The continuous rise in the demand

R. Kumari (✉) · M. Pandit · K. S. Sherpa
Electrical and Electronics Engineering Department, Sikkim Manipal Institute of Technology, Sikkim Manipal University, Sikkim, India

M. Pandit
e-mail: moumi.p@smit.smu.edu.in

S. Dhar et al. (eds.), *Advances in Communication, Devices and Networking*, Lecture Notes in Electrical Engineering 776,
https://doi.org/10.1007/978-981-16-2911-2_34

for sustainable energy production, has led to the interfacing of renewable energy sources with power electronics. The major drawback of using renewable resources is that the power extracted from individual renewable energy varies continuously. This leads to an imbalance between the input and the output, thus affecting the efficiency of the system. Most of the renewable energy systems that are currently available rely on solar and/or wind energy [1]. Out of the various renewable energies available, the researcher has also opted for solar and wind energy as the input due to its various advantages over the other resources [2].

A hybrid system proposed in this paper is the combination of wind and solar energy as the input source. This kind of setup works more efficiently during the summer and winter seasons. During summer the intensity of sunlight is more due to which the output generated by the PV panel will be more. Whereas, during winter the wind turbine would generate more output. The combination of these two renewable energy resources, i.e., energy from solar and wind together would be able to generate better results as compared with the stand-alone models.

A. Solar Energy: The availability and reliability of solar energy are in abundance. However, efficiency is the major concern for PV panels. The efficiency of these PV panels can be determined by the generation of electric power from each panel. Most of the PV panels have the efficiency to generate 15–20% of the sunlight into electric power. The power output generated from the solar PV panel is a fluctuating DC output due to the instability factors from the environment [3].

B. Wind Energy: Wind energy involves high initial cost and large space [5], whereas the maintenance and running cost of these types of systems is very less as compared to other renewable resources. The average wind efficiency of turbines is between 35 and 45%. A major challenge with this system is its low power generation capacity of an individual wind turbine [1]. Thus, for practical applications, a number of various wind turbines needs to be connected to produce a sufficient amount of output [1].

This paper is divided into six sections. Sect. 2 discusses the related works done in this field. Section 3 describes the proposed converter designed for hybrid system configuration. Section 4 has included a detailed explanation of the software and hardware results. The last section includes the conclusion of the work.

2 Related Works

In [6, 7], the authors have highlighted the integration of discussed cascaded boost converter with different photovoltaic modules. The paper presents the optimization techniques of using 'n' number of converters for 'n' number of solar cells connected in series. The optimal output is achieved by using the supervisory algorithm in MATLAB software. In [4], the author has discussed the optimization techniques for the cascaded boost converters for PV applications for achieving maximum efficiency. Maximum efficiency is achieved by optimizing the number of PV cells and by reducing the size of the cascaded converters. Apart from that, the author Hafez in [6] highlighted other techniques, like multi-level cascaded buck and boost converters, respectively, in series and parallel connections designed for PV applications. The system is analyzed in both dynamic and static conditions.

The output of the cascaded converters is very much dependent upon the type of connection, i.e., whether the converters are connected in series or parallel. Within this context, a comparison is done on the basis of loads, input voltage and output voltage at different duty ratios. Chang et al have discussed the two-switch cascaded buck and boost inverters using H-bridge unfolding circuits used for both step-up and step-down voltages. Apart from cascading DC- DC converters, the literature also highlights the cascading of DC-AC, AC-DC and AC-AC converters. Hossain et al. [8] presented cascading of a non-isolated converter with the isolated converters. Using the Cockcroft Walton principle, a full-bridge converter is cascaded with the boost converter. The topology includes various advantages like cost-effective, higher voltage gain, less requirement of components and smaller size of the inductor.

The major challenge with renewable resources is the low power generation of an individual source. For sufficient power generation, the solar PV panel and wind turbine needs to be connected together before they are fed to DC-DC converters. But the existing DC-DC converters are not sufficient enough to meet the requirements and this diverted the concern of the researcher to use multi-input renewable sources with cascaded converters. The proposed topology receives inputs from two sources, i.e., a solar cell and a NANO wind turbine designed in the laboratory. This topology can be easily applied like residential areas, aerospace and portable electronic devices where there is the need of using more than one energy source.

The simple DC-DC converters like boost or buck can be used for different purposes but they are not acceptable for high voltage gain or high efficiency. These requirements cannot be satisfied using conventional converters. Also, the use of step-down or step-up converter will lead to difficulties like switching surges, losses and increasing the complexity of the system. A technique that can provide larger conversion ratios without using a transformer is a cascade connection. Using cascaded converters with multi-inputs will increase the voltage gain compared to conventional converters and also less components will be used. This paper intends to fill the gaps and proposes a topology for collecting the power from distributed systems i.e. solar and wind for cascaded boost and buck-boost converters.

The paper presents the following work contribution:

- A non-isolated DC-DC cascaded boost and buck-boost converters with dual inputs. The proposed topology comprises a minimum number of components resulting in reduced circuit complexity and making the system cost-effective.
- The solar PV panel and the designed Wind Energy system are the dual in puts for the cascaded converters.
- The proposed converter is implemented and tested using a hardware setup in the laboratory.

The proposed work involves two stages, i.e., stage I and stage II. Figure 1 presents the block diagram of the proposed system. Stage I represents the hybrid renewable energy system. The inputs from the distributed energy sources (solar and wind) are

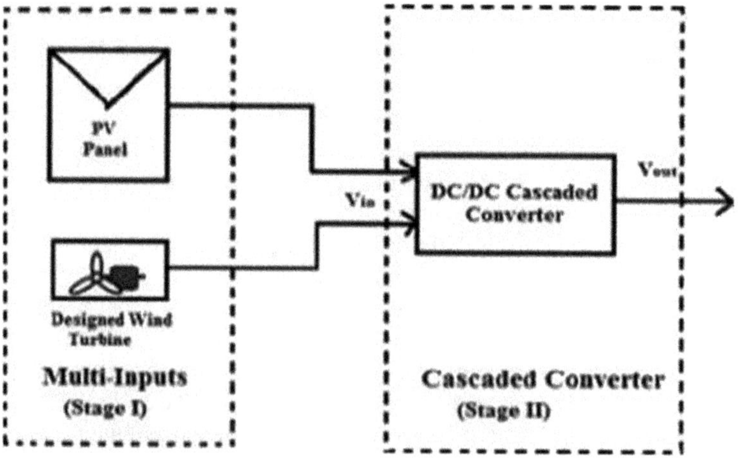

Fig. 1 Block diagram of the proposed system

connected in series. The outputs of stage I are fed to stage II i.e. cascaded DC-DC converter. Stage II of the proposed work consists of cascaded DC-DC converters. Boost and Buck-Boost are the two converter topologies designed for cascading.

3 Proposed Converter Design

After a thorough literature survey on various cascaded converters, this paper has proposed the cascading of boost and buck-boost converter. Figure 2 shows the circuit diagram of the model. It comprises a single switch. The key features of the proposed converters are as follows: higher voltage gain, minimum number of components used

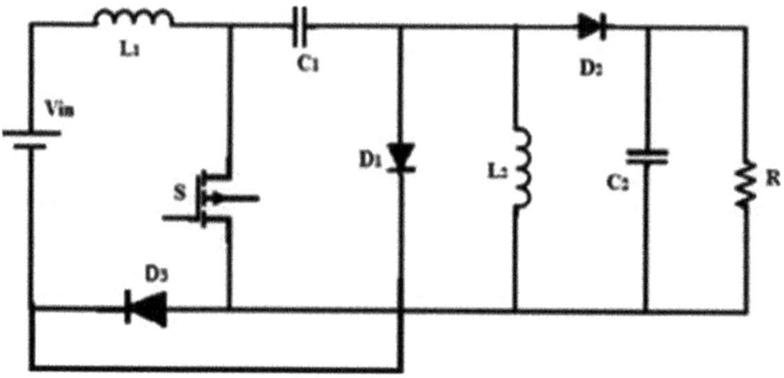

Fig. 2 Circuit structure of the cascaded model

to result in simplified structure and, higher efficiency. The voltage conversion ratio of the proposed structure is given by $D/(1-D)^2$.

The circuit includes two inductors $L1$ and $L2$. $C1$ is acting as a coupling capacitor and C_2 is the output capacitor. The cascaded converter also comprises one switch, diodes $D1$, $D2$ and $D3$. The working operation of the circuit is divided into three modes:

Mode 1: During the turn on condition of the switch, the diodes $D1$ and $D2$ are in reverse-biased condition due to voltage across the capacitor $C1$, whereas diode $D3$ is still in on condition. The voltage across inductor $L1$ is the same as the input voltage V_{in} due to which inductor current IL_1 increases linearly. The voltage near inductor $L2$ is VC_1 and the capacitor $C1$ is discharging its energy to inductor $L2$ through switch S. The output current across inductor L_2 rises linearly.

Mode II: Now, the switch is in turned-off condition. The path followed by the inductor current is through diode $D1$, inductor $L1$, input voltage Vin and capacitor $C1$. Due to this, IL_1 falls down linearly. For the same time period, current IL_2 across output inductor is forced to transfer its energy to diode $D2$. The energy stored in the output inductor L_2 is shifted to the output capacitor C_O and resistive load R. Therefore, inductor current IL_2 also decreases linearly and the diode $D3$ gets reverse-biased due to source voltage V_{in} [4].

Mode III: The diode $D2$ becomes reverse-biased as soon as the inductor current IL_2 becomes zero. This blocks the current across the inductor to rise linearly again. When diode $D1$ is turned ON and switch S and diodes $D2$, $D3$ are turned OFF the capacitor $C1$ transfers energy with the help of DC source and inductor $L1$.

4 Result Analysis

4.1 Simulation Results

The projected modified cascaded Boost and Buck-Boost Converter has been designed using the parameters mentioned in Table 1. The input voltage applied to the proposed converter is 12 V. The switching frequency used is 50 kHz. The proposed converter is simulated in MATLAB R2014a. The inductor values are L_1 0.4 and L_2 0.15 mH. However, the capacitor value is C_1 470 μF and the output capacitor C_2 is 100 μF respectively. The load for the proposed structure is 100 Ω. The output voltage is 59.2 V and a 60% duty ratio is achieved in MATLAB simulation. The proposed converter is simulated for 79.8 W of output power and 1.157 A of the output current.

Figure 3 shows the MATLAB Simulink model of modified cascaded Boost and Buck-Boost Converter. The switching frequency fed to the converter is 50 kHz. Figure 3 shows the regulated 12 V DC as the input voltage for the converter. The output voltage obtained is 49.2 V at a 60% duty ratio for the proposed converter, as shown in Fig. 4.

Table 1 Design parameters of the converters (simulation)

S. no.	Parameters	Value
1	Input voltage	12 V
2	Inductor L1	0.4 mH
3	Inductor L2	0.15 mH
4	Capacitor C1	470 μF
5	Capacitor C2	100 μF
6	Duty ratio	60%
7	Output voltage	49.2 V
8	Output power	79.8 W
9	Switching frequency	50 kHz
10	Output current	1.157 A
11	Load	100 Ω

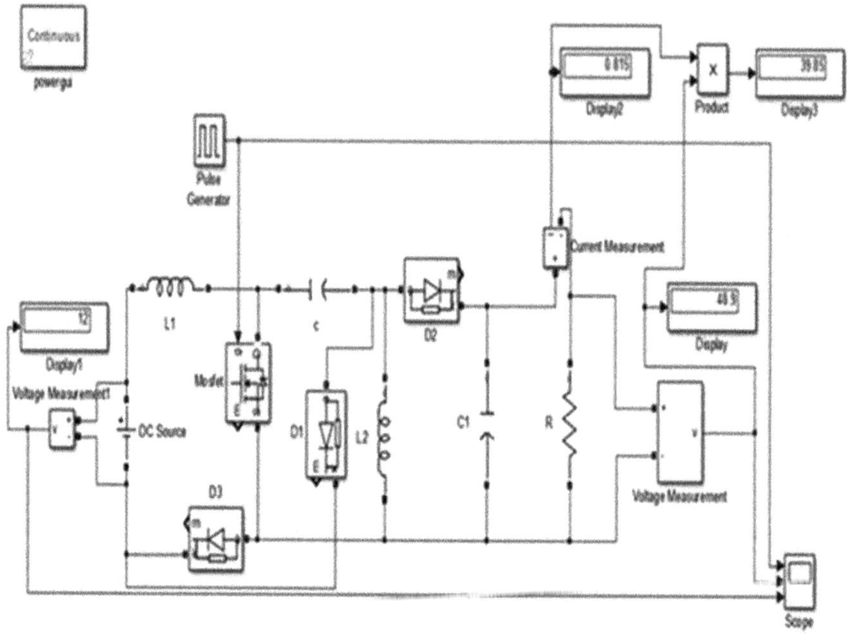

Fig. 3 MATLAB Simulink model of the cascaded buck-boost converter (at 50%)

4.2 *Hardware Results*

Figure 5 shows the circuit diagram of the proposed cascaded converter comprising a controller. The controller used here helps to provide multiple inputs in order to get constant output voltage. The Arduino Uno used here acts as a controller to which two inputs are fed, i.e., a PV cell and a NANO wind turbine generator. The cascaded

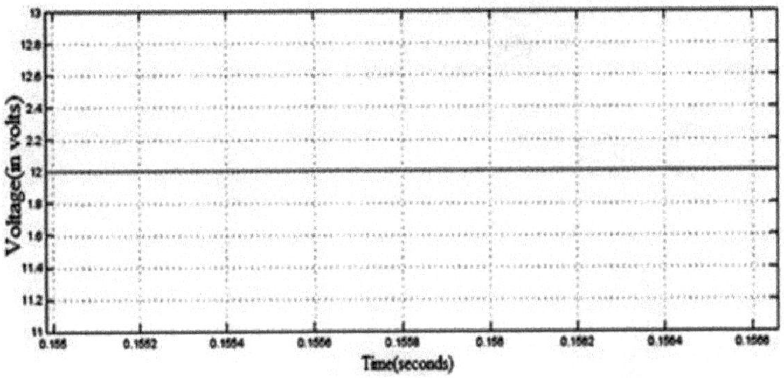

(a)

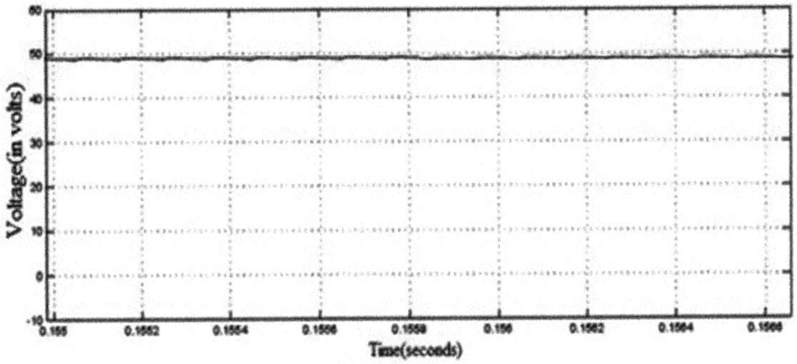

(b)

Fig. 4 Waveform of cascaded converter (50% duty ratio). **a** Input voltage, **b** Output voltage

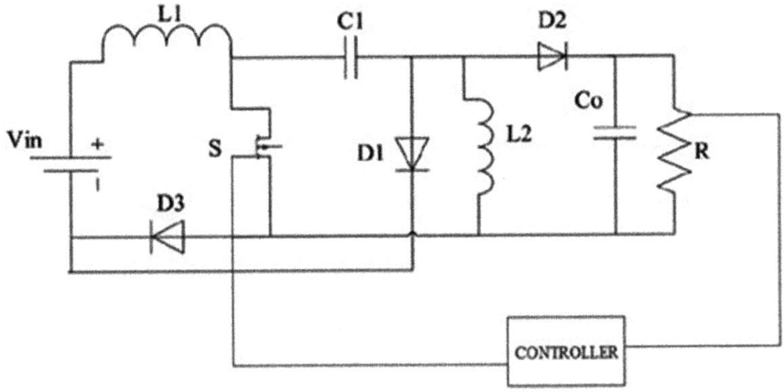

Fig. 5 Cascaded converter circuit for closed-loop system comprising controller

Fig. 6 Laboratory setup of the dual-input cascaded converter

converter includes a switch MOSFET (IRF 540 N). The input voltage fed is 9.8 V DC. The inductor L_1 is 0.4 mH and L_2 is 0.15 mH. Moreover, the capacitor C_1 is 470 μF and output capacitor C_O is 100 μF. The resistive load of 10 KΩ is used for which the output voltage obtained is 34 V at 53% duty ratio. Relay has been used for controlling the double inputs (wind and solar) fed to the converter. A display is added to show the voltage range, duty cycle and output voltage.

Figure 6 shows the experimental setup of the multi-input cascaded converter. The input of the Solar Panel is 7.08 V, to which the output voltage obtained is 32 V at 63% of duty ratio. Similarly, the input to the wind turbine is 12.6 V, to which the output voltage obtained is 36 V at 43% of duty ratio. Averaging both the value together, the input for the cascaded converter is 9.8 V and the output voltage obtained by the cascaded converter is 34 V at 53% of its duty ratio. The output from the experimental setup has been shown in Table 2. The proposed cascaded converter delivers high voltage gain. For example, the output voltage is 37.3 at 9.8 V of input voltage and 50% duty ratio. The conventional circuits of boost or buck-boost would be unable to reach that much of the output voltage for the same duty cycle.

Table 2 Output results of hardware model of cascaded converter using hybrid inputs

Solar energy			Wind energy			Cascaded converter		
Input	Duty ratio	Output	Input	Duty ratio	Output	Input	Duty ratio	Output
7.08 V	63%	32 V	12.6 V	43%	36 V	9.8 V	53%	34 V

5 Conclusion

In this paper, a DC-DC cascaded converter with dual-inputs has been proposed for the renewable energy system. The inputs fed to the structure are solar and wind energy. The circuit is simple comprising one switch and provides better voltage gain as compared to the conventional converters, and, it is cost effective. The proposed converter works with two inputs: solar and wind. Solar energy is received through a photovoltaic cell, whereas wind generator represents the wind source. Voltage from both the supplies is fed to the cascaded structure. Moreover, the unavailability of any one input source will have the back-up by the second source making the system applicable for the hybrid renewable systems. In the future, other than these two renewable resources other sources can also be implemented with cascaded converters to make the electric power generation free from the conventional energy resources.

References

1. Zhang N, Sutanto D, Muttaqi KM (2016) A buck-boost converter based multi-input DC-DC/AC converter. In: 2016 IEEE international conference on power system technology (POWERCON)
2. Sernia PC (2002) Cascaded DC-DC converter connection of photovoltaic modules. In: 33rd Annual IEEE power electronics specialists conference proceedings (Cat No 02CH37289) PESC-02
3. Morales-Saldana JA, Guti EEC, Leyva-Ramos J (2002) Modeling of switch-mode dc-dc cascade converters. IEEE transactions on aerospace and electronic systems
4. Krishnamurthy H Stability analysis of cascaded converters for bidirectional power flow applications. 30th International Telecommunications Energy Conference, 09/2008
5. Ortiz-Lopez MG, Leyva-Ramos J, Diaz-Saldierna LH, Carbajal-Gutierrez EE (2007) Multi-loop controller for N-Stage cascade boost converter. IEEE International Conference on Control Applications
6. Hafez AAA (2015) Multi-level cascaded DC/DC converters for PV applications. Alex Eng J
7. Kumari R, Haider A, Pandit M, Sherpa KS (2020) Dual input cascaded converter for high voltage gain. Students Conference on Engineering and System (SCES)
8. Zakir Hossain M, Jeyraj A / L Selvaraj, Rahim NA (2018) High voltage-gain full-bridge cascaded dc-dc converter for photovoltaic application. PLOS ONE

Simulation and Analysis of 11T SRAM Cell for IoT-Based Applications

Saloni Bansal and V. K. Tomar

Abstract The rapid growth in e-marketing demands the IoT (Internet of things)-based devices, which interconnect several devices to each other and can efficiently operate on ultra-low-power supply voltage. In this context, an 11T static random access memory (SRAM) cell has been simulated on the Cadence Virtuoso tool on a 45 nm technology file. The obtained simulated results are analyzed and compared with conventional 6T and read decoupled 8T SRAM cells with varying the supply voltages from 0.6 to 1 V. As a consequence, excellent results have been observed in terms of write power, write ability, and read delay of 11T SRAM cell as compared to considered SRAM cells. It is observed that read power dissipation in the 11T SRAM cell is decreased by 1.10× as compared to the conventional 6T SRAM cell. Furthermore, write power dissipation is decreased by 1.05×/ 3.38× as compared to conventional 6T and RD 8T SRAM cells. The write ability of the 11T SRAM cell also improved by 1.17×/1.1× as compared to conventional 6T and RD 8T SRAM cells at 0.6 V supply voltage. The write delay of 11T SRAM is 2.91× improved as compared to the conventional 6T SRAM cell.

Keywords IoT · SRAM · Read/write power · Read/write stability · Read/write delay

1 Introduction

The excessive advancement in the field of IoT-based devices has changed daily lives [1]. IoT describes the structure of sensual objects that consist of information, of different software, and several technologies, which help devices to interconnect with each other and sharing the data among them [2]. Cloud computing needs to

S. Bansal (✉) · V. K. Tomar
GLA University, Mathura 281406, Uttar Pradesh, India
e-mail: Saloni.bansal@gla.ac.in

V. K. Tomar
e-mail: Vinay.tomar@gla.ac.in

S. Dhar et al. (eds.), *Advances in Communication, Devices and Networking*,
Lecture Notes in Electrical Engineering 776,
https://doi.org/10.1007/978-981-16-2911-2_35

work on CMOS devices with very strong realizations while edge computing requires devices having very-low-power consumption because these have to cope with plenty of devices available in the virtual space. To facilitate high-speed sharing of data, among the IoT devices low-power memory circuits are of primary importance. Today's SRAM cells occupy 90% of total space on silicon-on-chip (SoC) devices [3] and high-performance VLSI circuits. It also becomes the major contributor to the total power dissipation. Therefore, the design of low-power SRAM cells becomes an essential requirement for IoT-based devices due to limited available energy resources. In SRAM cells, supply voltage scaling is one of the adequate methods to mitigate the power dissipation. The reduction in power consumption is beneficial for the entire system and also increases the yield and upgrades the reliability of SoC [4].

2 Related Work

In this section, various recently reported SRAM topologies [5–13] are discussed. Nabavi et al. [5] present a 6T SRAM bit-cell. In this cell, PMOS transistors are utilized as access transistors, which results in lower zero-level degradation that further increases the read stability of the cell. Sachdeva et al. [6] reported a single-ended read/write 11T SRAM cell. In this cell, read/write access time is improved along with enhancement in read/write stability. A loop-cutting approach [7] is utilized to improve both read static noise margin (RSNM) and write static noise margin (WSNM) of 10T SRAM cell. This cell is also free from half select issues. In addition to this, I_{on} to I_{off} current ratio is also improved as compared to the conventional 6T SRAM cell. Kumar et al. [8] reported a 12T SRAM cell operated in the sub-threshold region and has reduced power consumption with improved stability. In this cell, both read and write access time is improved with a reduction in power dissipation. Another [9] 12T SRAM cell is presented with a low read power delay product. This cell utilized a Schmitt-trigger-based design to achieve the improvement in various parameters. Ensan et al. [10] reported an 11T SRAM cell with improved write ability as well as write delay by using Fin-FET technology. So by keeping all the trade-offs under consideration, we have simulated and analyzed 11T SRAM cells for IoT-based applications.

3 SRAM Topologies

In this section, we have simulated the considered SRAM cells such as conventional 6T, RD 8T, and 11T SRAM cells on 45 nm technology node using the Cadence Virtuoso tool.

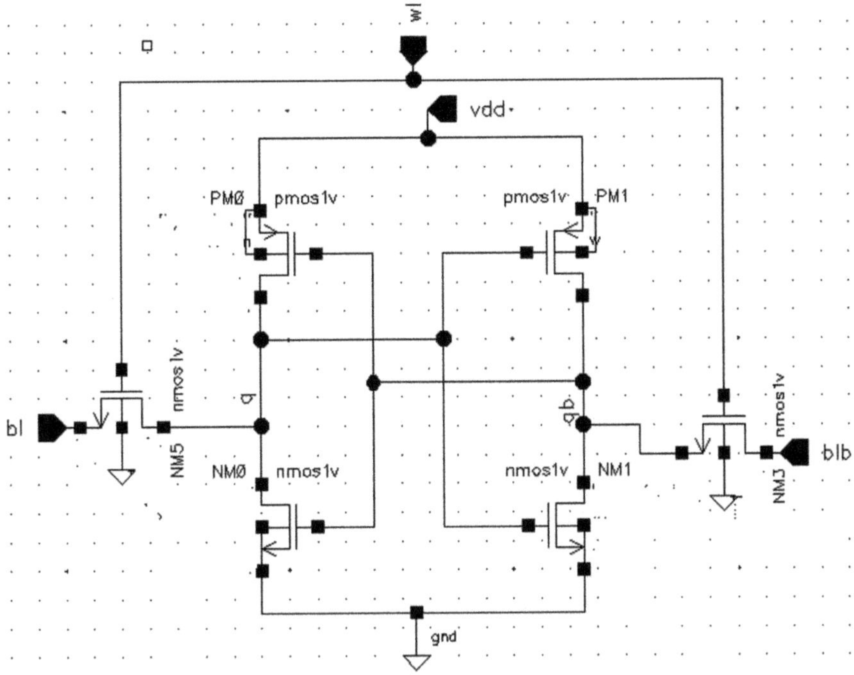

Fig. 1 Conventional 6T SRAM cell

3.1 6T SRAM Cell

Figure 1 shows the conventional 6T SRAM cell. It is formed by two back-to-back inverters used to form latch type structure that works as storage nodes. To perform a write operation, bit-line (bl) is raised to be logic high. To perform the write '1' operation, the word-line (wl) signal is kept at a high logic level and generally connected to $V_{DD.}$ Current passes through access transistors NM3 and NM5 and logic '1' is stored at node Q and logic '0' stored at node qb that turn on the PM0 and NM1 transistors. For a read operation, wl remains at a logic high by which both access transistorsNM3 and NM5 become turn on. Bit-line bar (blb) tends to ground throughNM1transistor while bl stands with V_{dd} following the NM0 transistor to be turn off. This operation concluded to read '1' operation.

3.2 Read Decoupled 8T SRAM Cell

In Read Decoupled structure of 8T SRAM cell, transistors PM1, PM2 and NM1, NM2 form the cross-coupled circuitry and works as core memory of the circuit. NM11 and NM12 work as access transistors. This circuit has a single-ended discrete

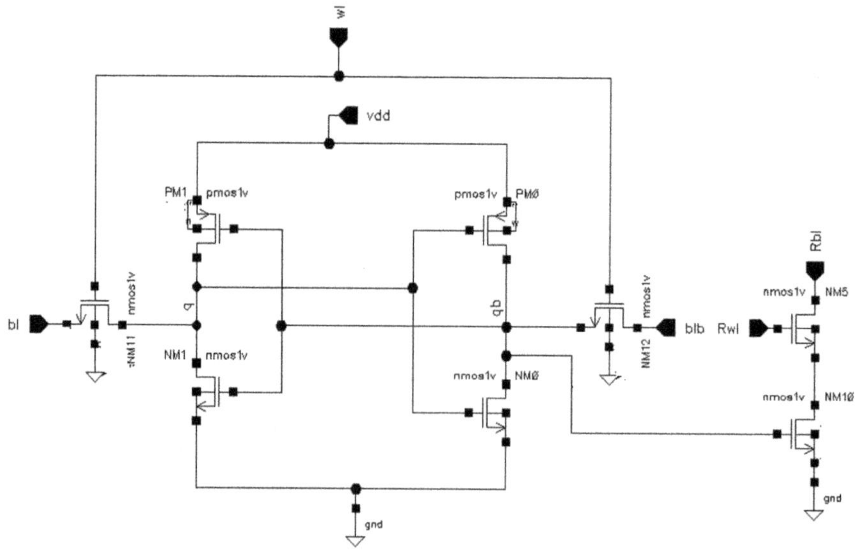

Fig. 2 Read-decoupled 8T SRAM

read terminal to enhance the read operation and improves the read stability. In this cell, the differential write operation is performed. The schematic of the RD 8T SRAM cell is shown in Fig. 2.

3.3 11T SRAM Cell

Figure 3 shows the schematic of the 11T SRAM cell. This cell consists of 11 transistors. T1, T2, T3, and T4 form the core latch part of the cell. T5 and T6 are the access transistors. Here, VT1 and VT2 both NMOS transistors are connected with different voltage sources having voltage below the supply voltage provided for the circuit to perform write operation. Both [14] source terminals of NMOS transistors VT1 and VT2 are connected with bL and blb. T8 and T9 transistors are connected in series and act as resistors, whereas T7 transistor behaves as a switch. The circuit works with a self-controllable voltage technique in which the cell reduces the leakage current. The 11T SRAM cell works in three modes of operations as discussed below.

RL signal is kept at a logic low during read operation while word-line (wl) is signal kept at a logic high level. This turns on both the access transistors. The bit-lines bl and blb are connected to the pre-charge circuit. Subsequently, read operation is performed as similar to conventional 6T SRAM cell. The read transfer characteristic curve of the 11T SRAM cell is shown in Fig. 4. During write operation, bit-line (bl) rises to logic high or stores logic '1' and bit-line bar (blb) is connected with low logic or stores '0'. Due to this, Transistors VT2 is on and subsequently, VT1 is off. As a

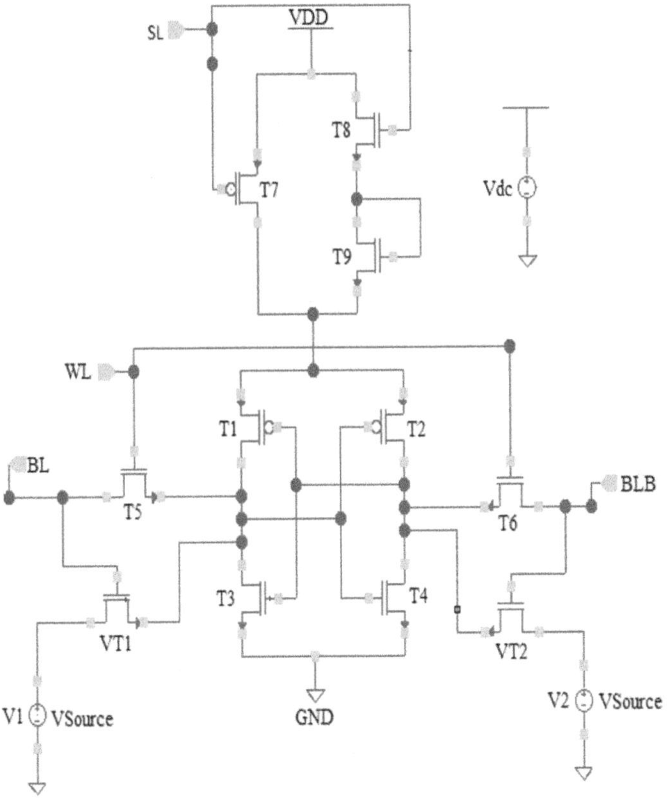

Fig. 3 11T SRAM cell

consequence, the fluctuations at the output node get reduced due to voltage source V2 connected with the source terminal VT2. At the time of hold operation, the SL signal rised with a logic high. VT1 and VT2 transistors are connected to voltage sources with equal voltage. Due to activation of the RL signal, NMOS transistors N8 and N9 find the resistive path. Thus, it reduces the leakage current. The write transfer characteristic curve of the 11T SRAM cell is shown in Fig. 5.

4 Result and Discussion

In this section, the obtained simulated results such as read/write power, read/write stability, and read/write access time of all considered SRAM cells are compared and analyzed. All comparisons are performed at 0.6 V supply voltage.

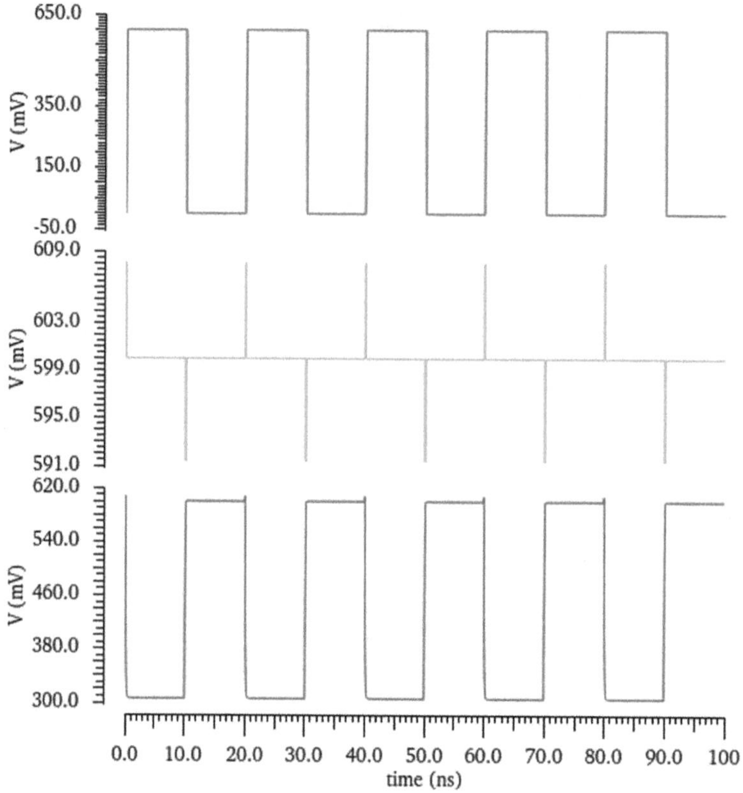

Fig. 4 Read waveform of 11T SRAM at 0.6 V

4.1 *Read/Write Power*

The total power dissipation is the sum of dynamic and static power dissipation. Dynamic power dissipation occurs due to swapping of voltages while static power dissipation occurs when the circuit is in standby mode.

The obtained simulated results of read power for 11T SRAM cell and considered SRAM cells are shown in Figure. It can be observed from Fig. 6 that the 11T SRAM cell consumes 1.10× less read power than the conventional 6T SRAM cell. Further, RD 8T SRAM cell consumes the lowest read power as compared to conventional 6T and 11T SRAM cells at 0.6 V supply voltages. Figure 7 shows the write power dissipation of 11T SRAM cell and considered cells. It is observed that the write power of the 11T SRAM cell reduces by 1.05× and 3.38× as compared to conventional 6T and RD 8T SRAM cells at 0.6 V supply voltage.

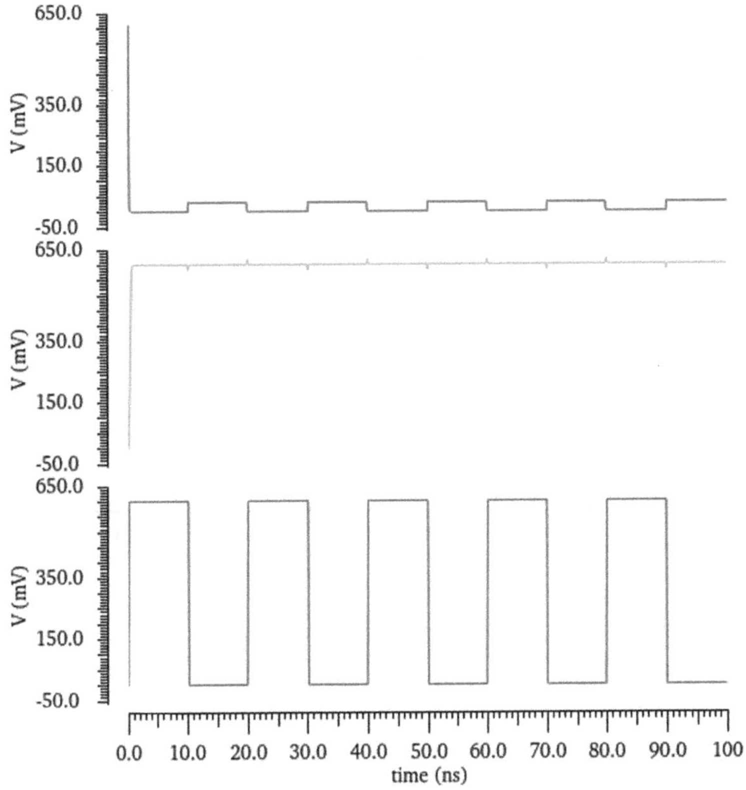

Fig. 5 Write waveform of 11T SRAM at 0.6 V

4.2 *Read/Write Stability*

The stability of any cell can be examined by the term SNM (Static Noise Margin). SNM can be defined [15] as the minimum amount of voltage required to flip the content of the cell. Stability of the cell depends on cell ratio, pull-up ratio, and supply voltage. There are basically two methods to measure the stability of the cell are butterfly-curve method and n-curve method. In this work, butterfly-curve method is used to measure the stability of the SRAM cells. In this method, an appropriate shape of square can be fixed into the smallest lobe of the curve and measured length can be called as Read Static Noise Margin. Figure 8 shows the read stability with variation in supply voltages. It can be observed that RD 8T SRAM cell has the highest value of RSNM at 0.6 V supply voltage as compared to conventional 6T and 11T SRAM cells. Moreover, smallest value of RSNM has been noticed in 11 SRAM cell.

The write static noise margin can be calculated by the integration of the read static curve (RTC) and write static curve (WTC) and then measure the width of the largest lobe of the curve [14]. Figure 9 shows the write ability of 11T SRAM cell along with

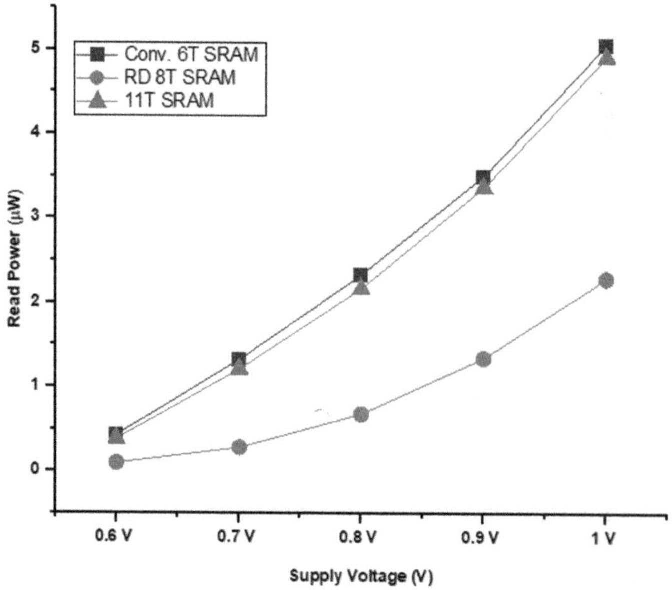

Fig. 6 Read power with variation in supply voltages

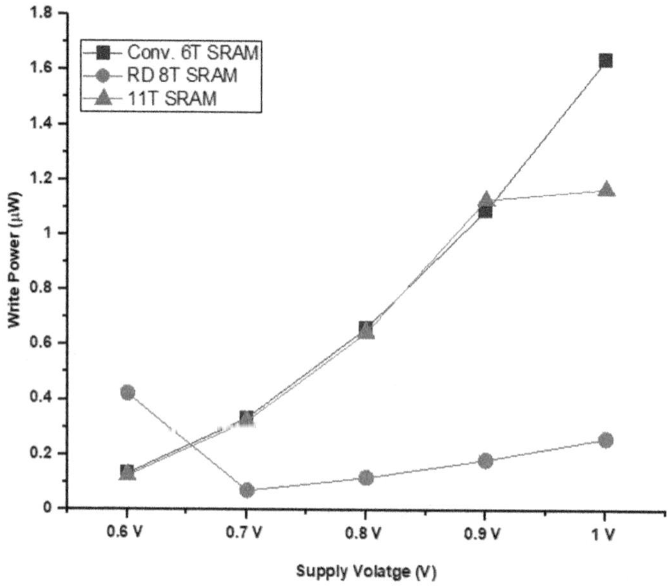

Fig. 7 Write power with variation in supply voltages

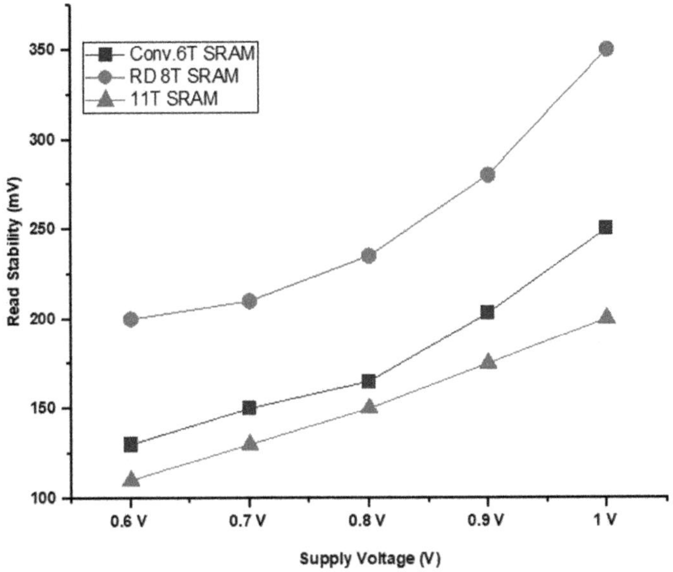

Fig. 8 RSNM with variation in supply voltages

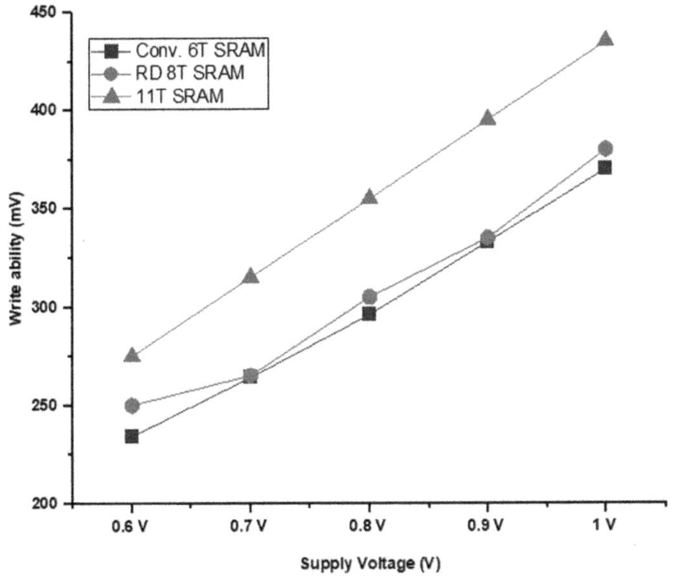

Fig. 9 Write ability with variation in supply voltages

the considered SRAM cells. It has been observed that 11T SRAM cell has far better write ability as compared to RD 8T and conventional 6T SRAM cells. It happens due to extra voltage sources V1 and V2 connected with the source terminals of VT1 and VT2 transistors. The write ability of 11T SRAM is 1.17×/1.1× better than the conventional 6T/RD 8T SRAM cell at 0.6 V supply voltage.

4.3 *Read/Write Delay*

Read delay of any SRAM cell for differential mode operation can be defined as the time duration to achieve a voltage difference of at least 50 mV among the bit-lines. Figure 10 shows the read access time of 11T SRAM cell along with considered SRAM cells. It reveals that 11T SRAM cell has the lowest value of read access time among other considered SRAM cells at 0.6 V supply voltage. Write delay of any cell can be defined as the time duration in which voltage of word-line (wl) goes to the 50% of maximal voltage of time in which one storage node from of two is getting pulled up from 0 to 50% of the supply voltage. In 11T SRAM cell, the write delay is improved by 2.91× as compared to the conventional 6T SRAM cell (Fig. 11).

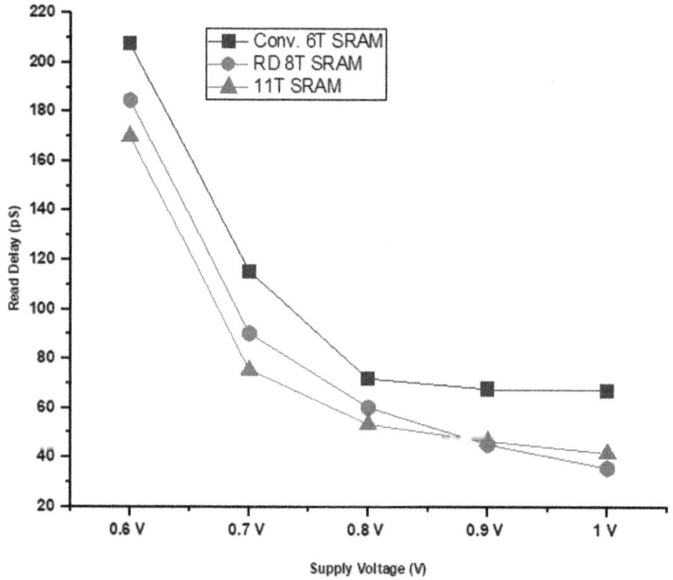

Fig. 10 Read delay with variation in supply voltages

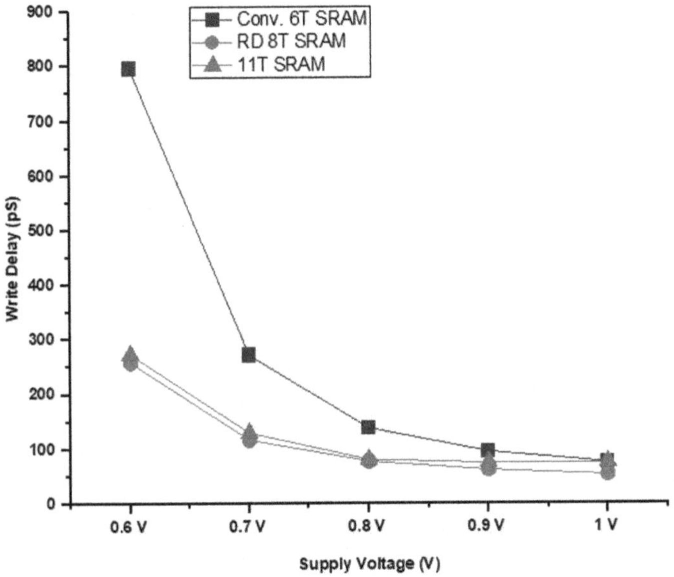

Fig. 11 Write delay with variation in supply voltages

5 Conclusion

In this work, 6T, RD 8T, and 11T SRAM cells are simulated on a 45 nm technology node with Cadence Virtuoso tool for IoT-based applications. The performance parameters such as read/write power, read/write stability, and read/write delay of all considered cells are compared and analyzed. The 11T SRAM cell shows a commendable reduction in read power and significant improvement in write delay. Further, the write ability of 11T SRAM cell also improved by 1.17×/1.1× as compared to Conventional 6T and RD 8T SRAM cells at 0.6 V supply voltage.

References

1. Pal S (2019) Transmission gate-based 9T SRAM cell for variation resilient low power and reliable internet of things applications. IET Circuits Devices Syst 13.5:584–595
2. Muntjir Mohd, Rahul Mohd, Alhumyani Hesham A (2017) An analysis of Internet of Things (IoT): novel architectures, modern applications, security aspects and future scope with latest case studies. Int J Eng Res Technol 6(6):422–447
3. Dasgupta Sudeb (2017) Compact analytical model to extract write static noise margin (WSNM) for SRAM cell at 45-nm and 65-nm nodes. IEEE Trans Semicond Manuf 31(1):136–143
4. Maroof N, Kong B (2017) 10T SRAM Using Half- $V_{\text {DD}}$ Precharge and row-wise dynamically powered read port for low switching power and ultralow RBL leakage. In IEEE transactions on very large scale integration (VLSI) Systems, vol 25, no 4. pp 1193–1203. https://doi.org/10.1109/tvlsi.2016.2637918

5. Nabavi M, Sachdev M (2017) A 290-mV, 3.34-MHz, 6T SRAM with pMOS access transistors and boosted wordline in 65-nm CMOS technology. IEEE J Solid-State Circuits 53.2:656–667
6. Sachdeva A, Tomar VK (2020) Design of a stable low power 11-T static random access memory cell. J Circuits Syst Comput 2050206. https://doi.org/10.1142/S0218126620502060
7. Sachdeva A, Tomar VK (2020) Design of low power half select free 10T static random-access memory cell. J Circuits Syst Comput 2150073. https://doi.org/10.1142/s0218126621500730
8. Kumar H, Tomar VK (2020) Design of low power with expanded noise margin subthreshold 12T SRAM cell for ultra low power devices. J Circuits Syst Comput. https://doi.org/10.1142/s0218126621501061
9. Sachdeva A, Tomar VK (2020) A Schmitt-trigger based low read power 12T SRAM cell. Analog Int Circuits Signal Process 1–21. https://doi.org/10.1007/s10470-020-01718-6
10. Ensan SS, Moaiyeri MH, Hessabi S (2018) A robust and low-power near-threshold SRAM in 10-nm FinFET technology. Analog Int Circuits Signal Process 94.3:497–506
11. Sachdeva A, Tomar VK *SPIN-2020* Statistical stability characterization of schmitt trigger based 10T SRAM cell design. Presented in 7th international conference on signal processing and integrated networks, https://doi.org/10.1109/spin48934.2020.9071365
12. Krishna H, Tomar VK (2019) Stability Analysis of Sub-threshold 6T SRAM cell at 45 nm for IoT application. Int J Recent Technol Eng 8(2):2434–2438. https://doi.org/10.35940/ijrte.b1989.078219
13. Kumar HK, Tomar VK A review on performance evaluation of different low power SRAM cells in nano-scale era. Wireless Personal Communications. https://doi.org/10.1007/s11277-020-07953-4
14. Gavaskar K, Ragupathy US (2019) Low power self-controllable voltage level and low swing logic based 11T SRAM cell for high speed CMOS circuits. Analog Int Circuits Signal Process 100(1):61–77
15. Singh J, Mohanty SP, Pradhan DK (2012) Robust SRAM designs and analysis. Springer Science & Business Media

Simulation and Analysis of Schmitt Trigger-Based 9T SRAM Cell with Expanded Noise Margin and Low Power Dissipation

Harekrishna Kumar and V. K. Tomar

Abstract In portable electronic devices, SRAM cells play a major role to decide the performance of gadgets. In this paper, a Schmitt trigger-based 9T (ST9T) SRAM cell is simulated at 45nm technology node through Cadence Virtuoso tool. ST9T SRAM cell shows a considerable improvement in read stability which is 2.8× and 2.25× higher in comparison to considered SRAM cells. ST 9T SRAM cell has the highest read stability as compared among the considered SRAM cells. The read/write power of the ST9T SRAM cell is 1.52× and 1.47×/3.12× and 5.39× lower as compared to conventional 6T and 8T SRAM cells, respectively. The process variation effects are also analyzed through Monte Carlo simulation under 5000 random samples. It has been found that the ST9T SRAM cell has the lowest variability among considered cells. Such improvements in the characteristics of ST9T SRAM make it suitable for ultra-low power applications. All the comparisons have been performed at 0.3V supply voltage.

Keywords Subthreshold · Stability · Power dissipation · Access time

1 Introduction

Nowadays, there is an increase in demand for battery-operated portable devices, like wireless sensor networks, gas analyzers, and blood glucose monitors. These devices require low power memory circuits to extend the battery life [1]. Static random access memory (SRAM) consists of back-to-back connected inverters that form a latch circuitry to store the data which occupies a large portion of the system on chip to improve the performance. Due to their large portion with high performance, power dissipation becomes the primary concern [2]. In a conventional 6T SRAM cell, four transistors are used to form core latch circuitry to store the data as shown in Fig. 1a. The read/write operation is performed through two common access transistors N1

H. Kumar (✉) · V. K. Tomar
GLA University, Mathura 281406, Uttar Pradesh, India
e-mail: harekrishnabgp@gmail.com

S. Dhar et al. (eds.), *Advances in Communication, Devices and Networking*, Lecture Notes in Electrical Engineering 776,
https://doi.org/10.1007/978-981-16-2911-2_36

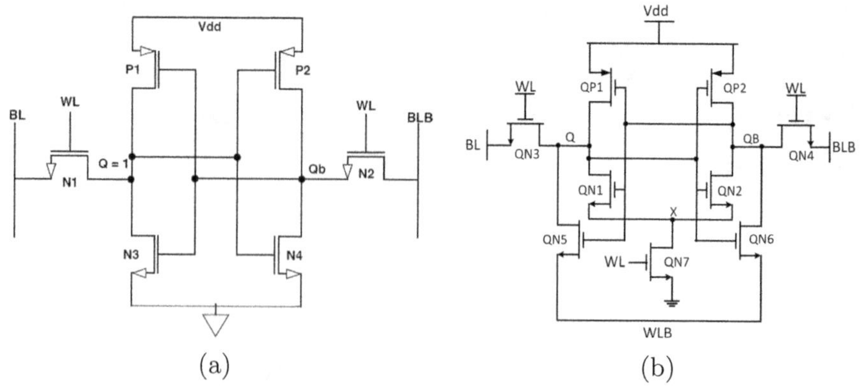

Fig. 1 Schematic of **a** 6T SRAM cell, **b** 8T SRAM cell

and N2. Due to the common access transistors during the read/write operation, access transistors conflict (ATC) is the major issue of conventional 6T SRAM cells in the sub-threshold region. Further, the stability of the cell is also degraded with ATC issue [3, 4].

To improve the SRAM cell characteristics like read/write stability, read/write delay, and power dissipation, a lot of SRAM cell designs are reported in the literature. An improvement in read stability is achieved by strengthening the pull-down transistors [5]. Besides, the read decoupled structure method is another approach to improve the read stability of the cell [6]. Pal et al. [7] presented a low-power 8T SRAM cell as shown in Fig. 1b which utilized a read-assist transistor to improve the read stability of the cell. It has been found that the isolated read/write structure improves the stability of the cell. In the conv.6T SRAM cell, write stability of the cell is increased by strengthening the access transistors with plenty of areas. A 12T SRAM cell has been reported in [3], in which the cell write stability is improved through a feedback loop cutting technique which declines the power consumption of the cell. Power dissipation becomes the primary concern in ultra-low power devices. In a differential structure, power dissipation of the cell is increased due to the higher value of the activity factor which takes minimum time to charge/discharge the capacitances. It improves the performance of the cell. To reduce the power dissipation, a single-ended SRAM structure is utilized [8, 9]. This structure minimizes the activity factor by half which results in reduction in power dissipation. A single-ended structure is utilized in a positive feedback control 10T SRAM cell to minimize the read power dissipation of the cell. Additionally, the leakage power of the cell is also reduced due to the stacking effect.

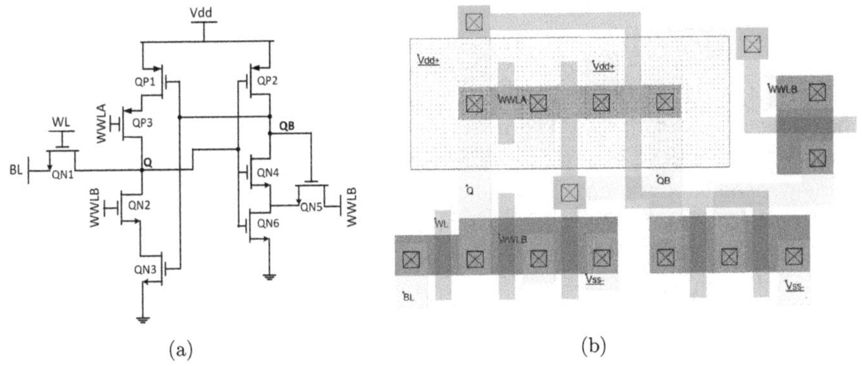

Fig. 2 **a** Schematic of ST9T SRAM cell, **b** Layout of ST9T SRAM cell

2 Schmitt Trigger-Based 9T SRAM Cell

The schematic and layout diagram of the ST9T SRAM cell is shown in Fig. 2. In this cell, during the read operation, WWL and WWLB signals are connected to logic ‘0’ and ‘1’ which turn on the transistors QP3 and QN2, respectively. At the same time, the WL signal is at logic ‘1’ which also turns on transistor QN1. Discharging of bit-line (BL) depends on the storage nodes (Q and QB). It is assumed that node QB stored logic ‘1’ which enables transistors QN3 and provides the discharge path of BL. It confirms the stored logic ‘0’ at storage node Q.

During the write ‘0’ operation, WWLA, WWLB, and WL signals are connected to logic ‘1’. Biline signal (BL) is connected to logic ‘0’. The WWLA signal disconnects the storage node Q from the VDD by turning off transistor QP3 while the WWLB signal provides the ground path to storage node Q. It results in logic ‘0’ at node Q. However, during write ‘1’ operation, WWLA, WWLB, and WL signals are connected to logic ‘0’ and logic ‘1’, respectively. The input signal is passed from BL to the storage node through transistor QN1. The WWLA signal maintains the data at node Q. During hold mode, signals WWLA and WWLB are connected to logic ‘0’ and logic ‘1’. The WL signal is connected to logic ‘0’ which turns off the transistor QN1 to preserve the data at the storage node.

3 Simulation Result and Discussion

The reported ST9T and considered SRAM cell (6T, 8T) are compared at 45 nm technology node with the same aspect ratio (width/length). The characteristics of ST9T SRAM cell such as power dissipation, access time, and stability are obtained at different supply voltages. The process variation effects are also analyzed at SRAM cell through Monte Carlo samples (5000 random samples).

3.1 Stability

Stability of ST9T SRAM cell is determined through the butterfly method [4, 10]. The obtained values of RSNM with varying supply voltages are shown in Fig. 3a. It is noticed that the RSNM of the ST9T SRAM cell is improved by 2.80×, and 2.26× in comparison to conventional 6T and 8T SRAM cells. Figure 3b shows the process variation effects on RSNM at 0.3 V supply voltage. It is evident that the mean value (RSNM) of the ST9T SRAM cell is highest among considered cells. It signifies that the variability in RSNM of ST9T is lowest as compared to other considered cells. It happens due to the Schmitt trigger structure which improves the noise handling capacity at the storage node. The write static noise margin (WSNM) is measured to define the write ability of the SRAM cell. To measure the WM, the bit-line swing method has been utilized [11]. Figure 3c shows the WM of SRAM cells at various supply voltages. The WM of ST9T SRAM cell is increased by 1.16× and 1.27× as compared to conventional 6T and 8T SRAM cells at different supply voltages. It happens due to the use of a write-assist transistor.

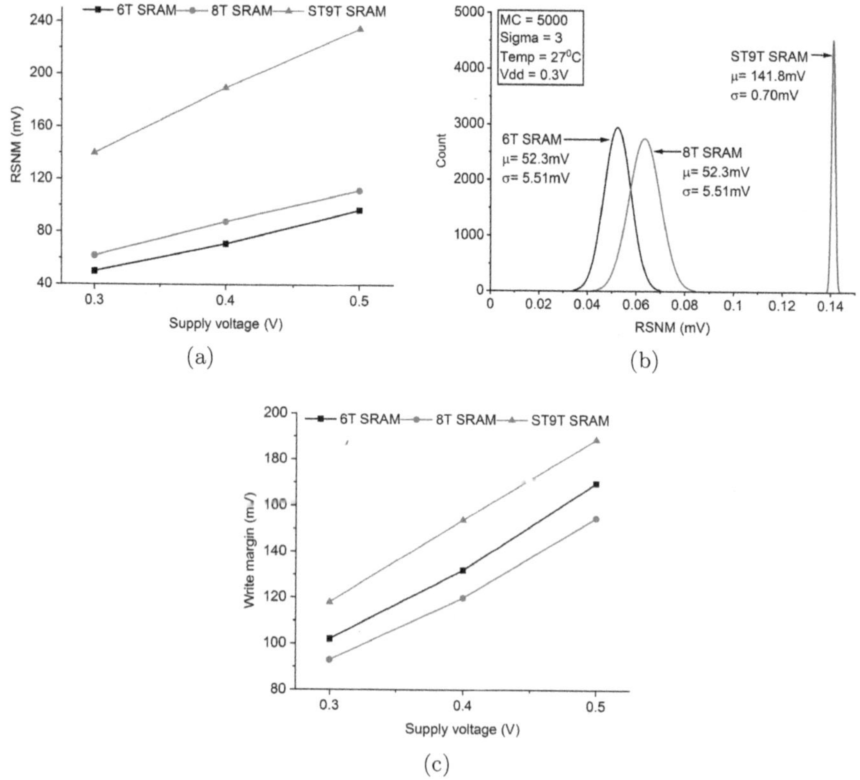

Fig. 3 **a** RSNM of SRAM topologies at different supply voltages, **b** MC simulation of RSNM at 0.3V supply voltage, **c** WM of SRAM topologies at different supply voltage

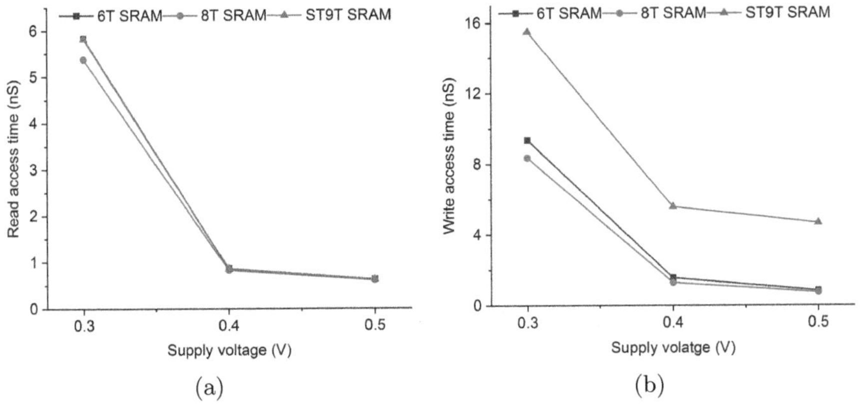

Fig. 4 **a** Read access time at different supply voltage, **b** Write access time at different supply voltage

3.2 Read/Write Access Time

The read delay is defined as the time duration required by the cell to develop a voltage drop of minimum required voltage (50 mV) on the read bit-line. Figure 4a shows the read delay with variation in supply voltages. It is worthy to note that the read access time of ST9T SRAM cell is approximately similar to 6T SRAM cell and highest to 8T SRAM cell. The write delay is expressed as the time difference between the activation of the WL signal and charging the storage node up to 90% Vdd. Figure 4b shows the write delay with variation in supply voltages. It has been observed that the write delay of the ST9T SRAM cell is highest among conventional 6T and 8T SRAM cells.

3.3 Read/Write Power Dissipation

Power consumption in SRAM cells becomes the primary concern for portable electronics devices. In this paper, average read and write power dissipation are calculated at different supply voltages as depicted in Fig. 5a, b. The read power consumption in the ST9T SRAM cell is decreased by 1.51× and 1.47× in comparison to conventional 6T [4] and 8T SRAM cells, respectively. Further, write power dissipation is reduced by 3.12× and 5.39× as compared to conventional 6T and 8T [2] SRAM cells at 0.3 V supply voltage. It happens due to a single-ended read/write structure which reduces the activity factor of the cell.

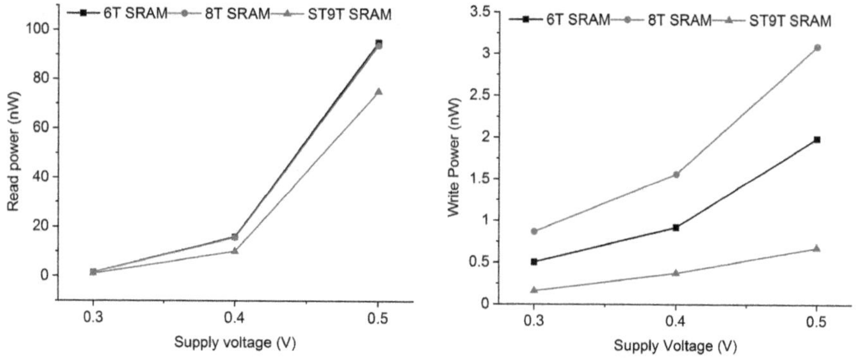

(a) (b)

Fig. 5 **a** Read Power of SRAM Topologies, **b** Write Power of SRAM topologies

3.4 Layout Design

In electronics devices, SRAM occupies a large portion of the system on chip (SoC). There is always some trade-off that exists between SRAM characteristics and the area of the cell. There are various steps involved during the layout process of SRAM cells. The layout of the SRAM cell is verified with design rule checks and Layout Versus Schematic. The layout of SRAM cells is shown in Fig. 6. It is observed that the area of the ST9T SRAM cell is the largest among considered cells. It happens due to more number of transistors.

The layout of the SRAM cell is verified with design rule checks and Layout Versus Schematic. The layout of SRAM cells is shown in Fig. 6. It is observed that the area of ST9T SRAM cell is the largest among considered cells. It happens due to more number of transistors.

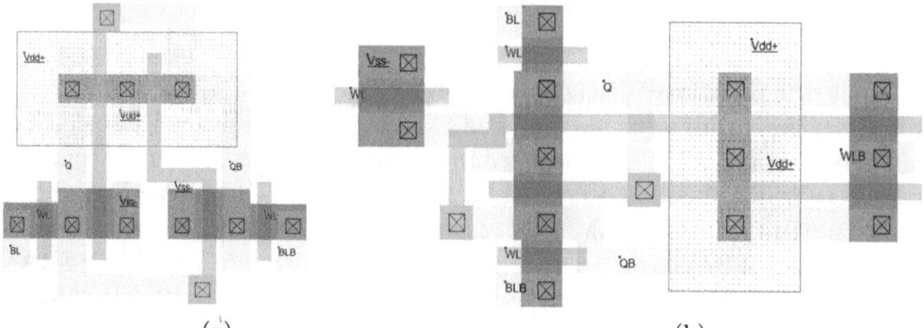

(a) (b)

Fig. 6 Layout of **a** 6T SRAM cell, **b** 8T SRAM cell

4 Conclusion

In this paper, the authors have utilized Cadence Virtuoso tool to simulate the Schmitt trigger-based 9T SRAM cell at 45 nm technology node. The obtained characteristics of 6T, 8T, and ST9T SRAMs are compared at 0.3 V supply voltage. It has been found that the read/write stability of ST9T SRAM cell is 140/118 mV which is better among other considered cells. The read and write power dissipation of ST9T SRAM is lowest than other considered cells due to the single-ended structure. The limitation of the ST9T SRAM cell is that it has the highest value of read/write delay. This 9T SRAM cell may be utilized in ultra-low power applications such as sensor-node processors and medical equipment.

References

1. Singh P, Reniwal BS, Vijayvargiya V, Sharma V, Vishvakarma SK (2018) Ultra low power-high stability, positive feedback controlled (PFC) 10t sram cell for look up table (LUT) design. Integration 62:1–13
2. Chien YC, Wang JS (2018) A 0.2 v 32-kb 10t SRAM with 41 nW standby power for IoT applications. IEEE Trans Circuits Syst I Regular Papers 65(8):2443–2454
3. Kumar H, Tomar V (2020) Design of low power with expanded noise margin subthreshold 12t SRAM cell for ultra low power devices. J Circuits Syst Comput. https://doi.org/10.1142/S0218126621501061
4. Kumar H, Tomar V (2020) A review on performance evaluation of different low power SRAM cells in nano-scale era. Wirel Pers Commun
5. Kumar H, Tomar VK (2019) Stability analysis of sub-threshold 6t SRAM cell at 45 nm for IoT application. Int J Recent Technol Eng (IJRTE) 8(2):2432–2438
6. kumar H, Tomar VK (2019) Single bit 7t subthreshold sram cell for ultra low power applications. Int J Adv Sci Technol 28(16):345–351
7. Pal S, Islam A (2015) Variation tolerant differential 8t SRAM cell for ultralow power applications. IEEE Trans Comput-aided Des Integr Circuits Syst 35(4):549–558
8. Surana N, Mekie J (2018) Energy efficient single-ended 6-t SRAM for multimedia applications. IEEE Trans Circuits Syst II: Express Briefs 66(6):1023–1027
9. Roy C, Islam A (2019) Power-aware sourse feedback single-ended 7t SRAM cell at nanoscale regime. Microsyst Technol 25(5):1783–1791
10. Sharma V, Vishvakarma S, Chouhan SS, Halonen K (2018) A write-improved low-power 12t SRAM cell for wearable wireless sensor nodes. Int J Circuit Theory Appl 46(12):2314–2333
11. Takeda K, Hagihara Y, Aimoto Y, Nomura M, Nakazawa Y, Ishii T, Kobatake H (2005) A read-static-noise-margin-free SRAM cell for low-VDD and high-speed applications. IEEE J Solid-State Circuits 41(1):113–121

Reduction of Power Fluctuation for Grid Connected qZSI-Based Solar Photovoltaic System Using Battery Storage Unit

Satyajit Saha, Pritam Kumar Gayen, and Indranil Kushary

Abstract In recent years, solar energy conversion system is adopting quasi Z-source inverter (qZSI) for connecting solar photovoltaic array to AC grid. But, output power of solar photovoltaic array is always of fluctuating profile. When this fluctuating power is injected into local grid or micro-grid, voltage at point of common coupling (PCC) experiences a flickering effect. This is an undesired impact on the local grid from the local consumer's point of view as it affects the performance of electrical gadgets as well as it creates irritation to people's eyes through illuminating devices. Sometimes, voltage flicker violates the permissible limit, which is specified in grid code. Therefore, this paper provides the solution to mitigate objectionable power quality issue in respect of local grid-connected qZSI-based solar energy conversion system. Here, fluctuating power at PCC is reduced by the usage of bidirectional converter-based battery storage unit. The charging and discharging operation of battery counteracts fluctuating power profile, hence it significantly smoothens out power profile at PCC and thus, the voltage fluctuation problem at the point can be avoided. The whole study of this paper is done using MATLAB-SIMULINK software.

Keywords Battery storage unit · Local AC grid · Quasi Z-source inverter · Solar energy conversion system · Power fluctuation

1 Introduction

In recent time, quasi Z-source inverter [1–4] is widely used for converting DC power output of solar photovoltaic array into AC power. This inverter has the following various advantages over traditional voltage source inverter:

S. Saha · P. K. Gayen · I. Kushary (✉)
Kalyani Government Engineering College, Kalyani, Nadia, West Bengal, India

I. Kushary
JIS College of Engineering, Kalyani, Nadia, West Bengal, India

S. Dhar et al. (eds.), *Advances in Communication, Devices and Networking*, Lecture Notes in Electrical Engineering 776,
https://doi.org/10.1007/978-981-16-2911-2_37

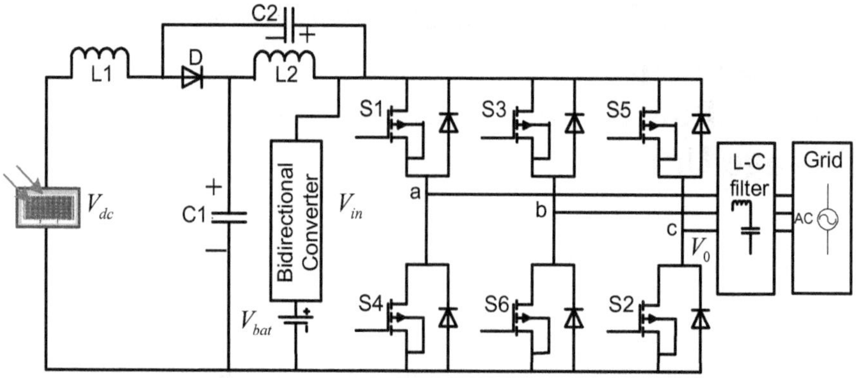

Fig. 1 Schematic diagram of grid-connected qZSI-based solar system with battery storage

i. It has voltage boosting capability,
ii. No dead time is required between pulse width modulation (PWM) of switches of the same leg, i.e. free from power distortion due to dead-time effect,
iii. False triggering of switches due to EMI does not cause harmful effect on the inverter,
iv. It draws steady input current from DC source.

These advantages make the single-stage inverter preferable choice in solar energy conversion system, and usage of this converter is increasing in practice. Many research works focus on the improvement of its performance, topology and control logic. Thus, for injecting output power of solar array into local AC grid, qZSI is used in the study of this paper. The scheme is shown in Fig. 1. The input-side solar power of qZSI fluctuates due to unsteady solar conditions in real-time application. This creates power oscillation at output of qZSI. In effect, voltage flicker [5] has definitely arrived at the point of common coupling (PCC) with the local AC grid. Thus, it is very essential to mitigate the voltage flickering effect so that it is kept within permissible limit as per standard. In this respect, this paper adopts the usage of a battery storage unit [6, 7] to smoothen out power profile via dynamic charging and discharging action. As a result, the voltage profile at PCC is improved and thus, it is expected that local user gets relief from voltage flicker problem.

2 Operation of qZSI-Based Solar Energy Conversion System with Battery Storage Unit

The overall system of this paper as shown Fig. 1 has three parts, which are discussed as follows:

2.1 Operation of qZSI

At first, the operating principle of qZSI has been presented. It is single-stage converter for converting low level DC voltage such as solar PV array or fuel cell to AC power system via intermediate DC voltage boosting action. It was previously done via a two-stage converter system (DC–DC boost converter and voltage source inverter). Nowadays, single-stage converter such as qZSI is replacing the two-stage converter due to the advantages [8] as stated before. However, the structure of qZSI has two parts—impedance network and inverter bridge. Here, impedance network prevents short circuit of terminals of input DC source in case of simultaneous switching of two power semiconductor switches of same leg of inverter bridge. Thus, the simultaneous switching of leg switches is the allowable state of qZSI unlike conventional voltage source inverter (VSI) and it is called as shoot-through state. In this state, no power is transferred to its output side.

$$V_{dc} = B.V_{mp} = \left(\frac{1}{1-D}\right).V_{mp} \tag{1}$$

$$B = \frac{1}{1-2D_{sh}} \tag{2}$$

where D_{sh} = shoot-through duty ratio, B = boosting factor.

On the other hand, power transfer through qZSI takes place via active state of its second part, i.e. inverter bridge. Here, the modulation index of the inverter bridge is controlled like conventional VSI to get controllable output voltage for achieving targeted AC-side power injection. The power transfer state is conventionally referred as non-shoot-through state. Finally, the peak value of output AC voltage is

$$\hat{V}_0 = MBV_{dc}/2 \tag{3}$$

where M = modulation index factor.

From Eq. (3), the overall voltage gain is defined as,

$$G = BM = \frac{1}{1-2D_{sh}}M \tag{4}$$

2.2 Operation of MPPT-Based Solar PV Array

It is well known that the output power of solar array is decided by solar conditions (solar irradiation and temperature). Maximum available power from solar PV array is extracted with the help of maximum power point tracking (MPPT) algorithm.

Fig. 2 Characteristics of PV Array

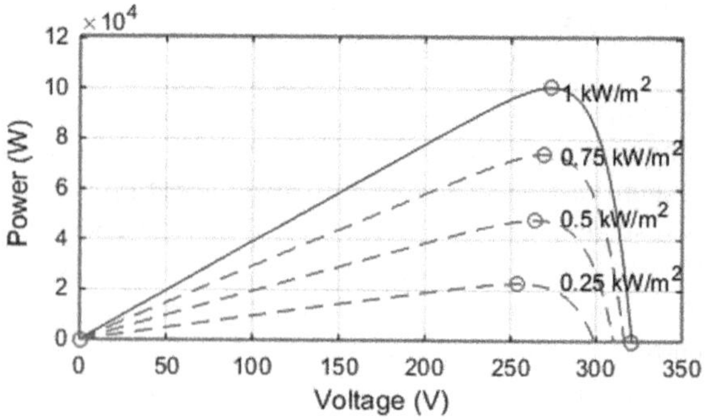

(a) Variable irradiations at 25°C

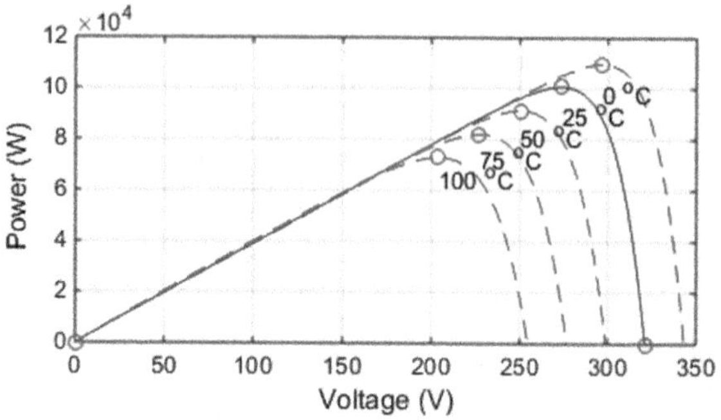

(b) Variable temperatures at 25°C

The P–V characteristics of solar PV array are given in Fig. 2, which are used in the simulation study of this paper. The perturb and observe (P&O) MPPT algorithm [8] is used in this paper, which is shown in Fig. 3.

2.3 *Operation of Bidirectional DC–DC Converter-Based Battery Stack*

The bidirectional converter controls required charging and discharging operations of battery for compensating power fluctuation. Here, the used control logic for bidirectional converter of battery is presented in Fig. 4.

3 Results

Various parameters of the overall system are provided in Table 1. The simulation study using MATLAB-SIMULINK software and observations are described as follows:

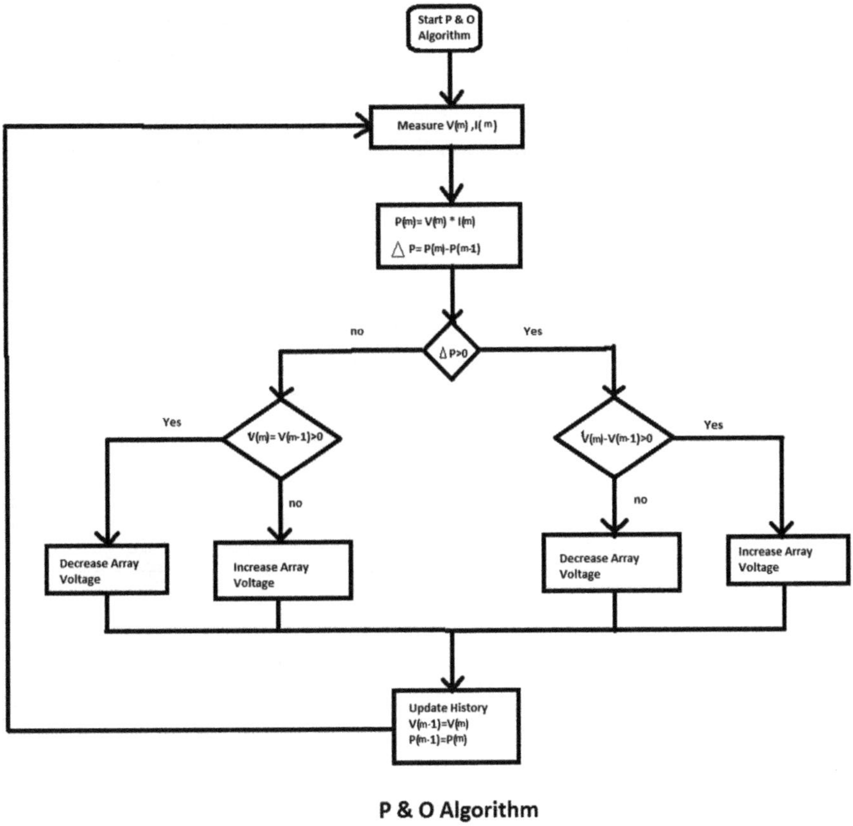

Fig. 3 Steps for P&O–MPPT Algorithm

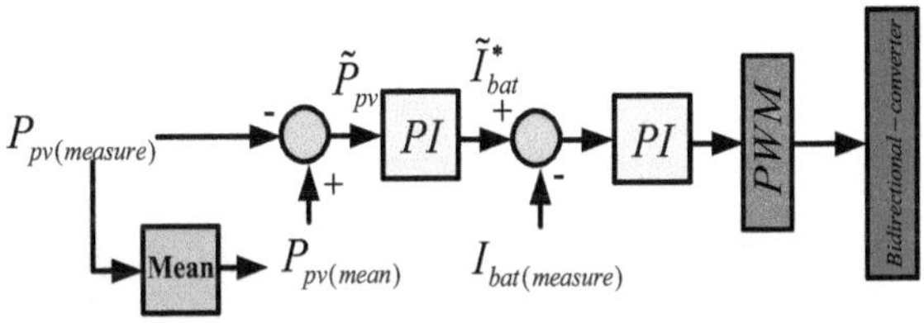

Fig. 4 Control Loop for DC–DC bidirectional Converter

Table 1 Parameters of the studied system

Components	Specifications
Rechargeable Battery	Lead acid 120 V Nominal voltage, 10Ah.
C_1, C_2	200 μF, 100 μF
L_1, L_2	500 μH, 8.0 μH
k_p, k_i	1, 20

The fluctuating temperature and solar irradiation patterns, which are set in the study, are given in Fig. 5. Figure 6 shows various power responses under *battery disconnected* mode. Here, the solar conditions of Fig. 5 cause fluctuation in output maximum power profile of solar PV array under the action of MPPT logic. This fluctuating solar power creates variations in power injection at PCC and this power response is shown in Fig. 6. The power responses at PCC under *battery-connected* mode are given in Fig. 7. It shows that grid-side injected power profile is improved with significantly lesser fluctuations due to usage of battery bank in grid-connected qZSI-based solar energy conversion system.

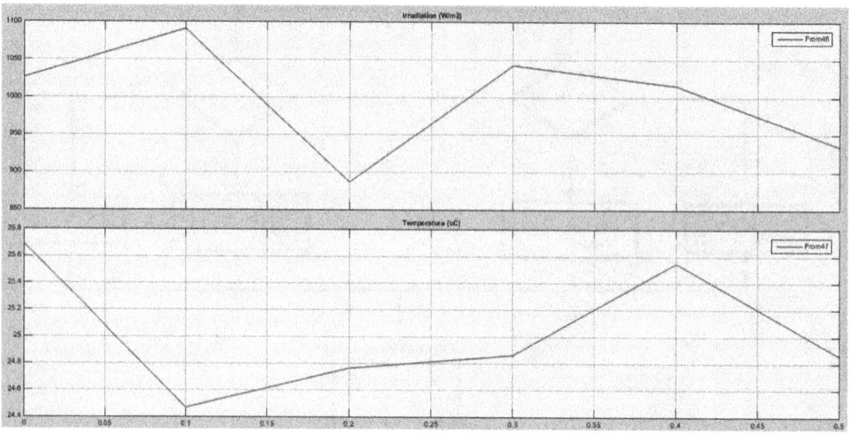

Fig. 5 Temperature, irradiation variations of solar PV array

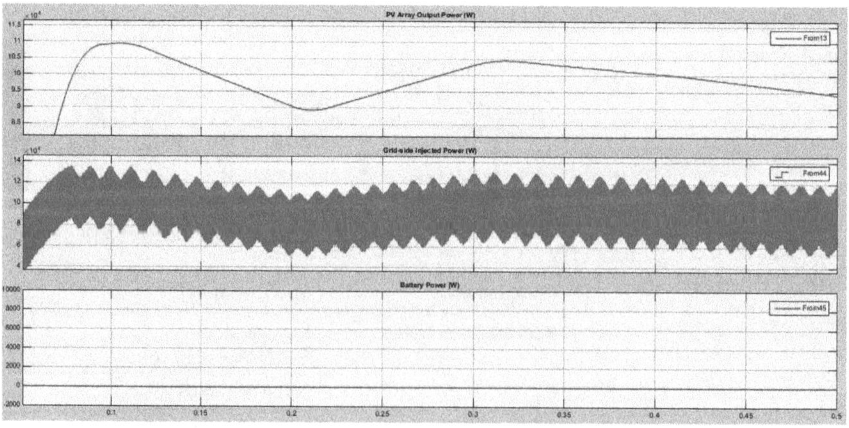

Fig. 6 Various dynamic power responses (solar PV array, grid-side injected power, battery power) at PCC under battery disconnected mode (battery power value of 0 W)

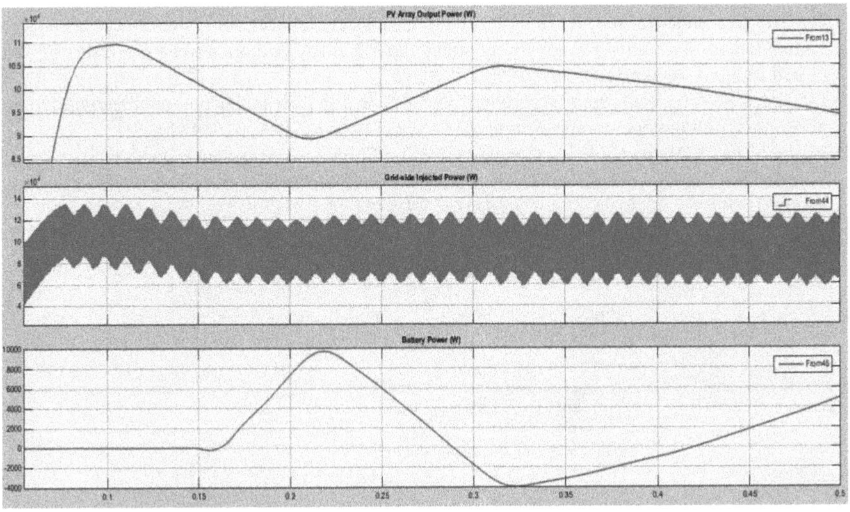

Fig. 7 Various dynamic power responses (solar PV array, grid-side injected power, battery power) at point of common coupling (PCC) under battery connected mode

4 Summary/Conclusion

The three-phase grid-connected qZSI-based solar energy conversion system is taken in the study of this paper. Under fluctuating solar conditions, there is a possibility of voltage flickering effect. To mitigate the problem, the battery storage unit counteracts the fluctuations in solar PV power to smoothen out injected power profile. The whole concept is validated using MATLAB-SIMULINK software.

References

1. Li Y, Jiang S, Cintron-Rivera JG, Peng FZ (2013) Modeling and control of quasi-Z-source source inverter for distributed generation application. IEEE Trans Ind Electron, 1532–1541
2. Padmapriya M, Raghavendiran TA (2020) Improved diode assisted voltage fed three phase quasi z-source inverter for photovoltaic application. J Ambient Intell Human Comput. https://doi.org/10.1007/s12652-020-02061-4
3. Siwakoti YP, Peng FZ, Blaabjerg F, Loh PC, Town GE, Yang S (2015) Impedance source networks for electric power conversion part II: review of control and modulation techniques. IEEE Trans Power Electron, 1887–1906
4. Ge B, Abu-Rub H, Peng FZ. Li Q, de Almeida AT, Ferreira FJTE, Sun D, Liu Y (2013) An energy stored quasi-Z-source inverter for application to photovoltaic power system. IEEE Trans Ind Electron, 4468–4481
5. Wu CJ, Fu TH (2003) Effective voltage flicker calculation algorithm using indirect demodulation method. Proc Inst Elect Eng Gen Transm. Distrib, 493–500

6. Dubarry M, Devie A, Stein K, Tun M, Matsuura M, Rocheleau R (2017) Battery energy storage system battery durability and reliability under electric utility grid operations: analysis of 3 year of real usage. J Power Sources, 65–73
7. Kusakana K (2015) Optimal scheduled power flow for distributed photovoltaic/wind/diesel generators with battery storage system. IET Renew Power Gener, 916–924
8. Zhang Y, Liu J, Zhang C (2012) Comparison of traditional two stage buck-boost voltage source inverter. In: Twenty seventh annual IEEE apply power electronics conference and exposition, pp 141–148

A Unique Developmental Study in the Design of Point-of-Care Medical Diagnostic Device for Kidney Health Care of Metastatic Brain Cancer Patients to Avoid Chemotherapy Side-Effects

Sumedha N. Prabhu and Subhas C. Mukhopadhyay

Abstract A brain cancer metastatic treatment involves chemotherapy treatment. The chemotherapy shows its negative side effect on kidney health care of cancer patients by raising the toxic waste content in their blood. Brain metastasis is a painful condition and a patient with brain cancer metastasis has a very less median rate of survival. Therefore, precise diagnosis, accurate treatment and careful monitoring of side effects of chemotherapy treatment is a must for improving the quality of life of patients. In this research, we are working on the development of creatinine as a toxic waste-specific highly selective Point-of-Care diagnostic prototype medical device system. This kidney health care management device aims to improve the quality of life of brain metastatic cancer patients.

Keywords Brain cancer metastasis · Chemotherapy · Creatinine · Point-of-Care

Metastatic cancer and medical devices definition & characteristics: Cancer is a disease caused by a mutation in DNA and results in uncontrolled body cell division. The metastasis stage initiates when cancerous cells begin their spreading within neighbouring tissues. Brain metastasis is observed when cancerous cells migrate through the lymph system or bloodstream from the site of the primary tumour and spread to the brain (secondary site). At the secondary site, the migrated metastatic cells begin to multiply. Metastatic cancer which spreads from its initial location is identified by the naming of the site of primary cancer. For example, if cancer has spread from the kidney to the brain, it will be called as metastatic kidney cancer and not brain cancer. Although there are multiple theories related to what are the roots of some cancerous cells to spread and why some cancerous cells migrate to the brain usually to the cerebral hemisphere or the cerebellum. It is also found that lung cancer metastasis to the brain is observed at early stages whereas breast cancer metastasis to the brain is observed at later stages. Overall, when compared with all cancers, lung cancer, colorectal cancer, kidney cancer, breast cancer and skin

S. N. Prabhu (✉) · S. C. Mukhopadhyay
School of Engineering, Macquarie University, Sydney 2109, Australia
e-mail: sumedha.prabhu@students.mq.edu.au

S. Dhar et al. (eds.), *Advances in Communication, Devices and Networking*,
Lecture Notes in Electrical Engineering 776,
https://doi.org/10.1007/978-981-16-2911-2_38

cancer have the highest tendency of metastasis to the brain. It is estimated that 20 to 40% of cancer patients develop metastatic brain cancer. In 2018, brain cancer was linked as the tenth most common reason for causing deaths in Australia and resulted in the loss of 1,410 lives out of which 845 males and 565 females. In 2020, the estimated Australian new brain cancer diagnosis case rate is about 1879 out of which 1113 are males and 766 are females. In 2020, about 1,518 people died of brain cancer in Australia and amongst them, about 921 are males and 597 are females. Minimum 1,200 deaths in Australia are linked to brain cancer each year. [1] The median rate of survival for metastatic brain cancer patients is between 3 and 6 months. [2] With accurate diagnosis and treatment, the life expectancy can be significantly raised with a 5-year survival rate of 22%. [1] Therefore, precise diagnosis, accurate treatment and careful monitoring of side effects of provided treatment are highly important to improve the quality of life. Brain metastasis is observed at a single spot in 30–40% of patients; whereas multiple spots metastasis is observed in most of the patients which cause meningitis. The symptom of meningitis is also observed with lymphomas, leukaemias, as well as also with advanced stages of different types of cancers, therefore precise diagnosis is important. During brain metastasis, symptoms are dependent on the location as well as the size and number of tumour growths within the brain, or the area covered with swelling. The metastasis can add up or produce the symptom of swelling in specific areas of the human brain causing area-specific symptoms. Not every individual having brain metastasis will show symptoms but the frequent population does show symptoms. In general, the symptoms are headache, memory issues, sleepiness, inability to move parts of the body such as forelimb and hindlimbs, frequent mood swings, hearing, seeing and swallowing issues, nausea and vomiting, poor communication, delirium, ataxia as well as seizures.

Various medical devices are used for the diagnosis and treatment of brain metastatic cancer. A medical device is also called as in vitro diagnostic medical device (IVD) if it is used in the form of a calibrator, reagent, kit, control material, software, apparatus, specimen receptacle, instrument, also it can be a piece of equipment or a complete patient care system, whether used unaccompanied or practiced in combination along with additional diagnostic materials for in vitro purpose.

Diagnostic medical devices and procedures for metastatic brain cancer detection: The metastatic brain tumours remain undiagnosed until the patient's body starts showing symptoms. There are ways with which doctors diagnose the occurrence of metastatic brain tumours including physical exam, neurological exam, computed tomography (CT or CAT scan), magnetic resonance imagining (MRI), diffusion tensor imaging (DTI) and biopsy.

(1) Physical exam: After collecting information about the patient's symptoms along with the patient's personal and family health care history, the doctor initiates vision check-up, physical examination and reflex test.
(2) Neurological exam: In the neurological exam, the doctor examines patients' vision, balance, hearing ability, strength, reflexes and coordination. A patient's difficulty in one or multiple areas acts as a clue in detecting which part/s of the brain could have been affected by the growth of the metastatic tumour/s.

(3) CT scan: A CT scan was performed for the diagnosis of brain cancer metastasis. It's a non-invasive imaging technique that utilises measurements of special X-rays for producing axial or horizontal brain images. It can provide in detail data regarding brain structures and tissue over a standard X-ray analysis of the head. It supplies more information regarding diseases and/or injuries associated with the patient's brain. During the CT scan of the brain, the X-ray beam moves around the body, thus permitting multiple views of the brain. The gathered information is sent to the computer which aids in the interpretation of the X-ray data and it is presented in a 2-dimensional form on the screen of the attached computer. The CT scan of the brain may be completed with or without "contrast". Contrast is a substance taken either by mouth or it can be injected via an intravenous route. The contrast aids in seeing the particular organ or tissue under a CT scan more clearly.

(4) MRI: MRI is a diagnostic device that utilizes a combination of a giant magnet, radio frequencies and the computer for producing clear images of the structures and organs inside the body. The MRI does not use any form of ionizing radiation as X-rays or CT scan. The MRI is a bulky tubular diagnostic device that generates a robust magnetic field surrounding the patient. The magnetic field and the radio waves change the hydrogen atom's natural alignment within the body of the patient. The radio wave pulses sent from a scanning machine hit the nuclei within patients' atoms outside of their ideal position. As the nuclei align again in an appropriate position, they start sending the radio signals. The signals are captured by the connected computer which processes and converts them in a 2-dimensional image of the body organ or structure under the process of examination. MRI is preferred over CT scan in the situation where soft tissues or organs are under the study, as MRI performs better while differentiating between normal and abnormal soft tissues. MRI is used for examining the brain and/or spinal cord for checking the incidence of structural anomalies, injuries or in case of the possibility of other conditions, including tumours, venous malformations, abscesses, congenital abnormalities, aneurysms, haemorrhage or haemorrhage within the brain or spinal cord. It is also used for detecting subdural hematoma, multiple sclerosis, degenerative diseases, encephalomyelitis, hypoxic encephalopathy, hydrocephalus, degeneration of discs of the spinal cord or herniation. MRI helps in planning surgeries on the spine, including decompression of a pinched nerve or spinal fusion. MRI is also helpful in identifying the exact position of a functional centre in the brain for supporting in the medication of an ailment of the brain. There are numerous other reasons with which doctors can recommend an MRI of the brain or the spine.

(5) DTI: The DTI scan permits the neurosurgeon and assisting hospital crew to imagine the brain circuitry of the brain, thus helpful in guiding during the process of surgery. The obtained images can then be loaded into the navigation systems. The navigation system is utilized within the operation theatre for aiding as a type of GPS. It also acts as a mapping system for the neurosurgeon.

(6) Biopsy: It is a surgical technique achieved for removing the cells or tissue out of the body for the process of investigation underneath a microscope. Few biopsies are performed in a doctor's clinic, whereas others require hospital admission. Besides, few biopsies can be performed by using a local anaesthetic by numbing the local area. Whereas remaining biopsies require sedation or even full anaesthesia that puts a patient entirely asleep during the entire procedure of biopsy. Biopsies are generally performed to detect if a tumour is cancerous. It also helps in finding the cause of a mysterious lesion, infection, mole, and also the process of inflammation. A process of biopsy can be performed in numerous ways. It rests on the type of specimen required. Tissue samples are typically minor in size and they are obtained from tissue that appears altered in the structure; for example, a tumour. An excisional or incisional biopsy is frequently utilized when a broader or deeper section of the tissue is required. By using a scalpel, a complete thickness of skin or all or part of a large tumour is detached for additional pathological examination. Post biopsy the wound is sewn closed by using the surgical grade thread. The biopsy can be also performed by using a needle. If the complete tumour is removed, then the technique is called an excisional biopsy. If only a certain area of the tumour is separated, the technique is called an incisional biopsy. Excisional or incisional biopsies can be performed by using local or regional anaesthesia. If the tumour is deep inside the body of the patient, general anaesthesia is utilized. Within a metastatic brain cancer biopsy circumstances, surgeons take an excisional or incisional biopsy which is instantly sent to the pathological clinic, and till the pathological results are obtained the patient remains under anaesthesia. The sample obtained by biopsy is viewed under a microscope. This helps in understanding if the sample is noncancerous (benign) or cancerous (malignant) in nature and also checked if the cells are from metastatic cancer origin or they are from a primary brain tumour. The process of biopsy also helps in the assurance of complete removal of a tumour from the patient's brain. The obtained data from the biopsy is critical to begin a diagnosis process. It also acts as a prognostic and a guiding step towards precise treatment.

Technological trends in advancements in medical devices: Newer addition of user preferences and indications for MRI has resulted in the advances in added magnetic resonance technology. Magnetic resonance angiography is a novel process utilised for estimating the flow of blood from the arteries present in a noninvasive method. It can also be utilized in detecting the vascular malformations and intracranial aneurysms. Magnetic resonance spectroscopy is an additional noninvasive procedure utilized in assessing the chemical abnormalities within body tissues of the brain. It may be useful in measuring disorders including stroke, head injury, HIV infection spread to the brain, coma, Alzheimer's disease, multiple sclerosis as well as tumours. Perfusion MRI is performed by using a particular MRI sequence. The collected databases are further post-processed for obtaining the perfusion maps having various parameters, such as blood flow and its volume along with time to peak and mean transit time. The brain's functional magnetic resonance imaging is utilized in determining the exact position of the brain areas responsible for a specific function, e.g. memory or speech areas. The universal locations of the brain where these functions happen are known, but the precise area may differ in every individual. In functional resonance imaging of the brain, the patient is requested for completing a particular work, e.g. reciting the national anthem, whilst the scanning is under process. This helps in investigating the precise position of the functional center within the brain. This is helpful to doctors in planning surgical or other treatments for metastatic brain cancer patients. Proton therapy is utilized in treating specific tumours in adults and children. Doctors combine advanced proton therapy technology along with the latest research while treating metastatic cancer tumours. Gamma knife is an innovative radiation treatment for children and adults having small to medium brain metastatic tumours, arteriovenous malformations, trigeminal neuralgia, epilepsy and additional neurological conditions. Gamma knife radiosurgery has developed as a vital treatment choice for metastasis brain tumours.

Treatments for metastatic brain cancer: This is vital for telling that metastatic brain tumours are frequently curable and they may be properly managed if accurate diagnosis and precise treatment is given to the patient on an immediate basis. Optimum treatment for metastatic brain tumours is decided differently for each patient as per their health care status. The neurosurgeon governs the utmost suitable treatment method, by calculating various factors. The list of factors includes the primary cancer subtype, patient's responses to the therapy and present status of their health, the position/s and number of metastatic tumour/s present inside the brain or spine, overall physical health status of patient and preferences concerning possible therapy options, present symptoms of the patient and risks associated around them, etc. Following are the treatment options used for metastatic brain cancer patients.

(1) Surgery: Surgery is a fast relief providing an option for "mass effect" and culminates into reducing the pressure within the patient's skull subsequently observed due to a growing tumour and swelling to the brain. The patients can understand enhancement in better feeling just a few hours of surgical procedure, in case the "mass effect" is triggering the symptoms associated with the tumour. Surgery aims to minimalize the tumour space taken up by bulking. It eradicates the maximum of the tumour as much as possible while upholding the neurological functions of the patient. Overall, doctors suggest surgery when a strong association of neurologic shortfalls with the tumour's location is observed. It is also suggested when the patient's primary cancer is curable as well as found to be under medical regulation. Surgery is prescribed when a patient has one or two metastatic brain tumours or a list of tumours that are near to each other and those can be carefully removed. The surgery which used to eliminate metastatic brain tumours is called a craniotomy, and it can be completed using different methods, such as the keyhole craniotomy. The neurosurgeon also prefers a microsurgical operation, by using the latest tools including minimally invasive endoscopy and image-guided surgery for ensuring the highest possibility for the best outcome [3].

(2) Radiation: This therapy is the anti-tumour procedure and it uses X-rays as well as other types of radiation, e.g. light energy for destroying cells from cancer. It prevents a tumour from increasing in its size. This treatment is also called radiotherapy. It is a non-invasive treatment that involves passing the beams of radiation through the body of the patient. It can treat those cancers of the brain which are hard to remove through a surgical procedure. Radiotherapy may comprise of just single or combination of the options such as whole-brain radiation, stereotactic radiosurgery, e.g. fractionated radiosurgery, external beam radiation therapy, liquid radiation, etc. All these radiotherapies can be also achieved post-surgery for preventing regrowing of tumours around the original location of the tumour elimination site and surrounding additional brain tissue. Precisely selecting radiation therapy type is a complicated procedure and requires a team decision. Depending on symptoms few patients receive a type of radiosurgery called stereotactic radiosurgery over traditional neurosurgery. Most patients receive whole-brain radiation therapy or a combination of multiple therapies. It is dependent upon a team approach and specific symptoms associated with each patient [3].

(3) Chemotherapy treatment: It is a little challenging for traditional chemotherapy treatment to pass through the blood–brain barrier. Therefore, a novel therapy called targeted therapy is also utilized as a subtype of chemotherapy for treatment of the metastatic brain tumours. These drugs potentially recognize as well as kill the cancerous cells with nominal harm to normal brain cells. Chemotherapy drugs also avoid development as well as the proliferation of cancerous cells. The targeted therapy may also be run post-surgery or together with radiation therapy for destroying the remaining cancerous cells from the tumour. Targeted therapies utilized for treating the metastatic brain tumours comprise drugs such as Trastuzumab and Erlotinib. Trastuzumab is used in HER2 receptor-positive breast cancer treatment which has metastasized to the brain. Erlotinib is used in treatment for the utmost common sub-type of lung cancer called the non-small cell lung cancer metastasized to the brain [3].

(4) Immunotherapy: Brain cancer immunotherapy uses vaccines, drugs and other therapies that activate the patient's immune system's innate capabilities to fight against cancerous growth. Immunotherapy is a specific type of cancer treatment that triggers the body's natural defences for fighting against cancer. It utilizes in vivo antibodies made by the body or in vitro substances manufactured inside the laboratory. It aids in improving the patient's natural immune system response works in finding and destroying the cancerous cells [3].

(5) Medication: High doses of corticosteroids may be administered for easing the symptoms of swelling in surrounding areas of the tumours. It helps in decreasing the neurological signs and symptoms in patients.

(6) Rehabilitation treatment: As brain tumours can grow in areas of the brain which regulate speech, motor skills, thinking and vision. Rehabilitation might be required as a step within the retrieval process. A doctor can suggest the patient for physical therapy which can support the patient in regaining motor skills or muscular stamina. The occupation therapy may support the patient in getting

into their usual everyday carry-ons, including work, post brain tumour illness recovery process. Speech treatment aid with speech-associated problems [4].

(7) Palliative care: It is a care which is dedicated medical health care aiming at offering psychological respite from pain and related signs of a serious sickness caused to the metastatic brain cancer patient for a better quality of life [4].

Challenges: Surgery helps in removing brain metastases but it carries multiple risks including infection and bleeding, neurologic deficits, etc. Other risks are dependent on the area of the brain where the patient's tumours are positioned. For example, surgery over a tumour nearby nerves connecting to the patient's eyes carries a major risk factor of vision loss. Side effects of whole-brain radiation (WBR) may include nausea, fatigue as well as hair loss. Long-term WBR is related to cognitive decline. Side effects of stereotactic radiosurgery may include headache, nausea, vertigo or dizziness and seizures. The danger of lasting cognitive decline after stereotactic radiosurgery is lesser than that with WBR. The chemotherapy carries major side effects by killing the non-cancerous cell, hair loss and results in the decline of kidney functioning [5–7].

Our research's active area: As chemotherapy shows significant side effects on kidney health care, it results in rising levels of toxic waste content inside the blood. Chemotherapy side effects will end up in rising normal blood levels of creatinine and blood urea nitrogen (toxic waste) of the patient. Therefore, monitoring kidney healthcare during chemotherapy treatment is vital for metastatic brain cancer patients [5–7].

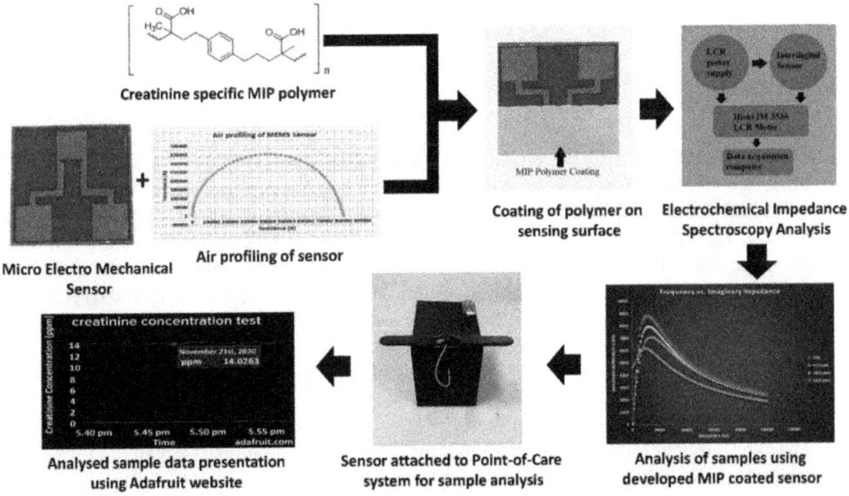

Fig. 1 A graphical representation of the development and functioning of PoC diagnostic device prototype for kidney healthcare management of metastatic brain cancer patients

At Macquarie University, we have developed a kidney healthcare aiding Point-of-Care (PoC) diagnostic device prototype system for management of creatinine levels [8–10]. Initially, micro electro-mechanical systems (MEMS) sensor was developed having 1-1-25 configuration. It was having 2 sensing electrodes between 1 excitation electrodes with a pitch length of 25 μm. In the second step, a selective polymer for creatinine was chemically synthesized using molecularly imprinted polymerization (MIP) technology. The air profiling of the MEMS sensor was performed and in the next step, the creatinine selective MIP polymer was coated on the sensing area of the sensor. This step helped in making functionalization of MEMS sensor for selective detection of creatinine from samples. In the consecutive step, the Electrochemical Impedance Spectroscopic analysis of functionalized MEMS sensor is performed followed by sample analysis by using the developed MIP coated MEMS sensor. In advanced steps, the MIP functionalized sensor was attached to the developed Long Range Wide Area Network-based PoC diagnostic medical sensing device and samples were tested for their levels of creatinine concentration. The analysed sample data was stored on the Internet of Things (IoT) based cloud server using Adafruit as a complementary website [8–10].

Conclusion: A MEMS-based creatinine-specific functionalized MIP sensor is used for the development and making of a PoC diagnostic medical sensing device prototype. The device was found to be successful in precisely measuring levels of creatinine from the samples and sending data to the IoT-based cloud server on the Adafruit website. The stored data can be easily accessible to any neurosurgeon. The medical device will be helpful in careful monitoring of kidney healthcare of patients undergoing chemotherapy treatment for metastatic brain cancer. Thus, the developed device can help to improve the quality of life of brain metastatic cancer patients.

References

1. BRAIN CANCER (2020) Brain cancer in Australia statistics. Available via the INTERNET. https://www.canceraustralia.gov.au/affected-cancer/cancer-types/brain-cancer/statistics#:~:text=The%20number%20of%20new%20cases,cases%20per%20100%2C000%20in%202016
2. Wong J, Hird A, Kirou-Mauro A, Napolskikh J, Chow E (2008). Quality of life in brain metastases radiation trials: a literature review. Current Oncol 5(5):25–45 (Toronto, Ont.)
3. JOHN HOPKINS MEDICINE (2020) Metastatic Brain Tumors. Available via the INTERNET. https://www.hopkinsmedicine.org/health/conditions-and-diseases/metastatic-brain-tumors#:~:text=Chemotherapy,for%20treating%20metastatic%20brain%20tumors
4. MAYO CLINIC (2020) Brain metastases. Available via the INTERNET. https://www.mayoclinic.org/diseases-conditions/brain-metastases/diagnosis-treatment/drc-20350140
5. Prabhu SN, Mukhopadhyay SC, Davidson AS, Liu G (2019) Highly selective molecularly imprinted polymer for creatinine detection. In: 2019 13th International Conference on Sensing Technology (ICST). Sydney, Australia, pp 1–5. https://doi.org/10.1109/icst46873.2019.9047696
6. Prabhu SN, Mukhopadhyay SC, Gooneratne C, Davidson AS, Liu G (2019) Interdigital sensing system for detection of levels of creatinine from the samples. In: 2019 13th International

conference on sensing technology (ICST). Sydney, Australia, pp 1–6. https://doi.org/10.1109/icst46873.2019.9047672
7. Prabhu S, Gooneratne C, Anh Hoang K, Mukhopadhyay S (2020) Development of a Point-of-Care diagnostic smart sensing system to detect creatinine levels. In: 2020 IEEE 63rd international midwest symposium on circuits and systems (MWSCAS). Springfield, MA, USA, pp 77–80. https://doi.org/10.1109/MWSCAS48704.2020.9184441
8. Prabhu SN, Mukhopadhyay SC, Gooneratne CP, Davidson AS, Liu G (2020) Molecularly imprinted polymer-based detection of creatinine towards smart sensing. Med Devi Sens. https://doi.org/10.1002/mds3.10133
9. Prabhu S, Gooneratne C, Hoang KA, Mukhopadhyay S IoT-Associated impedimetric biosensing for Point-of-Care monitoring of kidney health. IEEE Sens J. https://doi.org/10.1109/jsen.2020.3011848
10. Prabhu S, Gooneratne CP, Hoang KA, Mukhopadhyay SC, Davidson A, Liu G (2021) Interdigital sensors system for kidney health monitoring. In Interdigital sensors. Springer, Springer Nature Switzerland AG, pp 1–38

Reconfigurable Intelligent Surface (RIS)-Assisted Wireless Systems: Potentials for 6G and a Case Study

Chi-Bao Le, Dinh-Thuan Do, and Samarendra Nath Sur

Abstract As one of the key technologies for deployment of future wireless networks, the state-of-the-art reconfigurable intelligent surfaces (RISs) have rapidly gained a massive interest among researchers. In a specific case study, we study the outage performance of RIS-aided wireless systems in the presence of non-orthogonal multiple access (NOMA) scheme. In particular, different power factors are allocated to users which belong to a dedicated group. We the derive the exact outage probability of two users in a group. Specifically, it is assumed that the RIS is placed between the source and the users and the far user has better performance under the assistance of RIS. We also provide a comparative analysis to investigate the effect of the main parameters on the outage performance of our proposed system, such as the number of tunable elements of the RIS, power allocation factors, target rates, and the average signal-to-noise ratio at the base station. By using Monte Carlo simulation, we verify our analytical results via simulations. Our main results reported in this paper show the positive effect once we deploy RISs for guaranteeing fairness among NOMA users in wireless systems.

C.-B. Le (✉)
Faculty of Electronics Technology, Industrial University of Ho Chi Minh City, Ho Chi Minh City, Vietnam
e-mail: lechibao@iuh.edu.vn

D.-T. Do
Department of Computer Science and Information Engineering, College of Information and Electrical Engineering, Asia University, Taichung city, Taiwan
e-mail: dodinhthuan@asia.edu.tw

S. N. Sur
Department of Electronics and Communication Engineering, Sikkim Manipal Institute of Technology, Sikkim Manipal University, Majitar, Sikkim 737136, India

S. Dhar et al. (eds.), *Advances in Communication, Devices and Networking*,
Lecture Notes in Electrical Engineering 776,
https://doi.org/10.1007/978-981-16-2911-2_39

1 Introduction

Recently, one of attention for beyond 5G systems is the reconfigurable intelligent surface (RIS)-aided systems [1]. By exploiting the special structure of multiple low-cost reflecting elements in RIS-assisted planar array, incident signals are processed at such array using passive beamforming.

Regarding the architecture of RIS, several possible places are suitable to deploy the RIS in the real environment, such as situated by aerial platforms or coated on walls of houses/buildings. By employing RIS, the radio environment can be changed into a smart way that can assist information sensing, analog computing, and wireless communications [2]. In term of controlling, RIS-aided wireless systems will become more flexible along with the optimal management of the legacy RF transceivers to further assist diverse user requirements, such as extended coverage, increased data rate, reduce power consumption, and more secure transmissions [3, 5]. To conduct and transmit signals witha large size of scattering elements, the architecture of the RIS requires a control mechanism. As a result, RISs jointly change their EM behaviors and strength of signals on demand. For example, to deal with reconfiguration of RIS's EM behaviors, it alters phase control of individual scattering elements. To act locally with a scattering element and communicate to a central controller, RIS needs a joint design of the structure of tunable chips with the metasurface [6, 7]. In other work, a software-defined implementation of the control mechanism was studied as [8]. A field-programmable gate array (FPGA) and the tunable chips are designed along with typical PIN diodes to allow the RIS controller can be implemented in practical scenario [7]. By optimizing and distributing its phase control policies to all tunable chips, the embedded RIS controller can send and receive reconfiguration requests from external equipment. Especially, to reconfigure RIS's behavior relying on receiving the control information, each tunable chip varies its state and operates the corresponding scattering element.

Therefore, each element of IRS can be intelligently designed with emerging models as propositions in [9, 12]. The authors in [9] considered a model of the deployment and passive beamforming design for RIS by joint deployment, phase shift design, as well as power allocation in the multiple-input-single-output (MISO) to achieve the optimal energy efficiency associated with data requirements. They also studied machine learning approaches. By employing stochastic geometry, reference [10] considered the coverage of a RIS-assisted large-scale mmWave cellular network, and the authors derived the closed-form formulas of the peak reflection power expression of a RIS and formula of the downlink signal-to-interference ratio (SIR) coverage. The authors in [11] presented a novel RIS system including a simple controller associated with adjustable configuration, any number of passive reflecting elements, and a single Radio Frequency (RF) chain for baseband measurements. Further, they proposed an alternating optimization approach by considering sparse wireless channels in the beamspace domain for explicit estimation of the channel gains at the RIS elements. Reference [13] presented the analytic framework of the RIS-assisted systems to indicate the main performance metric, namely the ergodic capacity. Moreover,

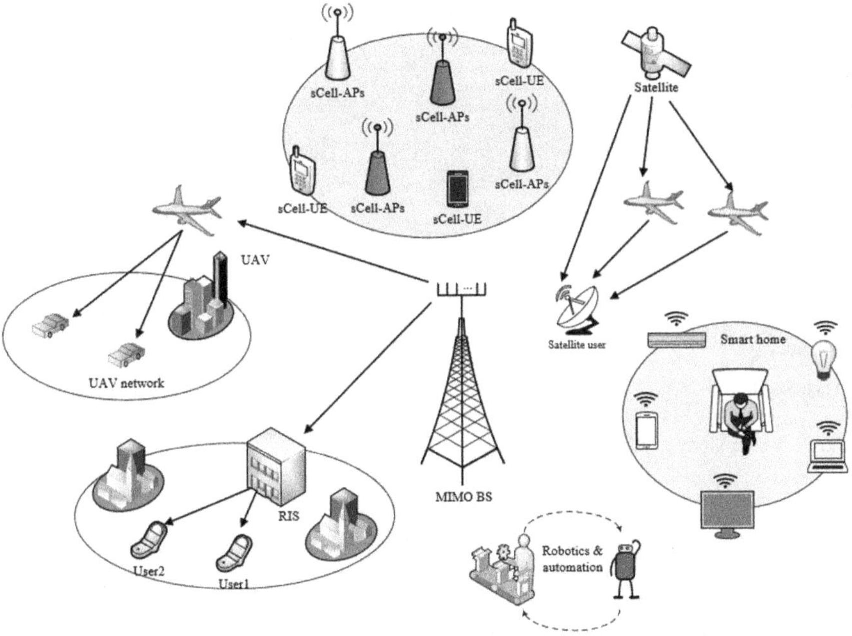

Fig. 1 Various emerging applications of future wireless systems

they also presented a high number of reflection units (RUs) approximations with high signal-to-noise ratio and studied ergodic capacity.

A wide range of possible applications of the future wireless network as well as state-of-the-art RIS-assisted wireless communications can be illustrated in Fig. 1. In this case study, we investigate how a RIS system can serve massive connections at the same time. Such an advantage can be implemented by exploiting NOMA.

2 A Case Study: System Model of NOMA-RIS System

To further increase spectral efficiency, one of the main objectives of 5G is non-orthogonal multiple access (NOMA) which is one of the promising techniques to address this issue [18, 27]. As the main concept, NOMA decides to serve which users following distance difference between users or power allocated to pair the users. The author in [26] studied the NOMA techniques with two scenarios including single-carrier NOMA (SC-NOMA) and multi-carrier NOMA (MC-NOMA).

Recent studies [28, 31] have presented NOMA-RIS systems. In [9], a novel frame-workstudied passive beamforming in the framework of NOMA-RIS. This paper aims to analyze the performance of two NOMA users who are benefited from the deployment of RIS. We consider a two-user NOMA scheme on downlink relying on RIS,

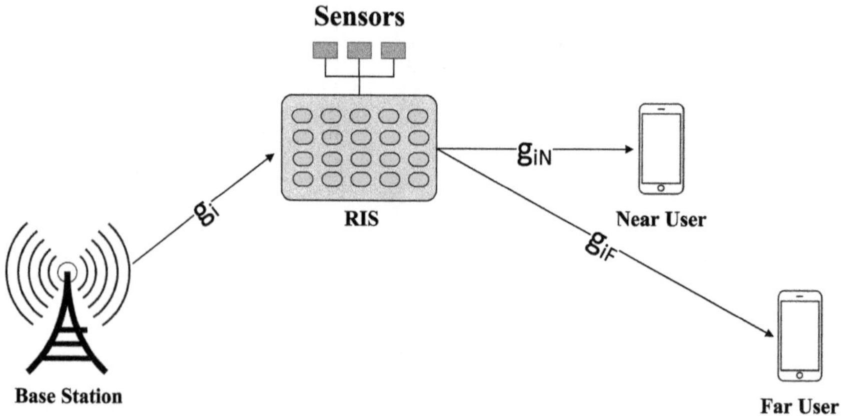

Fig. 2 A case study of RIS-NOMA system

as shown in Fig. 2. Two kinds of users are classified based on their locations, namely near user (NU) and far user (FU).

Regarding advances of NOMA, it is possible to extend the case of multiple users who serve by the proposed RIS-NOMA downlink transmission scheme, but the best performance for the two-user NOMA scheme is reported in the literature. Therefore, we do not consider degraded performance in such a system due to the deployment of multiple users, which is beyond the scope of this letter. In this circumstance, one could not be transmitted from the base station (BS) to mobile users directly due to heavy blockage or obstacles. The BS generates two beamforming vectors together with the technique of zero-forcing beamforming to serve two NOMA users. By grouping of paired users, RIS-NOMA satisfies different QoS requirements which are suitable to develop multiple services for mobile users in future wireless systems. Under the assistance of RIS equipped N reflecting elements. The RIS is also equipped with a controller associated with switching procedures including working modes. In particular, RIS operates in receiving mode for channel estimation and in reflecting mode for data transmission. Since the RIS is a passive reflecting equipment, we adopt a time-division duplexing (TDD) protocol for uplink and downlink transmissions and assume channel reciprocity for achieving the channel information acquisition in the downlink based on the uplink training sequence.

To enable NOMA mode, the superimposed signal $(a_1 s_1 + a_2 s_2)$ transmitting from the BS then is required to serve distant mobile users with the presence of RIS. It is noted that s_1 and s_2 correspond to two signals for users NU and FU, respectively. This study considers the NOMA concept to provide user fairness with a_1 and a_2 which are denoted as power allocation factors for user NU and FU, respectively. Due to less amount of power requiblack to supply for user NU, we have $a_1 < a_2$ and $a_1^2 + a_2^2 = 1$ [32].

We denote $A = \sum_{i=1}^{N} |g_i| |g_{iN}|$, $B = \sum_{i=1}^{N} |g_i| |g_{iF}|$ for ease in further computation. In particular, we can compute the signal-to-interference-plus-noise ratio (SINR) for the user NU to decode the FU's signal which is expressed by

$$\gamma_{NU,s_2} = \frac{A^2 a_2^2}{A^2 a_1^2 + \frac{1}{\rho}}, \tag{1}$$

where ρ represents the transmit signal-to-noise ratio (SNR).

By performing SIC, the user NU eliminates signal s_2, then it decodes its signal by computing SNR as [33]

$$\gamma_{NU,s_1} = A^2 \rho a_1^2. \tag{2}$$

Different from decoding signal at user NU, computing SINR to detect the FU's signal can be given by

$$\gamma_{FU,s_2} = \frac{B^2 a_2^2}{B^2 a_1^2 + \frac{1}{\rho}}. \tag{3}$$

The probability density function (PDF) of random variable (RV) T can be obtained as the following in [34, Eq. (24)]:

$$f_{T^2}(x) = \tfrac{1}{2b^{a+1}\Gamma(a+1)} x^{\frac{a-1}{2}} \exp\left(-\tfrac{1}{b}\sqrt{x}\right), T \in \{A, B\}, \tag{4}$$

where $a = \frac{N\pi^2}{(16-\pi^2)} - 1$ and $b = \frac{8}{\pi} - \frac{\pi}{2}$.

Next, the cumulative distribution function (CDF) of Y can be computed as [34, Eq. (25)]

$$F_{T^2}(x) = \frac{1}{\Gamma(a+1)} \gamma\left(a+1, \frac{1}{b}\sqrt{x}\right), \tag{5}$$

where $\gamma(.,.)$ is the lower incomplete Gamma function.

With the aid of [35, Eq. (8.350.1)], (5) can be obtained as

$$F_{T^2}(x) = \frac{1}{\Gamma(a+1)} \sum_{l=0}^{\infty} \frac{(-1)^l}{l!\,(a+l+1)\, b^{a+l+1}} x^{\frac{a+l+1}{2}}. \tag{6}$$

3 Performance Analysis

To look at system performance, we intend to derive new closed-form formulas for the outage probability, and performance gap among two NOMA users should be consideblack for such RIS-aided wireless systems.

3.1 Outage Probability at User NU

The outage behavior happens at the user NU once it fails to detect the FU's signal s_2 as well as its own signal s_1.

$$\begin{aligned} P_{NU} =& 1 - \Pr\left(\gamma_{NU,s_2} > \varepsilon_2, \gamma_{NU,s_1} > \varepsilon_1\right) \\ =& 1 - \Pr\left(A^2 > \varphi_2, A^2 > \varphi_1\right) \\ =& 1 - \Pr\left(A^2 > \varphi_{\max}\right), \end{aligned} \tag{7}$$

where $\varepsilon_1 = 2^{2R_1} - 1$ and $\varepsilon_2 = 2^{2R_2} - 1$ are the threshold of system, R_1 and R_2 are the target rate, $\varphi_2 = \frac{\varepsilon_2}{\rho\left(a_2^2 - a_1^2\varepsilon_2\right)}$, $\varphi_1 = \frac{\varepsilon_1}{\rho a_1^2}$, and $\varphi_{\max} = \max\left(\varphi_2, \varphi_1\right)$.

From (7), one can observe $P_{NU} = 0$ when $a_1^2\varepsilon_2 > a_2^2$.

Proposition 1 *The closed-form expression of outage probability to detect signal s_2 and s_1 at the NU is given, respectively, by*

$$P_{NU} = 1 - \frac{1}{\Gamma\left(a+1\right)}\Gamma\left(a+1, \frac{\sqrt{\varphi_{\max}}}{b}\right), \tag{8}$$

where $\Gamma\left(.\right)$ is the Gamma function and $\Gamma\left(.,.\right)$ is the upper incomplete Gamma function.

Proof See in Appendix A.

3.2 Outage Probability at User FU

Let us consider the outage probability of user FU as below

$$\begin{aligned} P_{FU} =& 1 - \Pr\left(\gamma_{FU,s_2} > \varepsilon_2\right) \\ =& F_{B^2}\left(\varphi_2\right). \end{aligned} \tag{9}$$

Applying (6) of the CDF, P_{FU} is calculated by

$$P_{FU} = \frac{1}{\Gamma\left(a+1\right)}\sum_{l=0}^{\infty}\frac{\left(-1\right)^l \varepsilon_2^{\frac{a+l+1}{2}}}{l!\left(a+l+1\right)b^{a+l+1}}. \tag{10}$$

4 Throughput Analysis

The system throughput of RIS-NOMA direct link is given by

$$\tau = (1 - P_{NU})\, R_1 + (1 - P_{FU})\, R_2. \tag{11}$$

5 Numerical Results

In this section, we simulate outage probability based on mathematical derivations and further verify with Monte Carlo simulation, and parameters used are summarized in Table 1.

In Fig. 3, due to the gap among power allocation factors for two users, the performance gap is expressed in terms of outage performance of user NU and FU at all values of the number of elements in the NOMA-RIS system. It is also the conclusion that the higher number of elements in RIS, the consideblack system shows

Table 1 Table of parameters for numerical results

Monte Carlo simulations repeated	10^6 iterations
The power allocation coefficient	$a_1^2 = 0.1, a_2^2 = 0.9$
The target rate	$R_1 = 3$, $R_2 = 1.5$ (BPCU)

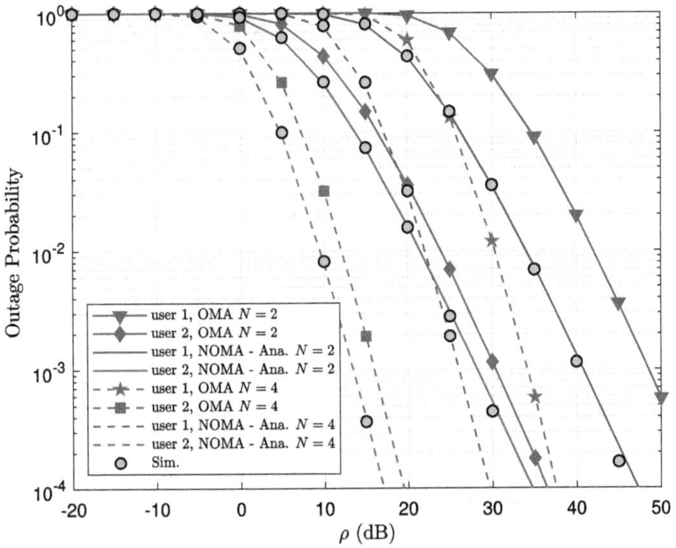

Fig. 3 Outage probability versus transmit SNR ρ with varying data rates

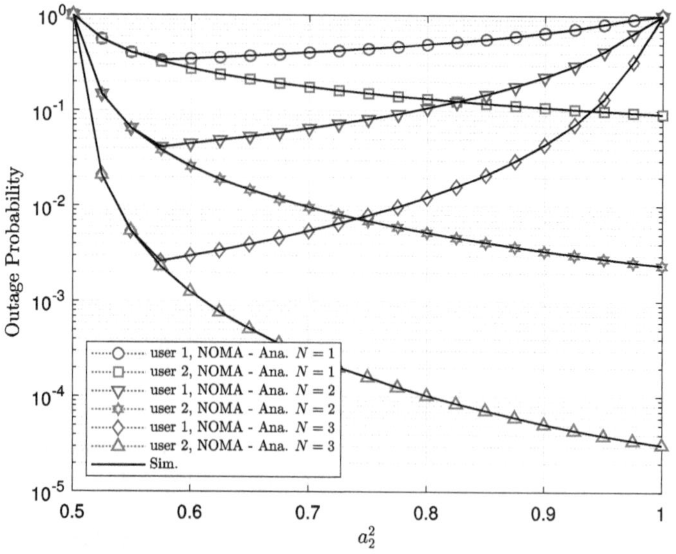

Fig. 4 Optimal outage performance of user NU, with $R_1 = 1$, $R_2 = 0.5$, and $\rho = 10$ (dB)

superiority in its performance. We can see tight matching between Monte Carlo and analytical simulations, especially in a high SNR regime. This figure provides slight improvement when we compare RIS deployed with NOMA and OMA schemes.

Figure 4 confirms that the optimal outage performance can be achieved once we vary the power allocation factor a_1. Unfortunately, it is likely difficult to find optimal outage probability for user FU. Of course, high SNR at source $\rho = 20dB$ is reported as a better case.

Furthermore, the target rates limit the outage performance of two users, shown in Fig. 5. By increasing the target rate, we can see trends of outage probability.

Such outage performance leads to throughput performance, shown as Fig. 6. In this scenario, different number of elements of RIS (N) demonstrates the difference in term of throughput.

6 Conclusion

In this study, we consider a case study for 6G wireless communications, namely RIS-NOMA-assisted wireless systems have been proposed to provide benefits of NOMA in terms of fairness. Considering the ideal condition of the optimal phases of the RIS elements over reciprocal channels, it can be achieved the exact outage probability for two users. Our main results indicate the impacts of the number of elements in RIS and power allocations factors on outage performance. Since the exact signal decoding procedure for the NOMA scheme is likely difficult to implement in

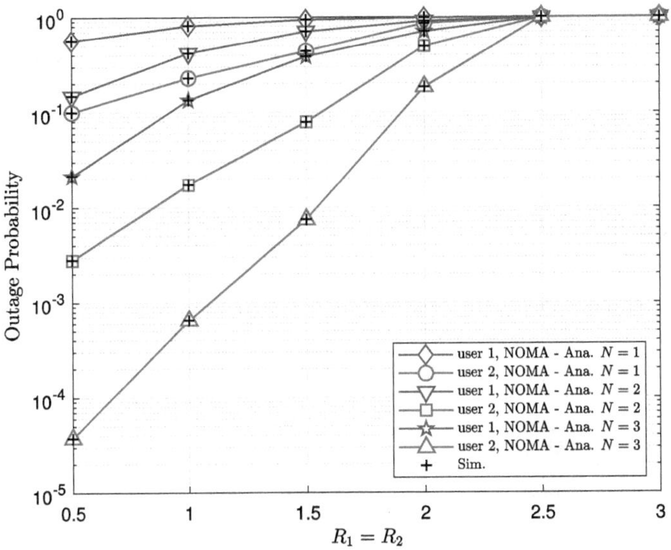

Fig. 5 Outage probability versus target rates $R_1 = R_2$, with $\rho = 10$ (dB), $a_1^2 = 0.05$, and $a_2^2 = 0.95$

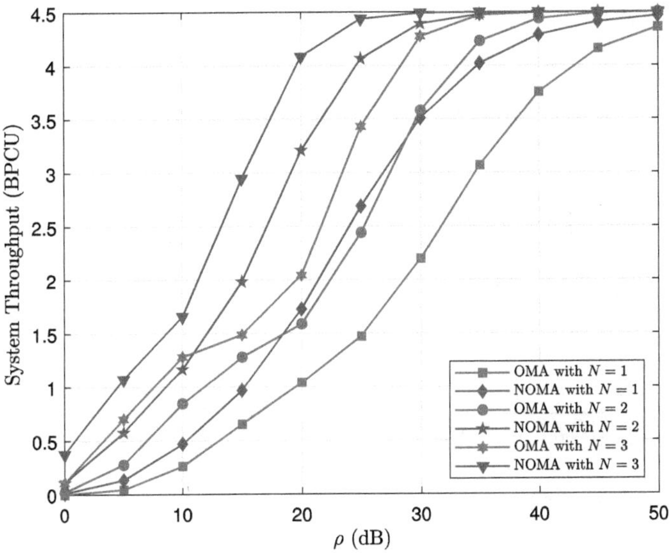

Fig. 6 Throughput performance

practice, we further consider imperfect SIC in such RIS-NOMA system. Moreover, the benefit of NOMA compablack with OMA is introduced in such system and this confirms joint deployment of NOMA, and RIS introduces a promising system for future development.

7 Appendix A

The outage probability P_{NU} can be further computed by

$$\begin{aligned} P_{NU} &= 1 - \Pr\left(A^2 > \varphi_1\left(\rho|h_I|^2 + 1\right)\right) \\ &= 1 - \int_0^\infty f_{|h_I|^2}(x)\left[1 - F_{A^2}\left(\varphi_1(\rho x + 1)\right)\right]dx \\ &= 1 - \frac{1}{\Omega_{h_I}}\int_0^\infty e^{-\frac{x}{\Omega_{h_I}}}\left[1 - \sum_{l=0}^{\infty}\frac{(-1)^l\varphi_1^{\frac{a+l+1}{2}}}{l!\Gamma(a+1)(a+l+1)b^{a+l+1}}(\rho x+1)^{\frac{a+l+1}{2}}\right]dx \\ &= \sum_{l=0}^{\infty}\frac{(-1)^l\varphi_1^{\frac{a+l+1}{2}}}{l!\Omega_{h_I}\Gamma(a+1)(a+l+1)b^{a+l+1}}\int_0^\infty e^{-\frac{x}{\Omega_{h_I}}}(\rho x+1)^{\frac{a+l+1}{2}}dx \end{aligned} \quad (12)$$

Let $t = \rho x + 1 \rightarrow \frac{t-1}{\rho} = x \rightarrow \frac{1}{\rho}dt = dx$, P_{NU} can be reformulated by

$$P_{NU} = \sum_{l=0}^{\infty}\frac{(-1)^l\varphi_1^{\frac{a+l+1}{2}}e^{\frac{1}{\rho\Omega_{h_I}}}}{l!\Omega_{h_I}\Gamma(a+1)(a+l+1)\rho b^{a+l+1}}\int_1^\infty e^{-\frac{t}{\rho\Omega_{h_I}}}t^{\frac{a+l+1}{2}}dt. \quad (13)$$

Using [35, Eq. (3.381.3)], (13) can be reformulated by (14)

$$P_{NU} = \sum_{l=0}^{\infty}\frac{(-1)^l\varphi_1^{\frac{a+l+1}{2}}\left(\rho\Omega_{h_I}\right)^{\frac{a+l+3}{2}}e^{\frac{1}{\Omega_{h_I}}}}{l!\Omega_{h_I}\Gamma(a+1)(a+l+1)\rho b^{a+l+1}}\Gamma\left(\frac{a+l+3}{2}, \frac{1}{\rho\Omega_{h_I}}\right). \quad (14)$$

This completes the proof.

References

1. Basar E, Di Renzo M, De Rosny J, Debbah M, Alouini M, Zhang R (2019) Wireless communications through reconfigurable intelligent surfaces. IEEE Access 7:116 753–116 773
2. Renzo MD et al (2019) Smart radio environments empoweblack by AI reconfigurable metasurfaces: an idea whose time has come. EURASIP J Wirel Commun Netw 129

3. Wu Q, Zhang R (2020) Towards smart and reconfigurable environment: intelligent reflecting surface aided wireless network. IEEE Commun Mag 58(1):106–112
4. Basar E, Renzo MD, de Rosny J, Debbah M, Alouini M-S, Zhang R (2019) Wireless communications through reconfigurable intelligent surfaces. IEEE Access 7:116 753–116 773
5. Huang et al C (2019) Holographic MIMO surfaces for 6G wireless networks: opportunities, challenges, and trends. arXiv:1911.12296
6. Liu F et al (2018) Programmable metasurfaces: state of the art and prospects. In: proceedings of the IEEE International Symposium on Circuits and Systems (ISCAS)
7. Yang H, Cao X, Yang F, Gao J, Xu S, Li M, Chen X, Zhao Y, Zheng Y, Sijia L (2016) A programmable metasurface with dynamic polarization, scattering and focusing control. Sci Rep 6(35692)
8. Liaskos C et al (2015) Design and development of software defined metamaterials for nanonetworks. IEEE Circuits Syst Mag 15(4):12–25. Fourthquarter
9. Liu X, Liu Y, Chen Y, Poor HV RIS enhanced massive non-orthogonal multiple access networks: deployment and passive beamforming design. IEEE J Sel Areas Commun. https://doi.org/10.1109/JSAC.2020.3018823
10. Nemati M, Park J, Choi J (2020) RIS-assisted coverage enhancement in millimeter-wave cellular networks. IEEE Access 8:188171–188185
11. Alexandropoulos GC, Vlachos E (2020) A hardware architecture for reconfigurable intelligent surfaces with minimal active elements for explicit channel estimation. In: ICASSP 2020 - 2020 IEEE international conference on acoustics, speech and signal processing (ICASSP). Barcelona, Spain, pp 9175–9179
12. C. H. et al (2019) Reconfigurable intelligent surfaces for energy efficiency in wireless communication. IEEE Trans Wirel Commun 18(8):4157–4170
13. Boulogeorgos AAA, Alexiou A (2020) "Ergodic capacity analysis of reconfigurable intelligent surface assisted wireless systems. IEEE 3rd 5G World Forum (5GWF). Bangalore, India, 395–400
14. Wu Q, Zhang R (2019) Intelligent reflecting surface enhanced wireless network via joint active and passive beamforming. IEEE Trans Wirel Commun 18(11):5394–5409
15. Guo H, Liang Y, Chen J, Larsson EG (2020) Weighted sum-rate maximization for reconfigurable intelligent surface aided wireless networks. IEEE Trans Wirel Commun 19(5):3064–3076
16. Chen J, Liang Y-C, Pei Y, Guo H (2019) Intelligent reflecting surface: a programmable wireless environment for physical layer security. IEEE Access 7:82 599–82 612
17. Guan X, Wu Q, Zhang R (2020) Intelligent reflecting surface assisted secrecy communication: is artificial noise helpful or not? IEEE Wirel Commun Lett 9(6):778–782
18. Do D-T et al (2019) Wireless power transfer enabled NOMA relay systems: two SIC modes and performance evaluation. Telkomnika 17(6):2697–2703
19. Do D-T, Le C-B, Le A-T (2019) Cooperative underlay cognitive radio assisted NOMA: secondary network improvement and outage performance. Telkomnika 17(5):2147–2154
20. Li X, Li J, Liu Y, Ding Z, Nallanathan A (2020) Residual transceiver hardware impairments on cooperative NOMA networks. IEEE Trans Wirel Commun 19(1):680–695
21. Li X, Li J, Li L (2020) Performance analysis of impaiblack SWIPT NOMA relaying networks over imperfect weibull channels. IEEE Syst J 14(1):669–672
22. Li X, Liu M, Deng C, Mathiopoulos PT, Ding Z, Liu Y (2020) Full-duplex cooperative NOMA relaying systems with I/Q imbalance and imperfect SIC. IEEE Wirel Commun Lett 9(1):17–20
23. Li Xingwang, Wang Qunshu, Peng Hongxing, Zhang Hui, Do Dinh-Thuan, Rabie Khaled M, Kharel Rupak, Cavalcante Charles C (2020) A unified framework for HS-UAV NOMA network: performance analysis and location optimization. IEEE Access 8:13329–13340
24. Do D-T, Thi Nguyen T-T (2018) Exact outage performance analysis of amplify-and forward-aware cooperative NOMA. Telkomnika 16(5):1966-1973
25. Do D-T, Le C-B (2018) Exploiting outage performance of wireless powered NOMA. Telkomnika 16 (5):1907–1917

26. Zeng J et al (2018) Investigation on evolving single-carrier NOMA into multi-carrier NOMA in 5G. IEEE Access 6:48268–48288
27. Liu G, Wang Z, Hu J, Ding Z, Fan P (2019) Cooperative NOMA broadcasting/multicasting for low-latency and high-reliability 5G cellular V2X communications. IEEE Internet Things J 6(5):7828–7838
28. Yang L, Yuan Y (2020) Secrecy outage probability analysis for RIS-assisted NOMA systems. Electron Lett 56(23):1254–1256
29. Elhattab M, Arfaoui MA, Assi C, Ghrayeb A Reconfigurable intelligent surface assisted coordinated multipoint in downlink NOMA networks. IEEE Commun Lett. https://doi.org/10.1109/LCOMM.2020.3029717
30. Li Y, Jiang M, Zhang Q, Qin J Joint beamforming design in multi-cluster MISO NOMA reconfigurable intelligent surface-aided downlink communication networks. IEEE Transactions on Communications. https://doi.org/10.1109/TCOMM.2020.3032695
31. Hou T, Liu Y, Song Z, Sun X, Chen Y, Hanzo L (2020) Reconfigurable intelligent surface aided NOMA networks. IEEE J Selected Areas Commun 38(11):2575–2588
32. Do D-T (2020) Anh-Tu Le and Byung Moo Lee, "NOMA in cooperative underlay cognitive radio networks under imperfect SIC. IEEE Access 8:86180–86195
33. Yue X, Liu Y (2020) Performance analysis of intelligent reflecting surface assisted NOMA networks. *arXiv*
34. Boulogeorgos A-AA, Alexiou A (2020) Performance analysis of reconfigurable intelligent surface-assisted wireless systems and comparison with relaying. IEEE Access 8:94463–94483
35. Gradshteyn IS, Ryzhik IM (2000) Table of integrals, series and products, 6th edn. Academic Press, New York, NY, USA

Analysis of the QoS Parameters of Different Routing Protocols Used in Wireless Sensor Networks

Prativa Rai, Nitisha Pradhan, and Kushal Pokhrel

Abstract There is a wide variety of sensor nodes that are capable of sending a wide range of parameters in the deployed environment. The sensed data in each of the cases are communicated by the Wireless Sensor Nodes (WSN) to the Base Station (BS) using one or the other wireless transmission mechanisms. During this type of transmissions, a proper selection of a suitable routing algorithm is seen to help in the overall performance enhancement of the network. This work aims to study the three of the popular routing protocols for WSN, viz: (a) Zone Routing Protocol (ZRP), (b) Fisheye State Routing (FSR), and (c) Landmark Ad hoc Routing (LANMAR). Comparative performance analysis of these three routing protocols has been done in QualNet 7.1 simulator based on the parameters like average jitter, throughput, end-to-end delay, and energy consumed during transmitting mode, receiving mode, and idle mode. The analysis shows that the LANMAR protocol offers a strong QoS foothold in the WSN framework compared to ZRP and FSR.

Keywords Wireless sensor network · Routing protocol · Energy · Jitter · Throughput · End-to-End delay

P. Rai (✉) · N. Pradhan
Department of Computer Science and Engineering, Sikkim Manipal Institute of Technology, Sikkim Manipal University, Gangtok, Sikkim, India
e-mail: prativa.r@smit.smu.edu.in

N. Pradhan
e-mail: nitisha.p@smit.smu.edu.in

K. Pokhrel
Department of Electronics and Communication Engineering Sikkim, Sikkim Manipal Institute of Technology, Sikkim Manipal University, Gangtok, Sikkim, India
e-mail: kushal.p@smit.smu.edu.in

S. Dhar et al. (eds.), *Advances in Communication, Devices and Networking*, Lecture Notes in Electrical Engineering 776,
https://doi.org/10.1007/978-981-16-2911-2_40

1 Introduction

A Wireless Sensor Network (WSN) comprises of a huge amount of sensor nodes, generally less complex and small in size [1]. The sensor nodes are capable of capturing the data that are crucial for checking the progress of certain objects like a man-made structure or even a human body and hence aid in the observation of the objects' lives under a particular environment. The data acquired from the sensor nodes is transmitted wirelessly to the Base Station (BS) for further processing, analysis, and retention. The data communication takes place under the governance of certain rules that specify how the routers should communicate to one another, some protocols allow the routers to be loaded with information concerning physical topology well in advance. However, other protocols allow the router to learn about the entire topology during the data sharing process.

Given the application of WSN, network layers have a growing demand for studies with the fundamental concept of the routing protocol to determine an optimal path in any network. The routing protocol is used to find a reliable and energy-efficient route. A lot of routing protocols are already available in the literature and can be categorized into three broad classes—Proactive Routing (also known as Table—Driven Routing), Reactive Routing (commonly known as On- Demand Routing), and Hybrid Routing protocols [2]. All these protocols differ in the way how the best path is selected. A comparative among different groups of routing protocols has been carried out in [3–6]. However, sufficient literature that directly compares Zone Routing Protocol (ZRP) [7], Fisheye State Routing (FSR) [8], and Landmark Ad hoc Routing (LANMAR) [9] protocols have not been reported so far.

Therefore, in this work, we focus on these three routing protocols designed for WSN and perform a comparative analysis of the same. The main goal of this paper is to assess the performances of ZRP, FSR, and LANMAR protocols for different scenarios of the variable density of nodes using the QualNet 7.1 network simulator. The remaining paper is ordered as follows: Sect. 2 gives a detailed description of ZRP, FSR, and LANMAR protocols. Related work has been presented in Sect. 3 followed by motivation in Sect. 4. Section 5 discusses the simulation environment along with parameters used for analysis. The simulated results and comparison of the performance of ZRP, FSR, and LANMAR protocols are presented in Sect. 6 with a summary of the results in Sect. 7.

Finally, we provide conclusions with future works directions in Sect. 8.

2 Background

In any network type, routing has always been a very challenging task because the route selected has to be energy efficient and produce less overhead and minimize the time required to find the best path. In this section, a brief explanation of how the best path is obtained in ZRP, FSR, and LANMAR protocols.

2.1 Zone Routing Protocol (ZRP)

The Zone Routing Protocol proposed in [7] incorporates the properties of both reactive and proactive protocols under consideration. Thus falls under the hybrid routing protocol category. This protocol works by creating zones for every node in the network. Each zone of a specific node considers entire nodes placed over a certain zone radius (*Zr*) which is defined by the number of hops. Thus, a zone includes all those nodes who are at most hops distant from the node considered. The nodes within a zone are divided into two types of nodes: interior and peripheral nodes. The nodes whose minimum distance is less than are called interior nodes and the nodes which lie on the circumference of the zone are called peripheral nodes.

The working of ZRP can be summarized as follows:

i. Routing within a zone: If the nodes within a zone need to communicate then it uses IntrA-zone Routing Protocol (IARP) [10]. IARP belongs to the proactive routing protocol family and is used when interior nodes want to communicate with each other. It identifies an optimal shortest routing path by eliminating redundant links and link failures.
ii. Routing outside zone: If the nodes falling in different zones needs to communicate then it uses IntEr-zone Routing Protocol (IERP) [11]. IERP is a reactive protocol and is used to find routes amongst the nodes lying in different zones. It has two phases: Route Request (RREQ) and Route Reply (RREP). ZRP also uses Bordercast Resolution Protocol (BRP) [12] to broadcast the RREQ packet to the peripheral nodes to find the destination node. The RREQ packet gets propagated in the network and the RREP packet with the path will be sent back to the source node.
iii. Route Maintenance: The local topology information is updated by IARP, so ZRP uses this information for route maintenance. In case of any link failures, an alternative multi-hop route is used.

ZRP minimizes the number of packet transmissions as compared to both proactive and reactive routing protocols [13, 14]. It is more reliable since it generates multiple path routes and it's a flat protocol so minimizes overhead and congestion compared to hierarchical routing protocols. But the performance of ZRP depends on the zone radius (*Zr*) parameter because it specifies the nodes distribution (node density) which directly affects the traffic in the network.

2.2 Fisheye State Routing (FSR)

Fisheye State Routing (FSR) [8] is a proactive protocol from the link-state protocol family. FSR introduces the concept of multi-layered fisheye possibility to decrease routing update cost in bigger network systems. In this approach, the deployed WSNs can exchange the respective routing information packets with the neighboring WSNs at a frequency that is dependent on their hop distance. It maintains the information of the topology at every node however avoids flooding the network with the information. It allows all nodes to exchange the topology data with their direct neighbors. It uses the fisheye technique and forms a scope or region. Each node gradually slows down the route update as the hop distance increases. The route information to the nearest neighbor nodes is updated with the highest frequency. The sequence numbers aids in identifying the change in topology.

The protocol comprises of three major tasks:

i. *Neighbor Discovery*: The nodes learn, establish, and maintain neighbor relationships.
ii. *Information Dissemination*: Link State Packets (LSP) are disseminated to other nodes that contain the neighbor link information.
iii. *Route Computation*: On receiving the LSPs, based on the information in the LSP, each node needs to compute the route to all the nodes in the network.

FSR is a hierarchical routing protocol and reduces routing updates and hence proves to be efficient for larger networks with low bandwidth and mobility [8].

Over the years, FSR has been used in MANET environments, because of its proven efficiency, scalability, and simplicity.

2.3 Landmark Ad Hoc Routing (LANMAR)

The Landmark Ad Hoc Routing (LANMAR) [9] protocol is based on a hybrid framework of landmark routing and the FSR scheme. As a result, this protocol has been seen to exhibit a sharp drop in the size of the routing table at layer-3 of the deployed WSN. This property of the LANMAR protocol leads to offload storage cost and line cost in the deployed WSN thus saving energy. It creates a two-tiered logical hierarchy with each landmark being the cluster head of a particular subnet. This protocol considers forwarding the data when the destination is known. In case the exact destination is unknown the data gets forwarded to the landmark of the destination subnet. Each subnet has a landmark elected. In each subnet a scope routing algorithm is available. While routing, the packets are restricted only within the scope while maintaining the correct routing information. The corresponding landmark of a subnet summarizes the routing information when the nodes to be communicated lies beyond the subnet's node's scope. These schemes facilitate LANMAR to reduce the routing table's size and hence hugely reducing the traffic overhead created during routing updates.

The phases of LANMAR protocol can be summarized as follows:

i. Node Identification: The protocol categorizes nodes based on common interest and forms a subnet with similar nodes
ii. Election: Each subnet holds an election to choose a landmark node. At the onset of the LANMAR algorithm, the landmark ceases to exist. The deployed sensor

node would first update about the other peer entities in the subnet in its scope. The subsequent growth in the routing computations leads the sensor nodes to learn about some of the peer WSNs in their scope. Thereafter, the sensor node declares the self as the landmark for this group. The sensor node also adds itself to the landmark distance vector. Now, all such landmarks would float the election weight to the neighbouring sensor nodes by way of broadcasting the distance vector update packets. As the landmark packets flood out, each of the sensor nodes in the subnet performs a winner competition procedure. In this way, at the end of the LANMAR algorithm, a subnet chooses only one landmark.

iii. Routing: Hierarchal addresses are assigned to each node. Each node then starts sharing its address to its immediate neighbor who shall further pass it forward to their immediate neighbors (in ripple fashion). On sharing the addresses each node discovers every other node in the network (outside their subnet as well). Hence each node uses the hierarchal address to find a route to another node.

3 Related Works

Noorani et al. [15] have analyzed the performance of proactive (Destination-Sequenced Distance Vector (DSDV [16]), reactive (Ad hoc On-Demand Distance Vector (AODV [17]) and hybrid (ZRP) routing protocols. An enhanced ZRP routing protocol has also been proposed that reduces the traffic flow using query detection methods and hence improves the performance compared to the standard ZRP protocol. Gasmi et al. [18] have enhanced the ZRP protocol and proposed a novel protocol called Stable Link-based Zone Routing Protocol (SL-ZRP) for routing on the Internet of Vehicles (IoV) networks. It ensures link stability in IoV applications by using three parameters: node priority, node speed, and delay.

Haglan et al. [19] have evaluated the performance of ZRP considering the size of the network and density to obtain the optimal value of zone radius. The protocol is simulated using the NS2.33 network simulator and they have reported that the optimal zone radius is two hops.

In [20], Dumala et al. have addressed the time series problem of LANMAR protocol using soft computing techniques. The fisheye scope parameter is configured dynamically using fuzzy-based rules. The work has been simulated using the EXata emulator with MATLAB and it is shown to achieve better performance than the standard LANMAR protocol.

A fuzzy logic-based enhanced version of ZRP has been proposed by Nithya et al. [21]. The zone radius is dynamically computed based on the remaining energy of the node, speed of a node, and density of neighboring nodes. The proposed protocol has been simulated in NS2 and proves to be better compared to the standard ZRP protocol.

The protocol considered in this work has various applications as reported in the literature because it suits networks with high node density and mobility of nodes.

4 Motivation

From Sect. 3, we can see that design of routing protocols for WSN is application-specific and a lot of research work is still being carried out. Many researchers have already proposed some good and stable routing protocols. With the emerging development of the Internet of Things (IoT) in the present time, we can try and find the suitable existing routing protocols that can be applied in some IoT-based application. Therefore, in this work, we have studied three already existing routing protocols (a) ZRP, (b) FSR, and (c) LANMAR. These protocols have been analyzed taking into consideration different performance metric parameters. Although various performance enhancements on these three protocols have been proposed over time. However, sufficient literature carrying out a complete performance comparison of these three protocols has not been reported so far.

5 Simulation Setup

We have simulated all three protocols using QualNet 7.1 simulator under various conditions. The nodes are positioned arbitrarily on the simulating environment and data traffic between source and destination is considered to be a Constant Bit Rate (CBR). We have considered a random waypoint mobility model in a rectangular field with various simulation parameters as shown in Table 1.

The performance metrics used for comparison of the three protocol under consideration is listed and described in Table 2.

Table 1 Scenario Factors

Parameter	Values
Routing protocol	ZRP, FSR, LANMAR
Size of region	1500 * 1500 Sq. Unit
Shape of region	Square
Number of nodes	20 to 200
Placement of nodes	Stochastic
Mobility model	Random waypoint
Antenna model	Omni directional
Traffic source	Constant bit rate
Size of the packet	512 bytes
Layer -2 protocol	802.11
Data rate	2 Mbps
Simulation time	300 s
Energy model	Generic

Table 2 Performance metric

Performance metric	Description	Measuring unit
End-to-End delay	It indicates the time consumed for a packet to be communicated across the network from source to destination	sec
Throughput	It is the ratio of successfully transmitted data per second, it is analyzed with CBR (Constant Bit Rate) data traffic	bits/sec
Jitter	It refers to the fluctuation in packet arrival time. It is also known as the variation in the latency occurred during packet transmission. It may be due to medium access route change, queueing delays. It is calculated as the difference between the two latency samples	ms
Energy consumed in transmission mode	It is calculated as the energy consumed by communicating nodes to transmit the packet. It directly depends on the size of the data packet being transmitted	mWh
Energy consumed in reception mode	It is calculated as the energy required by communicating nodes to receive data packets	mWh
Energy consumed in idle mode	The nodes which may not be transmitting or receiving at a particular moment however consume energy in idle mode as they are wirelessly connected and alive. This is known as energy consumed in idle mode	mWh

6 Results and Discussion

This work presents the study of the various performance of routing protocols for mobile nodes with different node densities. Figure 1 provides the snapshots of the QualNet 7.1 running simulation scenario of 100 nodes with a mobility speed of 10mbps for the ZRP routing protocol.

Figure 2 depicts the results obtained of the various end-to-end delays for ZRP, FSR, and LANMAR routing protocol for varying node densities. The simulation results reveal that WSN deployed with LANMAR protocol exhibits end-to-end delay not exceeding 0.06 s up to 200 nodes per simulation area. However, the same exhibited by WSN deployed using ZRP and FSR is observed to be more than 0.1 s and 0.3 s on an average respectively, in comparison. Therefore, WSN deployed using LANMAR protocol suffers the least end-to-end delay in comparison to that with FSR and ZRP protocols by a sufficient margin.

Figure 3 presents the variation of throughput for ZRP, FSR, and LANMAR routing protocol for different node densities. The simulation results reveal the following:

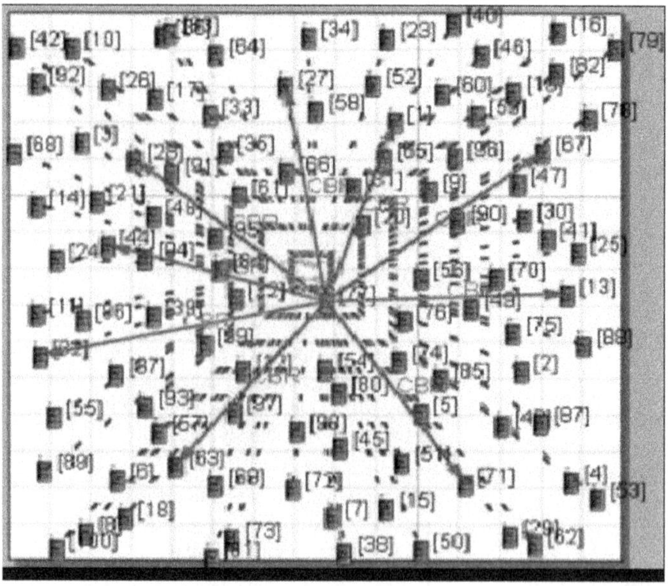

Fig. 1 Running simulation scenario

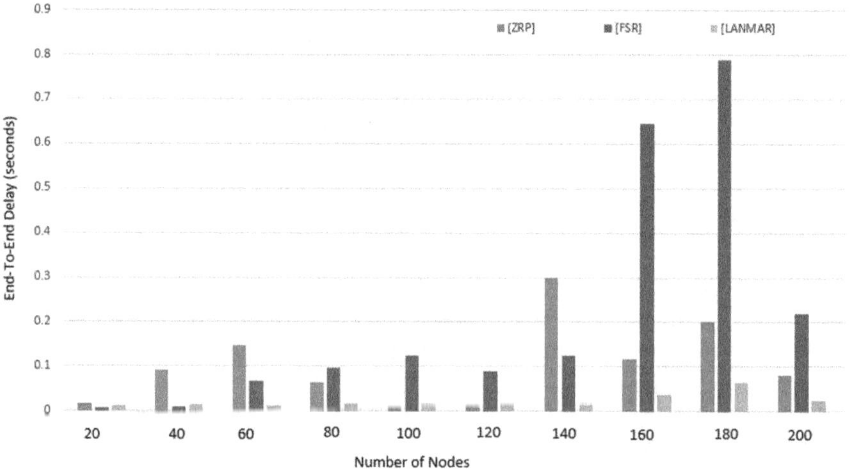

Fig. 2 End-To-End delay

i. For lower node densities (i.e., up to 60 nodes in the simulation area), the WSN deployed with ZRP protocol could deliver an average throughput of 2483 bits/sec whereas the same delivered by the WSN deployed with FSR and LANMAR is 57% and 64% more respectively than that of ZRP in comparison.

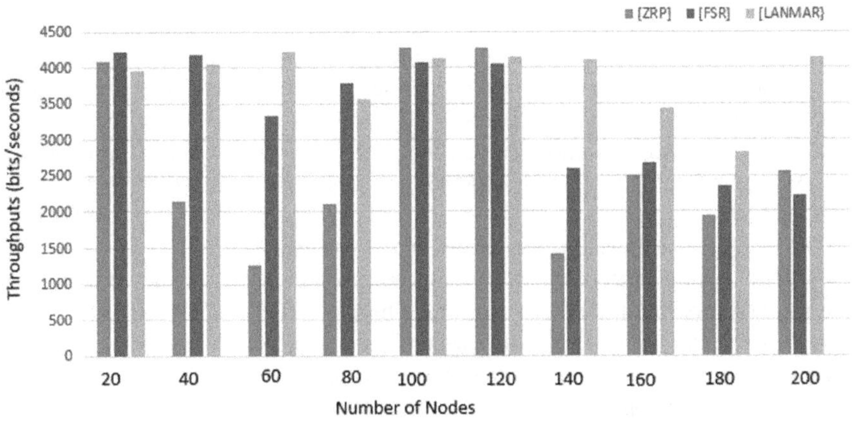

Fig. 3 Throughput

ii. For average node densities (i.e., from 60 to 140 nodes in the simulation area) the WSN deployed with ZRP protocol could deliver an average throughput of 3566 bits/sec whereas the same delivered by the WSN deployed with FSR and LANMAR is 12% and 10% more respectively, then that of ZRP in comparison.
iii. For higher node densities (i.e., from 140 to 200 nodes in the simulation area) the WSN deployed with ZRP protocol could deliver an average throughput of 2100 bits/sec whereas the same delivered by the WSN deployed with FSR and LANMAR is 17% and 71% more respectively than that of ZRP in comparison.

Therefore, the overall throughput exhibited by the LANMAR protocol is better than that compared with FSR and ZRP.

Figure 4 shows our simulation results for Jitter experienced by ZRP, FSR, and LANMAR routing protocol for different node densities. The simulation results show

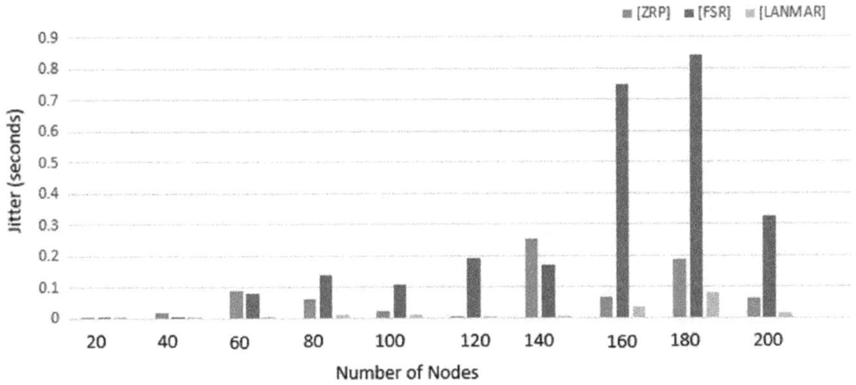

Fig. 4 Jitter

that the jitter experienced by the WSN deployed using LANMAR protocol does not exceed beyond 0.08 s for node densities up to 200 nodes in the simulation area. However, it is observed that the jitter suffered by the WSN deployed using ZRP protocol and FSR protocol exceeds 0.1 and 0.4 s on an average for the same node densities. Also, it is observed that the variation in jitter values is minimal for LANMAR protocol. Therefore, LANMAR protocol outperforms, FSR, and ZRP for jitter analysis.

The energy consumed by the nodes in the transmit mode of ZRP, FSR, and LANMAR routing protocol with different network densities is shown in Fig. 5. The result shows that the energy consumed by the nodes in ZRP does not exceed 0.2 mWh for various node densities whereas the same consumed for LANMAR is well above 0.8 mWh on an average for the same node densities. However, FSR performs poorly in comparison whose average energy consumption is well above 1 mWh comparatively. Therefore, ZRP outperforms FSR and LANMAR protocol for energy consumption in transmit mode.

The results obtained in Fig. 6 depict the energy consumed in the receive mode for the three protocols under consideration. The node densities have been kept the same for the sake of continuity. The simulation result shows that the energy consumed by WSN deployed considering the ZRP protocol does not exceed 0.4 mWh during the

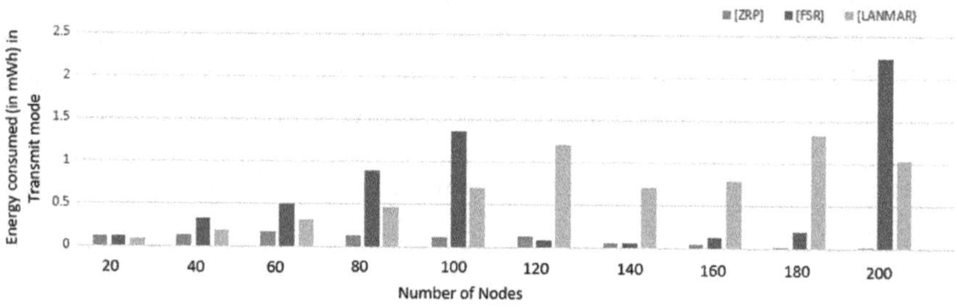

Fig. 5 Energy consumed in transmit mode

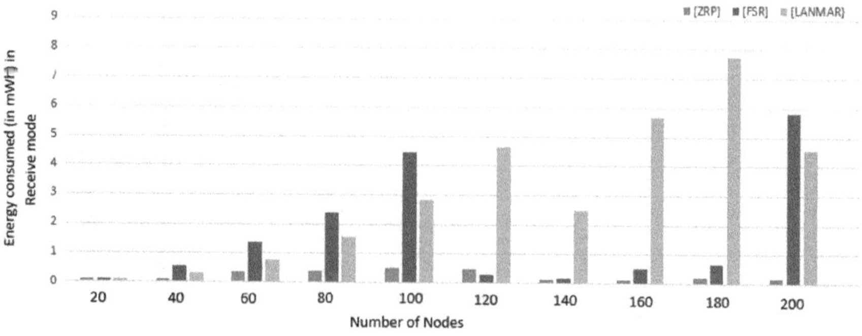

Fig. 6 Energy consumed in receive mode

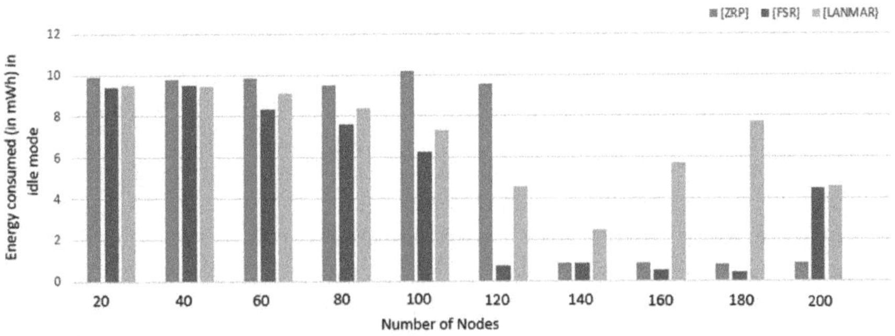

Fig. 7 Energy consumed in idle mode

entire simulation duration. However, the result also reveals that in comparison, the WSN deployed using FSR and LANMAR consumes average energy exceeding 2 and 3 mWh respectively. Therefore, the ZRP protocol outperforms FSR and LANMAR protocols for energy consumed in receive mode by a fair margin.

Figure 7 shows the variation of energy consumed in idle mode for various routing protocols measured for node densities up to 200 nodes in the simulation area. The simulation result shows that the average energy consumed by WSN deployed considering the ZRP protocol is approximately 6.35 mWh during the entire simulation duration. However, WSN deployed with FSR exhibits 27% lesser and that deployed using LANMAR is 7% more in comparison. The simulation result also reveals the following:

i. For node density up to 60 nodes in the simulation area, WSN deployed with ZRP consumes average energy of 10 mWh. In comparison, the same consumed by that of FSR and LANMAR is observed to be 14% lesser and 10% lesser respectively.
ii. Further, if we increase the number of nodes from 60 to 120 then WSN deployed with ZRP has been observed to consume lesser average energy at 9.6mWh. In comparison, the same consumed by that of FSR and LANMAR is observed to be 52% lesser and 31% lesser respectively.
iii. Also when the number of nodes was increased from 120 to 200, the energy consumption was observed to decline sharply. The average energy consumption for this case was found to be only 1.125 mWh for WSN deployed with the ZRP protocol. The same was observed to be 1.5 and 6.3 mWh for the WSNs deployed with FSR and LANMAR protocols respectively.

Therefore, WSN deployed with FSR protocol was observed to be the most energy-efficient during IDLE mode.

7 Summary

The summary of ZRP, FSR, and LANMAR protocol based on various parameters is listed in Table 3.

From the simulation results achieved in Sect. 6, the best performing protocol out of the three protocols has been listed in Table 4.

As we know that the attributes like end-to-end delay, throughput, and jitter are the main QoS parameters that have a strong impact on determining the quality of experience for end-users. If we ignore the energy consumption metric in our performance analysis, LANMAR protocol is seen to offer a strong QoS foothold in the WSN framework.

Table 3 Relative summary of ZRP, FSR, and LANMAR routing protocols

Parameter	ZRP	FSR	LANMAR
Routing approach	Hybrid routing	Table driven routing	Landmark routing
Routing structure	Plane structure	Flat structure	Hierarchical structure
The network is divided into	Zones	Fisheye	Landmark
Routing transparency	High inside zone	High	High
Routing computation	Broadcast	Distributed	Broadcast
Source routing	No	No	No
Compensation	Reduce retransmiss ions	Maintains accurate routing information for immediate neighbors	provide support for multiple hosts per router
Drawback	Overlappin g zones	The exchange frequency decreases proportionally to the distance	Does not provide support for unidirectional links

Table 4 Summary of performance analysis

Performance metric	ZRP	FSR	LANMAR
End-to-End delay			✓
Throughput			✓
Jitter			✓
Energy consumed in transmission mode	✓		
Energy consumed in reception mode	✓		
Energy consumed in idle mode		✓	

8 Conclusion

The performance analysis of the Zone Routing Protocol (ZRP), Fisheye State Routing (FSR), and Landmark Adhoc Routing (LANMAR) protocols for WSN was studied under varied node density using QualNet 7.1 network simulator. After experimental observations, it may be said that the LANMAR routing protocol might be best suitable for applications where average jitter, throughput, and end-to-end delay are unfavorable. In concerns to energy consumption in the sensor nodes on varied modes, ZRP consumes less energy as compared to FSR and LANMAR consumed maximum energy.

We can extend this work by enhancing these protocols and evaluating their applicability in IoT-enabled simple home-automation, health care, or wildlife monitoring applications.

References

1. García Villalba LJ, Sandoval Orozco AL, Trivino Cabrera A, Barenco Abbas CJ (2009) Routing protocols in wireless sensor networks Sensors 9(11):8399–8421
2. Bhushan B, Sahoo G (2019) Routing protocols in wireless sensor networks. In: Computational intelligence in sensor networks. Springer, pp 215–248
3. Raghunandan G, Lakshmi B (2011) A comparative analysis of routing techniques for Wireless Sensor Networks. In: 2011 national conference on innovations in emerging technology, pp 17–22
4. Yassine M, Ezzati A (2014) Performance analysis of routing protocols for wireless sensor networks. In: 2014 Third IEEE international colloquium in information science and technology (CIST), pp 420–424
5. Krishnaveni P, Sutha J (2012) Analysis of routing protocols for wireless sensor networks. Int J Emerg Technol Adv Eng 2(11):401–407
6. Tabbana F (2020) "Performance comparison and analysis of proactive, reactive and hybrid routing protocols for wireless sensor networks. Int J Wireless Mobile Netw (IJWMN) 12(4)
7. Beijar N (2002) Zone routing protocol (ZRP). Netw Lab Hels Univ Technol Finl 9:1–12
8. Pei G, Gerla M, Chen T-W (2000) Fisheye state routing: a routing scheme for ad hoc wireless networks. In: 2000 IEEE international conference on communications. icc 2000. global convergence through communications. conference record, vol 1, pp 70–74
9. Gerla M, Hong X, Pei G (2000) Landmark routing for large ad hoc wireless networks. In: Globecom'00-IEEE. Global telecommunications conference. conference record (Cat. No. 00CH37137), vol 3, pp 1702–1706
10. Haas ZJ, Pearlman MR, Samar P (2002) Intrazone routing protocol (iarp), IETF Internet Draft Draft-Ietf-Manet-Iarp-02 Txt
11. Haas ZJ, Pearlman MR, Samar P (2002) The interzone routing protocol (IERP) for ad hoc networks, Draft-Ietf-Manet-Zone-Ierp-02 Txt, pp 1–14
12. Haas ZJ, Pearlman MR, Samar P (2002) The bordercast resolution protocol (BRP) for ad hoc networks, IETF MANET Internet Draft, pp 13801–14853
13. Pearlman MR, Haas ZJ (1999) Determining the optimal configuration for the zone routing protocol. IEEE J Sel Areas Commun 17(8):1395–1414
14. Schaumann J (2002) Analysis of the zone routing protocol
15. Noorani ZY (2013) Performance analysis of DSDV, AODV, and ZRP routing protocol of MANET and enhancement in ZRP to improve its throughput. Int J Sci Res Publ 3(6)

16. Narra H, Cheng Y, Cetinkaya EK, Rohrer JP, Sterbenz JP (2011) Destination-sequenced distance vector (DSDV) routing protocol implementation in ns-3. In: Proceedings of the 4th international ICST conference on simulation tools and techniques, pp 439–446
17. Chakeres ID, Belding-Royer EM (2004) AODV routing protocol implementation design. In: 24th Proceedings international conference on distributed computing systems workshops, pp 698–703
18. Gasmi R, Aliouat M, Seba H (2020) A stable link based zone routing protocol (SL-ZRP) for internet of vehicles environment. Wirel Pers Commun 1–16
19. Haglan HM, Yussof S, Al-Ani KW, Jassim HS, Jasm DA (2020) The effect of network size and density to the choice of zone radius in ZRP. Indones J Electr Eng Comput Sci 20(1):206–213
20. Dumala A, Setty SP (2020) LANMAR routing protocol to support real- time communications in MANETs using soft computing. Data Eng Commun Technol 231
21. Nithya B, Mala C, Thivyavignesh R (2020) Performance evaluation of dynamic zone radius estimation in ZRP for multihop adhoc networks. Wirel Pers Commun 1–25

An Efficient Two-Wheeler Anti-Theft System Based on Three-Layer Architecture

Ranjit Kumar Behera, Mohit Misra, Amrut Patro, and Diptendu Sinha Roy

Abstract Amidst the notable growth in the purchase of modern velocipede, society has witnessed an increase in the number of vehicle thefts revealing an increase in the illegal means for bypassing the inbuilt security standards provided by the vehicle manufacturers. Some of the bypassing activities include theft by ignition wire tampering and disconnecting various parts of the vehicle. Also there has been a perceptible increase in fuel theft. Hence, an efficient three-layer architecture-based Iot system is proposed in this paper to solve the universal problem. The proposed model time to time collects data from sensors attached to the vehicle and uploads it into a real-time database present in the cloud. Any illegal activity or tampering with the vehicle is detected by the model and the same is reported to the user in the registered mobile number. The user also has an access to his/her vehicle's location and can forward the same to the authorities in case of emergency or help. This three-layer architecture-based model also has the potential for taking preventive measures against theft by blocking the flow of fuel. Another attention-gaining contribution of the system is that it can communicate with the owner both with and without an internet connection.

Keywords IoT · LoRa · RFID · Bluetooth · Face recognition

1 Introduction

In recent years, there has been a vigorous increase in the number of vehicle robbery cases in various parts of the world. According to the National Crime Records Bureau (NCRB) in 2018, there were 1,98,408 reported cases of two-wheeler theft in India out of which only 45,624 were recovered [1]. As a result of solving this universal problem, anti-theft vehicle security systems have gained a lot of attention from researchers.

R. K. Behera (✉) · M. Misra · A. Patro
National Institute of Science and Technology, Berhampur, India

D. S. Roy
National Institute of Technology, Shillong, Meghalaya, India

S. Dhar et al. (eds.), *Advances in Communication, Devices and Networking*,
Lecture Notes in Electrical Engineering 776,
https://doi.org/10.1007/978-981-16-2911-2_41

The alarm-based security systems provided by the manufacturers are often found to be ineffective in preventing thefts. Also, the use of immobilizers in vehicles is now a days outdated and is not able to ensure the safety of the vehicles as they can be easily hacked. According to the NCRB report it can be seen that the number of recoveries of the vehicle is found to be less than 23% of the actual number of robberies reported. The existing security systems are lacking with the potential of tracking the vehicles. As a result of this lack of a proper tracking system, once a vehicle is stolen it is very hard to be recovered again.

With these gaps in the existing security systems provided by the manufacturers there comes the need for an efficient vehicle security system which not only ensures the safety of the vehicle but also simultaneously interacts with the owner of the vehicle. A system that endlessly reports the status of the vehicle with the owner. A lot of studies have been carried out by researchers to fulfill the need for an efficient security system.

The authors in [2] have proposed a two-wheeler security system in which their main focus is to cover the maximum aspects needed for ensuring the security of the vehicle. Notifying the owner with the help of an SMS about a theft attempt, remotely having controls over the vehicle via SMS, tracking the location of the vehicle with the help of GPS, remote key system, locking system and parking mode indication are some of the aspects focused by the authors in this paper. Though the model covers most of the security aspects, still some limitations and imperfection can be noticed in the model. In case of a poor connection with the owner, there is no other possibility for communication even when a theft attempt takes place. Whereas in [3], a bio-metrics mechanism-based security model is proposed by the authors which also estimates the position of the vehicle on the basis of previously fetched data. The bio-metrics mechanism [4] for user authentication has been an efficient way of user authentication for many years. Still some drawbacks can be noticed which hinders their model from being a very secured and flexible system. In case, an obstruction layer is formed over the fingerprint scanner or the user's finger, the bio-metrics mechanism fails to verify the user. Also the proposed model does not provide an alternative for authentication. In case of an emergency, such kind of drawbacks may lead to a disastrous situation for the user. In [5] the authors have proposed an anti-theft tracking system that provides an all-round service for the vehicle owner. The vehicle can be controlled by an RFID [6] module for switching it switch on and off. Whenever, the vehicle is stolen, the vibration sensors mounted in the vehicle triggers and with the help of the GSM module the location is shared with the vehicle owner. Their proposed model is also limited to the Internet connection and also factors like fuel theft are not given a thought by the authors.

Hence, being mindful of these issues and challenges, a three-layered vehicle security system for two wheelers has been proposed in this paper which not only ensures an efficient and flexible way for user authentication but also has the potential for taking preventive measures against a theft attempt. On top of this, it also has an efficacious tracking system with competencies to communicate with the user both in online and offline mode. Even in case of poor network connectivity the system is able to communicate with the owner in offline mode within a defined geographical range.

The remaining sections of the paper are organized as follows. Section 2 describes the system architecture followed by the model design presented in Sect. 3. Section 4 presents the algorithms and their working. A discussion on the experimental setup and result analysis is done in Sect. 5. In Sect. 6, the paper concludes with a summary.

2 System Architecture

The overall system architecture considered in our proposed model comprises three different layers namely Surveillance Layer, Transmission Layer and User Interface Layer as shown in Fig. 1.

The surveillance layer consists of a Vehicle Safety and Surveillance System (V3S) embedded with a two wheeler vehicle. The V3S can be said as the heart of our model as it plays a very crucial role in the authentication of the user, detection of

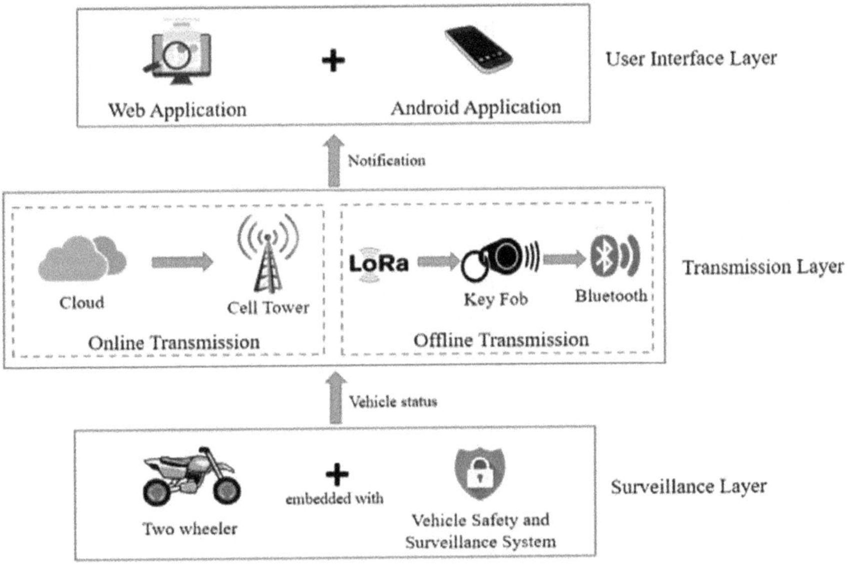

Fig. 1 Overall system architecture

any kind of illegal or tampering actions taking place with the vehicle, tracking the geographic location of the vehicle as well as controlling the ignition and flow of fuel in the vehicle. The V3S is a combination of two units. The first one is the Authentication Unit (AU) which is responsible for verifying the user trying to access the vehicle. The AU provides a multi-factor authentication facility for the user in which the authentication can be done by Radio-Frequency Identification (RFID) method, Personal Identification Number (PIN) or Face Recognition. However, the user has a control over choosing the factors for authentication with the help of a Web-based or Android-based Application. The second unit of V3S is the Tamper Detection and Communication Unit (TDCU) whose aim is to unceasingly detect any kind of tampering or damaging actions taking place with the vehicle and to simultaneously reflect the status of the vehicle in the user interface. The second layer in the architecture is the Transmission Layer, which is in charge of transmitting the data related to the vehicle status from the V3S to the user's smart phone. The data transmission in this layer is possible in two ways, i.e., online mode and offline mode. In situations, when there are no possibilities for offline data transmission, the data is endlessly uploaded by the TDCU into the real-time database present in Cloud [7]. The top most layer of the architecture which is the User Interface Layer that can also be termed as the Receiving End of the system. The owner of the vehicle is capable of fetching the status of the vehicle or receiving any kind of notification regarding the vehicle with the help of a Web-based or Android-based application defined in this layer. The applications are furnished with many other features such as enabling and disabling the system, selecting factors for authentication, tracking the vehicle's location as well as forwarding the location co-ordinates to the nearest police station for taking immediate actions against a theft. Even in the worst-case scenario, if the user is missing his/her smart phone, then also the vehicle status can be accessed anytime anywhere with the help of the Web-based application, proving a highly ubiquitous model.

3 Model Design

The systematic process of designing and implementation of the hardware part of the proposed model has been described in the following subsections along with the overall work flow taking place in the system.

3.1 Authentication Unit

This particular unit of V3S is in charge of authenticating the user trying to access the vehicle. It provides a multi-factor authentication facility in which the authentication of the user can be done by Radio-Frequency Identification (RFID) method, Personal Identification Number (PIN) or Face Recognition [8]. The AU becomes active as soon as the vehicle starting process is initiated. As shown by the block diagram in Fig. 2, the Ignition Switch is indirectly connected to the micro-processor with the help of a Voltage Converter. Whenever the ignition switch is turned on after switching on the ignition key the flow of current from the battery to the micro processor is possible, thereafter making the authentication unit active. Due to the lower working voltage of the micro processor compared to the working voltage in the circuits of the vehicle, it is mandatory to step down the voltage levels.

Hence, this is the reason for an indirect connection of the ignition switch to the micro processor with the help of a Voltage Converter. As soon as the authentication unit becomes active, it first checks the factors selected for authentication by the user and starts the authentication process thereafter. Whenever the Key-Fob of the vehicle is brought near the RFID reader present in the Multi-factor Authenticator, it verifies the user by receiving a unique id from the RFID tag present in the Key-Fob and matching it with the actual user id. Whereas the Keypad present in the authenticator helps the micro-processor for fetching PIN from the user and matching it with the actual user PIN provided to the system. User authentication by face recognition is done in the system by implementing an efficient object detection method based on 'Haar Cascade Classifiers' [9] proposed in [10]. However, the owner of the vehicle has control over disabling or choosing the factors for authentication by accessing the web-based application or smart-phone application. In a situation, if authentication of the user fails more than three times, the system notifies the owner for an unauthorized

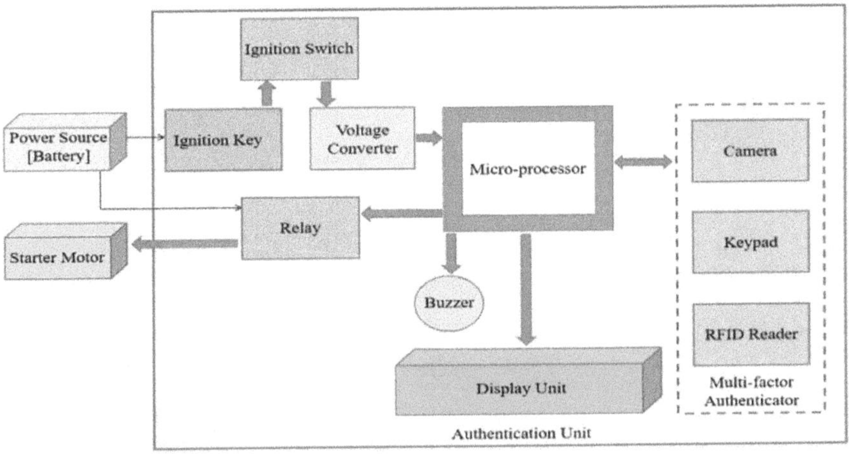

Fig. 2 Block diagram of authentication unit

access with the help of TDCU. Along with this, the AU takes some preventive measures such as alarming and avoiding the initiation of the Starter Motor with the help of a Buzzer and Relay connected to the micro processor. Therefore, from the design, it can be seen that the AU not only promises to be an efficient user authentication model but also proves to be a power-saving model as it becomes active only during the start process of the vehicle and remains inactive rest of the time.

3.2 Tamper Detection and Communication Unit

The TDCU whose main goal is to detect any kind of tampering activities taking place with the vehicle and ceaselessly reflect the status of the vehicle in the user interface, consists of a micro-controller that is connected to a number of sensors. The block diagram of TDCU, is presented in Fig. 3. The TDCU, periodically gathers data from the sensors connected to it. Data related to fuel level in the tank is endlessly fetched from the already existing fuel-level sensor present in the vehicle which helps the micro-controller in detecting a fuel theft. If the vehicle was in parking mode and if it is tried to be moved, then the change in speed of the vehicle is detected by the

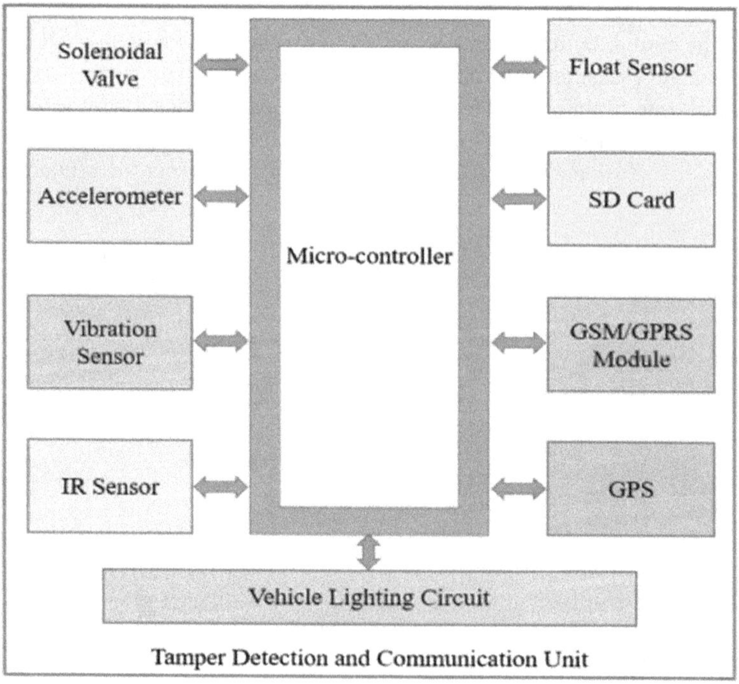

Fig. 3 Block diagram of TDCU

accelerometer. This feature further helps in acknowledging the owner even when the vehicle is towed from its resting place. If the vibration units of the motion exceed the defined threshold units, then the high voltage level is sent by the vibrational sensor to the micro controller. In this way, the micro controller is able to detect a damage or any kind of force against the vehicle taking place. The G.P.S module fetches the location coordinates of the vehicle and the IR sensor attached with the side stand of the vehicle transmits data to the MCU if there is any change in the parking mode of the vehicle. If the micro controller detects any kind of tampering activities taking place with the vehicle from the gathered data then it blocks the flow of fuel with the help of a solenoidal valve. This step taken by the system acts as a preventive measure against theft.

However, tamper detection is only done if the vehicle is in parking mode. Micro controller of this unit also receives forwarded data from AU to notify the user about any unauthorized access taking place. The TDCU parallely reports the status of the vehicle in online or offline mode. It first sends a data packet to the LoRa trans-receiver present in the Key-Fob. If it receives an acknowledgment packet in return from the Key-Fob that means the owner is within the offline mode communication range. With the possibility of offline mode, the status of the vehicle is forwarded to the Key-Fob by LoRa transmission [11]. The data received by the Key-Fob is further fetched by smart phone of the owner via Bluetooth. In case, if there is no possibility for an offline mode of communication then the General Packet Radio Service (GPRS) module uploads the status of the vehicle into the real-time database present in Cloud. The status from the Cloud real-time database is further fetched by the web-based or android-based application via the Internet. The second unit also could send SMS notifications to the owner as it is equipped with the Global System for Mobile communications (GSM) module. The owner also can remotely control the enabling and disabling of the sensors. Therefore, making the TDCU unit not only a strong tamper detection and communication model but also a system that focuses on power saving.

The Key-Fob is equipped with LoRa transceiver for the very purpose of receiving data packets from the V3S. The fetched data is stored in the EEPROM of Micro-controller. This stored data is further fetched by a smart phone application through Bluetooth. Moreover, the fob contains an RFID unique tag. The RFID tag when placed near the RFID receiver placed inside the Authentication Unit of V3S enables the system for RFID authentication of the user.

4 Algorithm Design

This section contains the high-level pseudo code design of AU and TDCU of the Vehicle Safety and Surveillance System (V3S).

Algorithm 1 Depicts the behaviour of the Authentication Unit present in the V3S.

Input: Facial Image set, Data from RFID Reader, Encrypted data from Pin
Output: Access Status

```
Ignition key turned on.
Step down of voltage level with the help of Voltage Converter.
Turn on micro processor.
authorizedAccess = False
RFIDverified = PINverified = FRverified = countFactors = 0
if RFIDauthentication == ENABLED then:
    countFactors = countFactors + 1
    for i = 1 to 3 do:
        Ask for input in display unit.
        Read data from RFID tag.
        if RFID == MATCHED then:
            RFIDverified = 1
            break
        end if
    end for
end if
if PINauthentication == ENABLED then:
    countFactors = countFactors + 1
    for i = 1 to 3 do:
        Ask for input in display unit
        Read data from Keypad.
        if PIN == MATCHED then:
            PINverified = 1
            break
        end if
    end for
end if
if FRauthentication == ENABLED then:
    countFactors = countFactors + 1
    for i = 1 to 3 do:
        Ask for input in display unit
        Capture image from Camera.
        if IMAGE == RECOGNIZED then:
            FRverified = 1
            break
        end if
    end for
end if
if countFactors == (RFIDverified + PINverified + FRverified) then:
    Initiate starter motor with the help of relay.
    authorizedAccess = True
else:
    Block started motor from initiating.
    Turn on alarm.
    authorizedAccess = False
end if
return authorizedAccess
```

Algorithm 1 Authentication Unit Algorithm

Algorithm 2 Depicts the real-time behaviour of the Tamper Detection and Communication Unit (TDCU).

Input: Feed from all the sensors
Output: Logs data to the cloud

```
while(TDCU == ENABLED) do:
    fuelTamper = motionTamper = locationTamper = False
    Fetch data from IR sensor attached with side stand.
    if parkingMode == ON then:
        if fuelTampering == ENABLED then:
            Fetch data from Fuel sensor.
            if fuel level == DECREASED then:
                fuelTamper = True
                Block flow of fuel using Solenoidal Valve.
            end if
        end if
        if motionTampering == ENABLED then:
            Fetch data from Vibration sensor.
            if force > vibrationThreshold then:
                motionTamper = True
            end if
        end if
        if accelarationTampering == ENABLED then:
            Fetch data from Accelerometer.
            if loaction == CHANGED then:
                locationTamper = True
            end if
        end if
    end if
    tamperingStatus = [fuelTamper, motionTamper, locationTamper]
    Send a packet to Key-Fob with the help of LoRa transreceiver.
    if acknowledgementPacket == RECEIVED then:
        Send [tamperingStatus, authorizationStatus from AU, current location] to Key-Fob via LoRa tranmission.
    else:
        Upload [tamperingStatus, authorizationStatus from AU, current loaction] to the real-time database in Cloud with the help of GPRS module.
        Send [tamperingStatus, authorizationStatus from AU, current loaction] as SMS to owner's smart phone with the help of GSM module.
    end if
end while
```

Algorithm 2 TDCU Algorithm

5 Experiment and Result Discussion

In this section, a discussion is made on how the setup for the simulation of our model was done and whether the model behaved as per the expectations or not. Figure 4. shows the hardware setup made for the working of our model.

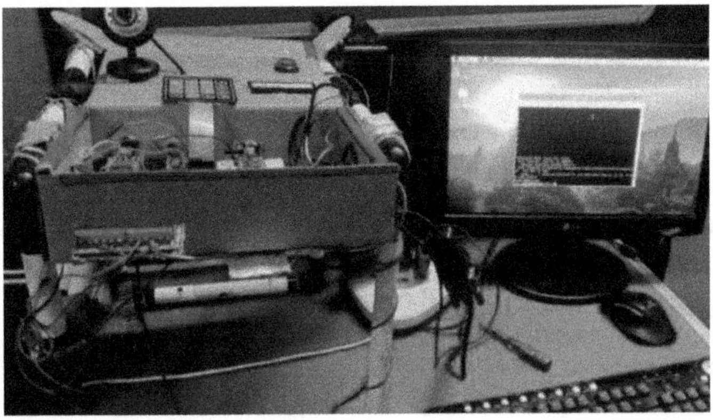

Fig. 4 Hardware setup of the proposed model

For designing the hardware model depicting the behaviour of Authentication Unit, a Raspberry Pi B3 Microprocessor was used. The Raspberry Pi B3 Microprocessor was connected to an MFRC522 RFID reader for verifying the RFID tags. Also, a camera was connected with the microprocessor to capture frames containing the user's facial features and a key-pad to read pins from the user. A piezo-buzzer was used for alerting people about theft and a 16 $\times$ 2 alphanumeric LCD Display for displaying purposes. Whereas for designing the Tamper Detection and Communication Unit (TDCU), Arduino Nano was used and a Neo-6 m for extracting the location co-ordinates of the vehicle. Due to the limitation of number of pins in Arduino Nano, two such micro-controllers were used, both communicating with each other through i2c communication. ADXL345 Accelerometer was connected to the Arduino for fetching the accelerometer data and for the long-range communication with the owner Lo-Ra SX1278 was used as a trans-receiver. The GSM module used here is SIM800L for the purpose of logging data into the Cloud. The Cloud portal used in the setup is Google Firebase which helped in creating a real-time database in Cloud. For the software simulation, the smart phone application was developed using Java Programming Language in Android Studio. The designed smart phone application ensures if there is a connectivity of the user to the Internet and fetches data from the cloud. The micro controller in Key-Fob maintains a buffer and as soon as all the data is retrieved completely, the data is written in the EEPROM present inside the micro controller which is further fetched by the smart phone application through Bluetooth. The Bluetooth module (HC-05) and the smart phone's built-in Bluetooth receiver are connected through serial communication with each other. The smart phone runs an asynchronous activity for collecting data through offline mode and provides the displays the data in text format as soon as a new data packet is retrieved. The data is transmitted in the form of Java Script Object Notation (JSON).

6 Conclusion and Future Work

With the rapidly increasing number of robbery cases in recent years, the security of vehicles has become a matter of concern. The security standards provided by the manufacturers have become inefficient and outdated. Though many studies have been carried to solve the universal problem still a lot of gaps were noticed in the existing models. Keeping in mind these gaps and challenges, an Efficient Three-layer Architecture based Two Wheeler Anti-theft System has been proposed in this paper. The model is proposed with a strong Authentication Unit and an efficient Tamper Detection and Communication Unit which not only ensures the safety of the vehicle but also takes preventive measures against theft. It is also capable of unceasingly communicating with the vehicle owner both in online as well as offline mode. For providing the owner with a ubiquitous model even the worst-case scenarios have been given a thought in this paper. The experimental and results analysis of our model has shown promising results and proved the efficiency of our proposed system. The future work of our paper includes, making the model more cost-effective and investigation of other possible technologies for widening the range of offline mode communication.

References

1. https://ncrb.gov.in/sites/default/files/Crime%20in%20India%202018%20-%20Volume%201.pdf. Accessed 18 June 2020 at 2:00 PM
2. Prashant Kumar R, Sagar VC, Santosh S, Nambiar S (2013) Two-wheeler vehicle security system. Int J Eng Sci Emerg Technol 6(3):324–334
3. Akinwole B (2020) Development of an anti-theft vehicle security system using gps and gsm technology with biometric authentication. Int J Innov Sci Res Technol
4. Campisi P (ed) (2013) Security and privacy in biometrcs. https://doi.org/10.1007/978-1-4471-52309-9
5. Liu Z, Zhang A, Li S (2013) Vehicle anti-theft tracking system based on Internet of things. In: Proceedings of 2013 IEEE international conference on vehicular electronics and safety. https://doi.org/10.1109/icves.2013.6619601
6. Roberts CM (2006) Radio frequency identification (RFID). Comput Secur 25(1):18–26. https://doi.org/10.1016/j.cose.2005.12.003
7. Furht B, Escalante A (eds) (2010) Handbook of cloud computing. https://doi.org/10.1007/978-1-4419-6524-0
8. Bruce V, Young A (1986) Understanding face recognition. British J Psychol 77(3):305–327. https://doi.org/10.1111/j.2044-8295.1986.tb02199.x
9. Cuimei L, Zhiliang Q, Nan J, Jianhua W (2017) Human face detection algorithm via Haar cascade classifier combined with three additional classifiers. In: 2017 13th IEEE international conference on electronic measurement & instruments (ICEMI). https://doi.org/10.1109/icemi.2017.8265863
10. https://www.hackster.io/mjrobot/real-time-face-recognition-an-end-to-end-project-a10826. Accessed 18 June 2020 at 2:00 PM
11. Augustin A, Yi J, Clausen T, Townsley WM (2016) A study of lora: long range & low power networks for the internet of things. Sensors 16:1466

Adoption of Robotics Technology in Healthcare Sector

Garima Bakshi, Anuj Kumar, and Amulay Nidhi Puranik

Abstract Healthcare professionals are facing a challenging time because of the COVID-19 pandemic. They are expected to perform their duties at the same time they need to take care of their own health. In such a disastrous time, robotics technology is emerging as the best support tool for healthcare professionals. In this paper, the researchers will discuss the role of robotics during the COVID-19 pandemic and propose certain adoption factors for robotics in health care. Robotics adoption is in the early stage in the healthcare industry, and there is further scope of primary research for the same. In this study, factors of adoption are taken from the Technology Acceptance Model and Diffusion of Innovation theory.

Keywords Health care · Robotics · TAM · DOI · Adoption · Benefits

1 Introduction

Coronavirus disease (COVID-19) is a communicable disease caused by a newly learned coronavirus. The best way to avoid and retard its spread is to stay aware of the causes of its transmission and take necessary precautions against it. The way the COVID-19 virus spreads, it is important to practice person-to-person distancing. At this point, there are no vaccines for COVID-19. However, a lot of research is being carried out all over the world for potential treatments or vaccines. Till the time vaccine is discovered, it becomes essential to practice social distancing [1]. To do so, robots and humanoids are playing a vital role in it [2]. Due to this pandemic, manufacturing, supply chains, restaurant business, medical care, education sector, etc., all are upside down today. The manual tasks which were carried out by humans are currently unsafe.

G. Bakshi
School of Engineering and Technology, Sushant University, Gurgaon, India

A. Kumar (✉)
Assistant Professor, Apeejay School of Management, Dwarka, New Delhi, India

A. N. Puranik
Amity Global Institute, Singapore, Singapore

S. Dhar et al. (eds.), *Advances in Communication, Devices and Networking*,
Lecture Notes in Electrical Engineering 776,
https://doi.org/10.1007/978-981-16-2911-2_42

Undertaking these tasks, along with no human-to-human contact, is the need of the hour. Robots offer the potential to carry out these jobs. The robotics sector has been relishing substantial development in current times, with several latest companies and fresh applications pouring greater than before acceptance in almost all sectors. In such testing times, the robots are offering two major opportunities, namely:

- To tackle the COVID-19 situation;
- To carry out tasks where socially distancing is required.

According to an analysis conducted by the experts, in the coming times robots are going to replace humans completely, and COVID-19 outbreak is fast-spreading. Because of this outbreak, consumer inclination has changed drastically and has led to many new possibilities for automation. Businesses, whether big or small, are working on how to make robots work to maintain social distancing as well as reduce the presence of the number of workers at work. Big retailers like Walmart, is utilizing robots to scrub its floor; likewise, in South Korea, robots are at work to distribute hand sanitizer and to measure temperature. There are many other technologies like artificial intelligence and machine learning that are also helpful to the healthcare segments, but the functionality is different. Artificial intelligence is linked with fast data diagnosis and X-rays. The combined algorithms based on artificial intelligence and machine learning are also helpful in countering the COVID-19 problem.

Robotics and artificial intelligence can come together as robotics can act as an interface between doctor and patient while artificial intelligence help in the diagnostic process and treatment. There are different functionalities of robotics that can help the healthcare segment. For example, through telepresence, videos can be taken from any remote location while teleoperation allows to operate robotics from any corner of the world [3]. Currently, the hospitals are facing a plenty of problems, hospitals have a huge number of audiences to handle. Robotics can help in mass-checking and maintaining social distancing too. Robotics can help in patrolling at regular intervals and it can also help at the reception in dealing with the number of patients [4]. Being a wireless device, robotics can connect with other devices and integrated technology can help further.

2 Objective

The goal of this study is to discuss the role of robotics technology in the healthcare industry during the COVID-19 pandemic. The second objective of this research is to propose certain factors of robotics technology adoption in the healthcare industry. Those factors will be proposed from the literature on robotics.

3 Insights from Literature Review

Reference [5] argued that the role of robotics in health care is much more significant now during the pandemic era. Robotics help in maintaining distance and performing all the tasks of the hospital include cleaning and sterilization. Medical robotics can provide high-end support to doctors in maintaining all the procedures. Reference [6] also argued about the increased role of robotics during a pandemic. The role of robotics is not limited to healthcare only; it is equally useful in restaurants, airports, transportation, and vehicles. In restaurants, robotics is playing the role of a waiter in taking orders. In all the industries, the life of frontline workers is at stake because they need to maintain a direct touch with the other party [7]. Robotics can be very useful in safeguarding the life of frontline workers. It can provide a shielding layer to frontline workers. It was impossible to treat patients without coming in physical contact with them, but robotics made it a reality [8]. Today, healthcare workers are evaluating and monitoring patients through robotics by maintaining distance [9]. Whenever any problem occurs, it asks for a creative solution. COVID-19 created challenges and problems all over the world [10]. The innovations in the form of robotics can help in dealing with this problem, and many countries are capitalizing on it [11]. Reference [12] also discussed the role of robotics in a hybrid framework for a safe working environment. Robotics and other 4.0 technology solutions have made technology adoption much faster during the COVID-19 pandemic [13]. Either it is health care or any other industry, they are looking for a feasible solution to safeguard their business from problems. For example, In Indonesia, healthcare workers, especially nurses, are ready to adopt robotics. They do not see robotics as their competitors. Instead, they are looking like technology partners in avoiding pandemic problems [14]. TAM theory [15, 16] and DOI theory [17] are the two most frequently talked about theories for technology adoption. TAM model is useful in studying technology adoption for an individual, while innovation theory (DOI) is relevant for studying adoption at the firm level. The basic factors of the TAM theory are perceived usefulness and perceived ease of use [18]. PU and PEOU are independent factors for behavioral intention to use technology. DOI theory's constructs are compatibility, complexity, relative advantage, trialability, and observability. There is no such literature available on factors available for the adoption of robotics in the healthcare sector based on the TAM or DOI model. Reference [19] tried to identify the factors responsible for robotics adoption in health care based on the UTAUT model. The findings suggest that facilitating conditions, influence by social circle, effort expectancy, performance expectancy, and safety concerns are positively contributing toward robotics usage [20]. In this paper, the researchers will try to identify which factors can contribute toward robotics adoption in the healthcare sector based on a combination of the TAM-DOI model.

4 Rise of Robots Amid COVID-19 Outbreak

The rise of robots that have replaced work without humans has brought significant changes in people's lives. It has resulted in developing robots that can provide both proactive and reactive measures that cannot be done by humans. A lot of changes have taken place in different industries. Of many tasks being carried out below, few tasks being delegated to robots during this pandemic outbreak are as follows:

- ***As Frontline Workers***: As frontline worker's robots work in the most infectious ward to reduce the risk of catching the disease by avoiding the direct contact of doctors with patients. Additionally, the use of robots also allows less protective clothing like PPE kits, which are currently in scarcity. Thus, the advantage is, therefore, double. Robots relay vital information like BP, heart rate, temperature monitoring between the doctor and patients remotely at any time of the day. Robots with UV light are disinfecting wards, thus keeping everyone safe even by reaching those areas which are hard to reach [21].
- ***As Logistics Robots***: Robots are being deployed to supply vitals like food and medicines to the patients. They are being put in warehouses and logistics facilities along with retail also where they disinfect, clean, and restock shelves. Also, robots are being employed to deliver food at short distances, thus maintaining human distancing [22]
- ***As Escort Robots***: In such a time of social distancing, loneliness is the major problem which most humans are facing, so these escort robots have found their utility in terms of companions as well as home care where human interaction is not possible [23].
- ***As Manufacturing Robots***: COVID-19 crisis has hard hit the manufacturing industry in a big way. But manufacturers have revamped and retooled their production lines. Now, robots are playing key roles to continue production that too in a safer and more efficient way [24].
- ***As Surveillance Robots***: Robots are in surveillance to ensure that none breaks lockdown by guarding the streets. Additionally, they are spreading a coronavirus awareness campaign in the areas where humans cannot go amid the current virus. A humanoid is being used to encourage patients in quarantine [25].

The battle against this novel virus is proving to be long and tiresome. So, many countries around the globe are moving toward automation. Millions of dollars are being invested in adopting robots to minimize human contact. In India, due to a shortage of personal protective equipment, the adoption to deploy robots has increased, and it will keep increasing in the future too. According to Rajeev Karwal, founder chairman, Milagrow, the Gurugram-based robotics company, the demand for robotics is increasing in Indian hospitals [26]. Many hospitals are talking to their company for robotics delivery. SMEs also exist in the healthcare segment, and those SMEs can adopt robotics technology for sustainability [27]. In China, a remote-controlled robot has been designed to take mouth swabs to carry out tests, perform scans. Robots have been stationed by the bedside to provide instant video connections between the patients and health workers. "Tommy" is name given to a robot that has been designed to help doctors in Italy. Tommy is helping doctors in taking care of patients. In Spain, around 500 people died in a day. Spain is planning to bring in a fleet of robots for conducting coronavirus tests. In Spain, most medical workforces have already got infected; thus, once the robot is deployed, it will reduce the risk for frontline fighters. Additionally, it is being planned that one of the robots will be stationed for early testing so as medical staff can put their efforts into treating the patients instead of conducting tests. In Korea, SK Telecom and Omron Electronics came together for the development of robots and those robots are powered by fifth generation technology to transmit real-time data.

Robots have been demonstrating and playing a vital role in fighting these difficult times of pandemic outbreaks across the globe. Robots are game-changers as they not only help in providing free contact alternatives but also are helping in the elimination of this virus from hospitals, public places. This contribution of robots shall make our health sector future-ready for any pandemic outbreaks in the future. As per the data published in world economic forums, not only the ground but also aerial robots are playing a key role in aspects relevant to manage and contain this pandemic. The below image is a glimpse of how intelligently and wonderfully robots have been globally integrated into the health care system to administer and manage the outbreak.

Healthcare professionals are being assisted by robots to perform interactions with infected and affected, thereby ensuring their safe well-being. Basic sanitization and disinfection of public spots, hospitals, healthcare clinics are being done by well-equipped robots like Xenex or Disinfection robot UVD. Former uses UV-C (ultraviolet-C) light to eliminate microorganisms. The robot disinfects at a rate of 12 rooms per hour. While later robots, although it is based on the same technology but can intelligently eliminate selected microorganisms. It has been programmed to work independently to disinfect infected and operation areas screening essential surfaces involving an appropriate quantity of light (UV-C) by eliminating selective pathogens. The robot disinfects a rate of 6 infected rooms per hour. Robots are being widely used in many countries of Asia, the United States, and Europe, where the pandemic outbreak has been virulent, robots have been a great help in managing timely logistics in sample collection from source patient to labs. Timely dispersal and availability of prescribed food, medicines, and medical equipment have led to the efficient management of operations.

Drones are being used to trace and identify potential patients and comply with quarantine and safe distance measures timely. Creating awareness adds another feather to the cap, timely dissemination of essential information to the public about updated protocols is contributing to the preventive measures. Robots have been involved in peer robot sanitation to ensure foolproof protocols to avoid any chance of disinfection. Telecommunication has brought another revolution in the health care industry, which led to a new phase in the industry post-COVID-19 era. Globe can be well prepared for any pandemic, epidemic, or disaster by utilizing robots without endangering precious lives due to safety reasons or workload. We can set emergency protocols in a systematic manner to aid our rescue teams during any such unforeseen circumstances. Re-deployment of these robots in building temporary and immediate

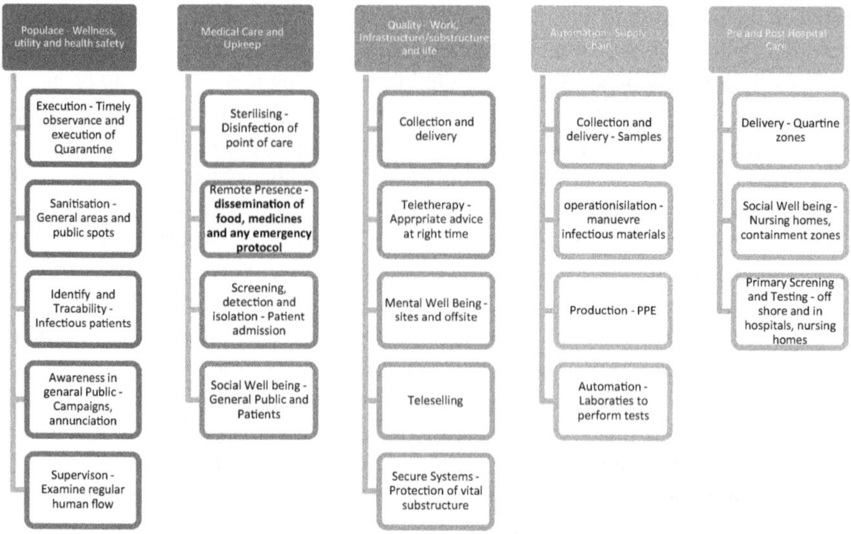

Fig. 1 Classification of details of robots for COVID-19 [28]

infrastructure is also one of the breakthroughs as it aids in the isolation of recuperating, symptomatic, and asymptomatic patients. Transforming existing infrastructure like train couches, cars to temporary treatment centers can also be one of the key areas of focus for robots (Fig. 1).

5 Requirements of Robots in Health Care

The requirement and demand for automated work styles are leading to an increased need for automatization and robotics. The International Federation of Robotics predicts the need for medical robots to be around 9.1 billion USD by the year 2022. The utilization of robots has not only reduced the workload of the medical and healthcare task force but also has improvised the productivity of entire facilities of health care.

1. *Kinematics and Dynamics*—Medical robot is dependent on the application for its kinematics and dynamics [29]. Both serial and parallel can perform all the tasks pertaining to the surgical and rehabilitation of service robots. Swiss robot—Delta was initially designed only for surgical operations, but it is now widely being used in the Food and Beverages industry. Variant robots are high payload robots with restricted DOM—Degree of Freedom. The surgical robots tend to be flexible, precise, and reliable if they have multi-DOF. Such robots can perform highly precise and complex surgeries with similar expertise and minimized error margin.

2. *Command and agility*—The biggest challenge is the command of medico robots to perform dynamic operations or tasks with the reverberation of external obstacles [30]. Sufficient DOF (degree of freedom) for end-effectors need to be provided to have movement in all directions. State-of-the-art technology is being utilized by medico robots to perform basic tasks like sterilization, cleaning, transportation, surgical applications, and patient care [31].
3. *Autoclave*—In order to prevent the spread of pathogens of communicable diseases, robots have been carefully designed to carry out massive sterilization. These robots can also multitask as spraying robots to disinfect the neighborhood surroundings.
4. *Cleaning Robots*—They are used to keep the general cleanliness of the hospitals; such examples are Roomba, UVD robots [32].
5. *Surgical Robots*—Minimally Invasive Surgery (MIS) is being performed by these robots with the same precision as human surgeons. Teleoperators are being used to perform remote operations and critical surgeries. The fourth generation of Da-Vinci surgical robots can be interfaced with other machines, and they help the hospitals and clinics in inventory control. KUKA LBR surgical robots have performed endoscopic and biopsy but also have performed highly precise surgeries like cranial and spine.
6. *Radiologist Robots*—These robots are a sigh of relief to mankind as the technicians no more need to get exposed to harmful radiations. Robots like twin robotic X-ray, a Siemens healthcare product are used in angiographies, fluoroscopies, 3D imaging. It can perform a multitude of X-rays in a single room without zero patient movements and with a real-time view for the doctor to investigate immediately. They are thereby saving the man-hours, delay in the diagnosis of the diseases. Many of these robots are contributing to chemotherapy treatments also.
7. *Rehabilitation Robots*—These robots assist and substitute human help to rehabilitate stroke or accident patients. They have made a breakthrough in the field of physiotherapy wherein they aid in assisting and treating various conditions of disabled, elderly, and inconvenient patients, thereby relieving the human from laborious work.

6 Findings

Based on the above study, it can be proposed that the following factors can contribute to robotics adoption in the healthcare sector. The above literature suggests that medical staff and individuals believe that robotics can enhance their performance. In fact, during the COVID-19 pandemic, when patients cannot be touched, robotics can be the most trusted partner of healthcare staff members. It will improve the job performance of healthcare sector employees. The usefulness can be a contributing factor in the adoption of robotics.

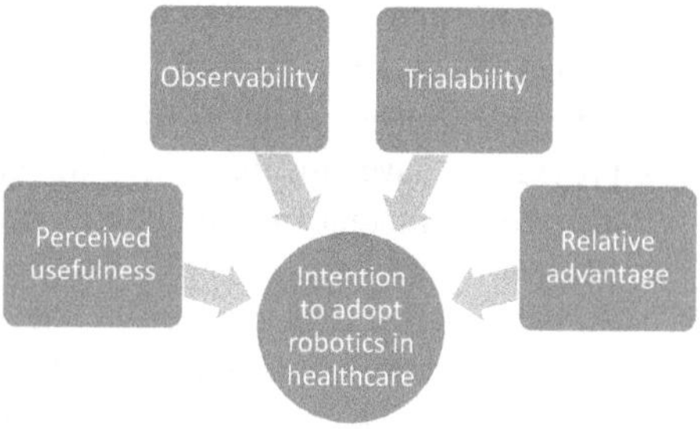

Fig. 2 A proposed model for adoption of robotics in the healthcare industry [15, 17]

- Ease of use can be justified when a person believes that he/she is completely free from effort. The adoption of robotics can be justified with ease of use because it can perform all the work of healthcare workers, but workers cannot reply to it completely. They need to monitor it. This factor cannot be suitable for adoption at this stage.
- Relative advantage refers to the degree to which innovation is better than the old one. No doubt robotics will help healthcare professionals in treating patients with proper physical distance, but robotics cannot replace healthcare professionals. The technology can be a supplement, not a replacement, in this case. But during COVID-19pandemic robotics can perform those complicated tasks which human hesitate to perform because it can be life risky. RA can be taken as a factor.
- Compatibility is also not a significant adoption factor because both healthcare professionals and patients are at a nascent stage in the usage of robotics. For patients, it is completely a new system, and they have no training as well.
- Complexity is defined as the degree of difficulty attached to the adoption of the innovation. Robotics is in the nascent stage in health care. It will take time before both patients and healthcare professionals will take it as a less complex solution.
- Trialability can be taken as a factor for robotics because different types of robotics have been passed the testing before reaching healthcare professionals.
- Observability can be taken as a factor because the usage of robotics is helping in taking care of the health of patients. The researchers have also mentioned the usage of robotics in improving the health of diabetic patients. So, results are visible (Fig. 2).

7 Conclusion

Based on the above discussion, it can be said that robotics technology emerged as prime innovation in dealing with patients without coming into personal contact with them. Different types of robotics are helping healthcare professionals in completing different types of treatment of patients, and this technology is also saving the life of healthcare professionals. The adoption factors responsible for robotics technology are RA, observability, trialability, and PU.

8 Limitations

The result of this study is based on the literature and knowledge gained about robotics from secondary resources. In the future, the researchers can collect primary data to test the significance of the proposed model suggested in this research.

References

1. AFP (2020) Robots and drones pick up delivery jobs in the West as humans face coronavirus risk. https://www.deccanchronicle.com/technology/in-other-news/160420/robots-and-drones-pick-up-delivery-jobs-in-the-west-as-humans-face-cor.html. Accessed 29 Nov 2020
2. Azeta J, Bolu C, Abioye A, Oyawale F (2017) A review on humanoid robotics in healthcare. In: MATEC web of conferences
3. Musa J (2020) How have humanoid robots been effective in supporting our healthcare workers during COVID-19?, SoftBank Robotics, 6 August 2020 https://www.softbankrobotics.com/emea/en/blog/news-trends/how-have-humanoid-robots-been-effective-supporting-our-healthcare-workers-during. Accessed 6 Dec 2020
4. Fagherazzi G, Goetzinger C, Rashid M, Aguayo G, Huiart L (2020) Digital health strategies to fight COVID-19 worldwide: challenges, recommendations, and a call for papers. J Med Inter Res 22(6):19284
5. Khan ZH, Siddique A, Lee CW (2020) Robotics utilization for healthcare digitization in global COVID-19 management. Int J Envrion Res Public Health 17:3819
6. Joshi AM, Shukla UP, Mohanty SP (2020) Smart healthcare for diabetes: A COVID-19 perspective. https://arxiv.org/pdf/2008.11153.pdf. Accessed 29 Nov 2020
7. Zemmar A, Lozano AM, Nelson BJ (2020) The rise of robots in surgical environments during COVID-19. Nature Mach Intell 2:566–572
8. Tavakoli M, Carriere J, Torabi A (2020) Robotics, smart wearable technologies, and autonomous intelligent systems for healthcare during the COVID-19 pandemic: an analysis of the state of the art and future vision. Adv Intell Syst 2:1–7
9. Zeng Z, Chen P, Lew A (2020) From high-touch to high-tech: COVID-19 drives robotics adoption. Tourism Geogr 1–11
10. Sahoo S, Bharadwaj S, Parveen S, Singh A, Tandup C, Mehra A, Chakrabarti S, Grover S (2020) Self-harm and COVID-19 pandemic: an emerging concern–A report of 2 cases from India. Asian J Psych
11. AlAttar A, Eissa M, AlHammadi O (2020) Robots versus COVID-19. Emirati Eng Abroad Lett 1(1)
12. Zhang H, Cai Y, Zhang H, Leung C (2020) A hybrid framework for smart and safe working environments in the era of COVID-19. Int J Infor Technol 26(1)
13. Clipper B (2020) The influence of the COVID-19 pandemic on technology. Adopt Health Care
14. Betriana F, Tanioka T, Locsin RC, Malini H, Lenggogeni DP (2020) Are Indonesian nurses ready for healthcare robots during the COVID-19 pandemic. Belitung Nurs J 6(3):63–66
15. Davis FD (1989) Perceived usefulness, perceived ease of use, and user acceptance of information technology. MIS Quart 13:319–339
16. Martín-García AV, Fernando M, David R (2019) TAM and stages of adoption of blended learning in higher education by application of data mining techniques. British J Educ Technol 50(5):2484–2500
17. Rogers EM (1995) Diffusion of innovations. Free Press, New York
18. Biucky ST, Harandi SR (2017) The effects of perceived risk on social commerce adoption based on tam model. Int J Electr Comm Stud 8(2):173–196

19. Vichitkraivin P, Naenna T (2020) Factors of healthcare robot adoption by medical staff in Thai government hospitals. Health Technol
20. Chao CM (2019) Factors determining the behavioral intention to use mobile learning: An application and extension of the UTAUT model. Front Psychol 10:1652
21. Breazeal C (2003) Toward sociable robots. Robot Auton Syst 42(3–4)
22. Lazzeri N, Mazzei D, Zaraki A, Rossi DD (2013) Towards a believable social robot. In: Conference on biomimetic and biohybrid systems, pp 393–395
23. Nagatani K, Kiribayashi S, Okada Y, Otake K, Yoshida K (2013) Emergency response to the nuclear accident at the Fukushima Daiichi nuclear power plants using mobile rescue robots. http://www.astro.mech.tohoku.ac.jp/keiji/papers/pdf/2013-JFR-Quince-online.pdf. Accessed 29 Nov 2020
24. Krueger V, Rovida F, Grossmann B, Petrick R, Crosby M, Charzoule A, Garcia G, Behnke S, Toscano C, Veiga G (2019) Testing the vertical and cyber-physical integration of cognitive robots in manufacturing. Robot Comput Integr Manuf 57:213–229
25. Witwicki S, Castillo JC, Messias J, Capitan J, Melo FS, Lima PU, Veloso M (2017) Autonomous surveillance robots: a decision-making framework for networked muiltiagent systems. IEEE Robot Autom Magaz 24(3):52–64
26. Ahaskar A (2020) Adoption of robots in India hospitals to grow during and post covid. https://www.livemint.com/news/india/adoption-of-robots-in-india-hospitals-to-grow-during-and-post-covid-11588139029183.html. Accessed 29 Nov 2020
27. Kumar A, Ayedee N (2013) Social media tools for business growth of SMEs. J Manag 5(3):137–142
28. Murphy RR, Adams J, Gandudi VBM (2020) Robots are playing many roles in the coronavirus crisis – and offering lessons for future disasters. [Online]. Available: https://theconversation.com/robots-are-playing-many-roles-in-the-coronavirus-crisis-and-offering-lessons-for-future-disasters-135527. Accessed 18 June 2021
29. Verma V, Gupta A, Gupta M, Chauhan P (2020) Performance estimation of computed torque control for surgical robot application. J Mech Eng Sci 14(3):7017–7028
30. Regmi S, Song YS (2020) Design methodology for robotic manipulator for overground physical interaction tasks. J Mech Robot 12(4)
31. Chiriatti G, Palmieri G, Palpacelli MC (2020) Collaborative robotics for rehabilitation: a multi-body model for kinematic and dynamic analysis. In: The international conference of IFToMM Italy
32. Kim J, Mishra AK, Raffaele L, Marco S, Nino C, Jose SV, Barbara M, Filippo C (2019) Control strategies for cleaning robots in domestic applications: a comprehensive review. Int J Adv Robot Syst 16(4)

Performance Analysis of Energy-Efficient Hybrid Precoding for Massive MIMO

Prashant Sharma and Samarendra Nath Sur

Abstract Massive multiple input multiple output (MIMO) is one of the main ingredients for the future generation (5G and beyond) communication system. Development of the green or energy-efficient communication system is the main concern of the future generation system. This work addresses this issue by designing an energy-efficient precoder for a millimeter-wave massive MIMO communication system. Here in this paper, the authors have proposed Lenstra–Lenstra–Lovász (LLL) algorithm-based lattice reduction (LR) aided hybrid precoder (HP) to have more energy efficiency. Through the simulation, in this work,the authors are able to show that the LR aided zero-forcing (ZF) and minimum mean square error (MMSE) precoders are much more efficient than conventional ZF and MMSE precoders.

Keywords Massive MIMO · Millimeter-wave communication · Energy efficiency · Spectral efficiency · Precoder

1 Introduction

Future generation communication system is looking for high data rate with low latency. As we are moving toward the implementation of a 5G technology to fulfill the user demand with a support that requires a huge amount of bandwidth. We know that bandwidth is a natural resource and is limited. As a 6 GHz spectrum is highly congested, so now we are looking for an alternative in which a millimeter wave(mmWave) can be implemented that can fulfill our requirement [1, 2]. Millimeter-wave communication technology is an emerging technology for an upcoming generation of cellular technology in which we can utilize the mmWave frequency resources [3–6]. It has a wide bandwidth with high spectral efficiency. One problem of mmWave upon its transmission is that it has a high path loss upon its propagation. As path loss in free space is inversely proportional to the square of the wavelength used [7]. To solve this issue, high engineering technique is required.

P. Sharma · S. N. Sur (✉)
Department of Electronics and Communication Engineering, Sikkim Manipal Institute of Technology, Sikkim Manipal University, Majitar, Rangpo, East Sikkim 737136, India

S. Dhar et al. (eds.), *Advances in Communication, Devices and Networking*,
Lecture Notes in Electrical Engineering 776,
https://doi.org/10.1007/978-981-16-2911-2_43

One way is that we can implement it by using a multiple antenna system in a small geographic area, which will improve the high traffic rate, transmission reliability, and quality of service [8, 9]. With the help of mmWave, adequate research work in multiuser MIMO setup has been going on all over the world.

Here massive MIMO system with a large number of array configurations can be implemented, which has higher spectral efficiency for wireless communications [10]. The capacity of MIMO can be increased by the interference alignment (AI) technique in which the interference among users can be used to obtain the intended information [11]. In such a mmWave MIMO system, a fully digital baseband precoder is applied in which each antenna individually requires a radio frequency (RF) chain. In such a system, hundreds of antennas are connected over a small geographic area. It is not feasible to connect each radio frequency (RF) chain into the individual antenna which leads to high power consumption and a complex hardware requirement [12]. We can remove these above limitations by designing a hybrid structure that contains analog processing and digital precoding to implement the mmWave MIMO system [13]. Here the digital filter is used at baseband for precoding or decoding and beamforming is done by the analog system. This system achieves high spectral efficiency and low hardware cost and less energy requirement. In [14], a novel algorithm has been developed which reduces the design complexity of the system and increases the spectral efficiency of different subcarriers. To manage energy consumption and hardware cost a user scheduling algorithm and resource allocation scheme has been proposed in [15], which will enhance the overall energy efficiency of the system. The hybrid structure allows a large antenna array to be connected in a small RF chain to provide reasonable array gain. To reduce the error between HP and digital baseband precoder alternate minimization algorithm was proposed [16]. A two-stage hybrid precoding multiuser MIMO system was designed in which the RF beamforming will perform in the first stage at the base station and user terminal to achieve maximum gain. And digital precoder with the ZF effect is applied in the second stage to reduce multiuser interference [17]. Based on the cluster-based channel model, some hybrid precoding technique was developed by combining local search technique and polarization matching in [18]. A hybrid precoding algorithm was developed in [19] to exploit the poor scattering nature of mmWave. An iterative precoding algorithm was developed in [20] where all precoding operations were done in the analog domain. In [21], the author assumes ful digital precoding at the transmitter side whereas on the receiver side with digital baseband RF beamforming has been employed. In [22], the author proposed a hybrid precoding sub-connected architecture with successive interference cancelation. MmWave MIMO system with hybrid precoding connected phase-shifting network exhibits power loss due to the presence of a huge number of the phase shifter, power divider, and combiner [23]. In [24], the author designs a HP and combiner with low complexity for both the downlink and uplink using multiuser mmWave MIMO systems.

Here, in this work, the authors have taken LLL algorithms for lattice reduction and proposed LP-based HP for mmWave massive MIMO (mMIMO) communication system. Finally, its performance has been compared with linear precoders like ZF and MMSE. Here, in this work, the authors have analyzed the performance of the LLR-MMSE and LLR-ZF-based HP for the mmWave mMIMO system.

The rest of this paper is organized as follows. The mathematical model related to the mMIMO system with hybrid precoding and mmWave channel model is presented in Sect. 2. All the analytical results are included in Sect. 3. In this section, the performance of different precoding schemes is presented in terms of energy efficiency. Finally, we have summarized the paper and presented it in Sect. 4.

2 System Model

2.1 mmWave MIMO System with Hybrid Precoding

Figure 1 shows a HP-aided massive mmWave MIMO System. In the above figure, the base station consists of a large number of antenna systems. Now, let the total number of antennas connected in the base station be N_T. In this model, the base station provides support for n number of mobile stations. We have designed this model in such a way that the mobile station has a single radio frequency chain and the base station consists of N_R radio frequency chain. The total number of antennas connected in mobile stations is N_M.

Here base station is communicating with each mobile station by a single radio stream and N_m represents the total number of streamed data transmitted from base stations. For, n number of users, base station can provide support which may be equal to or less than the total number of the radiofrequency chain ($n \leq N_R$) this can be achieved by proper implementations of the radio frequency chain.

The transmitted signal after baseband precoding and radio frequency precoding can be written as

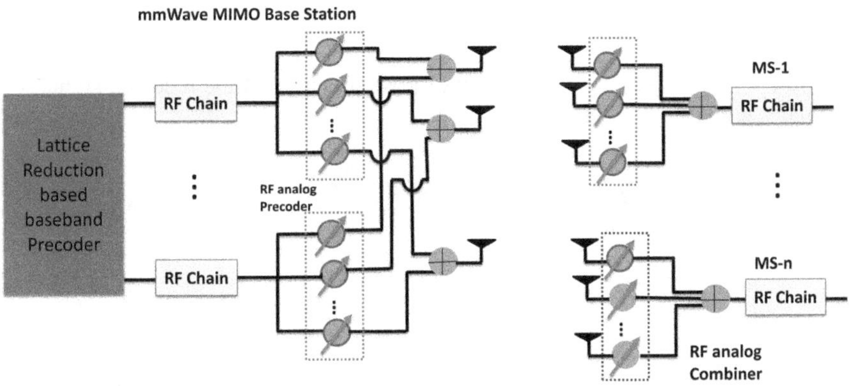

Fig. 1 mmWave MIMO System with Hybrid Precoding

$$y = \boldsymbol{P}_B \boldsymbol{P}_A \boldsymbol{m}$$

Where P_B denotes the baseband precoder of size $N_R \times N_m$ and P_A denotes the radio frequency analog precoder with the size of $N_T \times N_R$. The independent data stream which has been transmitted from the base station be m and having a size of $N_m \times 1$. Here we have assumed that the total power is equally distributed into an individual data stream which is given as $E\left[\boldsymbol{m}\boldsymbol{m}^K\right] = \frac{P_t}{M}\mathbf{I_n}$. Here P_t is the transmitted power. Due to noise or disturbance in the channel, some amount of fading will take place and the received signal at each mobile station after fading is given by

$$\boldsymbol{R}_j = \boldsymbol{K}_j \boldsymbol{P}_B \boldsymbol{P}_A \boldsymbol{m} + \boldsymbol{X}_j$$

Here K_j is the mmWave channel which is a correspondence between the path of mobile station and base stations, where K_j has the size of $N_T \times N_M$. Where X_j is the additive Gaussian noise which is corresponding to each channel.

The received signal at the mobile stations can be further simplified and can be written in the desired and interference component.

$$\boldsymbol{I}_j = \boldsymbol{K}_j \boldsymbol{P}_A \boldsymbol{f}_{Bj} \boldsymbol{m}_j + \boldsymbol{K}_j \sum_{i \neq j}^{n} \boldsymbol{P}_A \boldsymbol{f}_{Bi} \boldsymbol{m}_i + \boldsymbol{X}_i$$

Where $\boldsymbol{P}_A \boldsymbol{f}_{Bj}$ is the precoding metric which is corresponding to the nth the mobile station and $\boldsymbol{f}_{Bj}$ represents the column vector j of the baseband precoding metric $\boldsymbol{P}_B$. Here m_i represents the ith element of the symbol vector m. We already know that mobile stations usages only the radio frequency analog combiner, i.e., $\boldsymbol{W}_{Aj} = \boldsymbol{W}_j$. Now at the receiver side, the estimated symbols rate to each user is given as

$$\widehat{\boldsymbol{m}} = \boldsymbol{W}_j^K \boldsymbol{K}_j \boldsymbol{P}_A \boldsymbol{P}_B \boldsymbol{m} + \boldsymbol{W}_j^K \boldsymbol{X}_j$$

Where $(\cdot)^K$ represents the conjugate transpose. In terms of the received signal, the estimated signal can be written as

$$\widehat{\boldsymbol{m}}_j = \boldsymbol{W}_j^K \boldsymbol{K}_j \boldsymbol{P}_A \boldsymbol{f}_{Bj} \boldsymbol{m}_j + \boldsymbol{W}_j^K \boldsymbol{K}_j \sum_{i \neq j}^{n} \boldsymbol{P}_A \boldsymbol{f}_{Bi} \boldsymbol{m}_i + \boldsymbol{W}_j^K \boldsymbol{X}_j$$

2.2 mmWave Channel Model

In this paper, we have considered the geometric massive mmWave MIMO channel model, where we have assumed that there is L_j scattering associated with the jth mobile stations. The mathematical model mmWave channel can be expressed as

$$K_j = \sqrt{\frac{N_T N_M}{L_j}} \sum_{l=1}^{L_j} \alpha_{j,l}\, \beta_{MS}(\theta_{k,l}) \beta_{BS}^{K}(\phi_{j,l})$$

In the above equations, we have considered that scattering in the channel between the base station and mobile station. Where $\alpha_{j,l}$ represents the complex path gain between jth the mobile station and the base station. Here $(\theta_{k,l})$ represents the angle of arrival and $(\phi_{j,l})$ represents the angle of departure of the lth path. Here the angular range lies between $0 to 2\pi$. The antenna array response is represented by $\beta_{MS}(\theta_{k,l})$and$\beta_{BS}^{K}(\phi_{j,l})$, which is corresponding to the arrangement of the antenna in the mobile station and base station, respectively. And the array response of uniform linear array can be expressed as

$$\beta(\phi) = \frac{1}{N}\left[1, e^{j\frac{2\pi}{\lambda}\delta \sin\phi}, \ldots, e^{j(N-1)\frac{2\pi}{\lambda}\delta \sin\phi}\right]$$

Where δ is the distance between the antenna element and is the wavelength of signal used.

2.3 Digital Precoder Design

The effective channel between the base station and mobile station can be achieved by

$$K_{ej} = W_j^K K_j P_A$$

Now, the signal to interference ratio (SIR_l), which is corresponding to the lth user can be expressed as

$$SIR_l = \frac{\left(\frac{P_t}{n}\right)\left|K_{ej} f_{Bj}\right|^2}{\sum_{i \neq j}^{n} \left(\frac{P_t}{n}\right)\left|K_{ej} f_{Bj}\right|^2 + \varepsilon^2}$$

After calculating SIR_l, we can easily calculate the data rate between each user. The data rate between each user is given by

$$D_l = \log(1 + SIR_l)$$

The main motivation for using analog radio frequency precoders in the base station and mobile station is it increases the gain of an antenna array. We use radio frequency precoder P_A at the base station and a radio frequency combiner W_j at the mobile station and the baseband digital precoder P_B will help to enhance the overall system performance. In this system, a digital precoder has been designed based on the channel parameter ($K_{ef} = K_{e1}, \ldots, K_{el}$).

3 Results

This section represents the performance analysis of the mmWave mMIMO system. Here, we have considered 64 numbers antennas at the base station (BS) and 4 antennas at the user terminals. To create the multiuser scenario, we have considered 4 number of active user terminals for the simulation. To have a more realistic scenario, we have taken 10 multipath components in each cluster.

Figure 2 represents the spectral efficiency comparison between different HPs. As depicted, spectral efficiency is maximum in the case of a single user environment, where there is no interference effect. And fully digital precoder is much more spectral efficient in comparison to HP but a fully digital precoder is practically impossible for a mMIMO system as it is to associate a RF chain with each antenna element. By comparing the result, it is clear that out of HPs, LLR-MMSE based precoder performs much better than other precoders. Particularly, it is more effective in low SNR conditions. And also, it can be observed that variation in SNR has no effect on the spectral efficiency for analog beamsteering.

Figure 3 represents the energy efficiency comparison between different precoders. As presented, apart from the analog beamsteering, a fully digital precoder is the most energy inefficient precoder compare to HPs. Out of ZF, MMSE, LLR-ZF, and LLR-MMSE-based HP providers, LLR-MMSE based HP is the most energy efficient. As

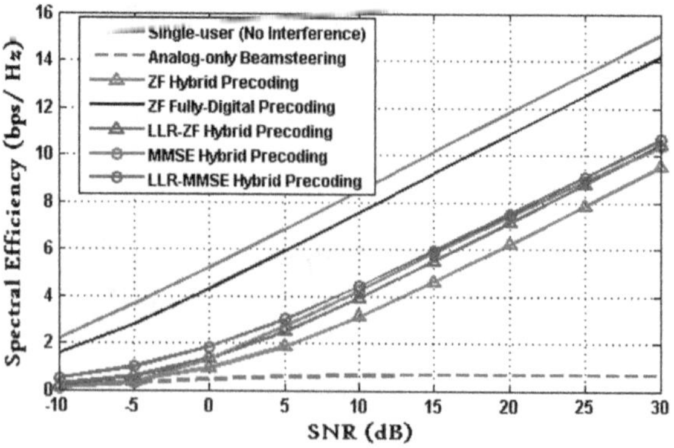

Fig. 2 Spectral efficiency comparison

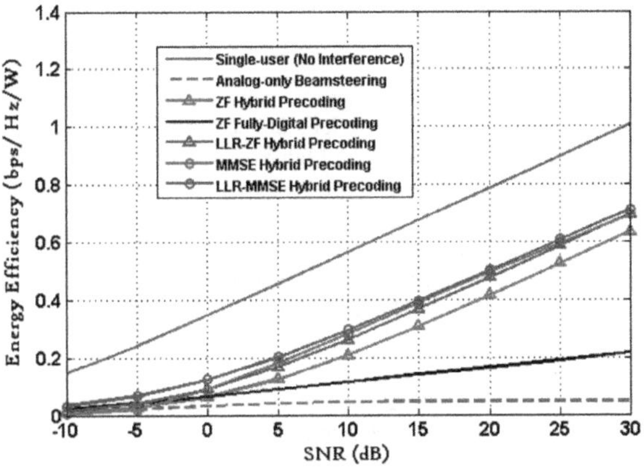

Fig. 3 Energy efficiency comparison

Table 1 Energy efficiency comparison between different precoders

Precoder	Energy Efficiency (bps/Hz/W)	
	At SBR = 0 dB	At SBR = 25 dB
Fully Digital Precoding	0.08958	0.1898
ZF- HP	0.0631	0.5241
LLR-ZF HP	0.09006	0.5854
MMSE	0.0895	0.5948
LLR-MMSE HP	0.1229	0.6065

presented in Table 1, LLR-MMSE HP is very effective under low SNR conditions and overall it outperforms the other precoders. As shown in Fig. 3, considering the energy efficiency of 0.4 bps/Hz/W, the SNR gap between LLR-MMSE HP and ZF HP is almost 5 dB and it is a significant improvement.

4 Conclusion

With the motivation of improving the energy efficiency of the mMIMO system, here, in this work, the authors have proposed LE-aided HPs. The proposed LR-aided HPs are much more energy efficient than the conventional ZF/MMSE precoders. Out of all precoders, LLR-MMSE-based HP outperforms all the HPs and particularly it is very much effective in low SNR conditions.

References

1. Wang C-X, Haider F, Gao X, You X-H, Yang Y, Yuan D, Aggoune H, Haas H, Fletcher S, Hepsaydir E (2014) Cellular architecture and key technologies for 5G wireless communication networks. IEEE Commun Mag 52(2):122–130
2. Andrews JG, Buzzi S, Choi W, Hanly SV, Lozano A, Soong ACK, Zhang JC (2014) What will 5G be? IEEE J Sel Areas Commun 32(6):1065–1082
3. Mark RG, Moody GB, Olson WH, Peterson PS, Schulter SK, Jr. Walters JB (1979) Real-time ambulatory arrhythmia analysis with a microcomputer. Comput Ardiol 57–62
4. Pi Z, Khan F (2011) An introduction to millimeter-wave mobile broadband systems. IEEE Commun Mag 49(6):101–107
5. Rappaport TS, Sun S, Mayzus R et al (2013) Millimeter wave mobile communications for 5 g cellular: it will work! IEEE Access 1:335–349
6. Akdeniz MR, Liu Y, Samimi MK et al (2014) Millimeter wave channel modeling and cellular capacity evaluation. IEEE J Sel Areas Commun 32(6):1164–1179
7. Roh W, Seol J, Park J et al (2014) Millimeter-wave beamforming as an enabling technology for 5 g cellular communications: theoretical feasibility and prototype results. IEEE Commun Mag 52(2):106–113
8. Constantine AB (2005) Antenna theory: analysis and design, 3rd ed. Wiley
9. Zhou F, Du M, Wang Y, Luo G (2016) Joint source-channel coding for band-limited backhauls in coordinated multi-point systems. IET Commun 10(13):1562–1570
10. Zhao R, Yuan Y, Fan L, He Y-C (2017) Secrecy performance analysis of cognitive decode-and-forward relay networks in nakagami-m fading channels. IEEE Trans Commun 65(2):549–563
11. Björnson E, Sanguinetti L, Hoydis J, Debbah M (2015) Optimal design of energy-efficient multi-user MIMO systems: is massive MIMO the answer? IEEE Trans Wirel Commun 6:3059–3075
12. Zhao N, Yu FR, Jin M, Yan Q, Leung VCM (2016) Interference alignment and its applications: a survey, research issues, and challenges. IEEE Commun Surv Tutor 18(3):1779–1803. (third quarter)
13. Heath RW (2016) Millimeter wave: the future of commercial wireless systems. In: Processing 2016 IEEE compound semiconductor integrated circuit symposium (CSICS), Austin, TX, USA, pp 1–4
14. Alkhateeb A, Ayach OE, Leus G, Heath RW (2013) Hybrid precoding for millimeter wave cellular systems with partial channel knowledge. In: 2013 information theory and applications workshop (ITA), San Diego, CA, USA, pp 1–5
15. Song Z, Zhang Z, Liu X, Liu Y, Fan L (2018) Simultaneous cooperative spectrum sensing and wireless power transfer in multi-antenna cognitive radio. Phys Commun 29:78–85
16. Zi R, Ge X, Tompson J, Wang C-X, Wang H, Han T (2016) Energy Efciency Optimization of 5G Radio Frequency Chain Systems. IEEE J Sel Areas Commun 34(4):758–771
17. Yu X, Shen J, Zhang J, Letaief KB (2016) Alternating minimization algorithms for hybrid precoding in millimeter wave MIMO systems. IEEE J Sel Top Sig Process 10(3):485–500
18. Alkhateeb A, Leus G, Heath RW (2015) Limited feedback hybrid precoding for multi-user millimeter wave systems. IEEE Trans Wirel Commun 14(11):6481–6494
19. Rusu C, Méndez-Rial R, González-Prelcic N, Jr. Heath RW (2015) Low complexity hybrid sparse precoding and combining in millimeter wave MIMO systems. In: Proceedings of the IEEE international conference on communication, London, UK, pp 1340–1345
20. Alkhateeb A, Ayach OE, Leus G, Heath RW Jr (2014) Channel estimation and hybrid precoding for millimeter wave cellular systems. IEEE J Sel Topics Signal Process 8(5):831–845
21. Ayach OE, Rajagopal S, Surra A, Pi SZ, Jr. Heath RW (2014) Spatially sparse precoding in millimeter wave MIMO systems. IEEE Trans Wireless Commun 13(3):1499–1513
22. Roth K, Nossek JA (2017) Achievable rate and energy efficiency of hybrid and digital beamforming receivers with low resolution ADC. IEEE J Sel Areas Commun 35(9):2056–2068

23. Gao X, Dai L, Han S, Chih-Lin I, Jr. Heath RW (2016) Energy-efficient hybrid analog and digital precoding for MmWave MIMO systems with large antenna arrays. IEEE J Sel Areas Commun 34(4)
24. Ribeiro LN, Schwarz S, Rupp M, de Almeida ALF (2018) Energy Efficiency of mmWave Massive MIMO precoding with low-resolution DACs. IEEE. https://doi.org/10.1109/jstsp.2018.2824762

Android-Based Mobile Application Framework to Increase Medication Adherence

Saibal Kumar Saha, Anindita Adhikary, Ajeya Jha, Vijay Kumar Mehta, and Tanushree Bose

Abstract Communication system has evolved enormously over the ages. With advancements, the scope of its integration with other fields has increased and these integrated systems may help to change the lives of people. The aim of this research is to design a framework using the process of communication, technology, and the theory of reinforcement to increase the medication adherence rate of patients. To accomplish the objective, an Android mobile application has been developed to capture the drug, diet, and exercise regime of the patients and store the same in a database. Based on the timing and instructions of the physician, reminders in the form of sound and text notifications would be sent to the patients to remind them of their medication regime. Tailored health tips will be received by the patient based on the usage habits of the application. A feedback system will help to graphically demonstrate the adherence behavior of the patient.

Keywords Communication · Health care · Medication adherence · Mobile application · Reminder · Drug · Diet · Exercise

1 Introduction

Communication plays a very important role in medication adherence among patients suffering from chronic diseases [1]. Several factors have been identified as causes of medication non-adherence: forgetfulness [2], beliefs [3], poor communication [4],

S. K. Saha (✉) · A. Adhikary · A. Jha
Department of Management Studies, Sikkim Manipal Institute of Technology–Sikkim Manipal University, Gangtok, Sikkim, India

V. K. Mehta
Department of Community Medicine, Sikkim Manipal Institute of Medical Sciences–Sikkim Manipal University, Gangtok, Sikkim, India

T. Bose
Department of Electronics and Communication Engineering, Sikkim Manipal Institute of Technology–Sikkim Manipal University, Gangtok, Sikkim, India
e-mail: tanushree.b@smit.smu.edu.in

S. Dhar et al. (eds.), *Advances in Communication, Devices and Networking*, Lecture Notes in Electrical Engineering 776,
https://doi.org/10.1007/978-981-16-2911-2_44

side effects [5], and depression [6] to name a few. Communicating with patients on a regular basis helps to develop a strong relationship with the patient and healthcare service provider and physician. Regular communication generates interest in patients, changes beliefs, and attitudes toward medication, side effects, treatment condition, disease criticality, and beliefs about his/her self-health. The theory of reinforcement [7] plays a vital role in changing beliefs and increases the odds of success. A well-amalgamated communication process with technology and reinforcement theory may help patients increase their medication adherence rate.

2 Literature Review

Adherence has been defined as the "active, voluntary, and collaborative involvement of the patient in a mutually acceptable course of behavior to produce a therapeutic result" [8, 9]. Following the medication, regime is a vital part of patient care and essential for attaining clinical goals. According to the medication adherence report published in 2003 by WHO, "increasing the effectiveness of adherence interventions may have a far greater impact on the health of the population than any improvement in specific medical treatment" [10]. Non-adherence to prescribed medication is common limiting the treatment process [11]. Studies report that nearly 50–60% of patients suffering from chronic diseases are non-adherent and fail to comply with the medication regime prescribed by their physicians [12, 13]. Studies also report that nearly 30% of medicine-related hospital admissions are because of medication non-adherence [14, 15].

Some of the studies have classified medication adherence as primary (initiation of pharmacotherapy) or secondary (implementation of the prescribed regime). The term primary non-adherence is given when the patient fails to fill prescriptions while initiating a new medication [16]. The term secondary non-adherence is given when the patient fails to take the medication as prescribed after the prescriptions are filled. This form of non-adherence is associated not only with poor health outcomes but also adds to the financial burden [17]. The concern for medication non-adherence has increased and has spread from clinicians, healthcare providers, physicians to payers. Medication non-adherence is associated with increased financial costs [18], disturbance in the family [19] and work life [20], depression, deterioration of patient health, and even death [21]. With the evolution of technology, a number of gadgets and techniques have been invented and used for increasing the medication adherence rate of patients [22]. Most of the electronic gadgets are costly and do not subsequently increase the adherence rates [23]. A number of mobile applications have also been launched in the market, but are limited in the features which are mainly confined to the timely reminder of the medicines.

From the literature, it is evident that there is no proper consensus about the right technique for increasing the medication adherence rate of patients. Although the rate of increase in adherence has been reported in most of the studies many suggest that there is a scope for improvement in the process with the use of technology.

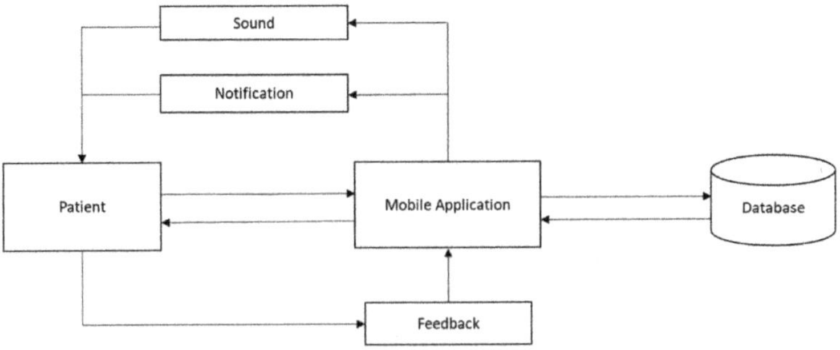

Fig. 1 Block diagram

3 Objective

The main objective of this study is to design a framework using the process of communication, technology, and the theory of reinforcement to increase medication adherence rates among patients suffering from chronic diseases.

4 Methodology

In order to accomplish the objectives of this research, a mobile application was developed using Android Studio and Java. The front end of the application consists of four interfaces: (a) main activity page (b) add reminder page (c) edit reminder page, and (d) health report page. Figure 1 shows the block diagram of the application which consists of the patient, mobile application, database, feedback mechanism, reminder in the form of sound, and notification outside the environment of the application.

Figure 2 depicts the flowchart of the reminder system. The patient, healthcare service provider, or the family member sets the drug, diet, and exercise details of the patient in the mobile application. The records get stored in the database. Reminder in the form of sound and notification is generated and notified to the patient. Based on the compliance status, the patient gets a provision to share the feedback which gets recorded in the database. With the help of this feedback data, the mobile application generates the adherence report for the patient.

5 Results and Discussion

The patient medication reminder mobile application consists of four pages: (a) main activity page (b) add reminder page (c) edit reminder page and (d) health report page. Figure 3a shows the blank main activity page which consists of upcoming reminders

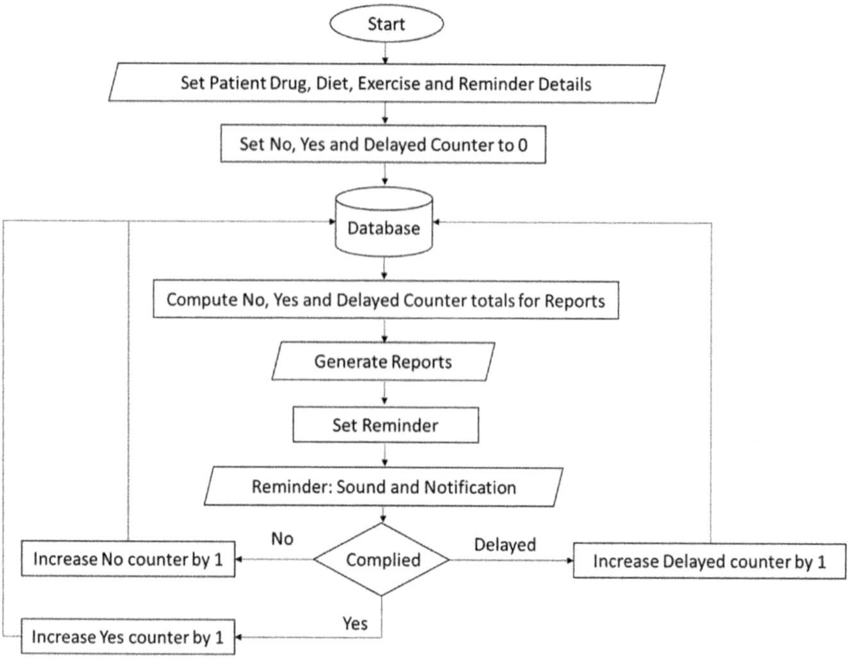

Fig. 2 Flowchart of alarm reminder system

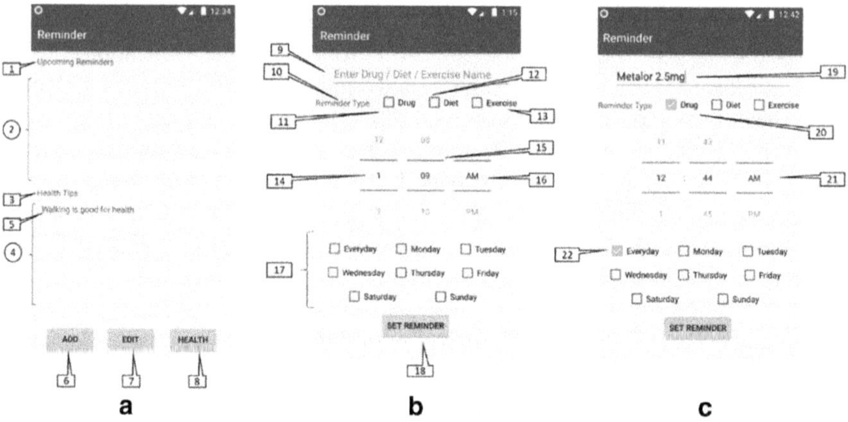

Fig. 3 Mobile application outputs **a** Blank main activity page. **b** Add reminder page **c** Add reminder page with data

(1), the section where all upcoming reminders will be shown (2). The second part of the page consists of health tips (3) and a section where health tips will be shown (4). A sample health tip (5) for the day is shown in figuers (6), (7), and (8) are the buttons for adding reminder, editing reminder, and viewing the adherence report of the patient. All these buttons lead to new pages.

Figure 3b shows the add reminder page. (9) is the field where drug, diet, or exercise name is to be entered. There are three types of reminder (10) drug (11), diet (12), and exercise (13). Time of the reminder in terms of hour (14), minutes (15), and AM or PM (16). The functionality frequency (17) of reminders helps in generating reminders for multiple days based on the instruction of the physician. A reminder is set by pressing the set reminder button (18).

Figure 3c shows a filled reminder page where a medicine name Metalor 2.5 mg (19), reminder type Drug (20), time of 12:44 am (21) and frequency of every day (22) is set.

When the reminder is set the application returns to the main activity page as depicted in Fig. 4d. The medicine name (23) appears and a small notification “Alarm is set” (24) is displayed for few seconds. Figure 4e shows that at the set time of 12:44 am (26) the remainder is triggered and a notification is displayed outside the environment of the application (25). A sound is also generated to notify the patient.

Figure 5f shows notification (27) outside the environment of the application. Three buttons Yes (28) No (29) and Delayed (30) are shown along with the notification. The patient can share the feedback whether he/she has complied with the medication on time or not or if it has been delayed. Figure 5g shows the edit page of the reminder. Instruction to delete the reminder is shown (31). The page also shows two sample reminders (33). The back button (32) navigates the page back to the main activity

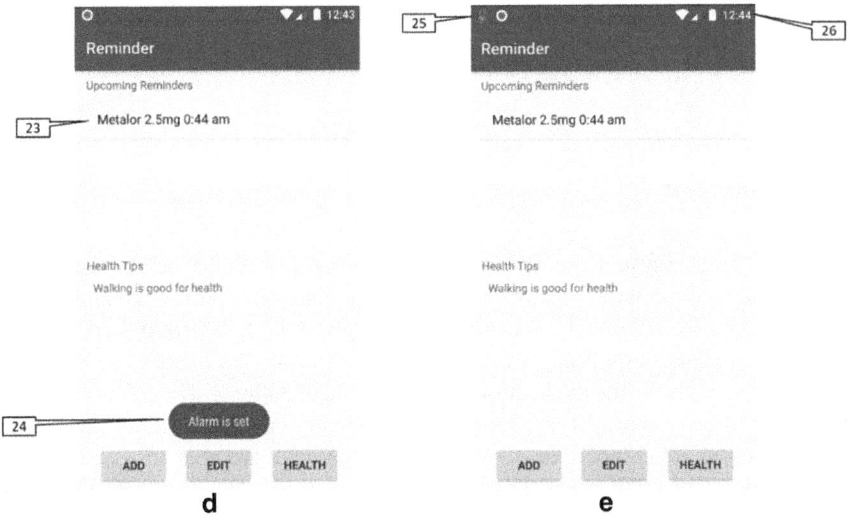

Fig. 4 Mobile application outputs **d** Filled main activity page. **e** Notification Generation

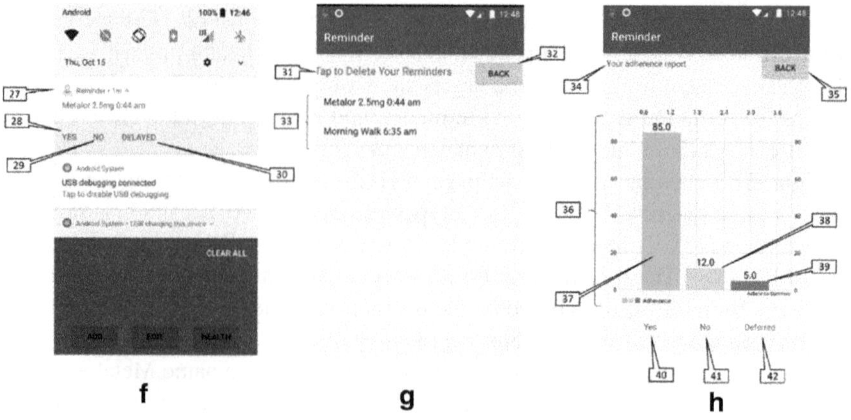

Fig. 5 Mobile application outputs **f** Notification **g** Edit page **h** Health report page

page. Figure 5h shows the medication adherence report (34) of the patient. (36) represents the area of the graph. (37) represents positive compliance, yes (40). (38) represents non-compliance, no (41) and (39) represent deferred (42). The back button (35) takes the user back to the main activity page.

Reports of medication adherence will be generated based on the set timing of the reminder and the timing of its compliance. The cumulative timing of the difference between the set timing and the actual timing will be computed using an algorithm and adherence scores will be given upon fulfillment of 80% and above criteria [24]. The scores of various events will be captured by the system and showed in the form of graphs. The visual appeal with different color combinations will engage the patient and provide encouragement and motivation to increase the score. Based on the type of reminder set, i.e., drug, diet, or exercise, health tips will be given with the aim to motivate the patient and increase adherence.

6 Conclusion

For the medication system to work properly it is very important that patients accept the disease and recommended treatment [25]. They should have knowledge and faith in the treatment, interaction with the clinicians and physicians, and follow a routinized therapy [26]. This study proposes a mobile application based medication reminder framework to increase the medication adherence rate of patients. The framework uses the theory of reinforcement, process of communication and technology to increase the medication adherence rate of patients. The patients are able to set reminders for three types of medication: Drug, Diet, and Exercise. They are able to set the time and frequency of occurrence. Patients also get an option to delete the reminder and share their feedback. The framework triggers tailored health related information that a patient should read and follow. In the proposed framework, the theory of reinforcement is not just confined to the set time of reminder for medicine or diet or exercise but it also takes care of the benefits, risks, care for self and others, motivation and self-control, which is sent to the patient from time to time in the form of health tips. This will motive the patient to open the mobile application and make it a part of their life, work with it, enjoy with it, and most importantly get benefitted from it.

The research is limited to Android application and the usage of few instructions from the patient. Reasons for missing the reminder/dose/diet or exercise are not captured. Future research can be done with attitude, beliefs, social and financial costs of non-adherence and could be integrated into the mobile application to study its effect on adherence.

References

1. Schoenthaler A, Chaplin WF, Allegrante JP, Fernandez S, Diaz-Gloster M, Tobin JN, Ogedegbe G (2009) Provider communication effects medication adherence in hypertensive African Americans. Patient Educ Counst 75(2):185–191. https://doi.org/10.1016/j.pec.2008.09.018
2. Judson MA (2020) Causes of poor medication adherence in sarcoidosis: poor patient-doctor communication and suboptimal drug regimens. Chest 158(1):17–18
3. Salama HM, Saudi RA (2020) Effect of patients beliefs about medications on adherence to drugs in diabetic patients attending family medicine outpatient clinic in Ismailia, Egypt. J Diabetes Metab Disord 1–8. https://doi.org/10.1007/s40200-020-00587-0
4. Toelle BG, Marks GB, Dunn SM (2020) Psychological and medical characteristics associated with non-adherence to prescribed daily inhaled corticosteroid. J Pers Med 10(3):126. https://doi.org/10.3390/jpm10030126
5. Ross XS, Gunn KM, Suppiah V, Patterson P, Olver I (2020) A review of factors influencing non-adherence to oral antineoplastic drugs. Supportive Care Cancer 28(9): 4043–4050. Springer. https://doi.org/10.1007/s00520-020-05469-y
6. Gu D, Shen C (2020) Cost-related medication nonadherence and cost-reduction strategies among elderly cancer survivors with self-reported symptoms of depression. Popul Health Manag 23(2):132–139. https://doi.org/10.1089/pop.2019.0035
7. Grossberg S (1987) A psychophysiological theory of reinforcement, drive, motivation, and attention. Adv Psychol 42(C):3–81. https://doi.org/10.1016/s0166-4115(08)60905-x
8. Meichenbaum D, Turk DC (1987) Facilitating treatment adherence: a practitioner's guidebook. Plenum Press
9. Delamater AM (2006) Improving patient adherence. Clinical Diabete Am Diabetes Assoc 24(2):71–77. https://doi.org/10.2337/diaclin.24.2.71
10. Sabaté E, Sabaté E (2003) others: adherence to long-term therapies: evidence for action. World Health Organizat
11. Fischer MA, Choudhry NK, Brill G, Avorn J, Schneeweiss S, Hutchins D, Liberman JN, Troyen AB, William HS (2011) Trouble getting started: predictors of primary medication nonadherence. Am J Med 124(11):1081.e9–1081.e22. https://doi.org/10.1016/j.amjmed.2011.05.028
12. Lavsa SM, Holzworth A, Ansani NT (2011) Selection of a validated scale for measuring medication adherence. J Am Pharm Assoc 51(1):90–94. https://doi.org/10.1331/japha.2011.09154
13. Svarstad BL, Chewning BA, Sleath BL, Claesson C (1999) The brief medication questionnaire: a tool for screening patient adherence and barriers to adherence. Patient Educ Counst 37(2):113–124. https://doi.org/10.1016/S0738-3991(98)00107-4

14. Osterberg L, Blaschke T (2005) Adherence to medication. N Engl J Med 353(5):487–497
15. McDonnell PJ, Jacobs MR, Monsanto HA, Kaiser JM (2002) Hospital admissions resulting from preventable adverse drug reactions. Ann Pharm 36(9):1331–1336. https://doi.org/10.1345/aph.1A333
16. Fischer MA, Margaret RS, Joyce L, Christine V, William HS, Brookhart MA, Joel SW (2010) Primary medication non-adherence: analysis of 195,930 electronic prescriptions. J Gen Intern Med 25(4):284–290. https://doi.org/10.1007/s11606-010-1253-9
17. Solomon MD, Majumdar SR (2010) Primary non-adherence of medications: lifting the veil on prescription-filling behaviors. Springer
18. Newman-Casey PA, Salman M, Lee PP, Gatwood JD (2020) Cost-utility analysis of glaucoma medication adherence. Ophthalmology 127(5):589–598. https://doi.org/10.1016/j.ophtha.2019.09.041
19. Keyser HH, De Ramsey R, Federico MJ (2020) They just don't take their medicines: reframing medication adherence in asthma from frustration to opportunity. Pediatr Pulmonol 55(3):818–825. https://doi.org/10.1002/ppul.24643
20. Desalegn D, Girma S, Abdeta T (2020) Quality of life and its association with current substance use, medication non-adherence and clinical factors of people with schizophrenia in Southwest Ethiopia: a hospital-based cross-sectional study. Health Qual Life Outcomes 18:1–9
21. Gerard R (2020) telehealth is tele-easy: a telehealth modality to improve medication adherence in older adults. University of San Diego
22. Mata J et al (2020) A mobile device application (app) to improve adherence to an enhanced recovery program for colorectal surgery: a randomized controlled trial. Surg Endosc 34(2):742–751. https://doi.org/10.1007/s00464-019-06823-w
23. Arulprakasam KC, Senthilkumar N (2020) To evaluate the impact of patient education on self-reported adherence, and management behavior of children with asthma. Int J Res Pharm Sci 11(1):581–588
24. Pietrzykowski Ł et al (2020) Medication adherence and its determinants in patients after myocardial infarction. Sci Rep 10(1):12028. https://doi.org/10.1038/s41598-020-68915-1
25. Elger BS, Harding TW (2002) Terminally ill patients and Jehovah's Witnesses: teaching acceptance of patients' refusals of vital treatments. Med Educ 36(5):479–488. https://doi.org/10.1046/j.1365-2923.2002.01189.x
26. George J, Kong DCM, Thoman R, Stewart K (2005) Factors associated with medication nonadherence in patients with COPD. Chest 128(5):3198–3204. https://doi.org/10.1378/chest.128.5.3198

Cross-Layer Optimization Aspects of MANETs for QoS-Sensitive IoT Applications

Nadine Hasan, Ayaskanta Mishra, and Arun Kumar Ray

Abstract Internet of things has introduced various network classifications that aim to achieve seamless connectivity. Flying Ad hoc Networks (FANETs), Vehicular Ad hoc Networks (VANETs), Wireless Sensors Networks (WSN), Airborne Ad hoc Networks (AANETs), and Mobile Ad hoc Networks (MANETs) are prototypes for future networking and IoT Architecture. This paper aims to study the adaptability and mobility of MANETs for different applications, along with the working mechanism of the cross-layer optimization concepts for Quality of Service (QoS) sensitive Internet of Things (IoT) application framework.

Keywords Mobile ad hoc network (MANETs) · Cross-layer optimization. quality of service (QoS) · IoT · Flying ad hoc networks (FANETs) · Vehicular ad hoc networks (VANETs) · Wireless sensors networks (WSN) · Airborne ad hoc networks (AANETs)

1 Introduction

IoT presents the future as a platform to be connected to everything with the internet. Things refer to smart physical objects that can be connected to the Internet to process, access, and monitor information. The information that is collected can be stored for future use to give suggestions and take important decisions. IoT represents the expansive field that allows many researchers to implant their ideas and contribute to the production of what drives technology. IoT comprises various technologies such as Wireless Sensor Networks (WSN), Mobile Ad Hoc Networks (MANETs), and Vehicular Area Networks (VANETs). MANETs are a group of nodes, self-organized, free to move without any restriction which forms a decentralized architecture. MANETs are a sub-classification of the wide range of wireless networks. Mobility, Availability, Networked, and Efficient Terminals are combined under the terminology of

N. Hasan · A. Mishra (✉) · A. K. Ray
School of Electronics Engineering, Kalinga Institute of Industrial Technology, Deemed to Be University, Bhubaneswar 751024, India
e-mail: ayaskanta.mishrafet@kiit.ac.in

S. Dhar et al. (eds.), *Advances in Communication, Devices and Networking*, Lecture Notes in Electrical Engineering 776,
https://doi.org/10.1007/978-981-16-2911-2_45

MANETs. This wireless network has emerged toward enabling various services to be available over the Internet seamlessly. Mobile nodes in MANETs act as routers and transceivers to forward the data packets and support mobility management. The wireless nodes move freely at different mobility models and in different directions as per a specific scenario requirement of a related application. This network model is one of the most preferable architecture suggested for many IoT applications. It will be a challenge to manage the network topology to provide the best routing between nodes and achieve a good Quality-of-Service (QoS) with better connectivity. The working mechanism of MANETs depends on the TCP/IP model. Routing protocols work to manage the shortest path to route and forward data packets between nodes and ensure the reliable forwarding mechanism of the data packets. They are set to increase the speed and the reliability of packet delivery in the considered network (wired or wireless). MAC layer is responsible for selecting the wireless medium to transfer the data to the allocated bandwidth. To mitigate the issues related to this protocol stack, a new term has been proposed within this stack known as a cross-layer. The design of the cross-layer gives rise to many solutions, to improve and increase the overall performance of the networks. The cross-layer is used to process different predominant technical issues. Implementation of a cross-layer in MANETs can come with benefits such as follows:

- Manage the transmission process by caching the link information and sensing the status of the nodes and save the nodes in the transmission state in the cache so that it increases the Packet Delivery Rate (PDR).
- AODV-SPF [1] mechanism is proposed to avoid a collision in the MAC layer to improve the performance of the network layer.
- Managing energy consumption in the communication process between the nodes.
- Control the number of hops to set the link within the shortest route in the network between two nodes. Figure 1 gives a provision of MANETs and some deployment domains.

The paper is constructed as follows: Sect. 2 describes related works, Sect. 3 provides the MANETs developing domain and the two layers of the protocol stack, Sect. 4 describes the general concept of the cross-layer, Sect. 5 describes the utilization of cross-layer models in MANETs, their applications, and Sect. 6 provides result analysis, and Sect. 7 gives the conclusion.

2 Related Works

Early research focused on evaluating and comparing the performance of several routing protocols in different working conditions and simulation environments using simulator tests. Many simulators were developed for MANETs simulation, the most used were NS2 in its versions, NS3 which was developed from NS2, OPNET, OMNET ++, and GloMOsim. Also, some real environments were performed to

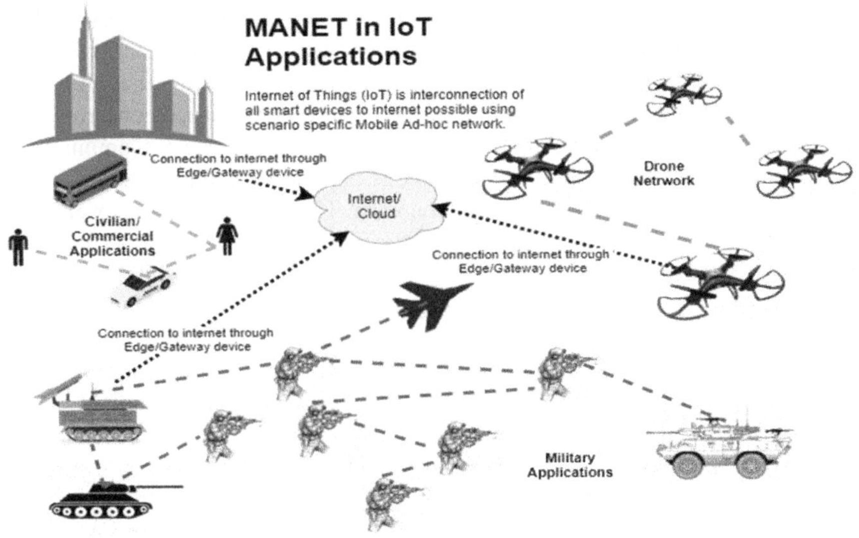

Fig. 1 Mobile ad hoc network architecture

test the performance level of the network. Evaluating MANETs has been done by comparing the parameters in various proposed circumstances and with different algorithms. Throughput, End-to-End Delay, Power consumption, Packet Delivery Ratio (PDR), Packet loss Ratio, and link stability are examples of these parameters. Most papers focused on parameters related to the first three layers. This is due to the major role these layers play in initiating, protecting, directing, and tracking the connection in MANETs as the same as controlling the power consumption. For meeting different goals in simulations and real test beds, many variables and algorithms were used. The evolution requirements, the seek for more flexibility, level up the QoS requirements [2], and adapt the requirements of high data applications have contributed to the cross-layer proposal. An improvement in work results has been observed in the routing protocol performance and the other parameters from different layers when applying the cross-layer concept. Flexibility and robustness for multimedia applications in MANETs were the concern in [2]. The authors performed some modifications on AODV, distance calculation between the nodes was done. A base station has been included within MANETs nodes. The adaptive cross-layer used with the implementation of MAODV has improved network performance. To reduce energy transmission and extend the network lifetime, a cooperative communication algorithm coupled with a cross-layer among the nodes [3]. Depending on the MAC layer protocol DEL-CMAC, AODV as the routing protocol, and using spatial diversity between nodes, the authors presented a cross-layer cooperative diversity-aware routing algorithm. In the simulation provided, this implementation has led to prolong the network life span comparing with other MAC protocols DEL-CMAC and DCF without cross-layer. In [4], the authors implemented a cross-layer scheme between routing protocols and

IEEE 802.11 DCF after modifying the later protocol. Based on the LEMO algorithm, the authors studied the effect of the cross-layer using AODV and DSR in multihop MANETs. A comparison between the two gave an increase in DSR performance that cross-layer achieved more than in AODV performance concerning the performance metrics used. While in [5], the authors worked on the optimization techniques to build the cross-layer that exploits the information in the physical and MAC layer and deploys it in the application layer. The simulation results were used to compare both real-time and best-effort traffic in the applications layer. The results have shown a clear and good improvement in overall network performance in delay reduction, throughput, and energy efficiency. For optimal service selection and direction, a cross-layer model in MANETs was suggested [6]. The authors compared the routing protocols such as DSR and DSDV rendition implemented with the cross-layer approach and their performance in the absence of a cross-layer utilized on the Server Location Protocol (SLP). They have proved that this approach increased the network utilization for providing fast service selection and periodic updating for redirection to the closest server when it's available. Throughput was approved for this comparison. An assessment of the cross-layer routing protocols designed for FANE was reported in [7]. The report provided an evaluation of transport layer protocols in the presence of the cross-layer approach. Different routing metrics and different propagation models were deployed in the simulation. The authors concluded through their simulation that TCP has outperformed the UDP, and this result is due to its reliability and retransmission process for the lost packets that TCP performs. All the experiments concluded have reported a satisfying performance with the presence of the cross-layer implemented for different goals. But for the security issue, there was no deep investigation among the presented works. Performing the handoff in MANETs with the help of the power wave received from the neighbor nodes is possible [8]. Depending on the power wave (the signal quality received from the neighbor node), the sending node can switch to another node before waiting for any link breakage. Anyway, this was done in very limited network size. Therefore, applying handoff in MANETs with high no. of nodes can be provisioned for stable connectivity by increasing the no. of nodes in the network. Evaluating the routing protocols (AODV, DSR, DSDV, TORA, and OLSR) performance was conducted in a variety of simulation environments without deploying any cross-layer approach. Different propagation models, different network sizes, different mobility models, variable number of hops, and different data metrics were used to compare and evaluate the previously mentioned protocols. The [9–12] protocols showed different performance levels in different network conditions. No stable and static characteristic was shown for any protocol included in different situations. Each protocol has outperformed the others for specific conditions.

3 MANETs Developing Domain (Research in MANETs)

Mobile Ad Hoc Networks, MANETs, are based on the wireless connection among many devices where wired medium doesn't exist. In the case of wireless connection, the design requires precision and consideration of different variables when using TCP/IP protocol stack. The key point of MANET architecture highly depends on the first layers. These layers will be responsible for allocating, managing, controlling the route of the link in the wireless channel, arranging the proper description for the data to be transmitted, facing and solving the problems that might intercept connection stability. MANETS acts as a self-control network and the expected system is to achieve connection with the Internet among the mobile nodes and to the other types of networks. The design of the network must be adapted dynamically with the mobility of the nodes. Different wireless parameters, i.e., availability of the wireless channel, stability of the link, different propagation models, the ability to shift between the intermediate nodes when both the source and the destination change their coordination will affect the connection placement. Each layer in the TCP/IP stack has its main privilege as well as its deficiencies when working in different networks. This imposes taking many aspects when designing MANETs. Different propagation models and disturbing factors such as fading and interference affect the stability and the reliability of the link. These considerations form the cross-layer for the overall network enhancement principle and elevate the level of their performance. Some of the parameters in protocol stack layers are provided in Fig. 2.

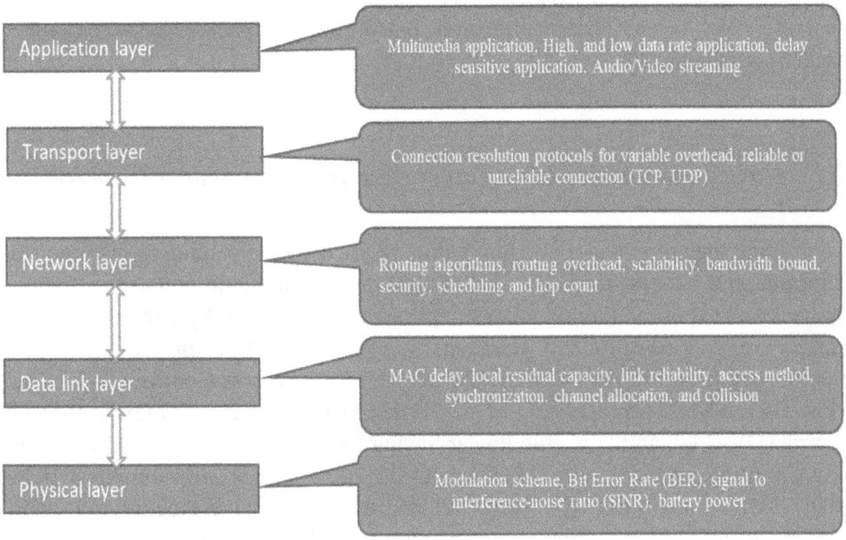

Fig. 2 Parameters of protocol stack layers

3.1 Network Layer and Routing Protocols

Network layer provides routing protocols, routing algorithms, scalability, security, and scheduling. As mobility is a good merit in MANETs, achieving routing to provide good QoS, flexibility, and security is a high requirement. Routing protocols in MANETS are classified into three types based on the method of creating the route of the link: Table Driven (Proactive), Source Initialized (Reactive or on-demand), and Hybrid protocols. The reactive protocols differ from proactive protocols in the way of establishing the link to direct the communication and differ in the bandwidth consumed. While the hybrid protocols combine the characteristics of the two types and operate according to parameters operate in the network. Some proactive protocols, OLSR, and DSDV. AODV and DSR are reactive protocols. ZRP and ZHLS are hybrid protocols. From the energy aspect, the dynamic mobility of the power constraint nodes makes the energy reservation an important topic in MANETS. Hence, energy-aware protocols are another classification of protocols [13] in both routing and MAC protocols. At the routing level, most of the energy is consumed by the retransmission of RREQ messages that caused flood over the network. Some of the energy-aware routing protocols are LBPC (Location-Based Power conservation) that use GPS for calculating the distance between the nodes to select the best route. SPAN is an energy-efficient coordination algorithm for topology maintenance. This algorithm depends on periodic instances to activate and deactivate the nodes when required. These instances are known as Stay awake and Sleep periods distributed among the nodes and depend on the adjacency between the nodes. EELAR (Energy Efficient Location Aided Routing) divides the network into small sectors depends on a reference node as a base station and controls both the packet floods and reduces packet over the head.

3.2 Data Link Layer: Link Layer and Medium Access Control

Controlling the errors due to the link failure or breakage, allocating the channel, and determining the access method to the wireless medium to set the connection between the nodes. At this level, energy consumption can be reduced by controlling errors, link failure, and the number of retransmissions. Some of MAC protocol that serves energy-saving Traffic-Aware Multi-Channel Medium Access Control (TAMMAC), Flow-Driven MAC protocol (FD-MAC), Dynamic Channel Assignment with Power control (DSAPC), and Dynamic Power Saving Mechanism (DPSM). Other energy-aware routing protocols are explained in [13]. MAC layer allocates the wireless channel which determines the bandwidth and the packet delay. According to IEEE, MAC layer 802.11 $\times$ can be operated either by [14] Distributed Coordination Function (DCF) or Point Coordination Function (PCF). These two models have been developed into enhanced models (EDCF and EPCF). DCF operates on CSMA/CA by which the sender listens to the channel and waits until the busy time of the channel ends. Distributed coordination function Inter-Frame Spacing (DIFS) is the time to wait after detecting a free channel. Then exchanging Request and Clear to Send (RTS/CTS) to set the initial transmission process. PCF depends on point coordination which manages and controls the access of the nodes to the channel. This is done by synchronizing the access with the help of local times of the nodes using Target Beacon Transmission Time (TBTT).

4 Cross-Layer General Concept

The importance of the MAC layer comes from its role to allocate the wireless channel which in turn affects the determination of the bandwidth and the packet delay in the transmitter. This helps the routing layer to select the wireless link to the destination. A cross-layer is proposed to simplify and mitigate the problems related to these layers and other different layers in the TCP/IP model and to provide a seamless connection in wireless networks as the same as between layers of the TCP/IP stack. The cross-layer is set to enhance the overall performance of the network such as minimizing error rate, node-to-node delay, and maximizing the throughput and reliability. Supporting the variety of applications starting from simple file transfer, ending to end-to-end delay-sensitive applications such as real-time video streaming. Security, quality of service (QoS), and mobility issues are also considered as cross-layer goals in wireless networks. For security issues in multi-hop wireless networks, the cross-layer achieves less routing overhead and reduces acknowledgment packets. It improves the QoS by overcoming the different shortness related to the routing protocols and MAC protocols used. This leads to the effective usage of the bandwidth available and good allocation of the channels. To provide stable and continuous communication and to avoid the side effects of node mobility, the cross-layer has been introduced many different solutions to overcome these problems. The working mechanism of the cross-layer is classified based on the area working in [15] Cross-layer at the node level communication and cross-layer at layers connection level where again each category is classified as shown in Fig. 3.

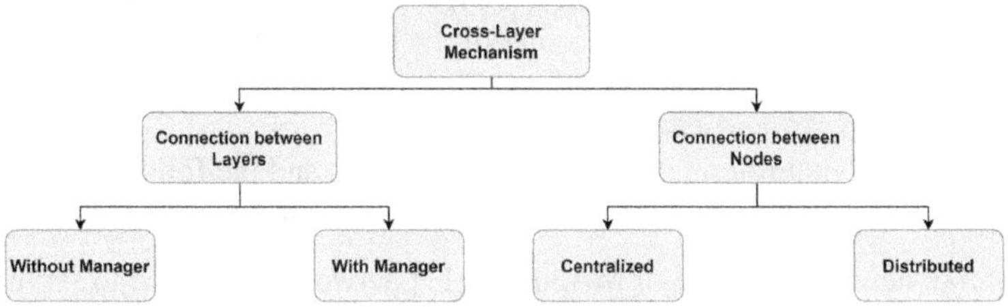

Fig. 3 Cross-layer classification

5 Utilization of Cross-Layer Models in MANETs and Applications

The align architecture supported by protocol stack, a water flow like architecture, is static and each layer has a fixed data format that passes to a higher or a lower layer. The optimization of the cross-layer overcomes different problems related to the static mechanism. The implementation of the cross-layer represents the managerial role of the TCP/IP stack implemented in MANETs. This provides integrity between the different functions of the five layers. New algorithms and protocols have to be implemented, or some modifications in the conventional protocols have to be done through cross-layer to reduce some of these problems.

The protocols of the cross-layer are proposed to support different parameters from multiple layers. The connection within the layers can be done using logical tunnels or interfaces connecting these layers under one target when needed. It is [7] possible to keep all the related parameters saved on cross-layer cache. The cache can be used as a database for future use by changing the variables according to deployment conditions. Each supposed model needs to satisfy different goals. Some of the goals to be satisfied are the following:

- Manage resource allocation [16], provide stability for link connectivity in the presence of dynamic changes in the network topology, the limited availability of ISM bandwidth, and to mitigate different propagation problems, and contribute to the reliability of the links.
- Providing a scheduling policy [16] to control the congestion issue. Detecting the discontinuity of the link by tracking the received signal strength of the packet in the physical layer. Based on the threshold, a new route discovery is implemented to maintain the connectivity between different connecting nodes.
- In the case of broadcasting, until the route is set with the reverse link between source and destination, there will be a loss of time, flooding many nodes that are not included in setting the supposed path. As a result, these excluded nodes will cause delays and more energy consumption as they are not included in the selected path/s. Load balancing can be optimized. Based on the session established between the layers, the required parameters to link between the nodes can be obtained. Algorithms to determine the type of coordination, which layers will be joined together, and how this joint will be highly required for any goal to be achieved. A provision of the cross-layer is shown in Fig. 4.

MANETs play a vital role in providing these services and the followings are some of MANETs implementation and application that work in these networks.

- Tactician networks are one of MANETS [17] networks improved to serve under mission-critical and latency-sensitive applications. Tactician networks depend on cognitive radio identification and Software-Defined Radio (SDR).
- Flying MANETS (FANETs) is the swarm of drones based on MANETS design and architecture. There is no grid in the mobility of the nodes, the design here

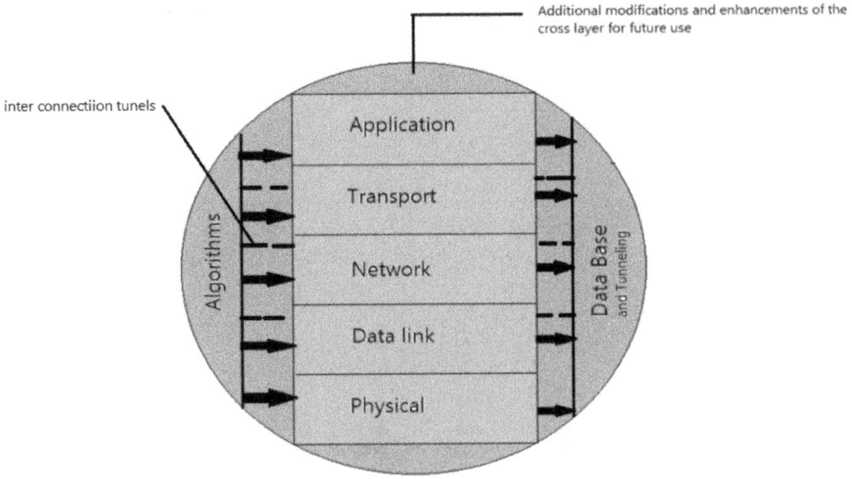

Fig. 4 A cross-layer design for MANETs

is built to work in the three-dimensional space. Drones in FANETs use wireless dongles supplied by 802.11 functionality that perform highly for different situations.

- Airborne Ad hoc Network (AANETs)are an extension of MANETs networks [18] that have been developed to include many applications in space as well.

Some **features of MANETs can be summarized as follows:**

- Research studies in tracking different types of wildlife and different health care domains [19].
- Flexibility to manipulate the emerged disaster such as floods, rescue processes, and sensitive applications such as military applications (tactician networks).
- In wireless networks applications such as metropolitan communication, broadband and internet services can be utilized using MANETs- based cross-layer.
- The dynamic and simultaneous infrastructure of MANETs gives these networks the importance to play a role as an intermediate link between servers and agents for different application services [6].

6 Analysis

A comparative analysis shows LAODV and LDSR: the LEss remaining hop More Opportunity (LEMO) algorithm based cross-layer optimization with IEEE 802.11 DCF outperforms regular AODV and DSR both in (a) Packet delivery ratio (PDR) % and (b) average end-to-end latency. Table 1 shows the PDR% and latency for with

Table 1 LEMO algorithm based cross-layer optimization QoS parameters

QoS Parameters	Packet delivery ratio (PDR %)				Latency (s)			
Scenario-1	AODV	LAODV	LDSR	DSR	AODV	LAODV	LDSR	DSR
Avg. value with varying speeds (0-35 m/s) with 20 nodes	14.86	17.46	9.98	9.59	5.23	5.04	47.88	50.25
QoS Parameters	Packet delivery ratio (PDR %)				Latency (s)			
Scenario-2	AODV	LAODV	LDSR	DSR	AODV	LAODV	LDSR	DSR
Avg. value with varying speeds (0–35 m/s) with 25 nodes	20.38	20.68	11.11	10.63	3.81	3.73	38.88	44.51

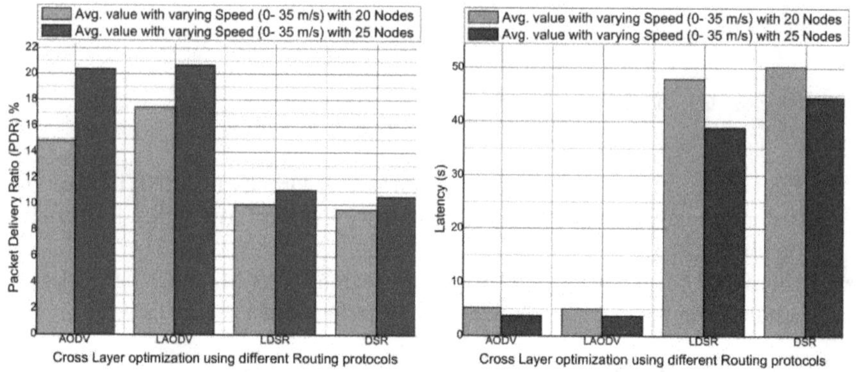

(a) (b)

Fig. 5 Comparative performance LEMO algorithm based cross-layer optimization. QoS parameters **a** Packet Delivery Ratio (PDR %) **b** Latency (s)

cross-layer LEMO variants (LAODV and LDSR) and without cross-layer variants of AODV and DSR implemented over two scenarios.

Figure 5 shows the QoS parameters (a) PDR % and (b) latency for two scenarios with 20 nodes and 25 nodes, respectively, implemented on mobile nodes (MANETs) with varying speeds of 0–35 m/s. The graph shows a comparative analysis of AODV and DSR and their cross-layer optimized variants LAODV and LDSR using LEMO and IEEE 802.11 DCF.

7 Conclusion

MANETs seek to take a step in developing many technological trends in the information era. With the ability to build a movable network connected to the Internet with this resilience, MANETs will be the futuristic step toward building a digital world.

In recent times, many things (machine to machine, factories, industrial, agricultural equipment, people, and natural entities.) are linked to the Internet. As expected, anything or everything will be connected to the Internet in the coming decades. MANETs are expected to be considered as the bedrock of the Internet's next generation. As the mobile nodes in MANETs could be any moving thing such as person, machine, drone, robot, or any expected device. The implementation of cross-layer concept in different MANETs applications has proved its high efficiency and good treatment of many problems faced by conventional MANETs (without cross-layer). In addition to that it, the advantage of being able to adapt to different environments faster with more flexibility. Cross-layer implementation in MANETs meets the different requirements of high QoS and the problems of using the ISM band. As a result, these networks are suitable for applications that do need remote control under high data rates. They enjoy flexibility and adaptability since their mobile nodes have directed and dynamic movement (decentralized architecture). Lastly, these networks proved to be highly efficient in terms of energy consumed. The data amount sent and received, controlling the congestion, the link reliability, the shortening of the path between the transmitter and the receiver, and other parameters that led to the improvement of the network performance significantly.

References

1. Sharma VK, Verma LP, Kumar M (2019) CL-ADSP: cross-layer adaptive data scheduling policy in mobile ad-hoc networks. Future Gener Comput Syst 97:530–563
2. Patil VR, Sandhya SV (2019) QoS aware manet routing protocol for multimedia traffic in an adaptive cross layer architecture. Int Res J Eng Technol (IRJET) 06(06):609–615
3. Priya ML, Saravanan K (2019) An efficient cross layer cooperative diversity optimization scheme together With DEL-CMAC for lifetime maximization and energy efficiency in MANETs. Int Res J Eng Technol (IRJET) 06:2140–2158
4. Walia M, Challa RK (2010) Performance analysis of cross-layer MAC and routing potocols in MANETs. Second Int Conf Comput Netw Technol 1–7
5. Xia X, Liang Q (2020) Bottom-up cross layer optimization for mobile Ad Hoc networks. IEEE paper, pp 1–7
6. Varshavsky A, Reid B, de Lara E (2005) A cross layer approach to service discovery and selection in MANETs. MASS, IEEE, pp 1–8
7. Bautista OG, Akkaya K (2020) Extending IEEE 802.11 s mesh routing for 3-D mobile drone applications in ns-3, Workshop on ns-3–WNS3, Gaithersburg, Maryland, USA, June 17–18, pp 25–32
8. Kwan WC, Judge J, Williams A, Kermode R (2002) Implementation experience with MANET routing protocols. ACM SIGCOMM Comput Commun Rev 32(5):49–59
9. Kanakaris V, Ndzi D, Azzi D (2010) Ad-hoc networks energy consumption: a review of the ad-hoc routing protocols. J Eng Sci Technol Rev 3(1):162–167
10. Kwan WC (2005) The behavior of MANET routing protocols in realistic environments. Asia Pacific Conf Commun 3–5:906–910
11. Divecha B, Abraham A, Grosan C, Sanyal S (2007) Impact of node mobility on MANET routing protocols models. J Digital Inf Manag 5(1):19–24
12. Tuteja A, Gujral RK, Thalia S (2010) comparative performance analysis of DSDV, AODV and DSR routing protocols in MANET using NS2. Int Conf Adv Comput Eng 330–333

13. Mohsin AH, Abu Bakar K, Adekiigbe A, Ghafoor KZ (2012) A survey of energy-aware routing and MAC layer protocols in MANETS: trends and challenges. Netw Prot Algor 4:82–107. ISSN 1943–3581
14. Suresh KC, Prakash S (2012) MAC and routing layer supports for QoS inMANET: a survey. Int J Comput Appl 60(8):40–46
15. Fu B, Xiao Y, Deng H, Zeng H (2014) A survey of cross-layer designs in wireless networks. IEEE Commun Surv Tutor 16(1):110–126
16. Sethi A, Vijay S, Saini JP (2019) optimized link stability and cross-layer design for secured ad-hoc network, pp 01–16
17. Nosheen I, Khan SA, Khalique F (2019) A mathematical model for cross layer protocol optimizing performance of software-defined radios in tactical networks. IEEE Access 2896363:20520–20530. Accessed 27 Feb 2019
18. Kumar P, Verma S (2020) Implementation of modifified OLSR protocol in AANETs for UDP and TCP Environment. J King Saud Univ Comput Inform Sci 1–7
19. Shimly SM, Smith DB, Movassaghi S (2019) Experimental analysis of cross-layer optimization for distributed wireless body-to-body networks. IEEE Sens J 19(24):12459–12509. Accessed 15 Dec 2019

A Review on Progress and Future Trends for Wireless Network for Communication System

Nira Singh and Aasheesh Shukla

Abstract Wireless Communication has a long history of development, which is also a means for exchanging information through free space or wireless media. This paper gives an introduction to wireless communication as well as tries to focus on some of the areas where recently the research is going on. A comparative analysis between wired and wireless communication systems has been shown. It has been found that wired communication is more reliable than a wireless communication system. However, a wireless system is portable and easy to install. Additionally, the generation of wireless networks is discussed in detail. It has been found that as the generation changes from 1 to 2G speed was up to 64 kbps, 2–3G speed was 144 kbps–2 Mbps, 3–4G speed is 100 Mbps–1 Gbps and still, the generation is changing with higher data speed. It is observed that as the data increases the energy consumption of devices increases proportionally. Besides, the authors have also discussed various trends in technology that are helpful for Wireless Communication.

Keywords Wireless communication · 6G · Reconfigurable intelligent surfaces · Visible light communication · Mm-Wave · Li-Fi

1 Introduction

Wireless freedom is about being removed from the telephone cord and having the ability to be anywhere you want to be. In normal words, the communication system can be defined as whenever we have two things that can be designated as a source and destination [1]. If we pass any information from source to destination, the process can be named as simple communication, that is, whatever must be transmitted can be understood for the receiver as described in Rappaport et al. [2] It can moreover communicate that the correspondence structure is a variety of individual transmission

N. Singh (✉) · A. Shukla
Department of Electronics and Communication, GLA University, Mathura 281406, India

A. Shukla
e-mail: aasheesh.shukla@gla.ac.in

S. Dhar et al. (eds.), *Advances in Communication, Devices and Networking*,
Lecture Notes in Electrical Engineering 776,
https://doi.org/10.1007/978-981-16-2911-2_46

interchanges network transmission systems, hand-off stations, associate stations, and terminal rigging ordinarily prepared for interconnection and interoperation to shape a sum. If analyzed from a channel point of view, communication can be divided into two broad categories: a wired system and a wireless system. Whenever there is a wired system, it means that it is a hardware connection between the transmitter and the receiver as explained by Zong et al. and Rappaport et al. [2, 3]. This means that it needs a physical medium. But a wireless system means that it does not require any hardware connection between the transmitter and the receiver. So in the end, it should be said that no physical medium is required to transmit information without any interconnection, or instead it should be said that without any physical connection. Table 1 is a simple comparison of wired and wireless communication systems. It compared these two systems to evaluate some parameters based on those parameters. So it can detect that transmission speed and quality are good in wired systems, but still said that wireless communication is very popular compared to wired systems.

When wired correspondence can do the greater part of the work that remote correspondence can do, for what reason do we need remote correspondence? The essential and significant advantages of remote correspondence are portability, adaptability, convenience, high throughput execution, simple to introduce framework with

Table 1 Comparison of wired and wireless system

Parameters	Wired system	Wireless system
Way of communication	Copper, Fiber, etc.	Air
Standard	IEEE 802.3	IEEE 802.11
Movable	Not much movable	Very highly movable
Speed/bandwidth	Higher than remote	Lower than cable network
Activity of investiture	Lumbering and labor intensive	Less work, concentrated, and simple
Time of investiture	Too much time taken	Less time taken
Cost of investiture	Greater	Lesser
Conservation cost	Greater	Lesser
Related equipment	Hub, Switch, Router	Wireless Router, Access point
Profit	• More noteworthy speed • Higher clamor invulnerability • High dependable • More noteworthy security	• No problems with links • Best for cell phones • More prominent portability • Simple establishment and the board

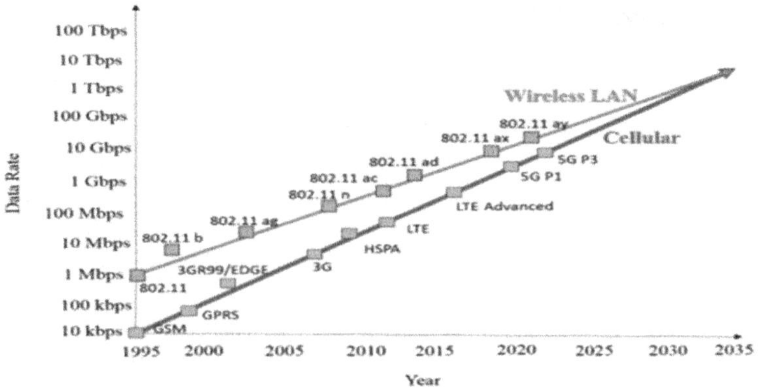

Fig. 1 Wireless roadmap outlook up to the year 2035

minimal effort, in crisis circumstances, and distant areas where wired correspondence is hard to set up, remote correspondence is a reasonable choice [1]. Wireless communication is very popular because of these things. It can also be said that although the wired system is not as good quality, even we prefer the wireless system because it makes the whole system moveable and with the help of this wireless communication concept not only communication between human beings is possible. [4] Goes but now in today's era, it can communicate with machines, vehicles, or anywhere around the world. Hence it can be said that nowadays wireless schemes are more popular.

Figure 1 shows the simple graphical representation that wants to improve wireless communication as like good performance in a wired system. With the help of this figure, it should be noted that in 2035 it can achieve a data rate with Terabit per second.

The objective of this paper is basically to cover the introduction of wireless communication as well as try to focus on some of the areas where recently the research is going on in the field of wireless. The rest of this paper is coordinated are as follows: In Sect. 2, wireless time-line and initial technologies are discussed. Section 3 focuses on trends technology in Wireless Communication. Section 4 concludes this paper.

2 Wireless Time-Line and Initial Technologies

This section will briefly survey the historical approach to wireless communications. Fundamentally, wireless communication aims at high quality and reliable communication. The growth of the history of wireless communication is divided into these six generations. It can be designated as an evolutionary journey of wireless communication. This section covers all generations with their merits and demerits in 1–6G wireless communication.

2.1 1G: Where It All Began

Start with 1G, it alludes to the original remote phone innovation, versatile media communications that were first presented during the 1980s and finished in the mid-1990s. Its speed was up to 2.4 kbps. The 1G network utilizes simple signs. The first 1G mobile system was launched in the USA. The drawbacks are very clear as this was the initial stage. So, there are many things that should be improved, for example, helpless voice quality, helpless battery life, huge telephone size, no security, restricted limit, helpless handoff dependability. So we can say that I was still the starting point, to do a lot of work to make good wireless communication or to shape wireless communication in a way so that it can be used practically, used by humans can go.

2.2 2G: The Cultural Revolution

Now with the concept of GSM (Global System for Mobile Communications) moving to 2G and it was launched in 1991 in Finland. The 2G network uses digital signals. This is a notable change from 1G to 2G. According to Alsharif et al. [5], its data speed was up to 64 kbps. As features are included, it enables services such as text messaging, picture messaging, and MMS (multimedia messaging). It provides superior quality and efficiency. There are drawbacks of 2G. 2G mobile phones require a strong digital signal to help work. If a specific area does not have network coverage, digital signals will be weak. These systems are unable to handle complex data such as video.

2.3 3G: The 'Packet-Switching' Revolution

There was a 3G transformation where we have moved from voice-level correspondence to information-level correspondence. It can likewise be known as the bundle exchanging unrest. 3G was dispatched in 2001 by NTT DoCoMo. The goal is to normalize the organization convention utilized by sellers. This implied that clients could get information from any areas on the planet as the 'information bundles' that drive web availability were normalized as explained in Gawas [1] This made world-wide meandering administrations a genuine opportunity unexpectedly. Information transmission speed expanded from 144 kbps to –2 Mbps. Generally, cell phones and highlights have expanded their transmission capacity. The disadvantages of 3G advancements are the costly charges for 3G permit administrations, it was trying to assemble a foundation for 3G, high transmission capacity prerequisites, costly 3G telephones, enormous phones.

2.4 4G: The Streaming Era

In 2009, 4G was first sent as a Long-Term Evolution (LTE) 4G standard in Stockholm, Sweden, and Oslo, Norway. It was later presented worldwide and made top-notch video in real time a reality for a huge number of purchasers. It can convey rates of 100 Mbps–1 Gbps [1]. Sorcery is one of the fundamental words used to depict 4G. Portable interactive media, whenever, any place worldwide portability uphold, incorporated remote arrangements, altered customized administrations. According to Zong et al. [3], highlights of 4G frameworks are intelligent interactive media, voice, video, remote Internet, and other broadband administrations, rapid, high limit and minimal effort per bit, worldwide versatility, administration convenience, adaptable portable organizations, consistent exchanging, variety of administrations. Nature of Service (QoS) prerequisites, planning and passage control innovation, call impromptu organizations, and multi-jump organizations. The disadvantages of 4G are high battery utilization, difficult to actualize, complex equipment required, costly hardware expected to execute cutting-edge organizations.

2.5 5G: The Internet of Things Era

5G is currently in the initial phase of deployment with carriers with limited 5G availability by 2020 and wide availability by the end of 2021. These network reforms will have far-reaching effects on the way people live, work, and play around the world. 5G technology promises to transform 2020 into a time of unprecedented connectivity and technological progress. With high capacity and speed, ultra-level latency, 5G power innovation that would be impossible under the 4G LTE standard. Full wireless communication with almost no limits [2] (Fig. 2).

2.6 6G: Next-Generation Wireless Technology

It caters to wireless networks without any limitations. It has an incredible transmission speed in the terabyte range. It will be available in the market in 2030 with greater reliability than previous mobile phones. This maximizes your data throughput and IOPS (input-output operation per second) [5]. The 6G network will be able to use higher frequencies than the 5G network. This will enable higher data rates to be achieved and greater overall capacity for 6G networks [4]. A very low latency level will almost certainly be a requirement. It would have to support a microsecond or even sub-microsecond latency communication, making communication almost instantaneous. Features of 6G are: It gives Frequency: 5.8 GHz, Bandwidth: 1 Gbps, 40 Mbps Connectivity Speed, Ultra-Broadband Internet Service, More Storage Capacity, Circuit and Packet Switching, and offers 3D Internet Concept.

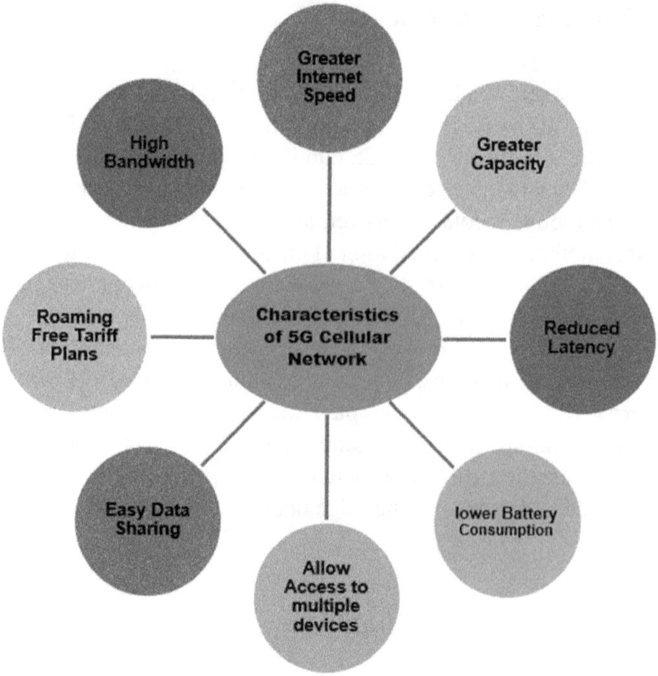

Fig. 2 Characteristics of 5G network

So it was a summary of the development of generations for wireless communication. In which the road from 1 to 6G was fully in the discussion, see its advantages and disadvantages.

3 Trends in Wireless Communication

Here we have discussed the various trends in the field of wireless communication. Like Communication with Reconfigurable Intelligent Surfaces, THz Communications, Visible-Light Communication, [6] Light Fidelity (Li-Fi), Mm Wave Wireless Communication, and Wireless Brain-Computer Interactions (BCI).

3.1 Communication with Reconfigurable Intelligent Surfaces

Reconfigurable intelligent surface (RIS) has as of late increased critical examination consideration as a method for empowering reusable dispersion conditions that can be tuned to give remote correspondence abilities is [7, 5]. We imagine an early

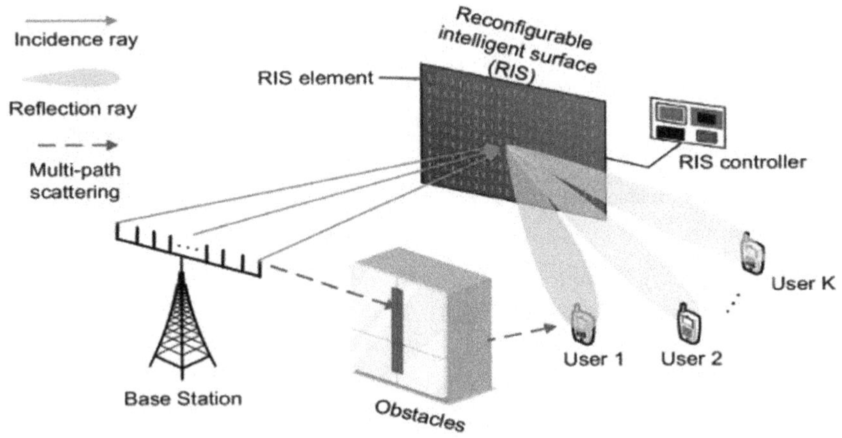

Fig. 3 Block diagram of reconfigurable intelligent surfaces

jump from customary enormous scope MIMO to huge clever surfaces (LIS) and savvy conditions that can give huge scope surface to remote correspondences and for heterogeneous gadgets. LISs empower inventive techniques for correspondence, for example, holographic radio recurrence (RF) and holographic MIMO [8]. RIS is made out of countless minimal efforts and energy-productive reusable intelligent components that can reflect electromagnetic waves with a controlled stage move with the assistance of a keen regulator (Fig. 3).

The upsides of RIS are that it very well may be conveyed all over the place. RISs are ecologically neighborly. RIS underpins full-duplex and full-band transmission, as they reflect just electromagnetic waves. What's more, RISs are practical as they don't expect simple to-advanced or computerized to-simple converters and force enhancers.

3.2 Visible-Light Communication (VLC)

Visible-light communication (VLC) is an information correspondence variation that utilizes obvious light somewhere in the range of 400 and 800 THz (780–375 nm). According to Khan [9], VLC is a subset of optical remote correspondence advances. History of 3.2 Visible light communication (VLC) can be traced back to the 1880s in Washington, D. C.. The innovation utilizes fluorescent lights to send signals at 10 kb/s, or for LEDs up to 500 Mbit/s over short separations. Frameworks, for example, RONJA can send a good way off of 1–2 km (0.6–1.2 mi) at full Ethernet speed (10 Mbit/s). The low transfer speed issue in RF transmission capacity has been explained in VLC because of the accessibility of enormous data transmission. According to Khan [9], VLC highlights are utilized to give both non-authorized channels, high

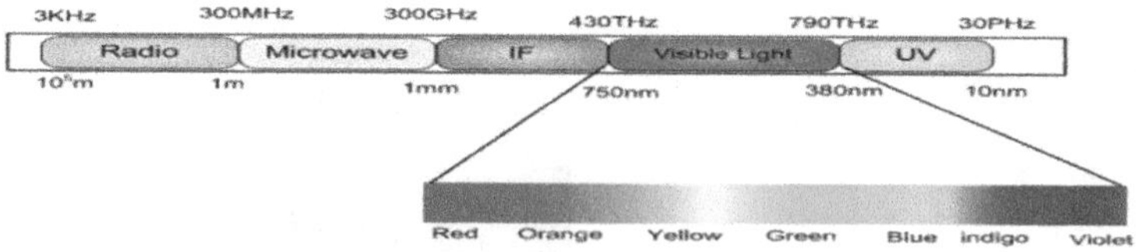

Fig. 4 Spectrum of visible-light frequency

transmission capacity, low force utilization, high transfer speed, insusceptibility to impedance, giving light and data (Fig. 4).

3.3 Li-Fi Wireless Technology

Li-Fi is a distant advancement that holds the best approach to understanding the challenges facing 5G. Li-Fi can send various gigabits, is more reliable, for all intents and purposes impedance free, and is regularly more secure than radio development, for instance, Wi-Fi or cell [6]. Li-Fi is an adaptable far-off advancement that uses light instead of radio frequencies to convey data. The development is maintained by an overall climate of associations embracing Li-Fi. Presently, the inquiry emerges how can it work. So the appropriate response is that the operational cycle is exceptionally straightforward. On the off chance that the LED is on, you send a digit 1 on the off chance that it is off you communicate a digit. 0. LEDs can be turned on rapidly and rapidly, which gives great open doors for communicating information. Everything necessary is a couple of LEDs and a regulator that codes the information in those LEDs. We simply need to shift the rate at which LEDs gleam dependent on the information we need to encode. Accordingly, each light source will fill in as a center for information transmission. The advantages of Li-Fi are rapid and transmission capacity, more solid, low dormancy, great security-wise, simple confinement, obstruction-free.

3.4 Mm Wave Communication

Millimeter-wave (mmWave) correspondence frameworks have pulled in huge enthusiasm for meeting the limited necessities of future 5G organizations. The mmwave system has a frequency range between 30 and 300 GHz, with a bandwidth of about 250 GHz available in total. It is also known as the Very High Frequency (EHF) band by the International Telecommunications Union (ITU). There are various telecommunications standards that specify carrier frequency in the MMW frequency range [10]. The advantages of mm-Wave are bigger data transfer capacity, higher transmission rate, dissemination rangeability and greater resistance to obstruction, amazingly high frequencies permit numerous short separations, need for slender shafts, decrease equipment size, High recurrence-radio wire is little. Burdens of MmWave are significant expense in development, incredibly high frequencies noteworthy lessening, millimeter waves significant distance applications, MmWave the infiltration quality of solid dividers is less known, likewise meddling with oxygen and downpour.

4 Conclusion

This paper basically gives a brief introduction to wireless communication. It also analyzed how wireless communication is needed in our society for development. Also, discussed various research areas such as reusable intelligent surfaces, VLC, mm-Wave, etc., on which any researcher and scholar can start their work. If it is said in the common language, the trends that are going on in the current time have talked about it in wireless communication, its benefits and what are the disadvantages are also explained in detail. This paper basically provides a guidance for our researchers and scholars.

References

1. Gawas AU (2015) An overview on evolution of mobile wireless communication networks: 1G-6G. Int J Recent Innov Trends Comput Commun 3(5):3130–3133
2. Rappaport TS et al (2019) Wireless communications and applications above 100 GHz: opportunities and challenges for 6G and beyond. IEEE Access 7:78729–78757
3. Zong B et al (2019) 6G technologies: key drivers, core requirements, system architectures, and enabling technologies. IEEE Veh Technol Mag 14(3):18–27
4. Zhang Z et al (2019) 6G wireless networks: vision, requirements, architecture, and key technologies. IEEE Veh Technol Mag 14(3):28–41
5. Alsharif MH et al (2020) Sixth generation (6G) wireless networks: vision, research activities, challenges and potential solutions. Symmetry 12(4):676
6. Bian R, Tavakkolnia I, Haas H (2019) 15.73 Gb/s visible light communication with off-the-shelf LEDs. J Lightwave Technol 37(10):2418–2424
7. Hu S, Rusek F, Edfors O (2017) Cramér-Rao lower bounds for positioning with large intelligent surfaces. In: 2017 IEEE 86th vehicular technology conference (VTC-Fall). IEEE
8. Basar E et al (2019) Wireless communications through reconfigurable intelligent surfaces. IEEE Access 7:116753–116773
9. Khan LU (2017) Visible light communication: applications, architecture, standardization and research challenges. Digital Commun Netw 3(2):78–88
10. Zhang X et al (2019) Phased-array transmission for secure mmwave wireless communication via polygon construction. IEEE Trans Signal Proc 68:327–342
11. Sun Y et al (2019) Blockchain-enabled wireless internet of things: performance analysis and optimal communication node deployment. IEEE Int Things J 6(3):5791–5802
12. Jafri SRA et al (2019) Wireless brain computer interface for smart home and medical system. Wireless Personal Commun 106(4):2163–2177
13. Lupu RG, Ungureanu F, Cimpanu C (2019) Brain-computer interface: challenges and research perspectives. In: 2019 22nd international conference on control systems and computer science (CSCS). IEEE

Development of an IoT-Based Smart Greenhouse Using Arduino

Nitesh Kumar, Barnali Dey, Chandan Chetri, Amrita Agarwal, and Aritri Debnath

Abstract The implementation of greenhouse farming is the need of the hour as it can benefit the farmers of the rural areas in a vast country like India to control and monitor their crops in the presence of various limitations like unfavorable weather conditions, lack of water supply management, etc. The implementation of a greenhouse is ideal for the production of crops, germination of seeds, and transplantation in the range of an affordable cost, as it provides all favorable conditions compared to the external environment. The greenhouse technique is nowadays used for large-scale production, off-season fruits, vegetables, improve existing agriculture methods, etc. This paper mainly focuses on greenhouse cultivation with the help of IoT (internet of things). The services and the possibility of continuous monitoring are better than regulating via manual methods. The operator can easily control, monitor, and view the real-time data to observe what is taking place inside the greenhouse via devices like a cell phone.

Keywords Greenhouse · Arduino · Sensors · IoT

1 Introduction

Our country, India, is a vast nation. Here, the production rates every year increases or decreases. Since all over the country, the soil condition is not very good throughout, it fails to meet the people's requirements. Some conditions like drought due to less rainfall, high humidity, invasion of insects, lack of fertility in soil, etc., affect the production rate and the farmer has to suffer a huge loss. Exchange of corps cultivation is possible but the overall outcome is less. Here the implementation of greenhouse came into existence. Greenhouse provides the solutions to overall problems [1].

N. Kumar · B. Dey (✉) · C. Chetri · A. Agarwal
SMIT, SMU, Gangtok, Sikkim, India
e-mail: barnali.d@smit.smu.edu.in

A. Debnath
CMRIT, Bangalore, Karnataka, India

S. Dhar et al. (eds.), *Advances in Communication, Devices and Networking*,
Lecture Notes in Electrical Engineering 776,
https://doi.org/10.1007/978-981-16-2911-2_47

A greenhouse is a dynamic closed trap of blockhouse, which provides favorable conditions to the inside plants for proper growth [2]. The sunlight can pass inside the house during the daytime and make the indoor environment, warm. A proper water supply management system is there for the plants, along with a drainage system so that excess water can go out. Bulbs and heater are also present, if in case plants do not get sunshine for a couple of days or weeks. Proper ventilation and a window system are there for the exchange of air. The greenhouse is effective for practical purposes because of no labor cost and can be done by self. Secondly, fewer expenditures and maintenance in terms of setting up the house. Thirdly, off-season production can be done along with new methods and techniques.

In this paper, the authors propose the implementation of a greenhouse method along with the Internet of Things (IoT) system. The system comprise various sensors like temperature, humidity, etc. that give their data to the Arduino-based microcontroller and then it forwards the information to a mobile set as SMS via a Wi-Fi module. If the system exceeds the given limitation values then a notification is sent to the owner.

2 Literature Review

Greenhouse technology has become a choice in many countries and has been optimally used in China and Japan for better yield of crops. The advantages of the greenhouse technology are manifold, which includes but not limited to reduced fertilizer waste, pests control in plants, better photosynthesis, optimal use of irrigation, vertical plantation, and planting of seedlings all through the year irrespective of the fact that the crops may be seasonal [3]. All these advantages of the greenhouse come from the continuous monitoring of microclimatic factors like temperature, humidity, intensity of light, moisture content of the soil, etc., within the greenhouse. Good monitoring and control of these will result in better productivity [4]. Traditional greenhouse systems are modified by the use of automation. In [5], the monitoring of the various parameters as mentioned above are done by an embedded system by round the clock checking for the threshold level. On reaching a certain set threshold level, it would further activate the sensors/actuators to keep a consistent atmosphere within the greenhouse. In [6], a wireless sensor network (WSN) monitors the microclimatic parameters within the greenhouse. The authors in [7] used a computer control system for controlling the parameters in the greenhouse system. In [8], the authors developed a digital signal processing based system for developing an environment monitoring system to monitor the climatic condition within the greenhouse system, which is vital for the growth of the plants. The authors in [9] used a microcontroller-based circuit for monitoring the parameters to achieve maximum yield in a greenhouse system.

3 Problem Definition

Apart from the soil condition, also waterlogging and water scarcity are continuous problems, which are faced by farmers in our country. Lack of knowledge about the quality of soil leads to cause both economic as well as permanent loss of quality of the soil. In rural and semi-urban areas, the problem of electricity is also a major hurdle that may limit the use of water pumps for irrigation in farmland. Over that attack of some phytophagous insects spoils most of the yield. The recent attack of locusts in the northern part of India during the spring season affected severely the yield of the crops. In this work, the problem of waterlogging and water scarcity is resolved by using a moisture sensor and a water pump. Soil condition is monitored by a pH sensor that can provide an accurate condition of the soil. The prototype system has been designed to work on a DC power supply so that the system can also function with a battery-operated power supply. For this also the system has a provision of outer covering made of plastic to protect the crops from unwanted insects. However, in the present work, the system has been designed without a covering while taking the measurements as it was tested within the closed environment of a laboratory.

4 Materials and Methods

4.1 System Architecture

The basic block diagram of the proposed system is shown in Fig. 1 and the flow diagram is also shown in Fig. 2. It can be understood from the diagram that the system has been designed in two parts. One is the hardware part of the system that comprise the Arduino and other sensors and actuators, and in the second part, there is a software part that helps in the implementation of the Internet of Things by use of a Wi-Fi module within the system. The Arduino is programmed using the Arduino software that is freely available.

The hardware unit comprises sensors for light, soil moisture content sensor, temperature and humidity sensor, a fan, bulb, DC motor, water pump, Arduino microcontroller board, Wi-Fi module, relay module, LCD, personal computer, and cell phone. The sensors play the most important part in the greenhouse system whose work is to sense the various microclimatic parameters. The data acquired by the sensors are given to the Arduino board, which computes the values and compares them with the input set threshold values. In case of any mismatch of the values of the parameters from the threshold values, the actuators are activated so that the ideal value is achieved automatically and the atmosphere is kept suitable for plant's growth. On the software side, these values are not only read and controlled but also they are transmitted via the Wi-Fi module to a remote monitoring station, in our case is the cell phone on an app called Blynk app that is freely available. Alternatively, an app may be developed exclusively for this system.

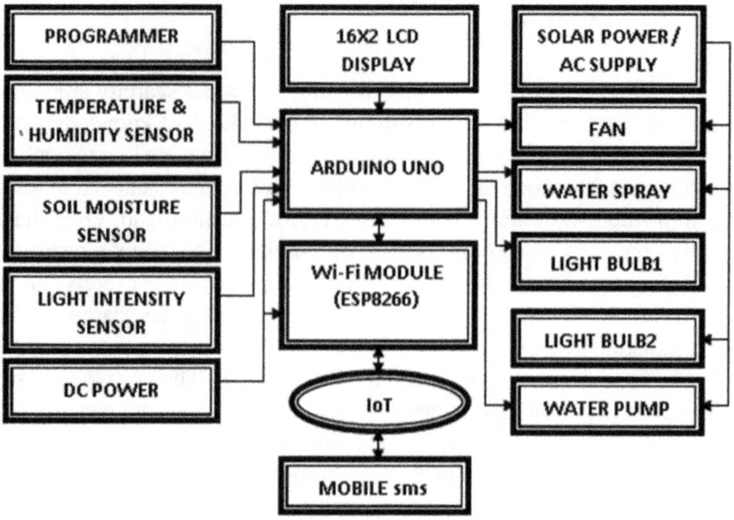

Fig. 1 Block diagram of the system architecture

4.2 *Materials Used*

The monitoring and controlling of the system is based on the parameters' values given by the following

A. *Arduino Microcontroller*: Arduino is an open-source electronics platform. It is an open source and is based on an easy to be used hardware and software module. A simple programming language may be used to send a set of instructions to the microcontroller for performing certain tasks.
B. *Relay Module*: The relay modules that have been used in the development of the smart greenhouse system is SRD-05VDC-SL-C. It runs on a 5 V power supply and can be controlled with any microcontroller. In our case, the microcontroller used is the Arduino Uno. It is a switch which when activated pulls the contact to make the high voltage circuit. It is activated with the help of a small voltage that is given by the microcontroller.
C. *LCD*: An LCD is used in the system in order to display parameter values. It has a parallel interface with the microcontroller and the interface has various pins like the register select (RS) which controls wherein the LCD memory a data is to be written. A read/write (R/W) pin, which is used to select between the reading and writing mode. An Enable pin that enables writing to the registers, display contrast pin, and power supply pins. The Liquid Crystal library compatible with the Hitachi HD44780 driver allows the user to control LCD display.

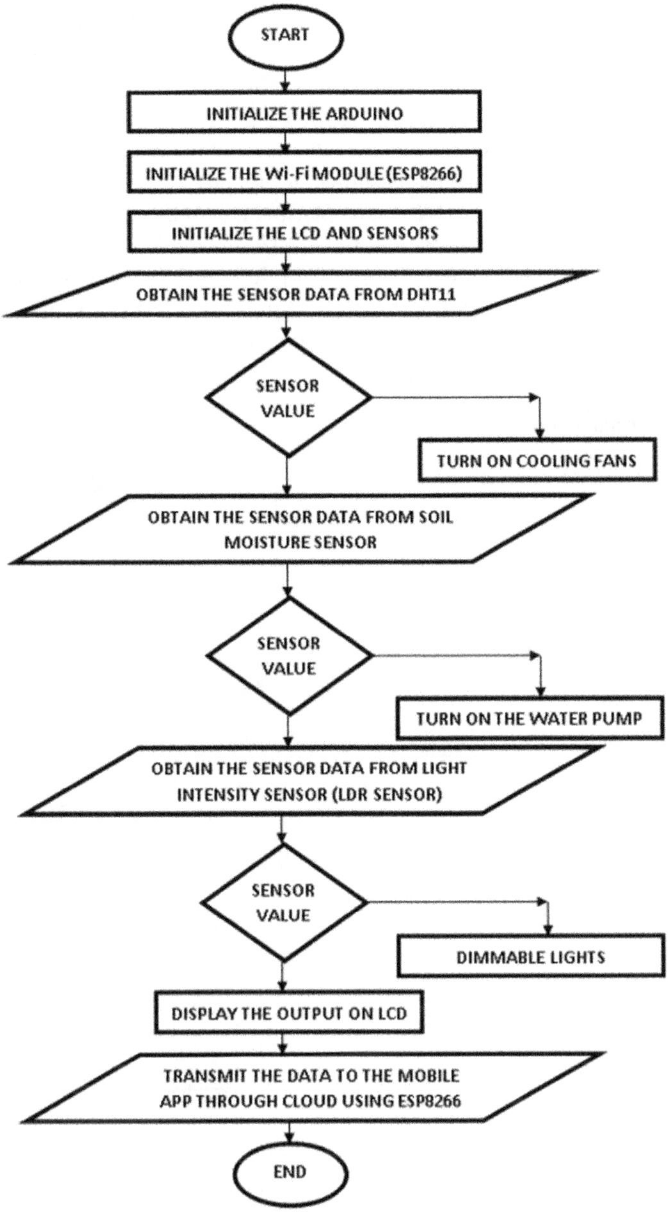

Fig. 2 Flow diagram of the system architecture

D. *NodeMCU ESP8266 Wi-Fi Module*: The **ESP8266,** which is a low-cost Wi-Fi microchip, is used in the system. It has a full TCP/IP stack and microcontroller capability. This small module allows the Arduino to connect wirelessly more than one kilometer using a Wi-Fi network [10]. A simple TCP/IP connection is established using Hayes-style commands. A general design of a Wi-Fi-based microcontroller is shown in Fig. 3.

E. *DHT-11 Temperature and Humidity Sensor*: Out of the various sensors that can detect temperature and humidity, DHT11 shown in Fig. 5, has been used in the system. According to the authors of [12], it is used to detect temperature and humidity. It also has anti-interference features, high level of accuracy, and ease of integration. It also uses low electricity and is small in size.

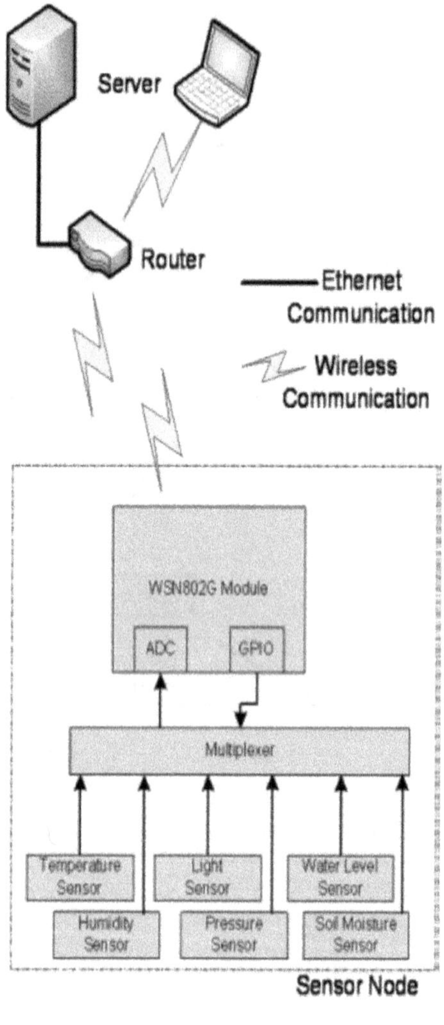

Fig. 3 The general design of a Wi-Fi-based microcontroller [11]

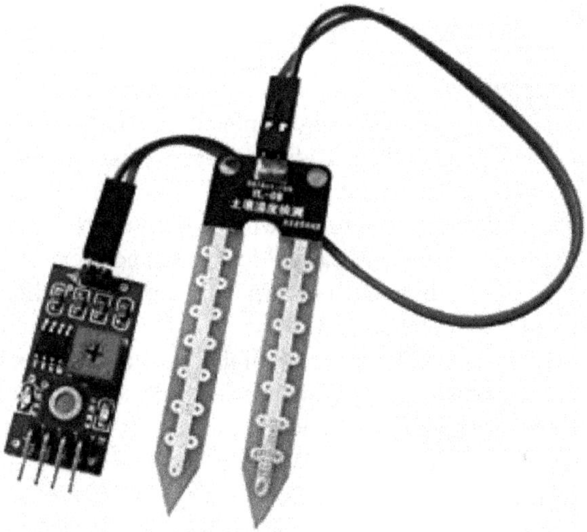

Fig. 4 Grove soil moisture sensor

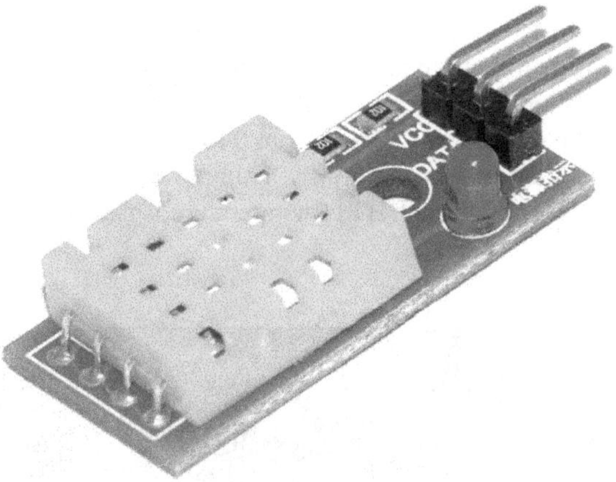

Fig. 5 DHT11 temperature and humidity sensor

F. *Grove Soil Moisture Sensor*: A soil moisture sensor shown in Fig. 4 can be used in two ways. It can be hand-made as reported by [13, 14] or one may use a ready-made sensor. In our system, a ready-made sensor has been used that measures the volumetric content of water inside the soil and gives us the moisture level as output. The soil moisture sensor FC-28 also contains a potentiometer that sets a threshold value. The LM393 comparator compares this threshold value. An output LED is activated according to the threshold value.

5 Results

A screenshot of the practical setup of the small-scale greenhouse system is shown in Fig. 6. An imperative component of the system is the control mechanism after the sensors give their readings. For example, an increase in temperature inside the greenhouse due to the plastic cover and heat outside may be controlled by turning ON the fan or opening the windows. Similarly, light can be controlled by turning ON artificial lights in case of less light and prevent the plants from being denied of sufficient daylight even in varying weather conditions and seasons. The outputs of the system are described below:

LDR: The Arduino output of the LDR is shown in Fig. 7 for low-light conditions. The circuit used for sensing light in the system uses a 10 kΩ fixed resistor, connected to a +5 V supply. The LDR's resistance is able to obtain 10 kΩ in dull conditions and 100 Ω in full brightness. While the optimal illumination value is given as 0–0.69 V, the transducer range is 0.7–2.5 V for dim light at a threshold of 500 lm of light. In case the intensity of light falls below the threshold, the relay module will turn ON an artificial light bulb.

DHT11 Temperature and Humidity Sensor: The sensor builds up a direct voltage versus RH yield. The point at which the supply voltage fluctuates, the sensor yield voltage follows to the same extent. It can work over a 4–5.8 supply voltage range. At 5 V supply voltage and room temperature, the yield voltage ranges from 0.8 to 3.9 V as the mugginess changes from 0 to 100% (non-condensing). The output voltage is converted to temperature by a simple conversion factor. The output of the sensor (screenshot of the DHT11 temperature is shown in Fig. 8) is taken according to the following formula:

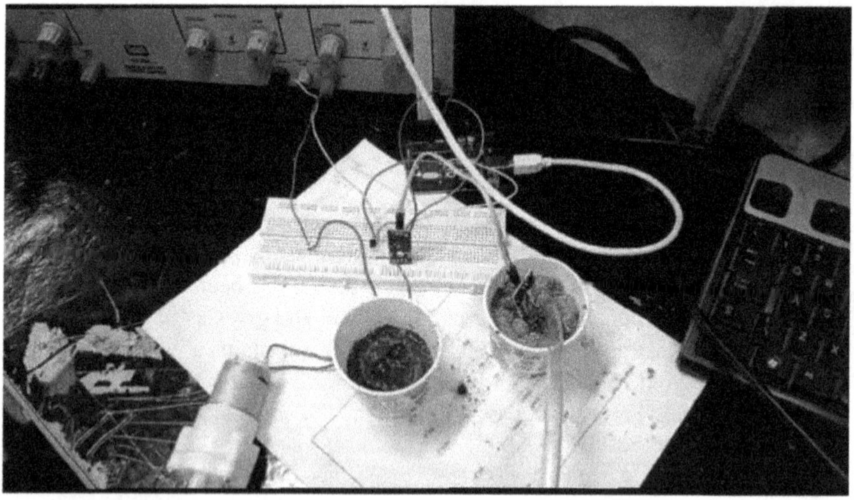

Fig. 6 Setup of the system using Arduino Uno

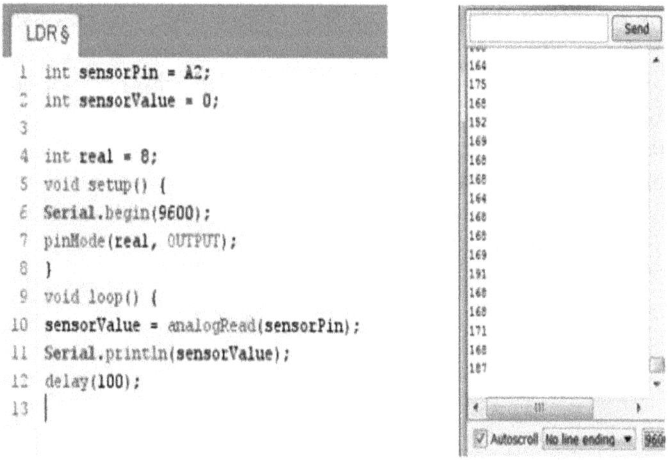

Fig. 7 Arduino output of LDR (Low-Light Condition)

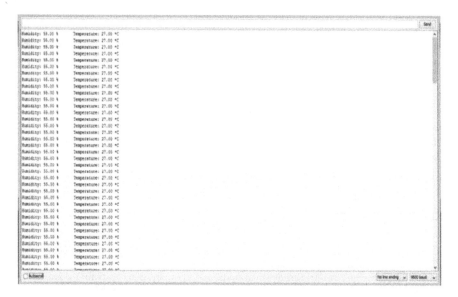

Fig. 8 Arduino output of DHT11

$$\text{RH} = ((\text{Vout}/\text{Vsupply}) - 0.16)/0.0062, \text{ typical at } 25\,^{\circ}\text{C}$$

The ideal temperature is considered to be 26–32 °C. As soon as the threshold value is crossed, the cooling fan in our case will be turned ON. Alternatively, exhaust fans can also be used which move a substantial volume of the heat through the outlet to freshen up and pull outside air in through the back vent. Similarly, the ideal value of the humidity is set to be between 40 and 50% beyond which the fans will be turned ON.

Soil Moisture Sensor: In this work, a soil moisture sensor is used to sense the condition of the soil to see whether it is dry, humid, or watery. If the soil condition is dry then a servo motor is automatically activated to turn ON the water supply that comes from a water pump. The water pump is attached to a tank. Here we have used bottles. Once the soil becomes moist enough, the sensor senses the moisture and it will close the water supply automatically.

6 Discussion

The aforesaid work has been designed and implemented as a prototype. The outputs were obtained and have been discussed in the results section in detail. All the sensors that have been discussed sense the parameters like temperature, humidity, and soil moisture. These outputs are relayed by the relays and the actuators are activated as and when required. The Wi-Fi module in this work is used to establish a communication link between the system and a mobile application so that information about the working of the sensors and actuators can be sent via SMS. However, as future work, an application can be developed as part of the system and all information can be sent as well as controlled by the developed app. Also, decision-making intelligence can be introduced in the same by the sensors into the system by use of machine learning techniques like Fuzzy Logic. If any controlling may be required, it can be done with the help of a cell phone through the app, for example, controlling the lights, motor, etc. Since the system has all access through the owner's cell phone, the owner can easily control and regulate the fans, valves of water system, bulbs, etc.

References

1. Modani V, Patil R, Puri P (2017) IoT based greenhouse monitoring system: technical review. Int Res J Eng Technol 10:2395–2456
2. Joudi KH, Farhan AA (2015) A dynamic model and an experimental study for the internal air and soil temperatures in an innovative greenhouse. Energy Conver Manag 91:76–82
3. Jha MK, Paikra SS, Sahu MR (2019) Protected cultivation of horticulture crops. Educreation Publishing
4. Shin JH, Son JE (1998) Development of a real-time irrigation control system considering transpiration, substrate electrical conductivity, and drainage rate of nutrient solutions in soilless culture of paprika. Eur J Hortic Sci 80(6):271–279. http://dx.doi.org/10.17660/eJHS.2015/80.6.2
5. Sumit K, Mohit K, Aswani K, Praneet G (2012) Monitoring and control of greenhouse environment. B Eng Project Depart Electr Electron Eng Bharat Inst Technol Meerut India
6. Mahmoud S, Ala'a I (2013) Greenhouse micro-climate monitoring system based on wireless sensor network with smart irrigation. Int J Electr Comput Electron Commun Eng 7(12):1072–1077

7. Melrolho JC, Serodio CMJA, Couto CACM (1999) CAN based actuation system for greenhouse control. Ind Elect 945–950
8. Kumar A, Singh A, Singh IP, Sud SK (2010) Prototype greenhouse environment monitoring system 2
9. Sahu K, Ghosh S (2012) Digitally greenhouse monitoring and controlling of system based on embedded system. Int J Scient Eng Res 3(1)
10. Brinkhoff J, Hornbuckle J (2017) Characterization of WiFi signal range for agricultural WSNs. In: 2017 23rd asia-pacific conference on communications (APCC), pp 1–6
11. Mendez GR, Md Yunus MA, Mukhopadhyay SC (2011) A WiFi based smart wireless sensor network for an agricultural environment. In: 2011 Fifth international conference on sensing technology, pp 405–410
12. Wang Y, Chi Z (2016) System of wireless temperature and humidity monitoring based on Arduino Uno platform. In: Processing 2016 6th international conferences instrument measurement computer communication control. IMCCC 2016, pp 770–773
13. Kumar MS, Chandra TR, Kumar DP, Manikandan MS (2016) Monitoring moisture of soil using low cost homemade Soil moisture sensor and Arduino UNO. In: 2016 3rd international conference on advanced computing and communication systems (ICACCS), vol 01, pp 1–4
14. Ardiansaha I, Bafdalb N, Suryadib E, Bonoc A (2020) Greenhouse monitoring and automation using arduino: a review on precision farming and internet of things (IoT). Int J Adv Sci Eng Inform Technol 10(2):703–709

Automatic Irrigation System with Rainwater Harvesting

Rajeev Sharma, Keshab Ch Gogoi, and Saikat Chatterjee

Abstract An automated irrigation system has been developed using sensor technologies interfaced with a microcontroller to efficiently utilize water for irrigation purposes. This paper generally discusses that how the accuracy of the sensors needs to be maintained so that the problem of false turning on and off of the motor can be eliminated. The resistive type soil moisture sensor which was used earlier does not provide accurate moisture reading with respect to the type of soil, the climatic condition, geographical area, and also the sensor is exposed to corrosion with time. So to overcome these problems, the implementation of capacitive soil moisture sensors has been done into the agricultural land. An algorithm has been built out with a proper range of operation as per the type of soil, geographical area, and climatic conditions so that the capacitive soil moisture sensor can control the water quantity in the soil as per the requirement of the crop. The sensor has an anti-corrosion layer that prevents it from corrosion and a water level sensor has been implemented to measure the water level in the tank. We also have a circuit that can forecast the status of rain and also a circuit that can monitor the rise of water level in the cultivated plot so that when the rise of water crosses the desired level in the field the siren turns on indicating that the field is likely to be overflowed with water. The excess water further can be made to flow through a proper channel to the water storage tank instead of wasting it.

Keywords Microcontroller · Capacitive soil moisture sensor · Water-level sensor · Rain sensor · Water level indicator

1 Introduction

India has plenty of agricultural lands and the basic livelihood of Indians depends on agriculture. The productivity of different kinds of crops does not depend on the constant water supply to the crops, but depends on better time matching of water supply as per the water demand by the crops and other uniform environmental

---

R. Sharma · K. Ch Gogoi · S. Chatterjee (✉)
EEE Department, Sikkim Manipal Institute of Technology, Sikkim Manipal University, Majitar, Sikkim 737136, India

S. Dhar et al. (eds.), *Advances in Communication, Devices and Networking*,
Lecture Notes in Electrical Engineering 776,
https://doi.org/10.1007/978-981-16-2911-2_48

conditions that are different for different crops. The crops that are cultivated cover a huge area of land and it is not manually possible to irrigate all the crops equally. Hence to fulfill the time to time water demand by a crop, we need an automatic and efficient irrigation system to improve the productivity of different crops by using water in an economic way [1]. Sensor technologies are being used to control the flow rate of water from the pump by accessing the moisture content present in a particular type of soil as for different types of soil the water intake capacity of the soil are different and also as per the types of soil suitable crops are grown in it. The heart of the system is the microcontroller and the accuracy of controlling the system using the microcontroller can be adjusted only if the values generated from the sensors are accurate. The water pump is operated as per the value sensed by the moisture sensor, which must be accurate enough so that problem of false turning on and off the motor can be controlled. In this paper, we will describe a prototype that can automatically irrigate the field using sensor technologies. The accuracy of the sensors is maintained so that false tuning on and off the motor can be controlled [2]. By developing this prototype, we can also control the misuse of water as when moisture content present in the soil is not found precise then the motor is turned on and when the moisture content sensed by the moisture sensor is adequate then the pump will remain off.

Rain is usually felt to be beneficial to crops, but we have to examine the ideal requirement of water needed for a crop. If the average rainfall is much lower or higher than the ideal, then it leads to significant problems and affects the quality of the grown crops. So, we are using a rain sensor, which can forecast the status of rain [2], and an analog temperature sensor that will give us information regarding the surrounding temperature of the agricultural field. We have used a water-level indicator that can monitor the rise of water level in the yard [3]. When the level of water crosses the desired level in the field, the siren turns on indicating that the field is about to be overflown by water and the excess water further can be made to flow through a proper channel to the water storage tank instead of wasting it. In this way, the rainwater can be reused in the crops or can be used after filtering it for domestic purposes like for drinking and for day-to-day activities [4]. The agricultural plots owned by the farmers are generally located in remote areas so that crops can be grown in a pollution-free environment. The plots owned by the farmers cover a huge area and the water pumps also have their own ratings.

Generally, for irrigating one acre of land, we nearly require 0.5–1 HP of water pump as per the geographical area and climatic conditions, so more the area of land more will be the horsepower and cost of the pump. So, for a heavy rating of the motor the power required will also be more and as a result problem of under voltage might come into the picture as many farmers will be using water pumps to irrigate their land. When under voltage is there, then the current drawn by the motor will increase which will heat up the motor and ultimately damage the windings of the motor, so to protect the motor we have used a voltage protection circuit to protect the motor from voltage fluctuations.

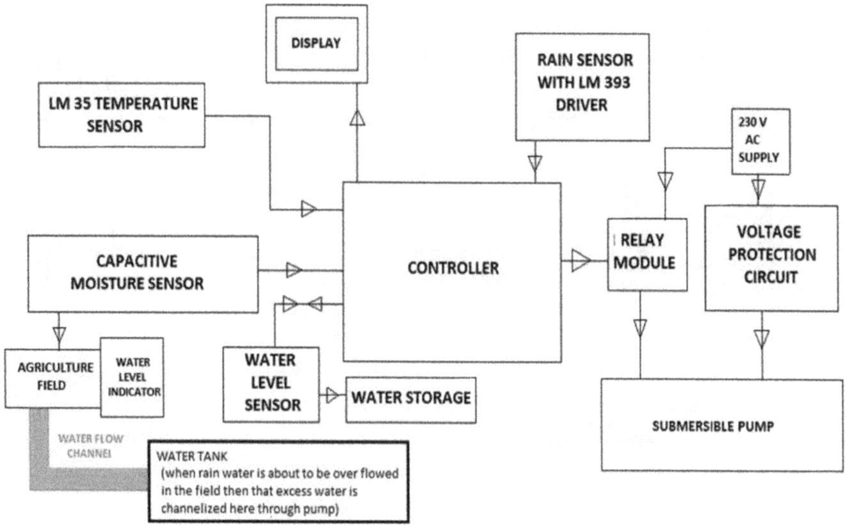

Fig. 1 Block diagram of the overall system

2 Block Diagram

See (Fig. 1).

3 Methodology

In an automatic irrigation system, there are many factors that need to be taken care of because ultimately our main objective for choosing an automatic irrigation system over manual irrigation is to minimize the wastage of water. Also, we know that the major source of water for irrigation is rainfall, so we must develop something which can store the rainwater and can be reused so that water is not wasted. Practically automatic irrigation works by accessing the moisture content present in the soil with the help of a resistive-type moisture sensor and with respect to that the pump will be operated via relay [5]. But here as per the type of soil, the range of operation for the sensor will also change so that the accuracy and stability of the system can be maintained [4].

The problem with the resistive-type moisture sensor was the accuracy of the range of operation of the sensor and with time the sensor is exposed to corrosion. So to overcome this we tried the same with a capacitive-type moisture sensor which gave us many accurate results compared to resistive-type moisture sensor and also the sensor is coated with an anti-corrosion layer which protects the sensor from corrosion [6].

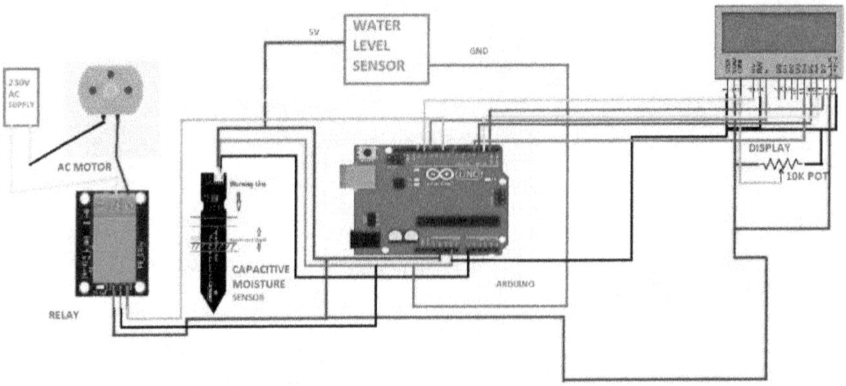

Fig. 2 Circuit diagram for automatic irrigation system

3.1 Monitoring of Moisture Content Present into the Agricultural Land Where Crops Are Being Cultivated

To adopt the actual moisture content present in the soil, a capacitive moisture sensor has been used. The operating range of the sensor has been implanted as per the climatic conditions and the type of soil where cultivation has been done. The microcontroller computes the status of soil as follows (Fig. 2):

- When the sensor finds out the soil to be in a dry condition, which means that the ions present in the soil are less so the generated analog value is high, then the availability of water is further checked by the water-level sensor and when everything is found precise then the motor is turned ON to irrigate the field and also it automatically turns OFF when sufficient amount of water is supplied to the field. And when the sensor detects adequate moisture in the soil or in other words adequate ions are present in the soil, then the motor will be in OFF condition.
- If the water is unavailable to irrigate the field, then the power supply is automatically turned OFF as the water-level sensor breaks the circuit and when everything is found precise then the circuit is automatically connected and the motor starts operating via relay module as per the moisture content detected in the soil by the sensor.

3.2 Predicting the Status of Rain Fall and Overall Outside Temperature

To predict the rainfall, a rain sensor with its driver circuit LM 393 has been used. The sensor works under the principle of resistance or in other words when the sensor is in

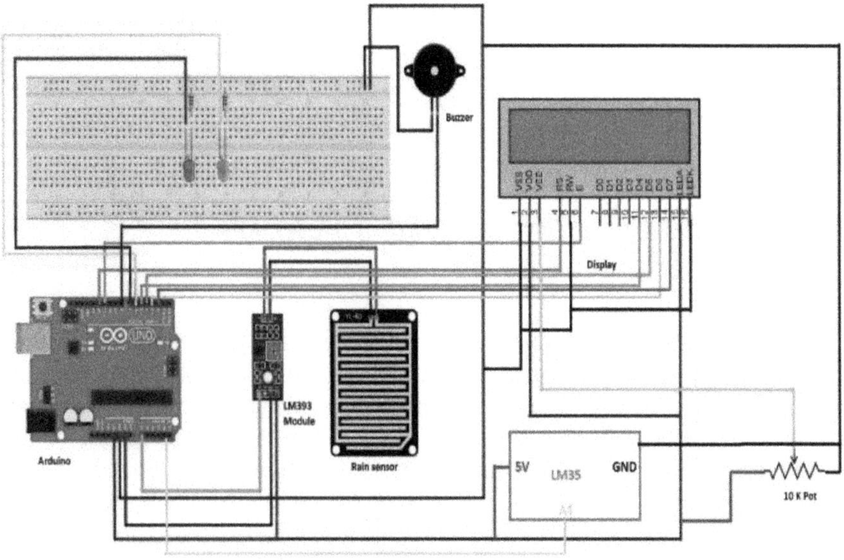

Fig. 3 Circuit diagram for the rain alarm circuit

dry condition, the resistance of the sensor is very high, and the sensor is in normally open condition and vice versa. The microcontroller computes the status of rainfall by comparing the preset threshold value and the generated analog value as follows (Fig. 3):

- If the analog data is greater than the pre-set threshold data, it denotes that it is raining and the same is indicated by a red colored LED and a buzzer.
- If the analog data is less than the pre-set threshold data, it denotes that it is not raining and the same is indicated by a white-colored LED. The white-colored LED will remain on in normal condition as it also denotes that the circuit is operating normally. The analog temperature sensor LM35 gives the outside temperature.

3.3 Motor Protection Circuit

We have used a step-down transformer which is stepping down the supply from 230 to 12 V (AC). After stepping down the AC voltage, the same is fed to the rectifier circuit which converts the AC voltage to DC voltage [7], since the household supply voltage is normally well synchronized, so we have arranged a provision to create the fault manually using potentiometers. After rectification from AC to DC, the output is not pure as it is affected by ripples which are bad for the load, to eliminate these ripples proper rating of the capacitor is used. Now the 12 V (DC) is fed to the voltage regulator IC which as an output gives a constant 5 V to the comparator IC. The comparator IC compares the value of voltage with the main supply voltage and if the voltage is more or less than the normal range then a signal is sent to the relay driver IC to trip the supply of the load via the relay. In this way, when the voltage is more or less than the normal voltage, then the relay trips the load for the safe and smooth operation of the system (Fig. 4).

3.4 Determination of Rise and Fall of Water Level in the Agricultural Field

To determine the rise and fall of water level in the field, we have used water-level indicator where the implementation of a transistor as a switch is done. We have connected four resistors at the base of transistors to avoid the transistors from being damaged by the high current across the LEDs. When there is no voltage applied at the base of transistors, there is no current flow through the collector and emitter terminal of the transistor, so as a result, the LED will remain OFF. The different levels of the water-level indicator are denoted as A (25%), B (50%), C (75%), and D (100%). When the water level reaches point A, a conductive path is formed between the base of the transistor (say T1) and the positive terminal of the battery. Therefore, the positive side of the battery and the base of the transistor T1 gets connected. As soon as the positive voltage is applied at the base of transistor T1, the transistor T1 gets turned ON and the LED 1 starts glowing indicating that the water level has reached 25% of the field. Similarly, when the water level reaches point B and C, it indicates the level of water has reached 50% and 75%, respectively (Fig. 5).

When the water level reaches point D, another transistor T4 (connected to buzzer) is turned ON due to the conductive path created, a positive voltage is applied at the base of it which in turn switches the buzzer ON indicating that the agricultural field is about to be overflown [8] and hence we need to make a provision to channelize the flow of excess water to a storage tank without wasting it.

4 Sensors and Its Parameters

4.1 Moisture Sensor Analog Value for Both Moist and Dry Condition

The above table represents the data recorded for both dry and wet conditions of the soil. The capacitive moisture sensor is manufactured in such a way that it can operate in a certain range of values sensed by it which is analog in nature. The capacitive moisture sensor is ranged from 0 to 1023, where the value sensed by the sensor near 0 is considered as the most wet condition and the value sensed by the sensor near 1023 is considered as the most dry condition. The range of the sensor depends upon the type of soil as well as the climatic and geographical condition of that area, so keeping

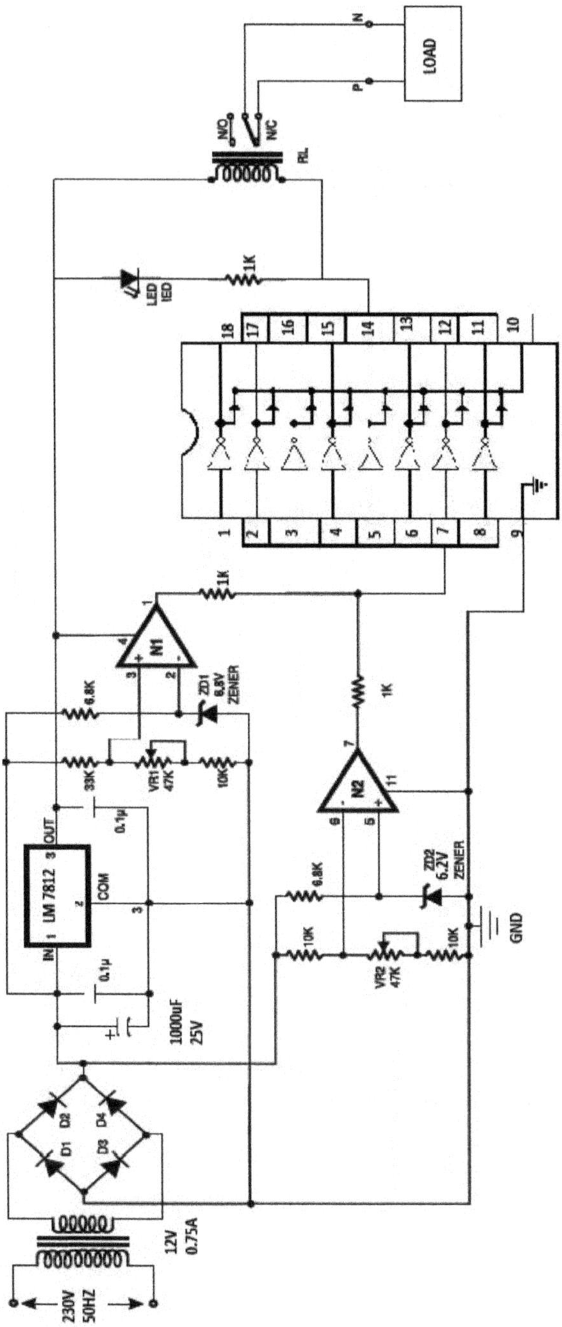

Fig. 4 Circuit diagram for voltage protection system

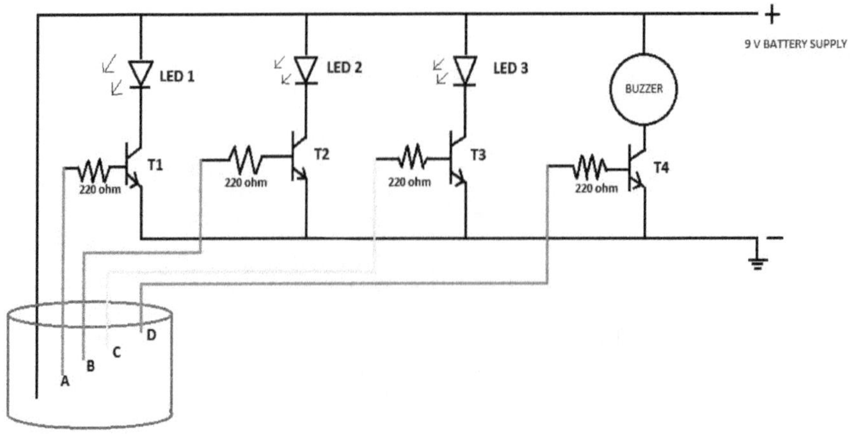

Fig. 5 Circuit diagram for water-level indicator

Table 1 Analog value for both wet and dry condition

Date	Analog value for wet condition	Analog value for dry condition	Average value for wet condition	Average value for dry condition
11/02/2020	319	606	320	610
12/02/2020	318	612		
13/02/2020	319	609		
14/02/2020	321	610		
15/02/2020	321	609		
16/02/2020	320	610		
17/02/2020	322	614		

these factors in observance we have tested the sensor for one week and obtained a set of analog values for both wet and dry condition. After attaining the values, we have considered the average of the sensed analog values. After testing we have fixed the operating range of the sensor to be from 320 to 610, where values generated near 320 is considered as most wet condition and values near 610 is considered as the most dry condition. The determined values are further processed by the microcontroller and is converted to their equivalent percentage value [9, 10] (Table 1).

4.2 *Rain Sensor Generated Analog Value*

When rain falls into the sensor then the resistance of the sensor decreases and voltage flows through the circuit because the circuit is complete, and it is in normally closed

Table 2 Threshold value for detecting rainfall

Date	Analog value	Average value
14/02/2020	501	500
15/02/2020	497	
16/02/2020	502	
17/02/2020	499	
18/02/2020	502	
19/02/2020	500	
20/02/2020	499	

condition. And when there is no rainfall the resistance of the sensor increases and the path of voltage supply breaks as the circuit is in normally open condition (Table 2).

5 Advantages

- The system is enhanced which reduces the wastage of water.
- The system is fast responsive, user-friendly, and better than the manual switching mechanism, which was earlier used by the farmers.
- It does not create any kind of nutrition pollution into the environment as the operation of the system is programmed and controlled via sensor technologies.
- A voltage protection system for the protection of the motor from voltage fluctuations is implemented and also we have provided a water-level sensor so that the motor is only under operation when there is an adequate amount of water available in the storage tank for the safe operation of the motor.
- The status of the rise of water level in the field due to rainfall is determined by a water-level indicator.
- The rainwater stored and collected can be further filtered out and used for domestic purposes or can be directly used to apply medicines and pesticides to the crops for better growth and production.
- It is quite economic and reduces the work of manual labor and saves a lot of time.
- This system gives us controlled and quality irrigation as it stops weeds and leaf diseases since water is directed to the roots.
- It also preserves soil structure and nutrients.

6 Future Scope

- We can use solar energy as a power source for the system.
- We can also use a GSM module, which can transfer all the final processed data like the status of the water pump, status of rain, outside temperature, condition of the

soil after checked by the moisture sensor to the contact number of the landowner through an SMS.

- The rain alarm circuit can be interfaced with the automatic irrigation circuit via a feedback loop to keep track of the operation of the motor. In other words, the data regarding the duration of working (ON/OFF) of the motor can be accessed whenever needed.
- The design and implementation of software parts for the smart controlled system need more improvement to meet various real demands of the crops grown.
- We can make a centralized database, where data regarding the maintenance of crops according to the atmospheric conditions can be preserved throughout the year.

7 Results

We have monitored the moisture content present in the soil for different crops for one week using the moisture sensor to analyze the water demand of the crop with respect to the type of soil in which it is cultivated for efficient and better production in a particular climatic and geographical condition.

- Paddy field located in Rupai Siding, Assam

  Type of soil–Clayey loam soil

The data below represents the amount of moisture content present in the soil of the paddy field. We have monitored the moisture content present in the paddy field soil for 1 week in both morning and evening shifts (Table 3).

- Black Mustard located near Talap, Rupai Siding, Assam

  Type of soil–Black cotton soil

Table 3 Moisture content present in paddy field

Date	Moisture content in percentage				Remarks (Average moisture content data monitored for one week)
	Analog value	Morning 8:30AM (%)	Analog value	Evening 5:30PM (%)	
09/03/2020	453	54	448	56	Morning–59.57% Evening–52% Overall–60.02%
10/03/2020	448	56	442	58	
11/03/2020	436	60	424	64	
12/03/2020	424	64	430	62	
13/03/2020	436	60	439	59	
14/03/2020	433	61	436	60	
15/03/2020	430	62	421	65	

Table 4 Moisture content present in black mustard crop

Date	Moisture content in percentage				Remarks (Average moisture content data monitored for one week)
	Analog value	Morning 9AM (%)	Analog value	Evening 5PM (%)	
09/03/2020	572	13	569	14	Morning–13.57% Evening–13.28% Overall–13.42%
10/03/2020	569	14	572	13	
11/03/2020	566	15	564	16	
12/03/2020	561	17	564	16	
13/03/2020	569	14	575	12	
14/03/2020	575	12	581	10	
15/03/2020	581	10	575	12	

The data below represents the amount of moisture content present in the soil of black mustard cultivated land. We have monitored the moisture content present in the black mustard cultivated soil for 1 week in both morning and evening shifts (Table 4).

8 Conclusion

By implementing this project, we can improve the earlier manual irrigation techniques used by the farmers and save the cost of daily labor charges by improvising the supply of the adequate amount of water to the crops and soil and also minimize the wastage of water. The automatic irrigation system allows cultivation in places with water scarcity thus improving sustainability. The system is incredibly versatile and economical as it does not need individuals on duty and is easy to operate and reliable. The agricultural fields which cover huge area can be monitored using highly sensitive sensors, which can access moisture covering a huge range of operation. This system can be used to irrigate agricultural lands, horticultural lands, parks, gardens, golf courses, etc. Thus, this system is efficient enough when compared to other types of automation systems. We know that if the average rainfall is more or less than the required amount of rainfall, this also leads to several kinds of problems in the growth and development of crops. So to avoid that, we have followed the process of rainwater harvesting using which we have accumulated and stored the rainwater for onsite use instead of allowing it to run off and create problems like soil erosion and flood. The rainwater is generally free from several types of pollutants and man-made contaminants and thus improves the quality and growth of the grown crops.

References

1. Anitha k (2016) Automatic irrigation system. YMCA, Connaught Place, New Delhi, pp 301–309. ISBN: 978–93–86171–10–8
2. Pavithra DS, Srinath MS (2014) GSM based automatic irrigation control system for efficient use of, resources and crop planning by using an android mobile. IOSR J Mech Civil Eng 11:49–55
3. Hade AH, Sengupta MK (2014) Automatic control of drip irrigation system & monitoring of soil by wireless. IOSR J Agricul Veter Sci 7:57–61
4. Seymour RM (2005) Capturing rainwater to replace irrigation water for landscapes: rain harvesting and rain gardens. department of biological and agricultural engineering, The University of Georgia–Griffin Campus, 1109 Experiment St, Griffin, GA 30223
5. Pavithra M, Priya MM, Bharathi GD, Annanayagi V, Keerthana MV (2017) An embedded based rain detection system in automatic irrigation. Int J Innov Res Sci Eng Technol 6. ISSN(Online). 2319–8753 ISSN (Print). 2347–6710
6. Mander G, Arora M (2014) Design of capacitive sensor for monitoring moisture content of soil and analysis of analog voltage with variability in moisture. RAECS UIET Panjab University Chandigarh
7. Barapatre P, Patel JN (2019) Determination of soil moisture using various sesnors for irrigation water management. IJITEE 8:2278–3075
8. Qu J, Fan J, Huang D (2014) The capacitive soil moisture sensor research. Appl Mech Mater 584–586:2142–2149
9. Rakshit D, Baral B, Datta S, Deb PB, Mukherjee P, Paul S (2017) Water level indicator. J Scient Res 1:181–182
10. Oyubu AO (2015) Design and implementation of a rainwater detector–alarm system. Int J Adv Res Technol 4:2278–2776

Home Automation: A Novel Approach

Jayant Singh, Kritika Garg, Nitesh Kumar, and Bikash Sharma

Abstract Until recently, automated central control of building-wide systems was found only in larger commercial buildings and expensive homes. Typically involving only lighting, heating, and cooling systems, building automation rarely provided more than basic control, monitoring, and scheduling functions and was accessible only from specific control points within the building itself. The work is a novel approach for low-cost solutions to home automation. Out of many prototypes developed, Solar Power System, Lights/Home Appliance Control, Air Quality Meter, Temperature & Moisture Meter, Home Surveillance, Water Tank Monitoring, and Password-Protected Automatic Door have been integrated and presented in this work. This system is integrated using two controllers Raspberry Pi and ESP8266 Board, where the Raspberry Pi is the main controller, which executes the commands, and the ESP8266 board is used as a Wi-Fi module to access the internet. The commands to control the overall system can be feed using Blynk App (open-source platform). The system also has a feature of manual control, in case of any failure the system needs to be a reboot and at that time the system can be controlled using manual switches. The prototype of the overall system has been tested and the results are promising.

Keywords Home automation · IOT · Low-cost

J. Singh
Department of Mechanical Engineering, Sikkim Manipal Institute of Technology, Sikkim Manipal Univeristy, Gangtok 737136, Sikkim, India

K. Garg
Department of Computer Science and Engineering, Sikkim Manipal Institute of Technology, Sikkim Manipal Univeristy, Gangtok 737136, Sikkim, India

N. Kumar · B. Sharma (✉)
Department of Electronics and Communiation Engineering, Sikkim Manipal Institute of Technology, Sikkim Manipal Univeristy, Gangtok 737136, Sikkim, India

S. Dhar et al. (eds.), *Advances in Communication, Devices and Networking*,
Lecture Notes in Electrical Engineering 776,
https://doi.org/10.1007/978-981-16-2911-2_49

1 Introduction

As we are in the era of new technologies, automation has played the main role in the development of the human race, and nowadays when we think about the future what comes to our mind is an easy way of life and all hard work to be done automatically with human commands, reducing manpower, avoiding accidents, and many more. But here, our work presents a home automation system including automatic lights with solar power, home appliance control, smart door lock system, water tank management and irrigation system, and surveillance cameras [1–3].

The system includes many features but it is very cheap as compared to recent automation, the system is made using programmable microcontroller boards like Arduino and Raspberry Pi and also many sensors for sensing purpose, it will help us in our day to day life and will make our life easy [4]. We will be able to control the home appliances using the mobile phone app and also we can monitor our home from anywhere in the world where we can access the Internet. It also has the feature of air quality meter, which will notify us if there is a leakage of LPG gas or any other harmful gases inside the home, in overall we can say that this system has every feature which will make our busy life easy and also a time saver and we can utilize our time in other important works. It will also provide us access to the latest technology and extending the ideas of automation further in a new stage [5].

The design and prototype module is one of the low-cost and robust systems. Bare minimum hardware has been so designed that the efficient control is to the optimum. The use of IoT further substantiates the low cost of the product and its robustness in controlling various devices and appliances remotely. Hence, the modules developed are highly promising to be graduated as a commercial product [6].

2 Proposed System Architecture/Circuit Design

The above Fig. 1 shows the overall system architectural design, the system consists of many blocks and every block plays a significant role to make an enhanced automation system. Raspberry Pi and ESP8266 (Wi-Fi module) have been programmed by a programmer in such a way that controlling range and system performance has been improved and also is been powered through a solar system, for controlling the home appliances mobile app Blynk is been used. We can give our command to the microcontroller through the cloud using this app and the controller will execute our command and will control the lights/home appliances present in our home [7]. Many sensors have been used to provide extra features to the system like gas sensor (MQ-TYPE) if there is any leakage of gas inside the home the sensor will provide an interrupt to the controller and an alert will be generated also will be indicated on the mobile app. Home surveillance and monitoring is been done by using an IP camera with Raspberry Pi and ESP8266 module, live surveillance and recording are

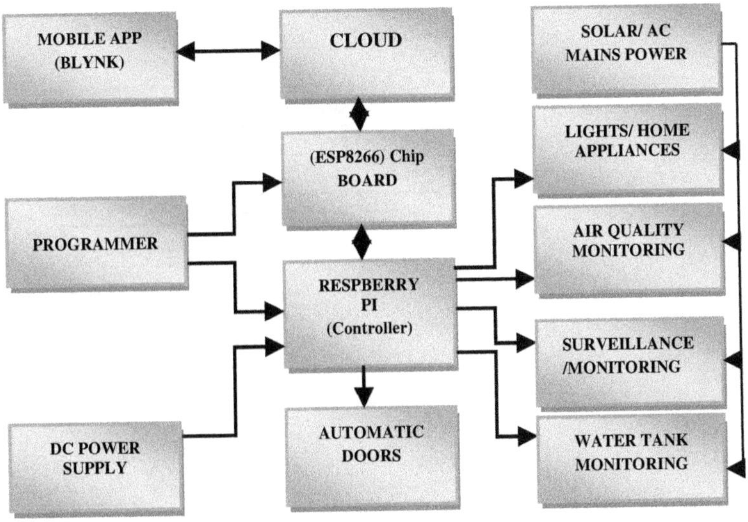

Fig. 1 Block diagram

provided on the mobile app for security purpose, the main door of the home is password protected, if anyone enters a wrong password then alert will be generated on mobile to secure the home [8]. The system also includes a water tank management system. There will be no overflow of water hence no wastage of water, if the tank is filled with water the interrupt will be generated on the controller and the valve will be closed.

3 Results and Conclusion

The proposed model has been successfully implemented and prototypes developed. Of many prototypes developed and integrated, we present the implementation of a few namely "Solar Power System, Lights/Home Appliance Control, Air Quality Meter, Temperature & Moisture Meter, Home Surveillance, Water Tank Monitoring and Password Protected Automatic Doors". This system is integrated using two controllers Raspberry Pi and ESP8266 Board, where Raspberry Pi is the main controller which executes the commands, and the ESP8266 board is used as a Wi-Fi module to access the Internet. The commands to control the overall system can be feed using Blynk app (open-source platform).

(a) **Flowchart**

See (Fig. 2).

Fig. 2 Flowchart

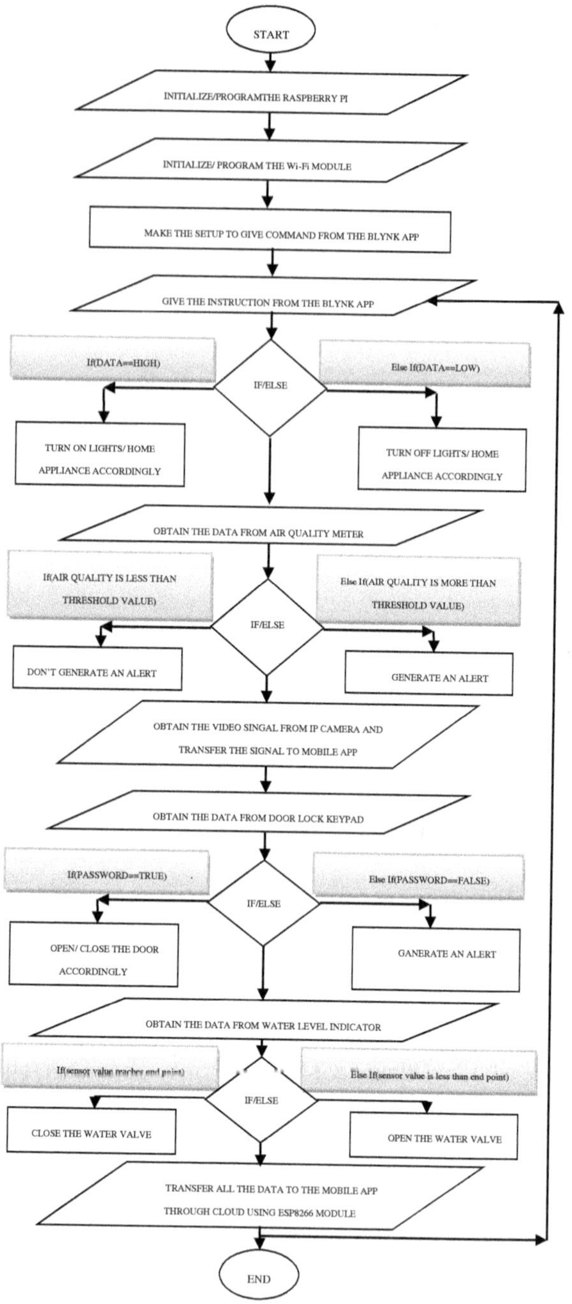

(b) **Algorithm**

i. Initialize/Program the Raspberry Pi module
ii. Initialize/Program the ESP8266 (Wi-Fi module)
iii. Initialize the mobile app Blynk
iv. Obtain the data from mobile app
v. Turn ON/OFF the lights and home appliances of the home according to the commands received from app
vi. Obtain the data from air quality meter
vii. If air quality is poor, generate an alert
viii. Obtain the video signal from the IP camera and transfer the signal to the mobile app
ix. Obtain the data from the door keypad, if the password is correct open the door else generate an alert
x. Obtain the data from the water-level indicator and close the valve if the level reaches maximum
xi. Transfer all the data to the mobile app through the cloud with the help of a Wi-Fi module (ESP8266)
xii. End

(c) **Prototypes Tested**

Figures 3 and 4 show the controlling of lights using IoT. The command is given from the mobile app to the controller through the Internet the controller executes the commands and turns ON/OFF lights accordingly.

Figures 5 and 6 show the prototype of password-protected door. To open the door we need to enter the password, if we enter a wrong password the alert will be generated on the mobile app, and also the buzzer will blow. Password can be changed by a user when required.

Figure 7 shows the surveillance system of the home. The IP camera is used to record the video signal and is been transferred to the mobile app through the Internet so that the live streaming can be done.

Figures 8 and 9 show the air quality and temperature/moisture meter.

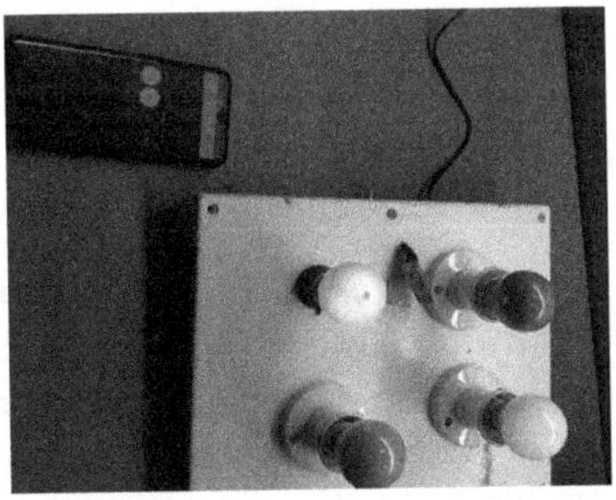

Fig. 3 Lights control

Fig. 4 Solar power

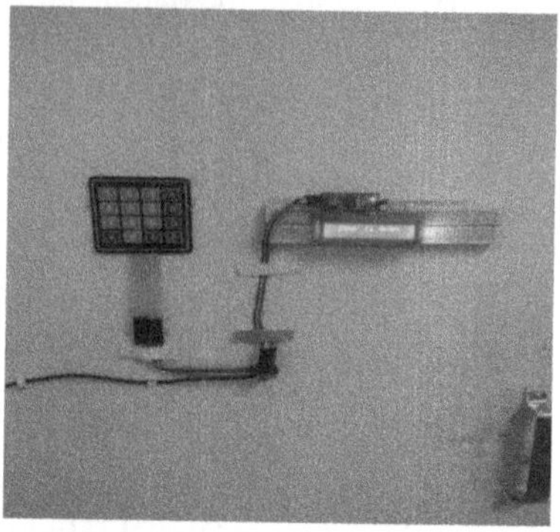

Fig. 5 Password-protected doors

Fig. 6 Password-protected doors

Air quality meter generates an alert when the air quality in the home gets poor or if there is any leakage of gases. Similarly, if the home temperature exceeds the threshold value alert will be generated.

The overall performance of the system integrated is promising with low-cost development. The individual modules prototyped are operating successfully and the future work is to bring all of them on one platform.

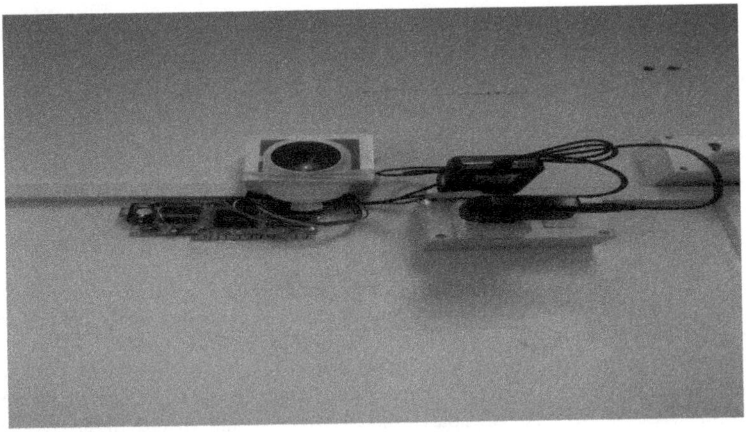

Fig. 7 Home surveillance

Fig. 8 Air quality meter

Fig. 9
Temperature/moisture meter

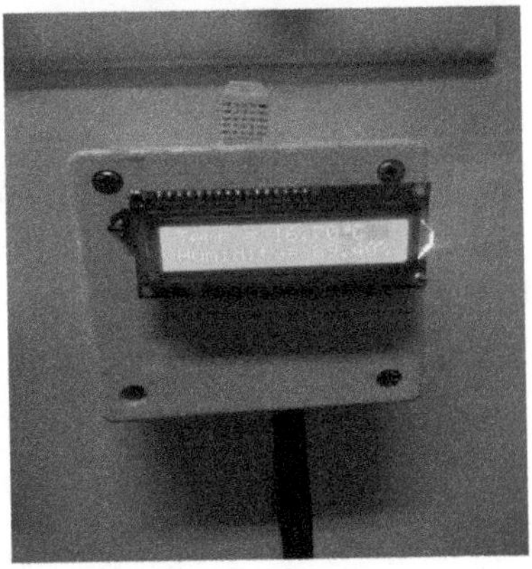

Acknowledgements This work was supported by Sikkim Manipal University (SMU) under TMA Pai University Research Fund (Minor (Students) Grant) vide ref.: Ref. No. 6100/SMIT/R&D/Project/13/2018 dated: 18th Dec. 2020.

References

1. Rai D et al (2019) Arduino-GSM interfaced secure and smart cabin for smart office. In: 2019 2nd international conference on intelligent communication and computational techniques (ICCT), Jaipur, India, pp 203–206. https://doi.org/10.1109/icct46177.2019.8969013
2. Kumar N, Dey B, Chetri C, Biswas A (2020) Surveillance robocar using IoT and Blynk App. In: Bera R, Pradhan PC, Liu CM, Dhar S, Sur SN (eds) Advances in communication, devices and networking. ICCDN 2019. Lecture notes in electrical engineering, vol 662. Springer, Singapore. https://doi.org/10.1007/978-981-15-4932-8_47
3. Chettri A et al Design and development of semi-automatic electrical incinerator using arduino. In: 2019 2nd international conference on intelligent communication and computational techniques (ICCT), Jaipur, India, pp 196–198. https://doi.org/10.1109/icct46177.2019.8969035
4. Rasaily D, Pradhan A, Kumar N, Rai D, Sharma BP (2019) smart houses for smart cities based on arduino and anroid app. Int J Res Anal Rev (IJRAR) 6(1):941–945. E-ISSN 2348–1269, P-ISSN 2349–5138
5. Gawli K, Karande P, Belose P, Bhadirke T, Bhargava A (2017) Internet of things (IoT) based robotic arm. Int Res J Eng Technol (IRJET) 04(03):757–759
6. Butkar Vinayak D, Devikar Sandip R, Jaybhaye Vikas B (2014) Android based pick and place. Int J Infor Futur Res 2(4):2347–1697. ISSN (Online)
7. Vanitha M, Selvalakshmi M, Selvarasu R (2016) Monitoring and controlling of mobile robot via internet through raspberry PI Board. IEEE. https://doi.org/10.1109/ICONSTEM.2016.7560864
8. Anusha1 S, Madhavi M, Hemalatha R (2015) Home automation using ATmega328 microcontroller and android application. Int Res J Eng Technol (IRJET) 02(06). www.irjet.net

Automatic Fire Detector

Saurabh Debabrata Das, Amrita Biswas, Rajdeep Bhattacharjee, Shivam Gupta, and Barnali Dey

Abstract In various places where the risk of fire is more, automatic fire-fighting systems are used. In this paper, an automatic fire-fighting system based on a micro-controller has been designed and implemented. The system will detect fire, inform about the fire and also sprinkle water on the fire. There is an indicator panel that will give visual information about the area under surveillance at the present time.

Keywords Arduino · Fire detector · Water sprinkler · Temperature sensor · Automatic

1 Introduction

Fire is a common occurrence. It is useful to us in various ways and is part of our daily lives. Cooking and heating are its common uses. Fire also serves as a source of energy in various applications. However, if the fire goes out of control it can wreak havoc and can cause huge destruction of life and property. There are four classes of fire: Class A which occurs on solids like wood and plastics, Class B type which occurs on liquids like oil and tars that are inflammable, Class C type fires occur on gases, and Class D type fires occur on metals. No matter which class the fire belongs to it will always create harmful smoke. Smoke is the suspension of carbon or other particles in the air emitted from the substance which is burnt. Very often indoor fires lead to the death of the victims due to the breathing in of harmful smoke which causes choking. The risk of such deaths increases at night. So it is of utmost importance to take steps for detection of fire at the initial stage. As a precaution against fire, it is always recommended that smoke detectors must be installed at all places. In case a fire breaks out, it can be tackled manually or automatically. However, an automatic fire sprinkler system that can detect fire and release water with the right pressure to

S. D. Das · A. Biswas (✉) · R. Bhattacharjee · S. Gupta · B. Dey
Electronics and Communication Engineering Department, Sikkim Manipal Institute of Technology, Gangtok, Sikkim, India
e-mail: amrita.a@smit.smu.edu.in

S. Dhar et al. (eds.), *Advances in Communication, Devices and Networking*,
Lecture Notes in Electrical Engineering 776,
https://doi.org/10.1007/978-981-16-2911-2_50

put the fire out will help in combating fire effectively and safely and decrease the losses involved.

The main purpose of firefighting systems is to provide timely information to the occupants of the building about the break out of the fire so that they can make a timely exit and also to stop the fire automatically.

Over the years lot of work has been done on automatic fire detection systems [1, 2]. Shandeep et al. [3] used a GSM Module to send an emergency notification to a mobile set when a fire occurs by sensing the infrared light that emits from the flame.

Khan et al. [4] designed an automatic fire-fighting system with locally available components, which can detect smoke and high temperature.

Shams et al. [5] proposed a fire-fighting system based on computer vision. Their system detects fire and also performs the essential steps needed to control the fire at the initial stage.

Sahoo et al. [6] studied the different types of water sprinklers that can be used in firefighting systems.

The primary aim of this work is to design a system that can detect fires early so that the occupants of the area get sufficient time to evacuate. This system is designed to ensure that it is more user-friendly and easy to operate. This work also involves minimal use of hardware and low-level processing is involved.

The main objective of this paper is to design a fire detection and water sprinkler system that would fulfill the following:

To indicate the room in which fire erupted.

To indicate the location where the fire has occurred.

To prevent fire and smoke.

To sound the alarm if a fire occurs.

To run the emergency exit servo motor and control the fire by supplying water to the remote area by a motor pump.

2 Work Done

In the proposed work the area surrounding the system is monitored using temperature sensors. It alerts the user if the temperatures go beyond the normal parameters. The sensor signals are fed to the (ATmega32) microcontroller for processing. The microcontroller polls the sensors to scan the environment and displays its status on the LED status indicator panel. The system also has a firefighting pump which is kept on standby. It is operated if the fire is detected by the sensors.

As shown in the basic block diagram in Fig. 1, the temperature sensor and the ultrasonic sensor provide the inputs, and motors (servo and DC motor), water pump with sprinkler, buzzer, fan and LED serve as the output. Arduino Uno microcontroller is used for processing and overall decision-making of the entire system. The chosen

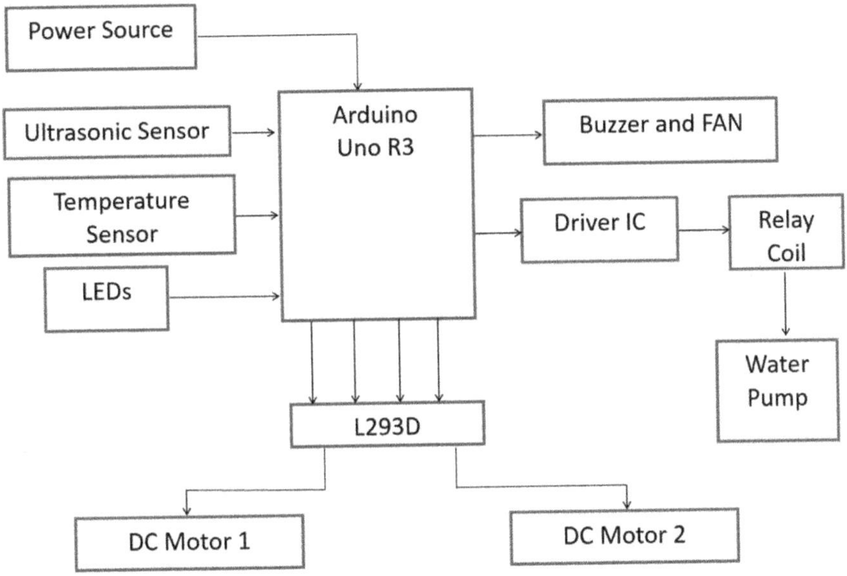

Fig. 1 Block diagram of the proposed automatic fire detector and water sprinkler system

microcontroller handles the computing task efficiently and software and hardware development tools for the Arduino Uno like compilers and assemblers are easily available.

Arduino code in the C programming language is used to interface the hardware and the software. The automatic fire-fighting robot can correctly detect the source of fire and put it off. The developed fire fighting robot finds the source of the fire by using a flame sensor and ultrasonic sensor. The flame sensor is used to detect fire and its location. The ultrasonic sensor detects the obstacles around the robot so that they can be avoided. The procedure of the robot in detecting is described clearly in the flowchart in Fig. 2. The temperature sensor detects the temperature of the area in order to have more accurate results from the incident area. All sensors are connected to Arduino Uno along with pump and DC motors to control the movement of the robot. If the five-channel flame sensor detects the fire, the DC motor will stop at 30 cm from the fire. The pump will start to react to push the fire terminator fluid in the source of the fire. Three fans have been installed which will keep the fire away from the robot and the robot can easily move forward to stop the fire.

The workflow has been shown in Fig. 2.

The Arduino IDE software has been used for writing the code.

For writing and compiling the code into the Arduino Module, the Arduino IDE has been used which is an open-source software. It makes the code writing and compilation task relatively simple and is readily available for operating systems like MAC, Windows, and Linux. The IDE platform has been used for creating the main code. A Hex File corresponding to the code is generated which is then transferred

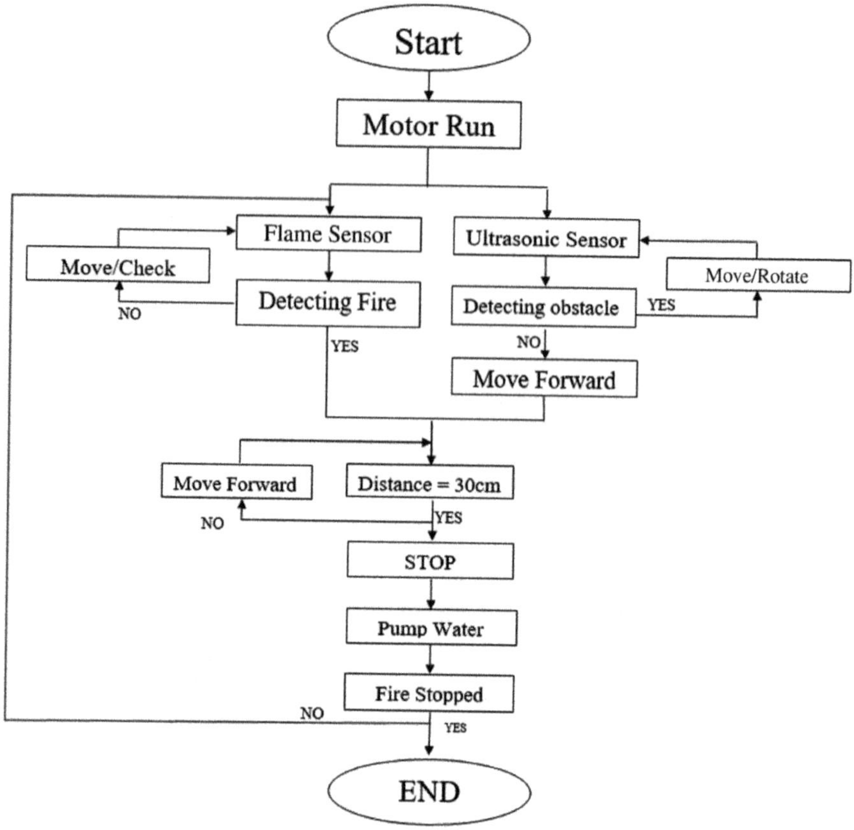

Fig. 2 Flowchart of the working of the automatic fire detector and water sprinkler

and uploaded in the controller on the board. The IDE environment mainly contains two basic parts: Editor and Compiler. The Editor is used for writing the code and the Compiler is used for compiling and uploading the code into the given Arduino Module. A picture of the developed model for the proposed system has been shown in Fig. 3.

3 Results

The proposed system performs two tasks. It detects fire and sets off an alarm using the buzzer and the robot moves towards the fire and sprays water. In our prototype system, we have not installed an actual water sprinkler. A relay is activated whenever the sprinkler is to be turned on. It also has a fan installed to keep the fire away from the robot so it can move forward.

Fig. 3 Model of the proposed automatic fire detector and water sprinkler system

The temperature sensor used is LM35. It can detect the temperature in the range of 55–150 °C. The threshold temperature is set at 40 °C. Whenever a temperature greater than 40 °C is detected, fire alarm is turned on. The buzzer and LED are turned on and the robot moves forward. The ultrasonic sensor detects obstacles in the path if any and the robot is steered toward the source of fire with its help. It uses a 58 kHz signal frequency. The resolution of measurement is 5 cm, the range is up to 11 m. Table 1 shows the results obtained.

Table 1 Results obtained

Temperature sensor	Ultrasonic sensor	Motors	LEDs	Buzzer	Fan	Relay
Not Activated	Not Activated	Not Activated	Not Activated	Not Activated	Not Activated	Not Activated
Not Activated	Activated	Activated and Robot turns/rotates	Not Activated	Not Activated	Not Activated	Not Activated
Activated	Not Activated	Activated and robot moves forward	Glows	On	On	Activated
Activated	Activated	Activated and rotate	Glows	On	On	Activated

4 Conclusion

The proposed system is reliable, portable, and cost-effective. Design is simple and can be operated at all places be it indoors or outdoors. Manual fire control systems will be soon replaced by such automatic fire control systems because they are faster and safer to use. Such systems can also be used in places where it will be difficult for humans to intervene. All construction and manufacturing firms should install such automatic fire control systems to ensure the safety of life and property.

References

1. Mohamed K (2018) Design and development of integrated semi-autonomous fire fighting robot. Int J Res Appl Sci Eng Technol 6:192–197. https://doi.org/10.22214/ijraset.2018.7025
2. Ahmed A, Mansor A, & Albagul A (2015) Design and fabrication of an automatic sprinkler fire fighting system. Lecture Notes Control Inform Sci 789. https://doi.org/10.1007/978-3-319-17527-0_5
3. Shandeep SR, Steeve Elisa Giftson V, Beny JR, Aishwarya C (2016) Design and implementation of fully automated fire fighting robot. Int J Scient Res Sci Eng Technol 2(6)
4. Khan MJA, Imam MR, Uddin J, Rashid Sarkar MA (2012) Automated fire fighting system with smoke and temperature detection. In: 7th international conference on electrical and computer engineering, December, 2012. Florida International University, Miami, 2003, Dhaka, Bangladesh, pp 20–22
5. Shams R, Hossain S, Priyom S, Fatema N, Shakil SR, Rhaman MK (2015) An automated fire fighting system. In: 12th international conference on fuzzy systems and knowledge discovery (FSKD), Zhangjiajie, pp 2327–2331. https://doi.org/10.1109/fskd.2015.7382316
6. https://www.scribd.com/document/402036321/AutomaticFireSprinklerSystem-pdf

CPSIA information can be obtained
at www.ICGtesting.com
Printed in the USA
LVHW080448140921
697767LV00002B/150